COSMIC RAYS, THE SUN AND GEOMAGNETISM:

The Works of Scott E. forbush

James A. Van Allen, Editor

American Geophysical Union

Library of Congress Cataloging-in-Publication Data

Forbush, Scott E., 1904-1984.
    Cosmic rays, the sun, and geomagnetismn : the works of Scott E. Forbush / James
    A. Van Allen, editor.
        p.   cm.
    Includes index.
    ISBN 0-87590-833-0
      1. Cosmic ray variations.   2. Solar activity.   3. Geomagnetism.
    I. Van Allen, James A.   II. Title.
    QC485.8.V3F67   1993   93-26144
    539.7223—dc20   CIP

ISBN 0-87590-833-0

Printed in the United States of America

American Geophysical Union
2000 Florida Avenue, N.W.
Washington, D.C. 20009

Scott E. Forbush (left) and James A. Van Allen on the occasion of the latter's 1977 visit to the Department of Terrestrial Magnetism of the Carnegie Institution of Washington.

CONTENTS

Selected Published Papers by Scott E. Forbush

Editor's Foreword

This monograph is a tribute to the character and achievements of Scott Ellsworth Forbush (1904–1984) who, almost single-handedly with only technical assistance, laid the observational foundations for an important part of the subject of solar-interplanetary-terrestrial physics. The heart of his research was the meticulous and statistically sophisticated analysis of the temporal variations of cosmic-ray intensity as measured by ground-based detectors at various latitudes and altitudes.

Forbush either discovered or put on a firm basis for the first time the following fundamental cosmic-ray effects:

- The quasi-persistent 27-day variation of intensity.
- The diurnal variation of intensity.
- The sporadic emission of very energetic (up to $\approx$ GeV) protons by solar flares.
- Worldwide impulsive decreases (Forbush decreases) in intensity followed by gradual recovery.
- The 11-year cycle of intensity and its anticorrelation with the solar activity cycle as measured by sunspot numbers.
- The 22-year cycle in the amplitude of the diurnal variation.

These effects became the infrastructure and inspiration for an immense variety of subsequent work by others.

His own work was concerned primarily with searching for effects of persistent significance within a bewildering array of fluctuations; and with patient and rigorous testing of their statistical validity. One of his central themes was the association of cosmic-ray variations with geomagnetic activity. The latter was attributed then, as now, to fluctuating solar corpuscular streams whose magnitude and frequency of occurrence were related to solar flare activity.

Forbush's central passion was to establish the facts.

He was quite clear about the identification of energetic particle events with solar flares. But he found that some impulsive decreases of cosmic-ray intensity were well correlated with the simultaneous signatures of geomagnetic storms whereas in other cases there was no apparent correlation whatever. In one of his relatively few excursions into theory, he discussed those facts as possibly attributable to a Störmer-Chapman ring current on geomagnetic cut-offs and concluded that such a geocentric interpretation was probably untenable.

But, for the most part, he did not venture into the realm of more far-ranging theoretical suggestions. In none of his papers did he propose that either a Forbush decrease or the 11-year intensity cycle was caused by the interplanetary medium; but later, with characteristic modesty, he welcomed and embraced this line of interpretation as established by others.

I first met Forbush in 1939 when I went to the Department of Terrestrial Magnetism of the Carnegie Institution of Washington as a postdoctoral fellow. At that time there were two distinct and quite different professional cultures within DTM. I became one of the "Young Turks" working in nuclear physics under the aegis of Merle A. Tuve. The other culture was represented by the DTM traditionalists or "old timers", led by the director John A. Fleming. The old timers were studying atmospheric electricity, earth currents, aurorae, geomagnetism, cosmic rays, and ionospheric physics. Scott Forbush and Harry Vestine were prominent members of this latter group. Also there were occasional visits by Sydney Chapman and Julius Bartels, who were then completing their great two-volume treatise *Geomagnetism*. I was one of the few individuals who crossed the DTM culture barrier, in either direction. As a result, I soon found myself more interested in cosmic rays and solar-terrestrial physics than in low energy nuclear physics. I was attracted especially by the quiet, thoughtful work that Forbush was doing. Tuve's view was quite the contrary. He regarded the old timers at DTM as hopelessly out of touch with modern science and he considered Forbush an outstanding example of a person who was filling massive notebooks with numbers that no one would ever find of interest or of importance.

These were the circumstances under which Forbush persisted with passionate and solitary (but not lonely) devotion to trying to unravel the meaning of the numbers in his notebooks of cosmic-ray and geomagnetic data. One is reminded of Kepler's devotion to the planetary observations in the notebooks of Tycho Brahe.

Scott was born on 10 April 1904 on his family's farm in Ohio but soon decided that he would rather be a scholar than a farmer. From a one-room rural schoolhouse, he moved to the Western Reserve Academy and then went to the Case School of Applied Science in Cleveland, from which he graduated with a B.S. degree in 1925. He attended graduate school at the Ohio State University briefly, then got a job at the National Bureau of Standards and, in 1927, transferred to DTM. His first assignment was as an observer at the Huancayo Magnetic Observatory in the Peruvian Andes and his second assignment was as a member of the staff of the famous nonmagnetic survey ship the *Carnegie*. He was on board the *Carnegie* when she was destroyed in November 1929 by an explosion and subsequent fire in Apia Harbor, Samoa. But, fortunately for cosmic ray physics, he escaped unharmed.

During subsequent years his home base continued to be DTM but he returned to Huancayo for two years, after which he came back to Washington and took graduate work in physics, statistics, and mathematics at universities in that area. His driving motivation was to apply everything that he learned to the analysis of geomagnetic and cosmic-ray data.

During World War II he worked at the Naval Ordnance Laboratory on the magnetic field of ships and related matters. Thereafter he continued his cosmic-ray work at DTM, had a prominent role in the program of the International Geophysical Year (1957–58), and lectured widely on the statistical analysis of observational data at the Peruvian National Universities of San Marcos, San Agustin, and Cuzco, and elsewhere.

In 1960–1961 Scott was a visiting professor at the University of Iowa. He wrote two valuable papers on the time variations of our Explorer VII data on particle intensities in the Earth's radiation belts and gave a series of seminar lectures entitled "Geomagnetism, Cosmic Radiation, and Statistical Procedures for Geophysicists". These edited lectures and related ones are reproduced in this monograph.

Nearly every cosmic-ray paper of Forbush's relatively short bibliography is a landmark in the subject.

In 1962 he was elected to the National Academy of Sciences and in 1966 he received the especially appropriate John A. Fleming Medal of the American Geophysical Union.

So much for Merle Tuve's disdain!

James A. Van Allen
University of Iowa

Acknowledgments

The editor is indebted to Evelyn D. Robison, Alice M. Shank, and John R. Birkbeck of the University of Iowa for skillful assistance in preparing this manuscript.

Editor's Addendum

In the process of selecting published papers for inclusion in this volume, I reviewed (a) Forbush's own comprehensive bibliography of 79 items for the period 1933–1970 (incl.) and (b) the open literature for the subsequent period 1971–1984 (incl.). I omitted the following two of his earliest papers because of their being outside of the central theme of his subsequent and more important cosmic ray studies:

> Apparent Vertical Earth-Current Variations at the
> Huancayo Magnetic Observatory
> — S. E. Forbush
> Terr. Mag., 38, 1–11, 1933

> Some Practical Aspects of the Theory of the Unifilar
> Horizontal-Intensity Variometer
> — S. E. Forbush
> Terr. Mag., 39, 135–143, 1934

The other items that I omitted were parts of internal reports of the Department of Terrestrial Magnetism of the Carnegie Institution of Washington, contributions to conference proceedings, short preliminary letters to the editor, and abstracts of oral papers. The papers that I selected include the substance of these miscellaneous items and are, of course, much more detailed and complete.

James A. Van Allen

Eos, Vol. 65, No. 33, August 14, 1984

Scott E. Forbush 1904–1984

Scott E. Forbush, a pioneer in cosmic ray research, was the quintessential geophysicist's geophysicist. Until, on the eve of his 80th birthday, he succumbed to pneumonia, he maintained an abiding interest in the continued reliable operation of the three remaining cosmic ray ionization chambers of the world-wide network that he had set up in the mid 1930's. No one could have predicted, when the first instrument at Cheltenham, Md. commenced operation in 1936, that Forbush was destined to discover most of the important multifarious time variations of cosmic rays that were accessible to his classic detectors: the first-generation instruments that were similar in principle to those with which a mysterious penetrating radiation, probably from an extraterrestrial source, had been discovered by Victor Hess in a series of manned balloon flights in 1912. The time scales of the effects which Forbush studied ranged from minutes to decades.

How did Scott Forbush get into a field in which he was to occupy an absolutely unique niche, assiduously pursuing a single unswerving goal, to derive from continuous observations with ionization chambers all of the statistically significant information that the data were capable of revealing? As he told it, ". . . around 1926 I wasn't overly in love with my job at the National Bureau of Standards, and I was offered the possibility of going to Peru to a Geomagnetic Observatory which was operated at Huancayo by the Department of

Terrestrial Magnetism (DTM) of the Carnegie Institution of Washington (CIW)".

In 1932, a committee set up by the CIW to consider a request by R. A. Millikan and A. H. Compton (who didn't agree very often) concluded that it would indeed be useful to have a network of cosmic ray detectors situated at "convenient places." These convenient places would be magnetic observatories because they already existed, and so the first detector in the network was installed at Cheltenham Magnetic Observatory, Md., in 1935. It is still in operation (at Fredericksburg). Forbush was put in charge of this program. The instrument (Figure 1) was called a Compton-Bennett meter or, alternatively, a model C meter. What you got, then and now, is a trace on a photographic bromide paper representing the combination of the ionization currents caused by the cosmic rays and any local radioactive material.

Forbush and one assistant laboriously scaled by hand and reduced all of these records which, of course, included barometric pressure readings. Subsequently, volumes containing the final results were sent to many workers throughout the world.

Forbush, by his detailed analyses of the many different cosmic ray intensity time variations, stimulated others to make more experimental observations and to propose theoretical explanations for these phenomena. The cosmic ray time variations cover a very large dynamic range. The shortest occurs during the onset of ground level enhancements (GLE's) associated with solar flares. There is a diurnal variation arising from the earth's rotation. There are transient events called Forbush Decreases, which are of somewhat longer duration. There are 27-day recurrences related to the rotation of the sun. There is an annual variation. There's a solar activity cycle effect (11 years), and a solar magnetic cycle effect (22 years). Each one of these was discovered (or put on a firm footing) by Scott Forbush in a one-man operation with the help of one dedicated assistant, Isabelle Lange until 1957, then Lisellote Beach until her retirement in 1975.

Forbush was very much influenced by Julius Bartels, who was an associate at the Carnegie Institution of Washington during the period 1931–1940. Actually, Forbush was somewhat of a professional statistician, who ". . . read on buses every book on the subject that I could get my hands on." Thus, he was able to benefit very greatly from Bartels' presence there, as exemplified in an early paper [*Forbush*, 1937a]: "The adequate characterization of the diurnal variation in any geo-physical phenomenon requires not simply a knowledge of its average value, but also a full knowledge of its variability. The latter, in general, is made up of an irregular (or random) part and a systematic part such, for example, as a systematic variation with season in the amplitude (or phase) of the diurnal variation. These facts, together with the methods of analysis used in this discussion, have been set forth clearly in numerous papers by J. Bartels, who, as a research associate for the Department of Terrestrial Magnetism of the Carnegie Institution of Washington, has made important applications to problems in terrestrial magnetism." This profound respect for and admiration of Bartels is evident throughout all of his writings, which frequently refer to Bartels. What Forbush, in his characteristic self-effacing modesty always claimed he was doing, was merely extending

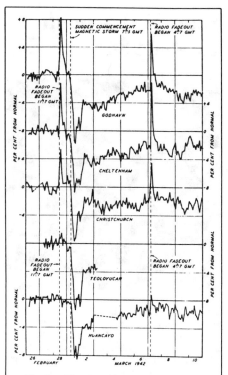

Fig. 2. The first observations of solar cosmic rays, 30 years after the discovery of galactic cosmic rays by Victor Hess [*Forbush*, 1942].

Fig. 1. Compton-Bennett meter, also known as model C meter, utilized in the worldwide network of the Department of Terrestrial Magnetism, Carnegie Institution of Washington, set up by Scott E. Forbush.

some procedures for which Bartels had (perhaps) given the basis but hadn't carried out to its full extent.

With his first 233 days of ionization chamber observations, Forbush did what was then a very elaborate statistical analysis, and found that the controversial diurnal variation was, indeed, real. In his paper on this subject [*Forbush*, 1937a] he stated, "To summarize the analysis of the data for Cheltenham demonstrates the existence of a physically real 24-hour wave in apparent cosmic ray intensity, which does not appear to be due to systematic instrumental effects but which may be due, in part at least, to variations in local radiation." He eked out of these data an exceedingly small vector which is the order of half a percent, and for the first time the probability that it is a real effect was rigorously and correctly evaluated. The now well understood semi-diurnal variation is smaller than the first harmonic and at this point Forbush had to decide its presence could not be established within the statistics of the available data.

Another phenomenon, that had been claimed by *Compton and Getting* [1935] on the basis of their analysis of ionization chamber data, was the effect of the motion of the galaxy: If there is a uniform distribution of cosmic rays coming from all directions, then the rotational speed of the earth and the galaxy should produce a net anisotropy in sidereal time. *Forbush* [1937b] also investigated this matter, and his conclusion was "Compton and Getting found from the data of Hess and Steinmaurer that the amplitude of the apparent 24-hour sidereal wave was nearly 10 times their estimate of its probable error. Their exact procedure in obtaining this estimate was not given." If there was anything that made Soctt Forbush angry (to put it mildly), it was failure to describe the statistical procedure that was used to obtain the claimed results. Quoting further, "Estimates of the probable errors in geophysical data are especially misleading." In this regard, Forbush then stressed a very important point. "If based on the departures of observed points from a fitted wave, they are invariably too small unless the departures for successive points are statistically independent. Tests on cosmic ray data from Cheltenham indicate that such departures are not independent. Our conclusions regarding the reality of the 24-hour sidereal

wave are based on a method of analysis which takes account of this. It is surely one of the most constructive recent developments in physics that such powerful tools have been evolved for evaluating the real or illusory nature of such interesting periodicities."

The first two observations of solar cosmic rays were made by Forbush in 1942 (Figure 2). Because of his cautious approach, Forbush waited for still another GLE to occur before publishing his discovery in a paper [*Forbush*, 1946] characteristically titled most cautiously. He concluded, "These considerations suggest the rather striking possibility that the three unusual increases in cosmic ray intensity may have been caused by charged particles actually being emitted by the sun with sufficient energy to reach the earth at geomagnetic latitude 48° but not at the equator."

In a later paper [*Forbush, et al.*, 1950], there is a conclusion, more or less in between the lines, from observations of a GLE for the first time on top of a mountain, that the spectrum of relativistic solar cosmic rays is very steep, indeed, and that what one is seeing is rather low-energy nucleons coming from the sun in this case. Actually, Forbush was very lucky because GLE's of sufficient magnitude to be detected with ionization chambers have not occurred since 1956.

The Forbush decrease [*Forbush*, 1937c] is the only discovery which carries his name (Figure 3). By comparing data from the

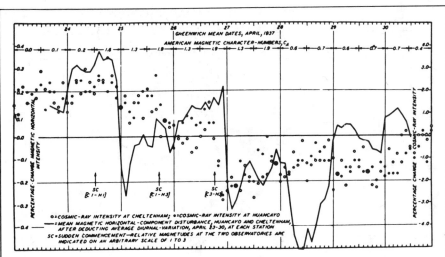

Fig. 3.   The discovery of the Forbush Decrease in 1937 [*Forbush*, 1937c].

worldwide network of stations that he had established, he was able to show for the first time that certain changes in cosmic ray intensity were worldwide [*Forbush*, 1938]. It was very common and natural in those days to associate those sudden decreases with changes in the geomagnetic cut off due to some ring current, for example. It was also natural if you were working at the Department of Terrestrial Magnetism, that you would think of this, and other effects—such as 27-day recurrences [*Forbush*, 1940], for example—as attributable in some way to geomagnetic field variations. So he found that there were events in which the cosmic ray intensity seemed to more or less track the horizontal intensity of the geomagnetic field. But then he found cases of a large geomagnetic storm during which the cosmic ray intensity didn't change at all [*Forbush*, 1955]. That remained a mystery for quite a while. An interesting point is brought out here. Relating the geomagnetic activity with the level of cosmic ray intensity *Forbush* [1938] stated, "Since the period of minimum values for the departures in cosmic ray intensity in this figure agrees roughly with that of maximum magnetic activity, and since we have also indicated the existence of a 27-day wave, probably quasi-persistent in cosmic ray intensity, it would not be unexpected to find, when adequate data are available, the 11-year cycle of sunspot activity reflected in cosmic ray intensity." That was really looking ahead! Forbush also noticed that during solar minimum, the variations in the intensity were very much less than when the sun was most active.

The last of the Forbush discoveries was the 22-year wave in the diurnal variation (Figure 4) [*Forbush*,1967, 1969, 1981; *Duggal et al.*, 1970a, b]. Although the solar cycle (11-year) variation was universally accepted, his claims for a 22-year wave were at first rejected by some members of the cosmic ray community but have since been vindicated.

The superimposed epoch technique was introduced by Sir Charles Chree (1913), and a medal bearing his name was established by the British Institute of Physics and the Physical Society. Forbush received this Chree Medal in 1961. He later remarked that he thought he probably got it because he ". . . was mad at Chree." The reason he was mad at Chree was that when Chree proposed this new way of doing things, he never told you how to do the statistics. It is very, very tricky, and the solution of this long-standing problem constituted Forbush's final contribution [*Forbush et al.*, 1982, 1983].

In 1966, Forbush received the American Geophysical Union's John A. Fleming award, the citation for which noted that his findings came through ". . . intricate development of statistical methods and the most erudite analysis of data."

Especially in his first detailed paper on the "Variation with a Period of Two Solar Cycles in the Cosmic-Ray Diurnal Anisotropy [*Forbush*, 1969], he developed an elegant albeit arcane analytical procedure and notation that

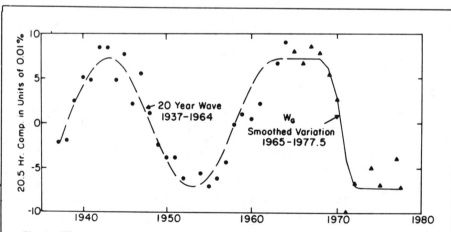

Fig. 4. The wave with a period of two solar cycles [*Forbush*, 1981]. The later portion of the curve was labeled W_G because it was called by Forbush "guessomatic."

somehow tended to conceal the significance of the results, the full understanding of which required great patience on the part of the reader. On the other hand, he was as demanding of others as he was of himself. He insisted that people should publish their data so that others could analyze them with their own procedures. He did not assiduously follow the literature because he felt that the cost is high. His disdain for laziness and sloppiness as he perceived it led him to ask "What can you believe?"

Scott Forbush was chairman of the Cosmic Ray Committee for the International Geophysical Year. He also served on the Visiting Committee of the Bartol Research Foundation, where he was appointed Distinguished Professor upon his retirement from D.T.M. in 1969. Forbush spent two happy periods at other institutions, one in 1959 at the University of Iowa, another at Imperial College, London, in 1968. While pursuing his research in Peru, he was named Honorary Professor and presented an award by the Universidad de San Marco, Lima. He was elected to the National Academy of Sciences in 1962, and was a fellow of AGU, American Association for the Advancement of Science, and APS.

I am proud to have been coauthor of a number of papers with Scott Forbush and Shakti Duggal, whose untimely death at the age of 50 in 1982, created the first gap in our long-term collaboration. The last two papers [Forbush et al., 1982, 1983] brought to the ultimate limit the quantitative implications of Scott's insight more than 40 years earlier. Scott had planned to spend a period at Bartol in March 1984, when he was struck down by a fatal illness. He had long enjoyed good health (few knew that he suffered from diabetes) and was an avid jogger many years before this form of exercise became popular. He was incensed when a younger person offered to carry his bag to his room in a hotel at Banff during the 10th International Cosmic Ray Conference (ICRC) in Calgary in 1967. He seldom missed these biennial meetings, but his failing sight precluded attending the most recent ICRC in India. He overcame this frustrating handicap by using a magnifying glass and by writing in very large letters.

Preparation of the last manuscripts was exceedingly difficult, for Forbush was meticulous about the format and even the choice of words. The statistical aspects of all cosmic ray papers emanating from Bartol were always examined critically by Forbush, and when the

Forbush Imprimatur was accorded a manuscript, we knew that we were right.

It is striking that he never succumbed to the "publish or perish" syndrome. His publication list comprises somewhat less than two dozen papers over a period of 46 years (a significant number appeared after his retirement). A review paper [Forbush, 1966], covering 30 years of work to that time, contains 12 Forbush references. But it is an undeniable fact that every single one of Scott Forbush's papers was a landmark result that will remain indelibly etched in the annals of science.

References

Compton, A. H., and I. A. Getting, An apparent effect of galactic rotation on the intensity of cosmic rays, Phys. Rev., 47, 817–821, 1935.

Duggal, S. P., S. E. Forbush, and M. A. Pomerantz, Variations of the diurnal anisotropy with periods of one and two solar cycles, Acta. Phys. Acad. Sci. Hungaricae, 29, Suppl. 2, 55–59, 1970.

Duggal, S. P., S. E. Forbush, and M. A. Pomerantz, The variation with a period of two solar cycles in the cosmic ray diurnal anisotropy for the nucleonic component, J. Geophys. Res., 75, 1150–1156, 1970.

Forbush, S. E., On diurnal variation in cosmic-ray intensity, Terr. Magn. Atmos. Electr., 42, 1–16, 1937a.

Forbush, S. E., On sidereal diurnal variation in cosmic-ray intensity, Phys. Rev., 52, 1254, 1937b.

Forbush, S. E., On the effects of cosmic ray intensity during the recent magnetic storm, Phys. Rev., 51, 1108, 1937c.

Forbush, S. E., On world-wide changes in cosmic-ray intensity, Phys. Rev., 54, 986, 1938.

Forbush, S. E., On world-wide changes in cosmic-ray intensity, Phys. Rev., 54, 987, 1938.

Forbush, S. E., On the 27-day and 13.5 day waves in cosmic-ray intensity and their relation to corresponding waves in terrestrial-magnetic activity, Trans. Wash. Meeting, Int. Union Geod. Geophys., Assoc. Terr. Magn. Electr. Bull., 11, 438, 1940.

Forbush, S. E., Three unusual cosmic-ray increases possibly due to charged particles from the sun, Phys. Rev., 70, 771, 1946.

Forbush, S. E., World-wide variations of cosmic-ray intensity, Proc. Int. Congr. Cosmic Rays, 5, 285–303, 1955.

Forbush, S. E., Time-variations of cosmic rays, Handbuck der Physik, 49, 159–247, 1966.

Forbush, S. E., A variation with a period of two solar cycles in the cosmic-ray diurnal anisotropy, *J. Geophys. Res., 72*, 4937, 1967.

Forbush, S. E., Variation with a period of two solar cycles in the cosmic ray diurnal anisotropy and the superposed variations correlated with magnetic activity, *J. Geophys. Res., 74*, 3451, 1969.

Forbush, S. E., Cosmic ray diurnal anisotropy, 1937–1972, *J. Geophys. Res., 78*, 7933, 1973.

Forbush, S. E., Cosmic ray diurnal anisotropy, 1937 to 1977.5, *Proc. Intl. Cosmic Ray Conf., 10*, 209–212, 1981.

Forbush, S. E., T. B. Stinchcomb, and M. Schein, The extraordinary increase of cosmic ray intensity on November 19, 1949, *Phys. Rev., 79*, 501–504, 1950.

Forbush, S. E., and Liselotte Beach, Cosmic-ray diurnal anisotropy and the sun's polar magnetic field, *Intl. Cosmic Ray Conf., 4* 1204–1208, 1975.

Forbush, S. E., S. P. Duggal, M. A. Pomerantz, and C. H. Tsao, Random fluctuations, persistence, and quasi-persistence in geophysical and cosmical periodicities: A sequel, *Rev. Geophys. Space Phys., 20*, 971–976, 1982.

Forbush, S. E., M. A. Pomerantz, S. P. Duggal, and C. H. Tsao, Statistical considerations in the analysis of solar oscillation data by the superposed epoch method, *Solar Phys., 82* 113–122, 1983.

This tribute was written by Martin A. Pomerantz, *Bartol Research Foundation of the Franklin Institute, University of Delaware, Newark, DE 19716.*

THE
LECTURES:

GEOMAGNETISM,
COSMIC RADIATION,
AND
STATISTICAL
PROCEDURES
FOR GEOPHYSICISTS

by
Scott E. Forbush

Geomagnetism, Cosmic Radiation, and
Statistical Procedures for Geophysicists

by

Scott E. Forbush

Preface

These six chapters were written in July 1959 as a basis for oral presentation of lectures for geophysicists given in Peru (September and October 1959) at the National Universities of San Marcos (Lima), San Agustin (Arequipa), and Cuzco. The lectures have been published (1960) in Spanish for distribution to the libraries and universities in Latin America.

Lectures I to IV provide an introduction to important aspects of geomagnetism and statistical procedures for geophysicists. The material of these four lectures was used as a basis for talks given at the Seminar on Solar and Terrestrial Physics at the Department of Physics and Astronomy of the State University of Iowa in the fall of 1960. For these talks some of the elementary mathematics of Lecture I was omitted as unnecessary for graduate students. For some of the talks the written material of the lectures was augmented with more mathematical detail and for other talks the material was presented from a different point of view.

References for more detailed discussion of various topics are given at the end of each lecture. However particular attention is directed to two comprehensive classical treatments of the subjects of the first four lectures. These are:

(1) *Geomagnetism*, Vols. I and II by S. Chapman and J. Bartels; International Monographs on Physics, Oxford at the Clarendon Press (1940).

(2) *Statistical Studies of Quasiperiodic Variables* by J. Bartels; reprints of three papers by J. Bartels from the Journal of Terrestrial Magnetism and Atmospheric Electricity (now the Journal of Geophysical Research), reprinted by the Carnegie Institution of Washington, Department of Terrestrial Magnetism, 5241 Broad Branch Road, Washington 15, D. C.

November 15, 1960

I. Earth's Main Field and Its Secular Variation

1.1 Introduction

The earth's main field is only roughly approximated by that of a magnetic dipole. Nevertheless it may be helpful to recall some characteristics of a dipole and to review briefly some essentials of scalar and vector fields.

1.2 Potential and Field Strength

In Figure 1 suppose that at the origin, O, there is a positive magnetic pole of strength m. At Q_2 the force on a unit magnetic positive pole is $H = m/r^2$. The work done to move this unit pole along the infinitesimal path $\overrightarrow{d\ell}$ from Q_1 to Q_2 is

$$dV = -\overrightarrow{d\ell} \cdot \vec{H} = -H \; d\ell \; \cos \; \alpha \; . \tag{1}$$

Or since $\overrightarrow{d\ell} \; \cos \; \alpha = dr$, then

$$dV = -\frac{m}{r^2} \; dr \; . \tag{2}$$

The potential at P is the work done in bringing a unit positive pole from ∞ to P or

$$V = -m \int_{\infty}^{P} r^{-2} \; dr = m/R \; . \tag{3}$$

Since, from (1), $-dV/\overrightarrow{d\ell} = H \cos \; \alpha$, it is evident that the potential decreases most rapidly in the direction of $\vec{H}$ or in the direction of increasing r; thus $\vec{H} = -dV/dr$.

1.3 Gradient

If with each point in space there is associated a scalar, V, which is a continuous and differentiable function of position, then the change in V for an infinitesimal change in position is

$$dV = \frac{\partial V}{\partial x} \; dx + \frac{\partial V}{\partial y} \; dy + \frac{\partial V}{\partial z} \; dz \; . \tag{4}$$

The form of (4) indicates that it can be written as a scalar product as follows:

$$dV = (\hat{i} \; dx + \hat{j} \; dy + \hat{k} \; dz) \cdot \left(\frac{\partial V}{\partial x} \hat{i} + \frac{\partial V}{\partial y} \hat{j} + \frac{\partial V}{\partial z} \hat{k} \right) \; . \tag{5}$$

2

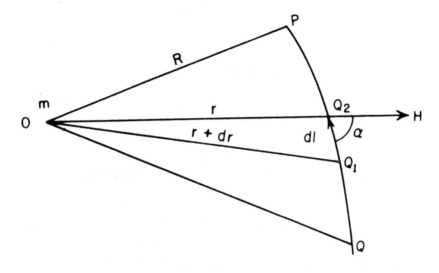

Fig. 1

The vector $(\hat{i}\ dx + \hat{j}\ dy + \hat{k}\ dz) = \vec{ds}$, the infinitesimal vector change in position. The second vector in (5) is the fundamentally important gradient of the scalar point function V; it is generally denoted by ∇V so that

$$\nabla V = \frac{\partial V}{\partial x}\hat{i} + \frac{\partial V}{\partial y}\hat{j} + \frac{\partial V}{\partial z}\hat{k} \ . \tag{6}$$

Thus the expression for dV in (4) can be written

$$dV = \nabla V \cdot \vec{ds} \tag{7}$$

wherein ∇ is defined as the operator

$$\nabla = \frac{\partial}{\partial x}\hat{i} + \frac{\partial}{\partial y}\hat{j} + \frac{\partial}{\partial z}\hat{k} \ , \tag{8}$$

which operating upon V produces the gradient of V. If V is the magnetic potential, the magnetic intensity $\vec{F} = -\nabla V$ and

$$dV = -\vec{F} \cdot \vec{ds} \ . \tag{9}$$

If $\vec{ds}$ is along an equipotential surface (V = constant), dV = 0 and $\vec{F}$ is normal to the equipotential surface in the direction in which V decreases most rapidly. If equation (9) is integrated along some curve between points 1 and 2, we have:

3

$$\int_1^2 dV = - \int_1^2 \vec{F} \cdot \overrightarrow{ds} \ . \tag{10}$$

Now

$$\int_1^2 \vec{F} \cdot \overrightarrow{ds} = \int_1^2 F_s \, ds$$

in which F_s is the component of $\vec{F}$ in the direction of the infinitesimal path length ds. When $\vec{F} = -\nabla V$,

$$F_s = -(dV/ds)ds \quad \text{and} \quad \int_1^2 F_s \, ds = -(V_2 - V_1)$$

or

$$V_2 - V_1 = - \int_1^2 \vec{F} \cdot \overrightarrow{ds} \ . \tag{11}$$

The line integral on the right of (11) thus has the same value for all paths that have the same initial point and the same end point. Consequently for any closed path the line integral of the gradient of a scalar is zero.

1.4 Air-Earth Currents

Some investigators found, from results of early surveys of the earth's magnetic field, that the calculated values of line integrals over large closed paths on the earth's surface did not vanish exactly. This was taken by some to indicate the possibility of vertical electric currents flowing from the atmosphere to earth, since the line integral around a closed path encircling such vertical currents does not vanish but is proportional to the current threading the path of integration, as may be seen readily from physical considerations or from the fact that for such a case the field is not derivable from a scalar potential. The (nonzero) line integrals over closed paths indicated the possibility of vertical currents of the order of 10^{-1} ampere km^{-2}. From subsequent measurements of atmospheric conductivity and potential gradient at several observatories the air-earth currents were found to be of the order of 10^{-6} to 3 × 10^{-6} ampere km^{-2} on about 10^{-5} of that required to explain the values found for line integrals around closed paths. Thus it seems probable that the failure of the line integrals to vanish exactly, around closed contours, is due to unavoidable errors in the magnetic charts.

1.5 Magnetic Dipole

In Figure 2 let m and −m be two magnetic poles separated a distance ℓ. The potential at P due to both is

4

$$V = \frac{m}{r_1} - \frac{m}{r_2} = \frac{m}{r - \frac{\ell \, \cos \, \theta}{2}} - \frac{m}{r + \frac{\ell \, \cos \, \theta}{2}} = \frac{m\ell \, \cos \, \theta}{r^2 - \frac{\ell^2 \, \cos^2 \, \theta}{4}} \, . \qquad (12)$$

For $r^2 \gg \ell^2$ we have

$$V = \frac{m\ell \, \cos \, \theta}{r^2} \, . \qquad (13)$$

The product $m\ell$ is called the magnetic moment M. By letting ℓ tend to zero and at the same time letting the pole strengths m tend to ∞ in such a way that $m\ell$ stays equal to M we have the concept of a magnetic dipole M · $\vec{M}$ is a vector along the axis of the dipole positive from $-m$ to $+m$. Thus

$$V = \frac{M \, \cos \, \theta}{r^2} \, . \qquad (14)$$

Or if $\vec{r}$ is the position vector of the point P (Figure 2), the potential

$$V = \frac{\vec{M} \cdot \vec{r}}{r^3} = \frac{\vec{M} \cdot \hat{r}_0}{r^2} \, , \qquad (15)$$

in which $\hat{r}_0$ is a unit vector in the direction of $\hat{r}$. Since the magnetic intensity (or force) is $-\nabla V$ we have from (8) and (14) that

$$-\nabla V = \frac{\partial}{\partial x} \frac{M \, \cos \, \theta}{r^2} \hat{i} + \frac{\partial}{\partial y} \frac{M \, \cos \, \theta}{r^2} \hat{j} + \frac{\partial}{\partial z} \frac{M \, \cos \, \theta}{r^2} \hat{k} \, . \qquad (16)$$

In Figure 2 let P be a point on the earth's surface with the dipole at the earth's center, O. With origin at P take an $\hat{i}$, $\hat{j}$, $\hat{k}$ set of rectangular right handed axes. ($\hat{j}$ is perpendicular to the plane of Figure 2 pointing inward.) With dx, dy, and dz measured respectively along $\hat{i}$, $\hat{j}$, $\hat{k}$ we have:

$$dx = r \, d\theta, \quad dy = r \, \sin \, \theta \, d\phi, \quad dz = dr \qquad (17)$$

and

$$\frac{\partial V}{\partial x} = -\frac{1}{r} \frac{\partial V}{\partial \theta}, \quad \frac{\partial V}{\partial y} = -\frac{1}{r \, \sin \, \theta} \frac{\partial V}{\partial \phi}, \quad \frac{\partial V}{\partial z} = \frac{\partial V}{\partial r} \qquad (18)$$

so that with $V = (M \, \cos \, \theta)/r^2$ we have

$$-\nabla V = -\frac{M \, \sin \, \theta}{r^3} \hat{i} + \frac{2M \, \cos \, \theta}{r^3} \hat{k} \, . \qquad (19)$$

Since along a parallel of latitude (in the direction of $\hat{j}$) the potential V is constant, it does not vary with the longitude ϕ, i.e., $\partial V / \partial \phi = 0$. In (19) the component along $\hat{i}$ is negative and that along $\hat{k}$ is outward along r.

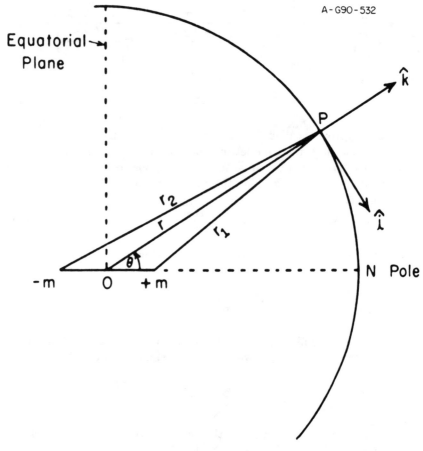

Fig. 2

The first of these is the horizontal component and the second, the vertical component. Since for the earth's field the horizontal component H is positive northward and the vertical component Z is positive downward in the northern hemisphere this means that the dipole approximating the earth's field is directed opposite to that in Figure 2. Even if the earth's main field were exactly that of a dipole this, of course, would not necessarily imply the physical existence of a dipole at the earth's center. For a uniformly magnetized sphere the intensity of magnetization I (I = magnetic moment per unit volume) has the same direction throughout the volume of the sphere. This uniform magnetization is equivalent to a dipole of magnetic moment

$$M = \frac{4 \pi a^3 I}{3} \tag{20}$$

with a = the radius of the sphere. From (19) the horizontal component H_0 of the earth's field (assuming this to be that of a dipole directed opposite to that of Figure 2 to which equation (19) applies) is

$$H_0 = \frac{M}{a^3} \qquad (21)$$

with a = earth's radius. From (20) and (21) we obtain

$$H_0 = \frac{4 \pi I}{3} . \qquad (22)$$

Since $H_0 \approx 0.30$, the value for I from (22) is about 0.07 gauss. Although cobalt-steel can be given an intensity of magnetization about 5×10^3 as great (i.e., 360 gauss) it is nevertheless quite certain that the material in the earth's solid crust could not have an intensity of magnetization sufficient to account for the earth's main field. Below the earth's crust the temperature is above the Curie point so that permanent magnetization therein is impossible. Also it can be shown theoretically that the dipole approximating the earth's main field could be produced by a current-system at the earth's surface. In this model the currents flow from east to west with a current density [1] of about 0.8 sin Θ amp cm^{-1}, wherein Θ = geomagnetic pole distance. Thus at the geomagnetic equator 0.8 amperes would flow westerly across each cm along a magnetic meridian.

Figure 3 shows the lines of force and the equipotential lines in a plane through the dipole axis. The direction of the dipole moment is not indicated but for the dipole approximating the earth's field the lines of force are directed downward at the north magnetic pole.

1.6 Dipole Approximation to Earth's Field

Bartels [2] showed that the best approximation to the earth's field (for 1922) with a magnetic dipole at the center of the earth was obtained when the axis of the dipole was aligned in the direction from a point A, near the south geographic pole, to the antipodal point B on the earth's surface, with B at 78°.5 north latitude and 291°.2 east longitude. The magnetic moment of the dipole was 8.1×10^{25} gauss cm^3 yielding a maximal horizontal intensity of 0.315 gauss at the earth's surface. Figure 4 shows the deviations of the actual horizontal vectors for the earth's field in 1922 from those obtained with the centered dipole approximation. From Figure 4 it will be noted that at many points on the earth the magnitude of the discrepancies amounts to as much as 0.1 gauss (or 10,000 γ). Bartels then determined the location of the eccentric dipole that best approximated the earth's field. This dipole was constrained to

7

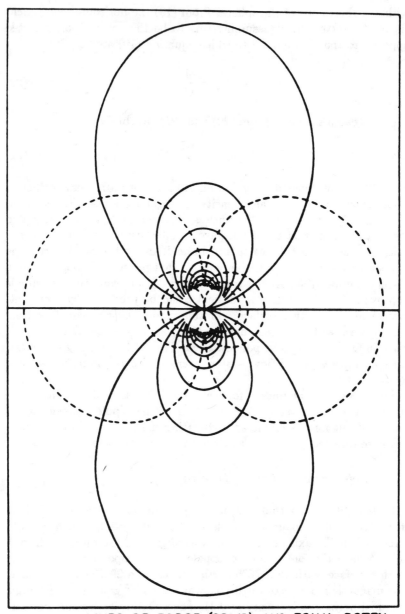

—LINES OF FORCE (SOLID) AND EQUAL POTEN-
TIAL (BROKEN) FOR MAGNETIC DIPOLE

Fig. 3

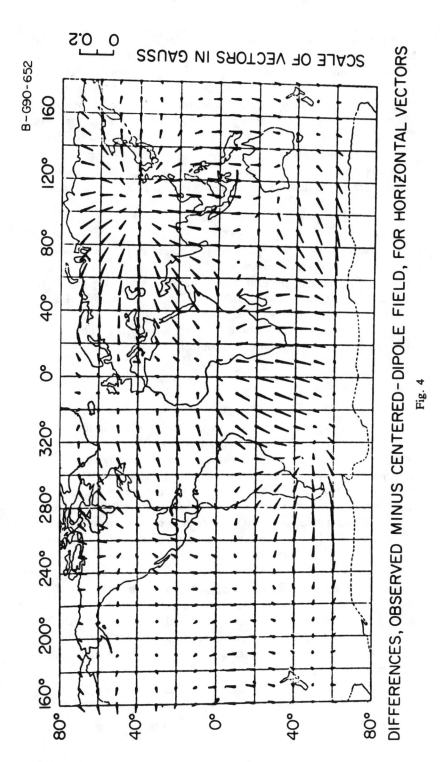

B–G90–652

SCALE OF VECTORS IN GAUSS

DIFFERENCES, OBSERVED MINUS CENTERED–DIPOLE FIELD, FOR HORIZONTAL VECTORS

Fig. 4

9

have the same magnetic moment as that for the centered dipole and to have a direction parallel to that of the best fitting centered dipole. This eccentric dipole had its center at a point C, 342 km from the earth's center, O, and in a direction towards a point C′ in latitude 6°.5 north and longitude 161°.8 east. Figure 5 shows the differences, observed minus eccentric dipole field, or regional anomalies, for horizontal vectors. The improvement in the approximation due to the shift of the magnetic dipole from the earth's center to the point C (magnetic center) is seen from Figure 5 to be scarcely noticeable for the field in the northern hemisphere but is distinct in the southern hemisphere.

1.7 Divergence and Laplace's Equation

Suppose that we have a vector point function $\vec{V}$ (the first space derivatives of which are finite, continuous, and single valued). Take a small element of volume of magnitude δ which contains the point P and is bounded by the small closed surface ω. Let the magnitude of a typical differential element of area of ω be denoted by $d\sigma$ and let $\hat{\mathcal{E}}$ denote a unit vector in the direction of the outward normal to ω. The divergence of $\vec{V}$ can best be defined [3] in the following way:

$$\text{div } \vec{V} = \mathop{\text{Lt}}_{\delta \to 0} \frac{1}{\delta} \int_\omega \hat{\mathcal{E}} \cdot \vec{V} \, d\sigma . \tag{23}$$

Thus divergence $\vec{V}$ at a point P is a measure of the outward flux of $\vec{V}$ over the surface ω or rather the limiting value of this flux as the boundary volume is shrunk towards zero. If the source of $\vec{V}$ is, for example, a positive charge of electricity of volume density ρ uniformly distributed throughout the volume, Δ, of a sphere with infinitesimal radius ϵ and with center at P, then $\vec{V}$ is the electric field intensity, $\vec{E}$, and it is directed normally outward over the surface of the sphere. Thus $\hat{E} \cdot \vec{V}$ is uniform over the small spherical surface and is equal to E. Applying the definition (23) we have div $\vec{E} = 4\pi\epsilon^2 \, E/\Delta$ and since $E = \rho\Delta/\epsilon^2$ then div $\vec{V} = 4\pi\rho$. If there are no sources of $\vec{V}$ (i.e., no charges, magnetic poles, etc.) within the volume δ in (23) then the normal flux of $\vec{V}$ over the surface of ω can only come from sources outside δ. The total flux normal to ω from outside sources can readily be shown to vanish since the normal flux into δ is just opposite to that out of δ. Thus everywhere in free space div $\vec{V} = 0$. In rectangular coordinates the divergence of $\vec{V}$ is given by:

$$\text{div } \vec{V} = \frac{\partial V_x}{\partial x} + \frac{\partial V_y}{\partial y} + \frac{\partial V_z}{\partial z} \tag{24}$$

in which V_x, V_y, and V_z are the components of $\vec{V}$. From the definition

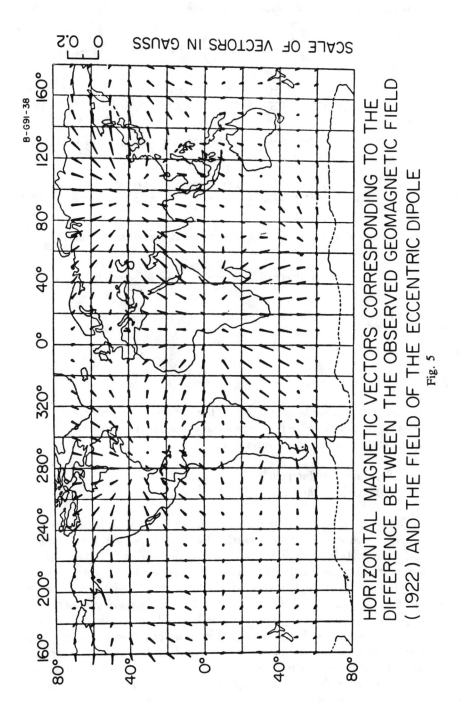

SCALE OF VECTORS IN GAUSS

0   0.2

B-G9I-38

HORIZONTAL MAGNETIC VECTORS CORRESPONDING TO THE DIFFERENCE BETWEEN THE OBSERVED GEOMAGNETIC FIELD (1922) AND THE FIELD OF THE ECCENTRIC DIPOLE

Fig. 5

11

of the operator ∇ in equation (8) we have

$$\nabla \cdot \vec{V} = \left(\frac{\partial}{\partial x} \hat{i} + \frac{\partial}{\partial y} \hat{j} + \frac{\partial}{\partial z} \hat{k} \right) \cdot (V_x \hat{i} + V_y \hat{j} + V_z \hat{k})$$

or

$$\nabla \cdot \vec{V} = \frac{\partial V_x}{\partial x} + \frac{\partial V_y}{\partial y} + \frac{\partial V_z}{\partial z} = \text{div } \vec{V} . \qquad (25)$$

If the vector $\vec{V}$ is the gradient of the scalar point function U then we have for the divergence of the gradient of U:

$$\text{div grad } U = \frac{\partial^2 U}{\partial x^2} + \frac{\partial^2 U}{\partial y^2} + \frac{\partial^2 U}{\partial z^2} . \qquad (26)$$

From the above discussion of divergence it can be seen that

$$\nabla^2 U = \text{div grad } U = \frac{\partial^2 U}{\partial x^2} + \frac{\partial^2 U}{\partial y^2} + \frac{\partial^2 U}{\partial z^2} = 0 \qquad (27)$$

for all points in free space. This is the celebrated equation of Laplace. In (27) ∇^2 is the operator

$$\nabla^2 = \frac{\partial^2}{\partial x^2} + \frac{\partial^2}{\partial y^2} + \frac{\partial^2}{\partial z^2} .$$

Equation (27) is fundamental to problems concerned with potential theory and, in particular, to problems arising in connection with the analysis of the earth's field and its variations.

1.8 Separation of Internal and External Potential for the Earth's Field

Equation (27) in rectangular coordinates is not directly suitable to problems concerned with the earth's field; for these problems a spherical coordinate system is not only more appropriate but also greatly simplifies the analysis in many cases. For problems concerned with the magnetic fields of ships or submarines a system of prolate spherical coordinates is most useful. Generalized techniques [3,4] exist for obtaining Laplace's equation in any system of orthogonal curvilinear coordinates, on the basis of which equation (27) may be obtained in spherical coordinates. Solutions of Laplace's equation in spherical coordinates are known as spherical harmonics. The potential V of the earth's magnetic field at or near the earth's surface may be expressed [5] as a series of spherical harmonic terms

$$V = \sum_n \sum_p V_n^p \qquad (28)$$

where V_n^p depends on θ the north polar distance (or co-latitude) and on λ the east longitude, solely through the surface harmonic factor

$$P_n^p(\cos \theta)(A \cos p\lambda + B \sin p\lambda) . \qquad (29)$$

$P_n^p (\cos \theta)$ denotes the associated Legendre function of degree n and order p; n and p are positive integers or zero and $p \leq n$. A and B are functions of r (the radial coordinate) of the type indicated in the following general expression for V_n^p:

$$V_n^p = \left\{ \left(E_{n,a}^p \frac{r^n}{a^{n-1}} + I_{n,a}^p \frac{a^{n+2}}{r^{n+1}} \right) \cos p\lambda \right.$$

$$\left. + \left(E_{n,b}^p \frac{r^n}{a^{n-1}} + I_{n,b}^p \frac{a^{n+2}}{r^{n+1}} \right) \sin p\lambda \right\} P_n^p(\cos \theta) . \qquad (30)$$

The terms in the potential which are of positive degree in r relate to the part of the field that has its origin above the earth's surface (r = a), while the other terms are associated with the part that originates within the earth. The factors (or coefficients) E_n^p, I_n^P, associated with these terms are chosen so as to suggest this internal and external character. The potential given by (30) may be differentiated to give spherical harmonic terms for the X (north), Y (east), and Z (downward) components of the magnetic field. These latter expressions indicate the type of functions which are used to fit the observed values of the earth's field as read from charts at points of a regular grid formed by uniformly spaced meridians and parallels of latitude. The coefficients of these functions are those in equation (30). In this way the ratio of the coefficients representing the part of the field which originates above the earth to that originating inside the earth can be obtained for terms of each order and degree. For the earth's main field practically all of the potential originates within the earth. The most recent analysis of the earth's main field was for the epoch 1945, made by Vestine, Lange, Laporte, and Scott [6]. In this analysis terms up to degree 6 and order 6 were used, which involved determining a large number of coefficients. Reference [6] gives further details concerning the procedure used to determine the coefficients, together with their numerical values.

1.9 Secular Variations

Figure 6 shows the total secular change in horizontal components of the earth's field (combined change in X and Y) during the interval 1885–

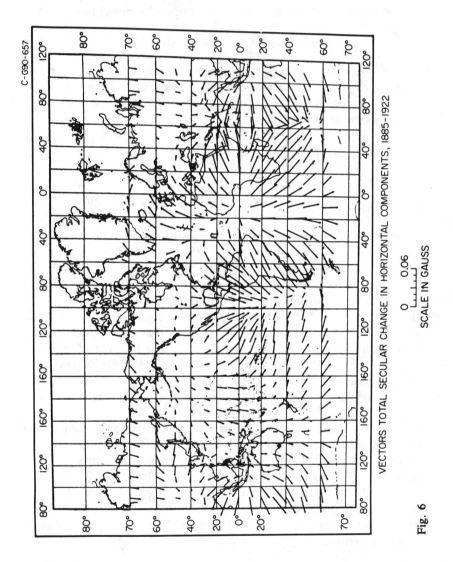

VECTORS TOTAL SECULAR CHANGE IN HORIZONTAL COMPONENTS, 1885-1922

0   0.06
SCALE IN GAUSS

C-690-657

Fig. 6

14

1922 as determined by McNish [7]. In some localities the magnitude of the change during these 37 years amounts to as much as twenty percent of the total horizontal force at the equator. Similarly the secular change in vertical intensity in some areas amounted in 37 years to as much as ten percent of the total vertical force at the poles. Figures 7 and 8 as derived by Vestine et al. [6] show respectively the contours for equal annual change in the northward (X) and eastward (Y) geomagnetic components for the epoch 1942.5. Results from charts like that in Figure 7 and in Figure 8 and from similar charts for the secular change in Z may be subjected to spherical harmonic analysis in the same manner as for the earth's main field. From the results of such spherical harmonic analysis the coefficients obtained may be used to describe the system of electric currents that would be required to produce the secular change, the source of which is inside the earth's surface. Vestine et al. [6] used the results of these spherical harmonic analyses to determine what these current distributions would be at depths of 0, 1000, 2000, and 3000 km below the earth's surface. Incidentally, the spherical harmonic coefficients facilitate the necessary extrapolations to various depths. Vestine et al. [6] thus found that the yearly changes in current rapidly increase in complexity with increasing depth and that this increased complexity would have been even greater if the observed secular change had been fitted with a greater number of harmonic terms than was actually used. They thus inferred that a major part of secular change does not originate in a region of greater depth than 3000 km, and that a lesser depth is probable by virtue of greater simplicity in concepts. On the basis of similar arguments Vestine concluded that if the main field of the earth were due to electric currents then the principal region of flow is likely to be between a depth of 1000 km and 3000 km below the earth's surface.

On the other hand Elsasser [8] believes that the currents that could account for the main field and its secular variation arise from fluid motions of the earth's core. Once a field is present he shows how this could be not only maintained but even amplified by the fluid motion of a highly conducting earth's core, and that the effects of the earth's rotation would tend to line up the eddies of current in the core in such a way as to provide an earth's field similar to that observed. Elsasser points out that some stars have magnetic fields which change rapidly and that, since stars have temperatures that certainly do not permit magnetization, a dynamic mechanism, such as that postulated for the earth's field, is required to explain their fields.

1.10   Drift of the Eccentric Dipole

The first few terms (or coefficients) from the spherical harmonic analysis of the earth's main field provide [2,9] the necessary data for locating

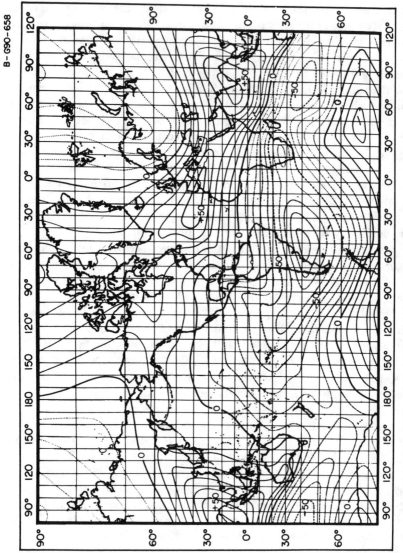

Fig. 7   GEOMAGNETIC SECULAR CHANGE IN GAMMAS PER YEAR, NORTH COMPONENT, EPOCH 1942.5

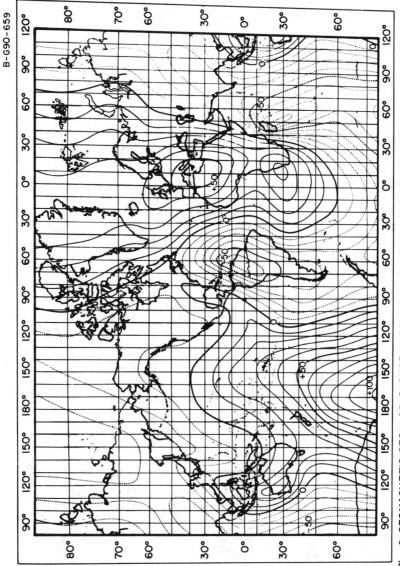

Fig. 8 GEOMAGNETIC SECULAR CHANGE IN GAMMAS PER YEAR, EAST COMPONENT, EPOCH 1942.5

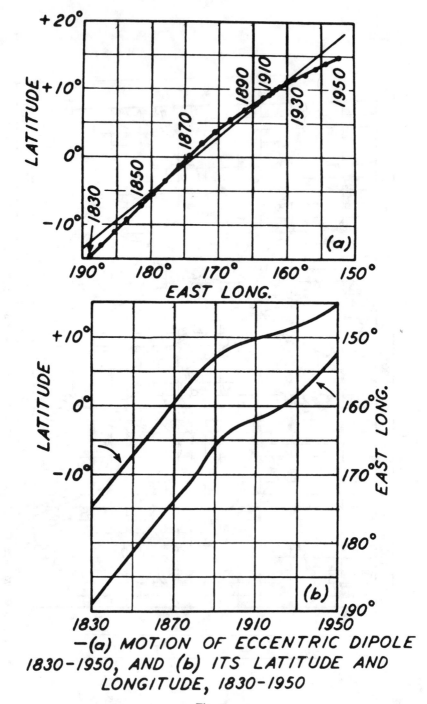

—(a) MOTION OF ECCENTRIC DIPOLE 1830-1950, AND (b) ITS LATITUDE AND LONGITUDE, 1830-1950

Fig. 9

the center of the eccentric dipole which best approximates the earth's main field. Figure 9(a) shows how the center of this eccentric dipole has apparently drifted westward and northward during the interval 1830–1950. Figure 9(b) shows how the latitude (upper curve) and longitude (lower curve) have changed with time during the same interval. Brouwer [10] has used astronomical data to estimate changes in the rate of rotation of the earth's crust during the period 1820–1950. Such changes are superimposed on the long term secular variation and are of much too great a magnitude to be explained by known effects in the atmosphere and oceans. Vestine [9] found that the rate of drift of the eccentric dipole corresponded in magnitude and direction to changes of the angular momentum of the core required to compensate those of the crust, i.e., in order to conserve total angular momentum. Hence, he concluded that the source of the geomagnetic field lies within a large-scale fluid circulation inside the central core of the earth and that this fluid circulation must be considered established as real since no other adequately large source needed to conserve angular momentum is apparently available. This conclusion is in general accord with Elsasser's theory. Vestine's interpretation is an example of how geomagnetic observations provide useful and important means of studying the earth's interior.

I. References

1. S. Chapman and J. Bartels, *Geomagnetism*, Vol. II, p. 645, Oxford at the Clarendon Press (1940).
2. J. Bartels, The Eccentric Dipole Approximating the Earth's Magnetic Field, *Terr. Mag.*, *41*, 225–250, 1936.
3. A. P. Wills, *Vector Analysis*, p. 93, Prentice-Hall, Inc., New York (1931).
4. M. Abraham (revised by R. Becker), *The Classical Theory of Electricity and Magnetism*, Hafner Publishing Company, Inc., New York (1932).
5. S. Chapman and A. T. Price, The Electric and Magnetic State of the Interior of the Earth, as Inferred from Terrestrial Magnetic Variations, *Philosophical Transactions of the Royal Society A, 229*, 427–460, 1930.
6. E. H. Vestine, I. Lange, L. Laporte, and W. E. Scott, *The Geomagnetic Field: Description and Analysis*, Carnegie Institution of Washington Publication 580, Washington, D.C. (1947).
7. A. G. McNish, The Earth's Interior as Inferred from Terrestrial Magnetism, *Trans. Amer. Geophys. Union*, 18th Annual Meeting, 43–50, 1937.
8. W. M. Elsasser, The Earth as a Dynamo, *Scientific American*, May 1958, pp. 44–48.

9. E. H. Vestine, On Variations of the Geomagnetic Field, Fluid Motions, and the Rate of the Earth's Rotation, *J. Geophys. Res.*, *58*, 127-145, 1953.
10. D. Brouwer, A New Discussion of the Changes in the Earth's Rate of Rotation, *Proc. Nat. Acad. Sci.*, *38*, 1-12, 1952.

II. Transient Geomagnetic Variations

2.1 Diurnal Variation

At each magnetic observatory, such as that at the Instituto Geofísico de Huancayo, variations in the earth's magnetic field are recorded continuously as daily magnetograms. On each magnetogram is recorded the magnitude of the horizontal component (H), its direction (D) or the declination, and the vertical component Z. Each day is assigned a character figure C on a scale 0, 1, or 2, according to whether the daily curves are smooth and regular (0), moderately disturbed or irregular (1), or very disturbed, showing large fluctuations and deviations from normal as on days of magnetic storms (2). Based on such reports from a large number of observatories, a central bureau assigns to each Greenwich day an average international character figure C. It also selects for each month the five "most quiet" and the five "most disturbed" days. Figure 10 shows the average diurnal variation for all of the 60 quiet days of the second International Polar Year (1932–1933) for eighteen observatories between geographic latitude $50°0$ S and $53°8$ N. For convenience in analysis the diurnal variation for each observatory, as shown in Figure 10, is that for the three geomagnetic components of the earth's field X, Y, Z which are the components referred to a set of axes through the centered dipole. ΔX, ΔY, and ΔZ are hourly mean departures from the daily mean of the northward, eastward, and downward geomagnetic components, respectively. These departures in Figure 10 are plotted according to local geomagnetic time, defined as the angle between the geomagnetic meridian through the observatory and the one opposite to the one through the sun. At stations in middle and low latitudes the geomagnetic time differs only slightly from the local geographic time [2]. Figure 11, similar to Figure 10, shows the diurnal variation of ΔX, ΔY, and ΔZ for several stations north of latitude $54°0$ N; these curves were derived by Vestine et al. [1] for the International Polar Year (1932–1933). Figures 10 and 11 show that, except for the polar regions, the diurnal variation of the three components varies in a rather systematic and regular manner with respect to latitude. Since any well behaved function on a sphere may be represented by a series of spherical harmonics, it is natural to consider the representation of the diurnal variation over the earth by such series.

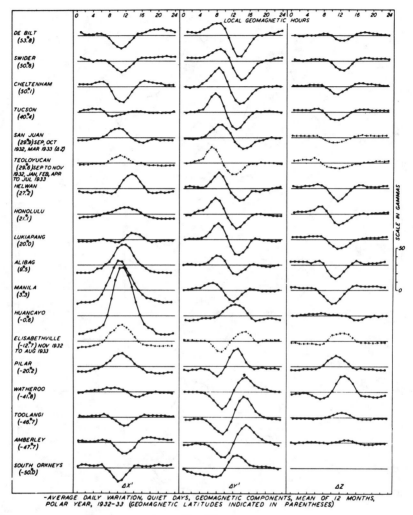

Fig. 10

Moreover, if the diurnal variation is derivable from a scalar potential and is a function only of latitude and of local time, the potential can be represented by a series of the type shown by equation (30), with only slight modification as follows:

$$V_n^p = \left[a\left\{ E_{na}^p \, (r/a)^n + I_{na}^p (a/r)^{n+1} \right\} \cos \, pt' \right.$$

$$\left. + a\left\{ E_{nb}^p \, (r/a)^n + I_{nb}^p (a/r)^{n+1} \right\} \sin \, pt' \right] P_n^p(\cos \, \theta) \qquad (31)$$

in which t' is the local time given by (t + k) where t is the time at

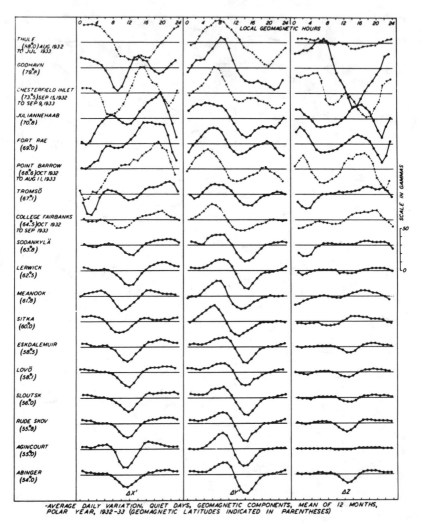

-AVERAGE DAILY VARIATION, QUIET DAYS, GEOMAGNETIC COMPONENTS, MEAN OF 12 MONTHS, POLAR YEAR, 1932-33 (GEOMAGNETIC LATITUDES INDICATED IN PARENTHESES)

Fig. 11

the meridian from which λ is measured. As outlined in connection with the spherical harmonic analysis of the earth's main field, the gradient of V_{np} from (31) in spherical coordinates provides expressions for the field components ΔX, ΔY, and ΔZ. For each of these components the dependence on longitude is given through the local time parameter t' as in (31). Thus the initial step in the spherical harmonic analysis is an ordinary harmonic analysis of the diurnal variation for each component and for each observatory. The next step consists in fitting the variation with latitude of the periodic components $p (p = 1, 2, 3$ etc.) with simple functions $P_{np}(\cos \theta)$ of the polar distance h or with functions simply

22

related to $P_{np}(\cos \theta)$ depending upon the field component being fitted. By proper combination of these results the coefficients E_{np} and I_{np} in (31) are obtained for each order p and degree n. The ratio of I_{np}/E_{np} for corresponding n and p determines the relative magnitude of the potential arising from causes inside the earth's surface to that from causes outside. This ratio is about the same for the first six or so terms required to fit the data and is about 0.4, showing that the external potential is about 2.5 times the internal potential. Consequently, the primary source of the diurnal variation is outside the earth. It should be mentioned that in the analysis it turns out that the quiet day diurnal variations, S_q, in X and Y are expressible in terms of the same spherical harmonic functions [3] and with the same numerical factors. This fact shows that the S_q field at the earth's surface has a scalar potential and thus that equation (31) may be used to describe it. Equation (31) shows that for the same p the coefficients for the potential of external (or internal) origin may be combined as follows:

$$E_{na}^p \cos pt' + E_{nb}^p \sin pt' = C_{ne}^p \cos(pt' + \alpha_{ne}^p) \qquad (32)$$

and similarly for I_{na}^p and I_{nb}^p. This permits a determination of the phase of each harmonic component (period = p) for the potential of external origin and for that of internal origin. The difference in phase between the two is summarized by McNish [4] on the basis of an early analysis by Chapman.

2.2 Electromagnetic Induction in the Earth

The separation of the potential for the diurnal variation S_q into parts S_q^e and S_q^i for the external and internal potential does not in itself explain either part. The similarity of the amplitude ratios and phase differences [3,4] for the four periodic components indicates a causal connection between the two. Since S_q^e is about two and one-half times greater than S_q^i, it is natural to consider S_q^e the cause and S_q^i the effect. A field like that from S_q^e which varies with time is certain to produce an internal field by electromagnetic induction since to some extent the earth is conducting. The internal field will arise from the currents induced within the earth. If the conductivity K were known everywhere inside the earth, the distribution of induced currents and their associated magnetic field could be calculated from the known S_q^e field. However, the earth's electrical conductivity, K, is not known below a depth of a few miles. Thus it is necessary to consider the internal fields that S_q^e would induce in a model earth of the same size as the actual earth, the model having some assumed distribution of conductivity. If the model is assumed [3] to have a constant conductivity throughout, calculation shows that the phase

23

difference between each periodic component of S_q^e and the induced field resulting from this periodic component depends only on K and not on the size of the sphere. For periodically varying external fields with periods 1, 1/2, 1/3, and 1/4 day (corresponding to the periodic components of S_q) the phase differences are all about the same and their sign is that found from the spherical harmonic analyses. These phase differences from the spherical harmonic analysis agree with those calculated from the model [3] if a value of $K = 3.6 \times 10^{-13}$ cgs is used. For comparison the value of K for sea water is 4.1×10^{-11}, for dry earth or rock $K = 10^{-15}$, and for copper $K = 6.1 \times 10^{-4}$. However, using the value $K = 3.6 \times 10^{-13}$ cgs for the whole model earth, the calculated ratios for the external and internal parts of S_q, for the four periodic components [3,4], are not in agreement with those derived from the spherical harmonic analyses. The results of calculation based on the model earth can be brought into agreement with those from spherical harmonic analysis if the model earth is assumed to have the same $(K = 3.6 \times 10^{-13}$ cgs) conductivity throughout a sphere of radius r_1, with r_1 about 160 miles less than the earth's radius, a; and if the spherical shell between radii a, and $a - r_1$ is assumed nonconducting. This outer shell is not absolutely nonconducting but its conductivity, if comparable to that for rock, would have an effect which would be small compared with that within the sphere of radius r_1 used for the model. In actuality the transition to the value 3.6×10^{-13} for the core is probably gradual and not abrupt as in the model. It should be remarked that the calculations based on the model earth show that about 90% of the field at the earth's surface that arises from induced currents is due to currents that flow within a fifth of the depth of the model core from its surface. Currents which flow at a depth greater than about half of the radius of this core have no appreciable influence on the surface S_q field. There is some evidence to indicate that K increases rapidly in the region from 160 miles depth to 400 miles depth. To obtain information on the conductivity of the central half of the core it is necessary to investigate magnetic variations with periods much longer than one day. The studies described above provide another example of how geomagnetic phenomena can be used to obtain information about the earth's interior that could be obtained in no other way.

2.3 The External Part of the S_q Field

Since there is no magnetized material in the earth's atmosphere and since the earth's atmosphere is practically nonconducting up to a height of about 70 km, the external part of the S_q field must arise from currents flowing in one of the ionospheric layers. It has been shown by

sending a rocket-borne magnetometer vertically through the lower ionosphere at the equator that the currents for the source of S_q^e flow in a
thin layer with height about 100 km. As indicated in connection with
the discussion concerning the origin of the earth's main field, the results
of spherical harmonic analyses afford a direct and convenient basis for
calculating the magnitude and direction of a system of electrical currents
which, flowing in a thin spherical shell, could produce the external part
of the S_q field. Figure 12 shows this current function for the daylight
side of the earth. This current function was derived from the observed
diurnal variation averaged for all days (1902) at each of several widely

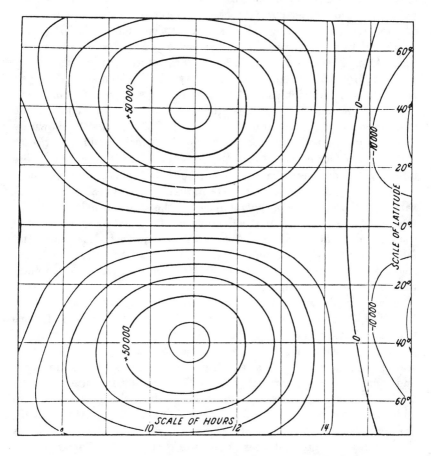

ISOMETRICS OF EXTERNAL CURRENT-FUNCTION, AVERAGE, ENTIRE
EARTH AMPERES ON DAYLIGHT SIDE OF EARTH AT INDICATED
LOCAL TIMES, ALL DAYS, 1902 DEDUCED FROM MAGNETIC DIURNAL-
VARIATIONS AT WIDELY SEPARATED STATIONS
(Original diagram by Bartels using Chapman's coefficients)

Fig. 12

separated stations (not including Huancayo which started operating in 1922). A total current of more than 60,000 amperes flows in each of the two circuits. The flow (as seen from above) is clockwise in the southern hemisphere and counterclockwise in the northern hemisphere. The current function for equinoctial months is similar to that of Figure 12 but in the months of northern summer the total current in the northern circuit is (see Figure 13) about one-third greater than that in Figure 12 while that in the southern circuit is about one-third less. In northern winter this asymmetry is reversed. The current system in Figures 12 and 13 is not fixed relative to the earth but would, if it could be seen, appear to be fixed (except for seasonal changes) to an observer on the sun. A similar but very much weaker current system flows over the night hemisphere of the earth. It will be noted in Figure 12 that at latitude 40° (north or south) the focus of the current system is overhead at 11 hours local time. After attention was called to the large diurnal variation in H (or X) at the Huancayo Magnetic Observatory, McNish [4] made an analysis of the diurnal variation using data for quiet days near the equinoxes of 1923 from five observatories near the 75° west meridian. The latitude of these stations ranged from 31°7 south to 43°8 north. As shown by the resulting current function, in Figure 14, which would give rise to the observed diurnal variations, the current system is very much more intense in the southern than in the northern hemisphere. More recently, and especially during the IGY (International Geophysical Year) observations of the variation of the amplitude of the solar diurnal variation in the neighborhood of the magnetic equator show that the amplitude of the diurnal variation of H (or X) is greatest at the magnetic equator and falls to about half this value at a distance of about 600 km north and south of the magnetic equator. In an IGY program carried out jointly by the Carnegie Institution of Washington, Department of Terrestrial Magnetism and the Instituto Geofísico de Huancayo the diurnal variation of H, Z, and D was observed in 1957 at about seventeen stations between Quito, Ecuador, and Mariá Elena, Chile. North and south of the magnetic equator the diurnal variation of Z was found to be opposite in phase and the largest amplitude occurred about 350 km north and south of the equator (magnetic). These results indicate a concentration of overhead eastward current flowing in a band about 1000 km wide with the center of the band over the magnetic equator. As for the normal diurnal variation, the current density in this band, known as the electrojet, is greatest near noon. If a current system analogous to that in Figure 14 were constructed on the basis of data from IGY stations operating at Talara, Chiclayo, Chimbote, and Yauru (all in Peru) together with data from Huancayo and the other stations from which data were used in obtaining Figure 14, this current system would be decidedly more concentrated near the magnetic equator (about 13° south

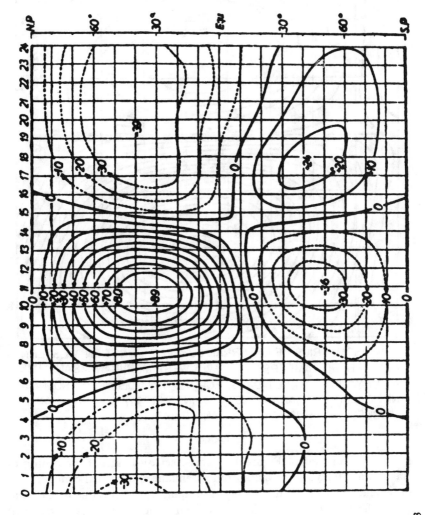

Fig. 13

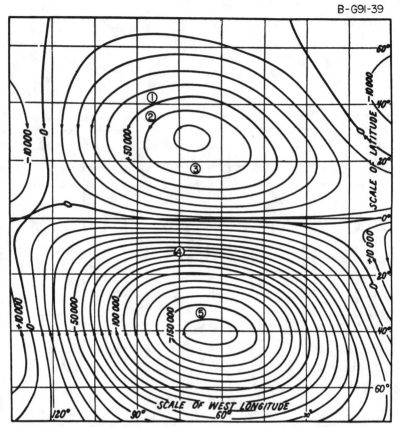

ISOMETRICS OF EXTERNAL CURRENT-FUNCTION WESTERN
HEMISPHERE, AMPERES AT II^h 75° WEST MERIDIAN MEAN TIME,
INTERNATIONAL QUIET DAYS, EQUINOX 1923 DEDUCED FROM MAG-
NETIC DIURNAL-VARIATIONS AT AGINCOURT (I), CHELTENHAM (2),
VIEQUES (3), HUANCAYO (4) AND PILAR (5)

Fig. 14

latitude) than that of Figure 14. This electrojet current, according to
Baker and Martyn [5], results from an enhanced electrical conductivity,
in this narrow band near the magnetic equator, which is due to the combined effect of the electric field which is driving the current system and
the Hall conductivity (conductivity perpendicular to the electric and
magnetic fields).

2.4   The Lunar Diurnal Variation

In some ways the lunar magnetic diurnal variation, L, is similar to
S_q. However the amplitude of the lunar variation is only about 7%

of S_q, although at Huancayo L at certain seasons may be nearly one-third of S_q. Special statistical procedures are required to separate L from S_q. This separation is difficult since on individual days, changes in amplitude or phase of S_q may appear, erroneously, as lunar effects. Basically, the lunar variation is a semidiurnal wave of constant phase but with amplitude greatly intensified during daylight hours. As in the case of S_q the potential for the lunar semidiurnal variations may be separated into external and internal parts and the phase of the periodic component determined for each. The phase difference is about the same as that for S_q and the ratio of the amplitude of the internal component to that of external origin is also about 0.4 as for S_q. It is thus reasonable to consider that the external part L^e of L is due to ionospheric currents and that the internal part L^i is due to currents induced in the earth. Figure 15 shows the current function [3] for L^e for the equinoxes and Figure 16 that for June. Chapman [2,3] showed that the magnitude of L varied with the inverse cube of the lunar distance. From apogee to perigee the lunar tide producing force increases about 37%; the increase in L was found [2] be about 33%. This is a clear indication that L^e is generated by the lunar atmospheric tides in the ionosphere.

2.5 Theory for Solar and Lunar Diurnal Variations

Of the various theories which have been suggested, that of Balfour Stewart proposed in 1882 is generally accepted as best explaining the solar diurnal variation S_q. This theory ascribed the S_q variation to horizontally flowing electric currents in the upper atmosphere [2,3,4]. Stewart suggested that these currents were induced in a conducting layer of the upper atmosphere by the systematic horizontal movements of this conducting layer across (i.e., at right angles to) the vertical component of the earth's field. The necessary air motions were attributed to tidal forces. From the analogy to current generation in a dynamo, this theory is called the dynamo theory. In 1889 Schuster first proved from spherical harmonic analysis that the primary cause of S_q (i.e., S_q^e) was outside the earth, and later still Kennelly and Heaviside showed that radio propagation around the earth required a conducting region in the upper atmosphere. The existence of this conducting layer was demonstrated directly by Breit and Tuve. Thus the conditions postulated by Balfour Stewart in the dynamo theory were later shown to be correct. The necessary tidal movements of the conducting layer involve principally a semidiurnal oscillation of the atmosphere which arises from solar gravitational and thermal effects. The latter are responsible for the well known 12-hour wave in barometric pressure, which, at the equator, attains an amplitude of about 1 mm Hg. The maximum of this wave occurs between 0930 and 1030 local time. This pressure variation can be

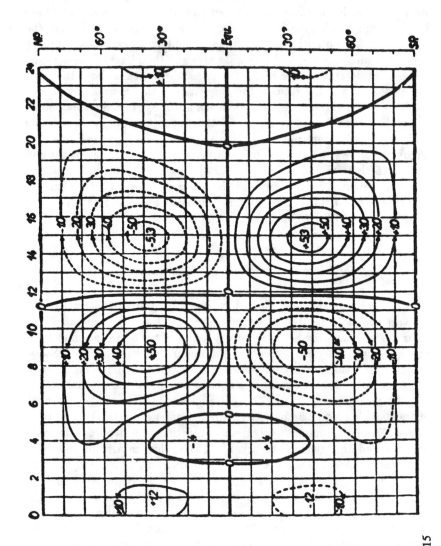

Fig. 15

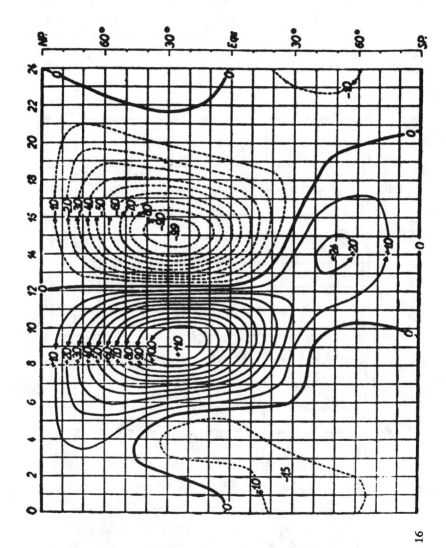

Fig. 16

closely approximated over the earth by means of a few surface spherical harmonics, which can in turn be conveniently used not only to obtain the air motions (at the earth's surface) but also to obtain the induced currents from similar air motions at ionospheric levels if the conductivity is known. Figure 17 shows the air motions derived from the semidiurnal atmospheric tides for the western hemisphere, and Figure 18 shows the induced electromotive forces arising from these motions of air at right angles to the earth's vertical field. In Figure 18 the emf's in the equatorial region are small. The current in Figure 14, for example, in the

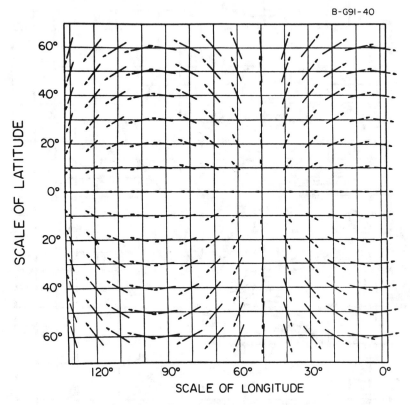

B-G9I-40

AIR-VELOCITIES ARISING FROM SEMI-DIURNAL ATMOSPHERIC TIDES IN WESTERN HEMISPHERE AT 11ʰ 75° WEST MERIDIAN MEAN TIME, TYPICAL OF THOSE OCCURRING ANYWHERE AT THE SAME LOCAL TIME
SCALE ⌐⊓⊓⊓⌐ CENTIMETERS PER SECOND
     O 50

Fig. 17

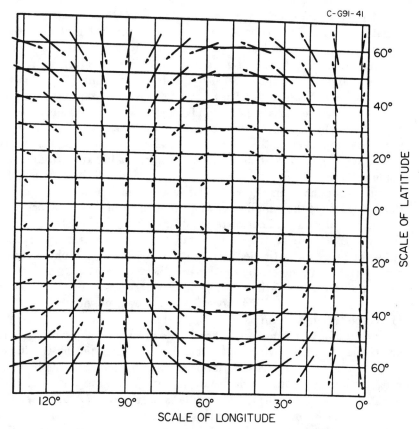

ELECTROMOTIVE FORCES IN ATMOSPHERE OF
WESTERN HEMISPHERE CAUSED BY SEMI-DIURNAL
ATMOSPHERIC TIDES AT 11^h 75° WEST MERIDIAN MEAN
TIME, ASSUMING VERTICAL MAGNETIC FORCE CON-
STANT ALONG EACH PARALLEL OF LATITUDE

SCALE ⌐⊤⊤⌐ VOLT PER KILOMETER
0   0.03

Fig. 18

neighborhood of the equator is driven by the electrostatic field set up to
preserve continuous current flow. Calculation of the current functions for
S_q from the air motions requires knowledge of the conductivity, which
in turn is a function of the solar zenith angle. In addition, it should
be mentioned that the earth's rotation modifies [2] air motions derived
from atmospheric semidiurnal pressure variations—and shown in Figure 17. When all of these effects are taken into account, the resulting
predicted current system does not agree in phase with the current sys-

33

tem derived empirically by spherical harmonic analysis of the magnetic variations. The discrepancy is in part due to the fact that both theory and observation indicate a phase change of the pressure wave with altitude. Recently, direct observations of the diurnal variation of wind direction and velocity have been made at Jodrell Bank (England) by Greenhow and Neufeld [6]. They used radio echoes from meteor trails to investigate the variations of wind velocity at altitudes between 80 and 100 km. In this region they found that the phase of the semidiurnal periodic component of the wind changes about $5°$ km^{-1} and that the amplitude increases by about 1 m sec^{-1} km^{-1}. At 100 km altitude they found that the amplitude of the semidiurnal wind velocities were 20 to 80 times that shown in Figure 17. Greenhow and Neufeld [6] indicated that the phase of the semidiurnal wind component in the region between $85-110$ km altitude did not correspond to that required for the dynamo theory of S_q. However, they pointed out that if they extrapolated their curve for phase vs. altitude to an altitude of 135 km, the phase would accord with that required by the dynamo theory. Measurements of wind velocities at other locations, using radio reflections from brief meteor trails ($0.2-1.0$ sec) and over long periods, would provide valuable information for understanding more completely the details of the dynamo theory. In addition, long series of such measurements might also permit determination of the winds at various heights generated by the lunar tide. The lunar semidiurnal variation is explained on the same basis as is S_q. Although the lunar tide producing force of the moon is twice that of the sun, the lunar atmospheric tide is only about 7% of the sun's. This difference is due to the resonant characteristics of the atmosphere. Mention should be made of the diagmagnetic and drift theories for S_q. The first of these, proposed by Gunn [7], ascribed the source of S_q to ions which, having thermal velocities, would in regions of long free paths spiral around magnetic lines of force to generate a current system. This theory gave a current system in close agreement with the empirical one but the number of ions required was greater than indicated from ionospheric measurements. The drift theory proposed by Chapman [8] was based on the fact that ions in a gravitational or electric field which is perpendicular to a magnetic field will describe paths which on the average are perpendicular to both and consequently generate a current system. But no detailed treatment of this theory was never made. The number of ions it required was more nearly in accord with the results of radio measurements than that for the diamagnetic theory. Neither of these two theories can explain the lunar variation. Since the only effect of the moon of consequence for explaining the lunar diurnal variation is the production of tidal atmospheric motions and since L_e/S_q^e is about the same as the ratio of the semidiurnal lunar and solar tides (at the ground), this is practically conclusive evidence that the dynamo theory is the correct one for S_q and for L. From the IGY electrojet project in Peru it may be possible to determine whether an electrojet effect exists for the lunar semidiurnal variation. Results from the newly established magnetic observatory at Arequipa, together with those from Huancayo, should prove most valuable in the future for studies of lunar variations as well as for investigations of variations in the electrojet effect over periods longer than the IGY. The presence or absence of electrojet effects in L should aid in determining whether the currents for L^e flow at the same level as do those for S_q^e. Systematic measurements of wind velocities from meteor trails at more places on the earth would provide invaluable data for a more thorough understanding of the details of the dynamo theory and for delineating the causes for the variability of the amplitude of S_q.

2.6   Solar Flare Effects

Figure 19 shows three photographs of the sun during a bright chromospheric eruption or solar flare on about April 8, 1936. At the top of Figure 19 three ionospheric records are shown, the center one of which shows the disappearance of radio echoes from the ionosphere at 16^h 46^m GMT. The upper right hand figure shows that echoes from the E layer (lower trace) and from the F layer (upper trace) did not appear until about an hour later. This radio fade-out is due to a large increase in ionization, probably by x rays in the solar flare radiation, in a region at the base of or below the E region. Such an increase of ionization in a region of high collision frequency increases the absorption of radio waves of the frequency used in detecting echoes from the ionospheric layers. After this absorbing layer has disappeared, the echoes from the E and F regions reappear in their normal state. The second record from the bottom of Figure 19 is a magnetogram from the Instituto Geofísico de Huancayo. At 16^h 46^m the horizontal intensity shows a rapid increase, and the Z record shows simultaneously a small decrease. Both records show a return to normal about half an hour later. The bottom record of Figure 19 shows effects simultaneously in earth-current potentials from Huancayo. In Figure 20 the light arrows show at several observatories the direction and magnitude of the horizontally flowing currents responsible for the normal diurnal variation S_q immediately preceding the solar flare effect. The heavy arrows show the additional currents that would account for the additional magnetic changes which began at 16^h 46^m. The heavy vectors in Figure 20 are all essentially parallel to the light ones, and at all the locations the amplitude of the heavy vectors is roughly half that of the light ones. This fact clearly indicates [9] that the effect of this solar flare was to enhance the strength of the current system for S_q by increasing the conductivity either in the region in which the S_q currents

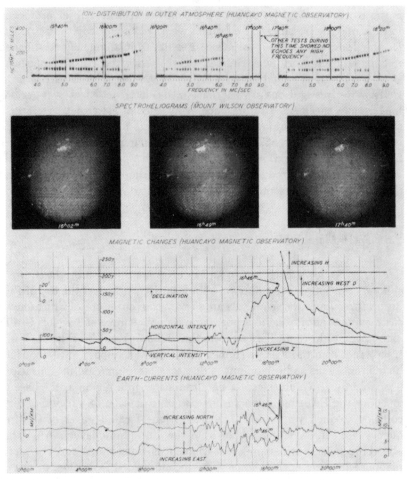

Magnetic, radio, and earth-current disturbances associated with brilliant solar eruption 1936 April 8: Greenwich mean time throughout (Carnegie Institution of Washington)

Fig. 19

flow or in a region in which the air motions are essentially the same as those in the S_q region. If the change of phase of the semidiurnal wind component with altitude is everywhere similar to that found by Greenhow and Neufeld [6], this would mean (using their figure of 5° km^{-1}) that the current system for the solar flare effects must flow at about the same altitude as that for S_q; otherwise the vectors in Figure 20 at each station would not be parallel. This geomagnetic effect from solar flares and the recent analysis by McNish [10] of geomagnetic effects from nighttime upper altitude nuclear explosions leave little doubt that the dynamo theory is the correct one for S_q (and L). It may be necessary to await measurements of flare radiation from satellites to determine with

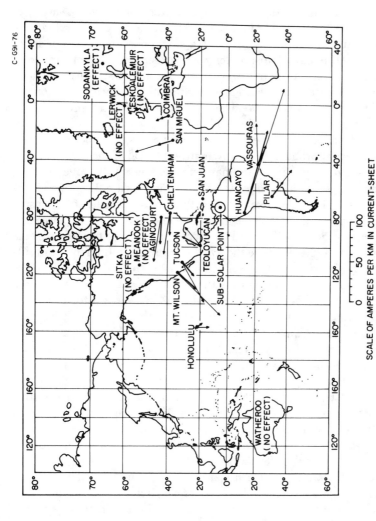

SCALE OF AMPERES PER KM IN CURRENT-SHEET

OVERHEAD CURRENTS NECESSARY TO PRODUCE MAGNETIC CHANGES AT TIME OF
SOLAR FLARE BEGINNING 16ʰ 46ᵐ GMT, APRIL 8, 1936 (HEAVY ARROWS) AND NORMAL
DIURNAL-VARIATION CHANGES AT SAME TIME (LIGHT ARROWS)

Fig. 20

37

certainty which component of the flare radiation causes the increased ionization.

2.7   Storm Time Variations

During days of magnetic storms magnetograms show such a variety of changes in the earth's field and such differences among storms that one would not anticipate from records at a single observatory for individual storms the remarkable simplicity and order that become apparent when storm data from several observatories are analyzed. Most magnetic storms start with an abrupt change or sudden commencement in one or more of the field components. These sudden commencements occur at the same time (at least within a minute or less) over the whole earth. Moreover the sudden commencements (from a large number of storms) occur with equal probability in any hourly interval of the day. This observation suggests averaging, for many storms, the hourly values for each magnetic component according to storm time. Figure 21 shows Chapman's [2] results for the average of forty storms of moderate intensity. This figure shows the initial rise in H for a few hours (initial phase) and the subsequent decrease followed by a gradual recovery towards normal. For stations at higher latitude the average decrease in H (right hand side of Figure 21) is less than that for stations near the equator (left side of Figure 21). Figure 21 shows that there is little change in D (declination) and that in the northern hemisphere Z (or vertical force V.F.) is positive during the main phase of the storm, that is, when H is below normal. For the main phase of the storm H is below normal in both the northern and southern hemispheres, whereas Z (or V.F.) is positive in the northern hemisphere and negative in the southern. The average storm time variation is essentially independent of longitude. Except for polar regions this fact and the simplicity of the storm-time variation mean that the potential (for a given hour during the main phase or for the daily mean components during the main phase) can be represented by a few zonal harmonics and that the ratio of the coefficients for the external and internal potential is readily determined for each term in the series. The largest coefficients are those that are equivalent to those which one would obtain for a dipole (or the equivalent current function). The ratio of the coefficient in this main term for the internal potential to that for the external potential is about 0.4 or about the same as that obtained from the S_q analysis, showing that the primary cause for magnetic storm time variations is outside the earth's surface. Figure 22 shows Chapman's [1,2] current system for producing the average magnetic storm time variation. In this figure, the total current between the northern and southern auroral zones is about 400,000 amperes. This value is for the average of moderate storms; for some great storms the

38

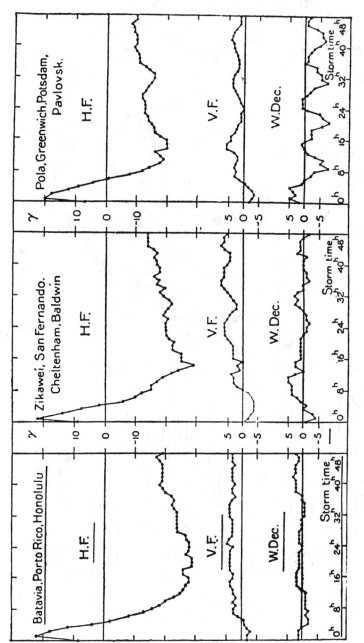

Fig. 21

39

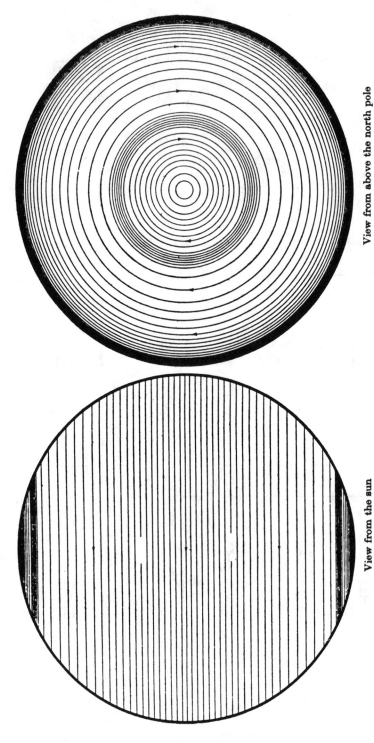

View from the sun

View from above the north pole

Views of the idealized overhead electric current-system that could produce the field of the average storm-time disturbance

Fig. 22

40

total current is several times as great. Although the source of storm time variations could be a current system in the upper atmosphere like that in Figure 22, it could also be (except for the part in the auroral zones which is known to flow in the upper atmosphere) a current which flows westward in a geocentric ring with the plane of the ring in the plane of the geomagnetic equator. The radius of such a ring must be at least twice that of the earth [11]. The correlation of geomagnetic activity with sunspot numbers, the more frequent occurrence of magnetic storms at the times of sunspot maxima, and the fact that magnetic storms often occur from 18 to 36 hours after a solar flare leave little doubt that they are caused by clouds of ionized gas emitted by the sun and arriving at the earth about 24 hours later. The physical mechanism involved in producing the observed effects is not yet clearly demonstrated. A likely possibility is that when such ionized clouds from the sun collide with the earth's magnetic field (or when the earth moves into such clouds) the charged particles in the cloud spiral round the lines of force in the earth's magnetic field. Since most of the charged particles will have a component of velocity along lines of force, they will follow the lines of force towards the auroral zones, where some of the particles will produce the aurora. Other particles are reflected from "mirror points" as proved in the Argus experiments and return to the opposite hemisphere along lines of force. As these charged particles oscillate between the northern and southern hemispheres they also drift around the earth (probably in the outer Van Allen belt) in such a way as to give rise to a westward flowing current system similar to that in Figure 22. This theory, yet to be established, has the merit of providing (1) the outer Van Allen radiation belt, (2) the aurora, and (3) the magnetic storm field. Further satellite (artificial) observations of the density and energy of particles and the magnetic field during magnetic storms will undoubtedly provide the information for determining whether this or other theories of magnetic storm mechanisms are correct.

2.8 Disturbance Daily Variation

Thus far we have discussed the solar and lunar diurnal variations and the storm time variation. The last was derived from averages for many storms of the hourly values arranged according to time (storm time) from the beginning of the storm. If now one averages the hourly values of magnetic components, for individual observatories, as a function of local time for the five most disturbed days of each month (60 per year) and from this average curve subtracts the average diurnal variation for the five quiet days, then one obtains the disturbance diurnal variation S_D; i.e., the additional diurnal variation on disturbed days. Part C of Figure 23 shows the current system for S_D; part B, the current system for

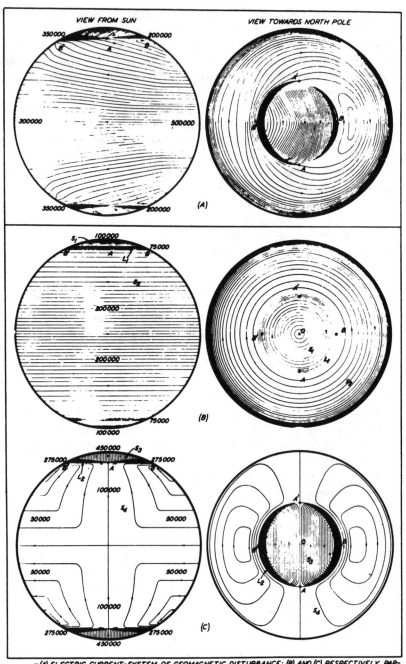

—(A) ELECTRIC CURRENT-SYSTEM OF GEOMAGNETIC DISTURBANCE; (B) AND (C) RESPECTIVELY, PAR-TIAL CURRENT-SYSTEMS D_{st} AND S_D COMPRISING (A)

Fig. 23

the storm time variation, D_{st} (as in Figure 22); and part A, the current system for S_D and D_{st} combined. It is evident in Figure 23C that the current system for S_D is concentrated near the auroral zone indicating a greatly enhanced conductivity there on disturbed days. The theory for the S_D variation is by no means complete though some investigators [12] have shown that a consistent wind system can account for both the S_q and S_D variations if the appropriate change of conductivity (especially in polar regions) is assumed for S_q. Vestine [13] also used S_D (and D_{st}) to derive a wind system that would explain both on the basis of the dynamo theory. Here it should be mentioned that the IGY cosmic-ray groups in a series of high altitude balloon flights to monitor cosmic-ray intensity discovered x rays during some aurora. A close correspondence in time was established between the occurrence of x rays and magnetic disturbances. There is no doubt that the large amount of IGY data in geomagnetism, aurora, ionosphere, cosmic rays, and other phenomena will prove of great value in clarifying theories for geomagnetic variations and related phenomena. Moreover, these data are available to anyone through the IGY World Data Centers.

II. References

1. E. H. Vestine, I. Lange, L. Laporte, and W. E. Scott, *The Geomagnetic Field: Description and Analysis*, Carnegie Institution of Washington Publication 580, Washington, D.C. (1947).
2. S. Chapman and J. Bartels, *Geomagnetism*, Vol. II, p. 645, Oxford at the Clarendon Press (1940).
3. S. Chapman, *The Earth's Magnetism*, Methuen's Monographs on Physical Subjects, John Wiley & Sons Inc., New York (1951).
4. A. G. McNish, On Causes of the Earth's Magnetism and Its Changes, *Physics of the Earth-VIII*, Terrestrial Magnetism and Electricity, pp. 308–384, McGraw-Hill Inc., New York and London (1939).
5. W. G. Baker and D. F. Martyn, Electric Currents in the Ionosphere. I. The Conductivity, *Phil. Trans. R. Soc.*, *246*, 281–294, 1953.
6. J. S. Greenhow and E. L. Neufeld, The Height Variation of Upper Atmospheric Winds, *Phil. Mag.*, *1*, 1157–1171, 1956.
7. R. Gunn, The Diamagnetic Layer of the Earth's Atmosphere and Its Relation to the Diurnal Variation of Terrestrial Magnetism, *Phys. Rev.*, *32*, 133–141, 1928.
8. S. Chapman, On the Diamagnetic Field of the Outer Atmosphere, *Terr. Mag.*, *34*, 1–16, 1929.
9. A. G. McNish, Terrestrial Effects Associated with Bright Chromospheric Eruptions, *Trans. Amer. Geophys. Union*, 18th Annual Meeting, 164–169, 1937.

10. A. G. McNish, National Bureau of Standards Technical News Bulletin, July 1959.

11. S. E. Forbush, On Cosmic-Ray Effects Associated with Magnetic Storms, *Terr. Mag.*, *43*, 203–218, 1938.

12. T. Obayashi and J. A. Jacobs, Sudden Commencements of Magnetic Storms and Atmospheric Dynamo Action, *J. Geophys. Res.*, *62*, 589–616, 1957.

13. E. H. Vestine and E. J. Snyder, The Geomagnetic Incidence of Aurora and Magnetic Disturbance, Southern Hemisphere, *Terr Mag.*, *50*, 105–124, 1945

III.   Solar Activity and Geomagnetic Effects

3.1   Twenty-Seven Day Recurrences in Magnetic Activity

It is well known from solar observation that sunspots often endure for several solar rotations though seldom do they last longer than six months. From recurrences of sunspot groups the period of the sun's rotation is found to be about 27 days. At the beginning of a solar cycle (duration about eleven years) sunspots appear midway between the sun's equator and its poles. For these spots the interval between successive crossings of the sun's central meridian is more than 28 days. After the period of maximum sunspot numbers has passed and these numbers approach a minimum, the spots occur nearer the sun's equator and the interval between successive solar central meridian passages of these becomes less than 27 days. The occurrence of aurorae produced by the impact of solar particles in the rarefied upper atmosphere and of the system of electric currents in the upper atmosphere that is required to account for magnetic disturbance S_D and for magnetic storm time variations lead to the idea that the sun ejects clouds or streams of charged particles. As will be shown later, certain changes or modulations of the intensity of cosmic-ray intensity also require the existence of such clouds or streams. Investigations of magnetic disturbances have led to the idea that these clouds or streams are ejected from well-defined areas on the sun.

The simplest measure of magnetic activity is the international character figure C. As already described, each magnetic observatory assigns for each Greenwich day a character figure: "0" for quiet, "1" for moderately disturbed, and "2" for very disturbed days—as judged from inspection of the daily magnetograms. The average of these figures for all participating observatories is the international character figure. The most extensive analyses of C have been made by Bartels [1] who has also developed rigorous statistical procedures for dealing with this as well as other geophysical phenomena, especially those involving periodic and quasi-periodic variations. Figure 24 shows his complete record of international character figures for each day from 1931 back to 1906 when the international character figure first came into use. It contains about 10,000 symbols. The day beginning each row is indicated by the date at the left. The key to this chart is shown at the lower left of the figure. Note that for days with C = 0.6 and 0.7 the space is blank and that there are about as many days with C < 0.6 as with C > 0.7. The two vertical charts are arranged so that intervals eleven years apart are side by side. These charts clearly show, by the distinct vertical columns of sequences in which the black and gray symbols occur, that quiet and disturbed conditions tend to recur after a 27-day interval or one solar rotation. The recurrent pattern of some of the sequences persists for a year or two. The fact that the black or gray symbols occur generally in vertical columns rather than in slanted columns shows that the recurrence period of 27 days predominates. The years around 1911 and 1923 are characterized by a prevalence of grey symbols and those around 1917 and 1928 show the effect of the eleven-year sunspot cycle since the former pair of years (near sunspot minima) was less disturbed than the latter pair (near sunspot maxima). However, a most remarkable fact is that in no row of the charts are the symbols all grey or all black. In fact pronounced sequences of quiet days persist, for example, near sunspot maximum in 1917 and 1918. Near sunspot minima in 1911 and 1923 conspicuous sequences of disturbed days occur. This, as Bartels [2] points out, clearly demonstrates that the sun's surface is never everywhere active nor everywhere quiet. The charts show that two or more sequences sometimes appear simultaneously. However, these never divide the 27-day interval into regular subdivisions, showing that the sun's surface does not exhibit a systematic pattern as has sometimes been assumed. There is a general correspondence between magnetic activity and the eleven-year sunspot cycle; also, there are occasions when magnetic disturbances appear as the consequence of activity in a particular sunspot group.

During the period covered by Figure 24 the greatest storm occurred from May 13–16, 1921. This storm was accompanied by the passage across the sun's central meridian on May 14 of a very large but short-lived group of spots. In general, however, the correlation between sequences of magnetically quiet (or disturbed) periods with recurrent spotless (or spotted) regions on the sun is poor. Magnetic sequences generally endure longer than spot groups, which seldom persist for more than four of five solar rotations. Also sequences of disturbed days sometimes persist for several solar rotations through periods of several successive weeks without visible sunspots, as, for example, the sequence in the first half of 1923 (Day 18) with well defined recurrences after

45

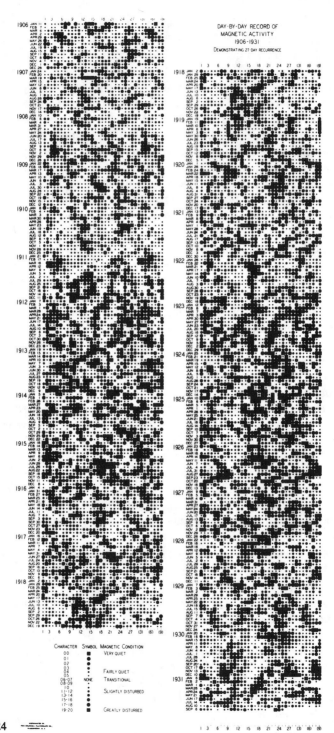

DAY-BY-DAY RECORD OF
MAGNETIC ACTIVITY
1906-1931
Demonstrating 27-Day Recurrence

CHARACTER	SYMBOL	MAGNETIC CONDITION
0.0	■	VERY QUIET
0.1		
0.2	●	
0.3		
0.4		FAIRLY QUIET
0.5		
0.6-0.7	NONE	TRANSITIONAL
0.8-0.9	·	
1.0		
1.1-1.2		SLIGHTLY DISTURBED
1.3-1.4		
1.5-1.6		
1.7-1.8		
1.9-2.0	■	GREATLY DISTURBED

Fig. 24

46

27 days. Figure 25 illustrates this effect for the years 1928–1930 on a larger scale than that in Figure 24. The middle and right-hand charts are, respectively, those for sunspot numbers and bright hydrogen lines observed on a central disc of half the sun's diameter, the region on the sun which is most likely to be effective. The bright hydrogen lines mark bright regions or markings on the solar surface as observed with the spectrohelioscope. The key to the three charts of Figure 25 is shown at the bottom. In Figure 25 it is readily seen that sequences in either of the two right-hand charts are evident in the other, indicating that sunspots and bright hydrogen clouds occur in about the same solar region. However, the pronounced sequences of magnetic disturbances in the left-hand chart are not recognizable on either of the other two charts even if account is taken of a possible time lag between solar and magnetic activity. Bartels [1] confirmed this fact for other solar phenomena and concluded [2] that geomagnetic activity reveals persistent solar influences which are distinctly revealed by recurrences of 27 days due to the sun's rotation. Bartels [2], therefore, attributed magnetic activity to some action from fairly definite regions on the sun—which he called M-regions. These regions were not then individually identifiable with any directly observable phenomena on the sun.

On the average, Bartels concluded that M-regions must vary in area as do sunspots since magnetic activity reveals the eleven-year cycle. More recently Babcock at Mt. Wilson has made regular measurements of magnetic fields over the sun's whole surface with a scanning solar magnetograph. His results have revealed "spots" where sunspots are not seen but which are detected from their magnetic fields. These spots may well be the M-regions postulated by Bartels as the source of geomagnetic activity, especially at times when no sunspots are seen. The charts in Figures 24 and 25 and the discussion concerning them indicate that an adequate interpretation of the relation between solar and geomagnetic activity can not be obtained from a short series of observations.

The 27-day recurrence tendency in magnetic disturbance, the duration of individual magnetic storms (a few days or so), and evidence that the clouds emitted by the sun travel to the earth in about a day (quite often solar flares are followed by magnetic storms after an interval of 18 to 36 hours), lead to the idea that active regions on the sun emit streams of particles more or less radially and that the streams rotate with the sun. A stream may be only a succession of individual clouds but as long as the emission continues the stream would sweep across the earth once in about 27 days. The behavior of these clouds as they approach the earth's magnetic field is complicated by the fact that these highly conducting clouds undoubtedly carry with them "frozen in" magnetic fields. The investigation of such phenomena embraces the new field of magnetohydrodynamics first introduced by Alfvén [4], who pointed out its importance to many cosmic and solar phenomena.

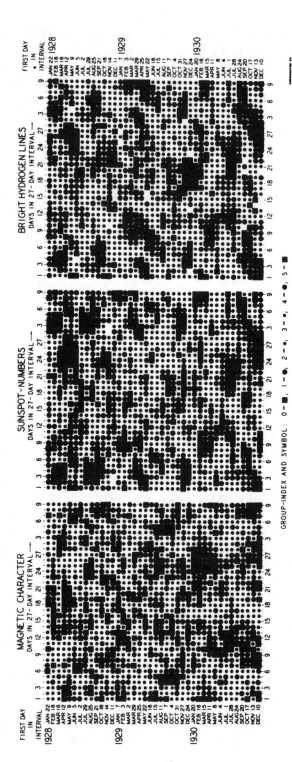

Fig. 25

48

3.2 Solar Cycle and Annual Variation of Magnetic Disturbance and the U-Measure of Activity

Figure 26 by Bartels [1] was prepared to disprove the contention that severe magnetic disturbances recurred on the same dates in successive years. This figure depicts the 392 days from 1906 to 1930 for which the international character figure $C \geq 1.6$; it shows no tendency for recurrence on the same days in successive years. The yearly totals of days with $C \geq 1.6$ at the right side of the figure exhibit the eleven-year cycle while the monthly totals indicate the annual variation (bottom of figure). The frequency of occurrence of disturbed days appears to be greatest near the equinoxes. To investigate the statistical reality of this semiannual variation in magnetic activity and for other purposes Bartels [1] derived a more objective measure of magnetic activity than that provided by C which, though very useful, suffered from the possibility of changing standards adopted by different observers in assigning character numbers for single days from a mere inspection of the records.

Figure 27 illustrates the basis for Bartels' u-measure of magnetic activity. It shows the worldwide nature of the depression in H (horizontal intensity) or in X (northward geomagnetic component) following magnetic storms or disturbances. The curves are drawn through means for consecutive 24-hour periods, six hours apart, with the means for the 24 hours centered at Greenwich midnight coincident with the vertical lines. The curve for Seddin (Germany) is for X and those for Watheroo (Australia) and for Huancayo (Peru) are for H. The similar curves show the worldwide nature of the typical depression of H (or X) during a disturbance and the gradual recovery to normal several days later. The u-measure (or interdiurnal variability) for a given day is the difference between the mean value of H (or X) for that day and the preceding day, taken without regard to sign, and then normalized before the results from several observatories are combined. The normalization [1,5] is done on the basis that the disturbance field can be regarded as a uniform field, say P, parallel to the earth's magnetic axis, like that which arises (except in high latitudes) from the current system shown in Figures 22 and 23(B) for the storm time variation. Suppose that at a certain observatory the interdiurnal variability identification (I.D.) of H or of X has been determined. The angle between the direction of H or of X and the earth's magnetic axis is easily calculated and may be called β. Thus the interdiurnal variability of P would be given by I.D./cos β which is the normalized u-measure. Consequently the u-measure is essentially the interdiurnal variability of the horizontal component at the equator, based on results combined from nine observatories. The unit of u-measure is taken as $10\,\gamma$ (10^{-4} gauss). Since the ideas involved should be useful in connection with other problems involving geophysical and solar

DAYS OF GREAT MAGNETIC DISTURBANCE DURING 1906 - 1930

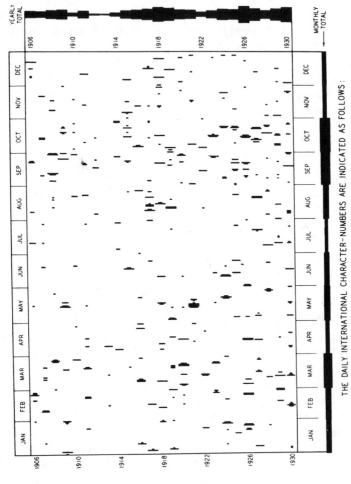

THE DAILY INTERNATIONAL CHARACTER-NUMBERS ARE INDICATED AS FOLLOWS:

CHARACTER-NUMBER	1.6	1.7	1.8	1.9	2.0
SYMBOL	·	▪	▬	▬	▬

Fig. 26

50

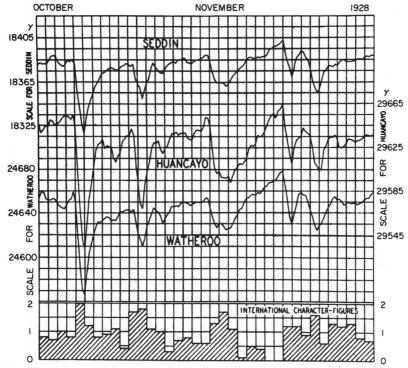

TYPICAL DEPRESSION FOLLOWING MAGNETIC DISTUR-
BANCES AND INTERNATIONAL MAGNETIC CHARACTER-FIGURES,
OCTOBER 14 TO NOVEMBER 19, 1928

Fig. 27

variables it is desirable to point out the defects of the u-measure and
the method used by Bartels [1] to correct them. In obtaining monthly
means of international character figures, the occurrence of a few greatly
disturbed days seldom has a very large effect on the monthly mean C
since C is never more than 2. On the other hand, the occurrence of
exceptionally great disturbance on a few days in a month may greatly
alter the monthly mean u from the value it would have had if such
exceptionally disturbed days had not occurred. For the purpose of investigating certain periodicities this effect is undesirable as it is also for
investigating relations between u and sunspot numbers, R. Bartels [1],
therefore, obtained a measure of magnetic activity which is just as well
defined as u but which, like C, mitigates the influence of the exceptionally great disturbances on u. This modified measure u_1 was defined as
a function, f, of the monthly mean value of u. f was chosen so that the
frequency distribution of monthly means of u_1 was similar to that for

sunspot numbers, especially for high values. This transformation from u to u_1 is essentially equivalent to adopting for u the variable scale of ordinates in Figure 28.

The relation between u and u_1 is shown in Figure 29. Lest this description of the derivation of u_1 from u lead to the conclusion that the transformation is just a refinement, let it be noted that for investigating the six-month wave in magnetic activity the statistical uncertainty in the average six-month wave from fifty-nine years of data using u_1 was about the same as it would have been from about one hundred sixty years of data using u.

Figure 28 shows the monthly means of magnetic activity, u, and sunspot numbers, R, 1900–1930 (lower part of figure) and the annual means of u and R, 1835–1930, and shows clearly the eleven-year variation in both. Figure 30 shows the correlation between annual means of u_1 and R for the years 1872–1930. The 117 points are annual means for January to December and July to June.

Figure 31 shows the annual variation of magnetic activity averaged for twenty years with high activity, nineteen years with medium activity, and twenty years with low activity. The monthly means in Figure 31 were smoothed according to the formula $b' = (a + 2b + c)/4$ in which a, b, and c are consecutive monthly means.

The maxima of activity near the equinoxes, March and September, and the minima near the solstices, June and December, are shown in all of the curves of Figure 31. One test of the significance of the annual variation, from a statistical point of view [1], proceeds as follows:

Let σ = standard deviation of monthly means within individual years from mean of year.

σ_m = standard deviation of average monthly means for all years in a group from the yearly mean for that group.

σ' = standard deviation of irregular variations superposed on the regular annual variation.

Using unsmoothed values (analogous to the smooth values in Figure 31) Bartels [1] found for the years of high, medium, and low magnetic activity the following values for σ_m (using u_1):

high 6.6; medium 5.8; low 3.2; all 4.7

and for σ:

high 19.5; medium 17.9; low 12.0; all 16.7.

If the irregular variations (with s.d. $= \sigma'$) are independent of the systematic yearly variation and are statistically random, σ' can be determined from the relation:

$$\sigma'^{\,2} = \sigma_m^2 + \sigma_i^2$$

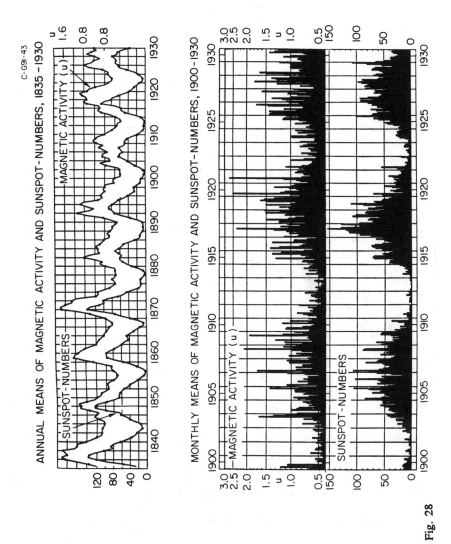

ANNUAL MEANS OF MAGNETIC ACTIVITY AND SUNSPOT-NUMBERS, 1835-1930

MONTHLY MEANS OF MAGNETIC ACTIVITY AND SUNSPOT-NUMBERS, 1900-1930

Fig. 28

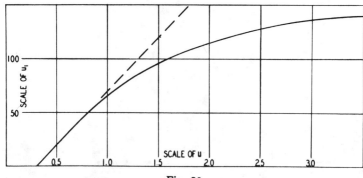

Fig. 29

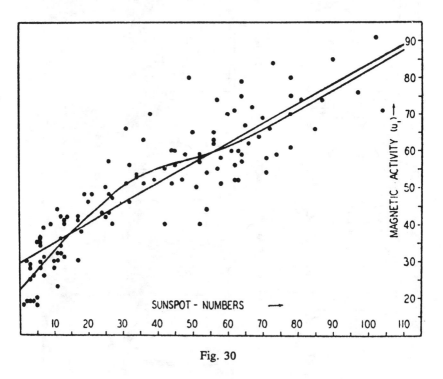

Fig. 30

since σ_i and σ_m are known. The resulting values of σ' are:

      high 18.3;   medium 16.9;   low 11.5;   all 16.0.

The ratios σ_m/σ' are as follows:

      high 0.36;   medium 0.34;   low 0.28;   all 0.29.

From these ratios Bartels concluded that the annual variation is not only

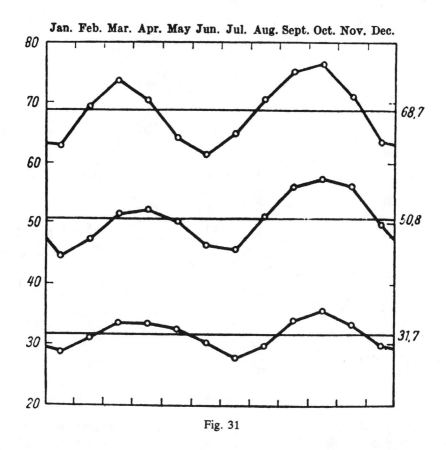

Fig. 31

more pronounced in an absolute sense in years of high activity but also more pronounced relative to the irregular fluctuations of simple monthly means. Actually the magnitude of the annual variation, expressed by σ_m, is about one-third that of the irregular variations expressed by σ'.

If we assume that σ does not contain a systematic annual variation but only random fluctuations, then the average monthly means for N years would be expected to have the standard deviation $\sigma/\sqrt{N}$, and the ratio $\sigma_m/(\sigma/\sqrt{N})$ can be taken as an index of the reality of the annual variation. Using the all-years group, this ratio has the value $4.7/(16.7/\sqrt{59}) = 2.2$. Since the expected value of the ratio is one in the absence of any systematic variation, the ratio 2.2 indicates the reality of the annual variation. That this annual variation does not arise from any annual variation in sunspot numbers was shown by Bartels [1] in a similar way to that above by the fact that all the critical ratios were less than unity.

Figure 32 shows Bartels' [1] results for the six-month waves of magnetic activity u, for the same data as in Figure 31. Distances of the

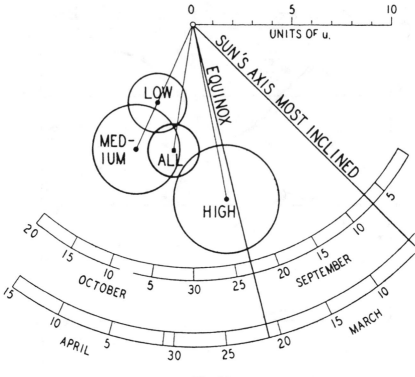

Fig. 32

points at the center of the circles from the origin O measure the wave amplitude on the scale shown, and the radii from O through the centers of the circles extend to cut the bottom circular arcs at the time of the wave maximum. While the details of general procedures will be discussed in Section IV, it will be pointed out here that the indicated probable error circles are such that the average wave from similar samples with the same number of data could fall with probability 0.5 inside (or outside) the circles. For the point marked "all" Bartels [1] showed that the probability of points, for similar samples, falling outside a circle with center at the point marked "all" and with periphery through the origin O is about 10^{-7}. This shows that the average six-month wave in u_1 is certainly statistically significant. Bartels computed the probability of points, from samples similar to that marked "all", falling within a ten-day interval centered on March 5, and within a similar one centered on March 21, with the result that it is highly improbable that the true six-month wave has its maximum within five days either side of March 5 but quite likely that it is within five days of March 21. On March 5 the sun shows us most of its southern hemisphere, and six months later, on September 7, it shows us most of its northern hemisphere. Since sunspots

56

occur most frequently in heliographic latitudes $10° - 15°$, and the sun shows comparatively few spots closer to its equator, it had been argued that if the solar streams leave the sun radially from the same belts in which the spots occur that the streams or clouds from them would more likely sweep across the earth near September 7 if they were emitted from the northern spotted belts and near March 5 if they were emitted from the southern belt. The results from Figure 32 argue strongly against this hypothesis and indicate instead that the real maxima of the six-month wave in magnetic activity are connected with the sun's crossing of the celestial equator on March 21 and September 23, i.e., at the equinoxes. This sharp distinction between times of maxima as close together as March 5 and March 21 would not have been possible without having used the u_1-measure instead of the u-measure. This example has been given not only for its own interest but to illustrate how judicious treatment of data has provided averages of the same reliability as would otherwise have been obtained only from nearly three times as many data (i.e., nearly 177 years instead of 59).

3.3 Variability of Wave Radiation from the Sun

The large amplitude of the diurnal variation S_q at Huancayo makes data from that observatory particularly valuable for investigating the variability of the amplitude of S_q and its relation to solar phenomena. For these studies Bartels [6,7] defined A as the excess of the five-hour average 9^h to 14^h ($75°$ WMT) of H (horizontal intensity) at Huancayo above the night level given by a straight line connecting the five-hour averages 0^h to 5^h. The lunar variation, L, appears in A as a wave with a half-month period. In the average for the months November to February near sunspot maximum (sunspot number R = 93), this lunar semimonthly wave has its maximum, $A = 149\gamma$, four days after new and full moon, and its minimum $A = 99\,\gamma$, a quarter month later [6]. Hence elimination of the lunar variation from A is required before using it to derive a measure for the amplitude of S_q. Bartels [8] used the results of his extensive analysis on the lunar variation to eliminate its effect on A. The seasonal variations of A are eliminated and then a value of A corresponding to R = 50 is computed. The normalized deviation ΔA from this value, expressed as a multiple of its standard deviation, is the daily measure ΔW for the amplitude of S_q. Figure 33 shows the correlation between quarterly means of sunspot numbers, R, and W ($= \Delta W$ except for scale). The correlation between ΔW and sunspot numbers is the closest yet found between phenomena on the sun and on the earth. Hence it is clear that R is a quite good measure of the wave radiation, W. The correlation coefficient between ΔW and R for eighteen Septembers is r = +0.966 and for no calendar month [6] is

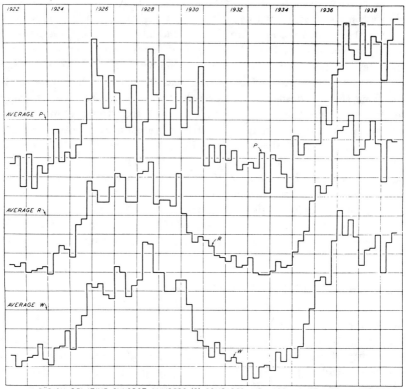

—ZÜRICH RELATIVE SUNSPOT-NUMBERS (R) COMPARED WITH GEOMAGNETIC MEASURES OF SOLAR CORPUSCULAR RADIATION (P) AND SOLAR WAVE-RADIATION (W); QUARTERLY MEANS, 1922-39 [UNITS CHOSEN SO THAT THE STANDARD DEVIATION FOR EACH CURVE EQUALS 2.5 SCALE-DIVISIONS]

Fig. 33

r less than +0.89. The upper curve in Figure 33 is a measure of solar corpuscular radiation, P, derived from the u_1-measure by eliminating the six-month variation of u_1. (Statistical tests similar to those discussed in the preceding section in connection with the six-month wave in u_1 show no statistically significant twelve-month wave in u_1.) Using ΔW Bartels [7] finds that the twenty-seven-day recurrence tendency in W is just as strong as in R. His statistical experiments show that the influence of "fast" variations of R (from one solar rotation to the next) is much stronger between R and W than between R and P. He finds that the fast variations of R are accompanied by similar variations of W lagging by not more than about one day. W is undoubtedly a solar radiation absorbed rather low in the ionosphere (near 100 km altitude), in or near the same level as the layer in which additional ionization is produced by ultraviolet radiation from a solar flare. Reference [7] contains tables of daily values of W for the period May 1922 to December 1937. These have been extended to 1947 and are available from Dr. John Manchley, Remington Rand Inc., 1624 Locust St., Philadelphia 3, Pa., U.S.A.

3.4   New Measure for Storm Time Variation

Walter Kertz has just published a new measure [9] for the worldwide storm time variation. This measure is derived as an average for every three hour interval from January 1939 through December 1945 (soon to be extended through the IGY period). Using values of H from Huancayo (Peru), Elizabethville (Africa), Watheroo (Australia), and Apia (Samoa), the storm time variation is derived by using only nighttime values. Since these stations are spaced at about six hour intervals of local time, the complete continuous storm time variation can be determined without using daytime values from any station. The night values of H are relatively unaffected by S_q and its variability. Moreover, S_D is much smaller at night than during the day and corrections are made for it. These new indices which Kertz [9] calls a new measure for the field strength of the geomagnetic equatorial ring current will be most valuable in many investigations such as those involving cosmic-ray variations, earth's conductivity, etc.

III.   References

1. J. Bartels, Terrestrial-Magnetic Activity and Its Relations to Solar Phenomena, *Terr. Mag.*, *37*, 1–52, 1932.
2. J. Bartels, How Changes on the Sun's Surface Are Recorded by the Earth's Magnetism, *Scientific Monthly*, *35*, 492–499, 1932.
3. J. Bartels, Random Fluctuations, Persistence, and Quasi-Persistence in Geophysical and Cosmical Periodicities, *Terr. Mag.*, *40*, 1–60, 1935.
4. H. Alfvén, *Cosmical Electrodynamics*, Oxford at the Clarendon Press (1950).
5. S. Chapman and J. Bartels, *Geomagnetism*, Oxford at the Clarendon Press (1940).
6. J. Bartels, Solar Radiation and Geomagnetism, *Terr. Mag.*, *45*, 339–343, 1940.
7. J. Bartels, Geomagnetic Data on Variations of Solar Radiation: Part 1 — Wave-Radiation, *Terr. Mag.*, *51*, 181–242, 1946.
8. J. Bartels and H. F. Johnston, Geomagnetic Tides in Horizontal Intensity at Huancayo, *Terr. Mag.*, *45*, 269–308, 1940.
9. W. Kertz, Ein Neues Mass für Die Feldstärke des Erdmagnetischen Äquatorialen Ringstroms, *Abhandlungen der Akademie der Wissenschaften in Göttingen Mathematisch-Physikalische Klasse*, Beiträge zum Internationalen Geophysickalischen Jahr, Heft 2, pp. 1–83, Gottingen Vandenhoeck & Ruprecht (1958).

IV. Random Fluctuations, Persistence, and Quasi-Persistence in Geophysical and Cosmical Periodicities

4.1 Introduction

The title of this section is exactly that of a famous paper by Bartels [1]. His paper is without doubt the sine qua non for putting the investigation of periodicities and related phenomena on a sound statistical basis. It emphasizes the pitfalls that exist in testing the statistical reality of periodicities, if proper account is not taken of the lack of independence (or randomness of sample) that generally prevails in geophysical data. Lack of statistical independence between successive individuals in a sample of geophysical data is of paramount importance in the application of most standard statistical techniques to geophysical problems.

The lack of independence (that is, autocorrelation) in a sequence of geophysical data generally arises in the sample when the sample consists of the values of a variable recorded in a time sequence, i.e., every hour, every day, every month, etc. Since many or most geophysical variables tend to have somewhat similar values between successive time intervals (hours, days, months, etc.), such sequences of data do not constitute a random sample as required for most statistical tests. Many of the established statistical procedures find interesting and important application in geophysical problems. A study of these procedures, including the ingenious methods used in biology and in agricultural experiments for searching out and testing the reality of real effects in a background of unavoidable variations, will not only familiarize the investigator with such techniques but will also suggest interesting problems which would appear insoluble without such techniques. However, standard textbooks on statistical methods are of little or no aid in connection with problems involving geophysical periodicities. References [l], [2 (Vol. II, Chapter XVI)], and [3] comprise the only available sources, in English, where the necessary statistical procedures for such problems are set forth. This section gives a summary description of the techniques of reference [1], which should be consulted for further details, proofs, and references.

4.2 General Procedure

As pointed out by Bartels [1], investigations of periodicities, cycles, recurrence tendencies, and related problems usually proceed in three stages: (1) Transformation, analytically, of the observational data by, say, harmonic analysis, (2) application of statistical treatment of these transformed data, including tests of significance, and (3) physical interpretation of the significant periodicities. In many geophysical problems the period (or frequency) of the variation is known in advance as in the case of solar and lunar diurnal variations. In other cases the length of the periods or recurrence-intervals may not be known in advance. The large number of periods and cycles in all kinds of geophysical phenomena (and even in business activity) that has been claimed without adequate statistical treatment emphasizes the need for using the sound statistical procedures which have been established for research involving periodicities.

4.3   Harmonic Analysis

Records of geophysical phenomena in general provide values of a variable, y, at equal intervals of time (i.e., hourly, daily, monthly, etc.). An individual record may cover the time $t = 0$ to $t = T$. It is convenient to introduce the time variable $x = 2\pi t/T$ so that the length of the record as measured by x is 2π. Let the number of values of y (ordinates) in the interval T be r, i.e., the times (or abscissae) $x_1, x_2, x_3 \ldots x_r$ divide the interval 0 to 2π into equal parts; and let y_ρ equal the ordinate at $x_\rho = 2\pi\rho/r$.

Consider sine and cosine functions of frequency $\nu = 0, 1, 2, \ldots k$; i.e., completing ν cycles in the interval 0 to 2π (length of periods $\rho_\nu = T/\nu$) and their sum

$$\phi_k(x) = a_0 + (a_1 \cos\, x + b_1 \sin\, x) + (a_2\, \cos\, 2x + b_2\, \sin\, 2x)$$

$$+ \cdots + (a_k\, \cos\, kx + b_k\, \sin\, kx) \ . \tag{33}$$

Harmonic analysis consists in determining the coefficients in (33) so that $\phi_k(x_\rho)$ gives the best fit in the least squares sense to the given ordinates y_ρ, i.e., so that s_k^2 is minimized with s_k^2 given by:

$$s_k^2 = \sum_\rho [y_\rho - \phi_k(x_\rho)]^2/r \ . \tag{34}$$

Inasmuch as (33) contains $(2k + 1)$ coefficients and is to be fit to r ordinates, it is useful to consider only values of k for which

$$(2k + 1) \leq r \ . \tag{35}$$

If $(2k + 1) = r$, then (33) fits the r ordinates exactly and $s_k = 0$. The coefficients in (33) are determined from the following equations:

$$a_0 = \frac{1}{r}\sum_{\rho=1}^{\rho=r} y_\rho; \quad a_\nu = \frac{2}{r}\sum_{\rho=1}^{\rho=r} y_\rho \cdot \cos\, \nu x_\rho; \quad b_\nu = \frac{2}{r}\sum_{\rho=1}^{\rho=r} y_\rho \cdot \sin\, \nu x_\rho \ . \tag{36}$$

For r even:

$$a_{r/2} = (-y_1 + y_2 - y_3 + y_4 - \cdots + y_r)/r \ . \tag{37}$$

Suppose that we have a second set of r ordinates y'_ρ for which the harmonic coefficients are a'_ν and b'_ν, then

$$\frac{2}{r} \sum_{\rho=1}^{\rho=r} (y_\rho + y'_\rho) \cos \nu x_\rho = a_\nu + a'_\nu$$

and similarly

$$\frac{2}{r} \sum_{\rho=1}^{\rho=r} (y_\rho + y'_\rho) \sin \nu x_\rho = b_\nu + b'_\nu \ ,$$

which show the additive property of the harmonic coefficients. In fact it is evident that

$$\frac{2}{r} \sum_{\rho=1}^{\rho=r} (Ay_\rho + A'y'_\rho) \cos \nu x_\rho = Aa_\nu + A'a'_\nu \ .$$

Thus any linear combination of ordinates gives rise to the same linear combination of coefficients. It should be noted in (36) that k does not appear so that any one of the coefficients a_ν or b_ν is the same regardless of how many other coefficients are determined.

4.4 The Harmonic Dial

For frequency ν the sine and cosine terms may be combined into a sine wave of amplitude c_ν and phase a_ν:

$$a_\nu \cos \nu x + b_\nu \sin \nu x = c_\nu \sin(\nu x + \alpha_\nu) , \qquad (38)$$

wherein

$$a_\nu = c_\nu \sin \alpha_\nu; \quad b_\nu = c_\nu \cos \alpha_\nu; \quad c_\nu^2 = a_\nu^2 + b_\nu^2, \quad \text{and}$$

$$\tan \alpha_\nu = a_\nu / b_\nu \ . \qquad (39)$$

Figure 34 represents these relations in the harmonic dial for frequency ν, in which the point P with coordinates a_ν and b_ν represents (38). The first of the ν maxima of (38) occurs for $(\nu x + a_\nu) = 90°$, i.e., at time $x_{max} = (90° - a_\nu)/\nu$. Thus $a_\nu = 90°$ corresponds to $x_{max} = 0$; $a_\nu = 0$, to $x_{max} = 90°/\nu$, etc. Thus on a circle about the origin O a scale may be constructed such that the vector $OP = \vec{c}_\nu$ points to the time of the wave maximum, i.e., towards t_{max} in the original time t. For a semidiurnal wave (time interval T from 0^h to 12^h), $t = 1^h$ corresponds to $x = 2\pi/12 = 30°$ and the scale for t_{max} becomes the ordinary dial of a clock, suggesting the name harmonic dial. Any number of sine waves

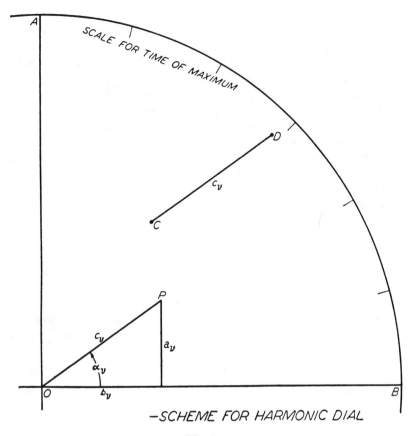

−SCHEME FOR HARMONIC DIAL

Fig. 34

of the same frequency may be represented by points in such a dial. Each point may be regarded as the end point of a vector from the origin. The average of all the sine waves represented by these points is represented by the average of all the vectors and has its end point at the mass center of all the points. Sums of several sine waves are represented by the vector sum of the individual vectors.

The calculation of harmonic coefficients can be done quite expeditiously, even by hand, if a proper schedule is adopted and especially if the computation is arranged for "mass production" when many analyses are to be made. In this arrangement each particular "operation" is carried out for, say, ten analyses before going on to the next operation. Similar operations on sums of ten sets of ordinates provide checks at each stage and greatly facilitate location of errors, if any. The footnote on page 8 of reference [1] lists references to several procedures. A very useful one of these is that of Bartels [1] in which noncyclic change (linear trend) in the ordinates is automatically removed by analyzing differences of successive ordinates so that the harmonic coefficients need not be corrected for the trend. Correction to the amplitudes when using ordinates that are means over finite intervals of time are derived in reference [1]. Figure 35 is a classical example of establishing a systematic variation in the presence of a much greater "noise" level, if sufficient data are available. The lunar tidal wave (semidiurnal) in atmospheric pressure at Batavia has an amplitude of 0.062 mm Hg. The consequent adiabatic compression gives rise to a theoretical lunar semidiurnal wave of temperature with amplitude only 0.0072° C, as shown below Figure 35. The harmonic dial of Figure 35 compares this theoretical value (as well as its phase) with that "observed", i.e., derived by Chapman from sixty-two years of temperature data at Batavia.

4.5   Twenty-Seven Day Waves in International Character Figure C

To illustrate the statistical principles needed in connection with the analysis of geophysical periodicities, Bartels [1] used the results of harmonic analysis of C, the international magnetic character figure, for the

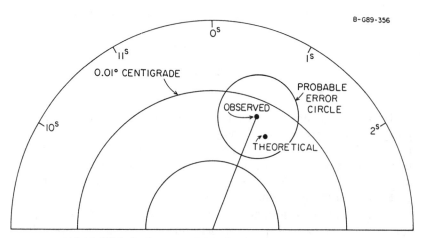

LUNAR SEMI-DIURNAL WAVE OF ATMOSPHERIC TEMPERATURE, BATAVIA, 62 YEARS (S. CHAPMAN)
PRESSURE-WAVE 0.062 SIN (2τ +68°) MILLIMETERS MERCURY
ADIABATIC COMPRESSION GIVES THEORETICALLY

$$\frac{\delta T}{T} = \frac{\gamma - 1}{\gamma} \frac{\delta P}{P} \; ; \; \delta T = 299° \times \frac{0.40}{1.40} \times \frac{0.062}{759} = 0.0072°$$

Fig. 35

period January 11, 1906 to December 30, 1933. The 10,206 days spanned 378 solar rotations which Bartels numbered 1 (beginning January 11, 1906) to 378. The same C values were used in the 27-day recurrence diagrams discussed in Section III. The harmonic dial of Figure 36 shows the results for the 378 27-day waves; the average wave is indicated by the cross near the center. The question of whether this average vector in Figure 36 is or is not so small that it can be ascribed to accidental causes is deferred until we have considered standard deviation of sine waves and residuals.

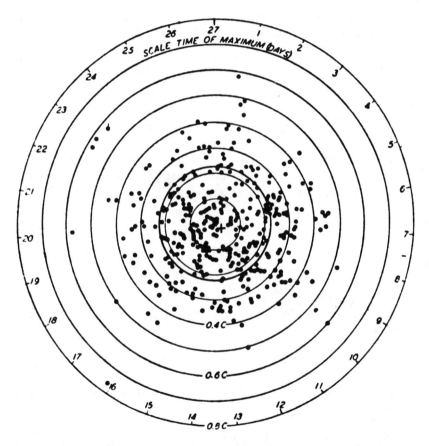

Harmonic dial showing the dial-points for the 27-day sine wave in each of 378 27-day sets of values of the international daily magnetic character-figures C, for the period 1906–33 beginning with 1906 Jan. 11

Fig. 36

4.6 Standard Deviations for Sine Waves and Residuals

From equations (34) and (35) it is evident that if fewer than r coefficients are used, then $\phi_k(x)$, defined in (33), is only an approximation to the given ordinates. To estimate the degree of this approximation let ζ be the standard deviation of the original r ordinates, y_r, from their mean, a_0. Then, as may be found in any standard text on statistics:

$$\zeta^2 = \sum_{\rho=1}^{\rho=r} \frac{y_\rho^2}{r} - a_0^2 . \tag{40}$$

It is shown in Appendix 1 of reference [1] that the average value of $\phi_k(x)$, defined in (33), is a_0, and that its standard deviation η_k is defined by:

$$\eta_k^2 = (a_1^2 + b_1^2 + a_2^2 + b_2^2 + \cdots + a_k^2 + b_k^2)/2 \tag{41}$$

and since $c_k^2 = a_k^2 + b_k^2$ [from (39)], then:

$$\eta_k^2 = (c_1^2 + c_2^2 + \cdots + c_k^2)/2 \tag{42}$$

except in the case of exact representation for which, when r is even, the last term in (42) is $2a_{r/2}^2$. The standard deviation of the residuals, defined in (34), is shown [1] to be given by the remarkably simple expressions:

$$s_k^2 = \zeta^2 - \eta_k^2 \tag{43}$$

or

$$s_k^2 = \zeta^2 - (c_1^2 + c_2^2 + \cdots + c_k^2)/2 . \tag{44}$$

Thus the variance s_k^2 of the residuals is reduced by half the squared amplitude of each additional harmonic term. For exact representation (and r even) $s^2 = 0$ and

$$(c_1^2 + c_2^2 + \cdots + c_k^2 + 2a_{r/2}^2) = 2\zeta^2 . \tag{45}$$

Suppose that amplitudes have been computed, for example, up to index k. Then:

$$c_{k+1}^2 + c_{k+2}^2 + \cdots + 2a_{r/2}^2 = 2\zeta^2 - c_1^2 - c_2^2 - \cdots - c_k^2 = 2s_k^2 . \tag{46}$$

Thus the square of the largest coefficient among those for frequency greater than k can not exceed $2s_k^2$. Even if one or two amplitudes are computed, an upper limit for any of the uncomputed amplitudes can be obtained similarly.

4.7 Statistical Procedures for Testing Periodicities

Imagine a set of ordinates (r in number) to be artificially constructed from:

$$\phi_k(x_\rho) = a_k \cos x_\rho + b_k \sin x_\rho \quad (\rho = 1, 2, 3, \ldots r) \qquad (47)$$

with $x_\rho = 2\pi\rho/r$). For the first set, let a_k and b_k each be a random sample drawn from normal populations with mean zero and respective standard deviations σ_{ak} and σ_{bk}. If this set of ordinates is harmonically analyzed, one obtains, of course, the same a_k and b_k that were used in computing the ordinates from (47). Suppose that (47) were written to include other additional harmonic terms with different frequencies, k_1, k_2, etc., in addition to the frequency k. Harmonic analysis of the ordinates would give the same coefficients a_k and b_k whether or not the additional terms for frequency k_1, k_2, etc. were included in computing the ordinates. Let a large number of pairs of coefficients a_k and b_k be similarly drawn at random from their respectively defined populations. For each pair of values of a_k and b_k let the amplitudes, $c_k = (a_k^2 + b_k^2)^{1/2}$, and phases θ_k be computed. We now inquire how these values c_k and θ_k are distributed. Since we are confining our attention to a particular frequency k, and since the results do not depend upon k, the subscript k will, for convenience, be omitted. The probability density for the a values, say f(a), is given in accord with our assumptions by the "normal" or Gaussian distribution; thus:

$$f(a) = \left(1/\sigma_a \sqrt{2\pi}\right) e^{-a^2/2\sigma_a^2} \qquad (48)$$

or f(a)da is the probability that a lies in the interval between a and (a + da). Similarly f(b) is

$$f(b) = \left(1/\sigma_b \sqrt{2\pi}\right) e^{-b^2/2\sigma_b^2} . \qquad (49)$$

The joint probability density function f(a,b) is:

$$f(a, b) = \left(1/2\pi\sigma_a\sigma_b\right) e^{-(a^2/2\sigma_a^2)-(b^2/2\sigma_b^2)} \qquad (50)$$

and f(a,b)da db is the probability that a falls in the interval (a + da), and b in the interval (b + db). Suppose now that $\sigma_a = \sigma_b$ and we put

$$M^2 = \sigma_a^2 + \sigma_b^2 . \qquad (51)$$

Then since $\sigma_a^2 = \sigma_b^2$: $M^2 = 2\sigma_a^2 = 2\sigma_b^2 = 2\sigma_a\sigma_b$ and we have from (50) for the joint probability function $f(c, \theta)$ the result

$$f(c, \theta) = (1/\pi M^2) e^{-c^2/M^2} \qquad (52)$$

67

and since this is independent of θ the distribution is circularly symmetric. The probability that c (or the end point of the vector $\vec{c}$ in the harmonic dial) lies in the element of area c dc dθ is $(1/\pi M^2)e^{-c^2/M^2}$ c dc dθ, integration of which with respect to θ from $\theta = 0°$ to $\theta = 2\pi$ gives the probability w(c)dc that the end point of the vector $\vec{c}$ lies between the distances c and (c + dc) from the origin with

$$w(c) = \frac{2}{M^2} c\ e^{-c^2/M^2} . \tag{53}$$

M^2 is called the expectancy, and it is convenient to express the distance from the origin as a multiple of the expectancy, i.e.:

$$c = K\ M \tag{54}$$

in which K is used for Bartels' Greek kappa. The probability that c lies between KM and (K + dK)M is w(K)dK with:

$$w(K) = 2K\ e^{-K^2} . \tag{55}$$

The probability that the end point of the vector $\vec{c}$ in the harmonic dial lies beyond a distance KM from the origin is found by integration of (55) between the limits K and ∞ giving

$$W(K) = e^{-K^2} . \tag{56}$$

Equations (52) through (56) are the same as those which govern [1] the distribution of distances from the origin reached after a random walk of n stretches of equal length ℓ if the expectancy M(n) is defined as $\ell\sqrt{n}$. Figure 37 shows one sample of such a random walk with 27 stretches, each equal to the radius of the circle in the upper left of the figure. Points on the circle indicate the direction of individual steps. Figure 38 shows a random walk with 125 stretches of unequal lengths; in this example the lengths were randomly sampled from a normal distribution of lengths with standard deviation 0.39 ℓ. Equations (52) through (56) apply also to the random walk with unequal lengths provided the expectancy M(n) = $\ell\sqrt{n}$ is computed with ℓ defined so that ℓ^2 is the average of the square of the individual step lengths taken over a large number of such walks each with n steps. The sampling distribution of step lengths is arbitrary within wide limits [1]. It should be noted that in applying equations (52) through (56) to the random walk, with the expectancy M(n) = $\ell\sqrt{n}$, that the equations apply to the total distance, from the starting point, that is reached after n steps, whereas the M defined in equation (51) and used in connection with our description of the harmonic dial applies to the distance of the end points of single vectors

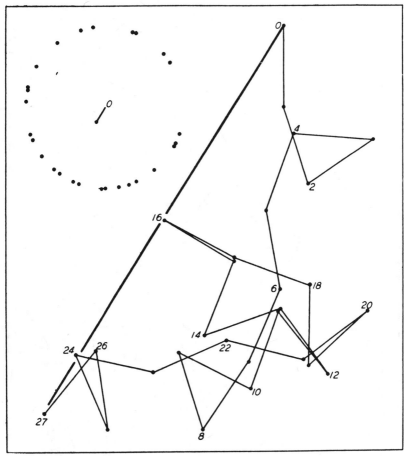

−RANDOM WALK WITH EQUAL STRETCHES

Fig. 37

(or steps) from the origin. The details of several analogies between the random walk and the harmonic dial are given in reference [1].

For the harmonic dial in Figure 36 Bartels finds M = 0.262 c with M^2 equal to the average of the 378 values of the squared amplitudes, (c^2), from Figure 36. This value of M is statistically equivalent to that which would have been obtained from equation (51) through the values of a and b for the 27-day wave provided this sample of 378 values of a and b was drawn from a population with mean values of a and b both zero (and consequently mean c = 0). Since one of the hypotheses to be tested is whether the mean value of c in Figure 36 can be regarded as a random sample from a population with mean c = 0, M may be derived in either way. In one case M is the expectancy for single amplitudes

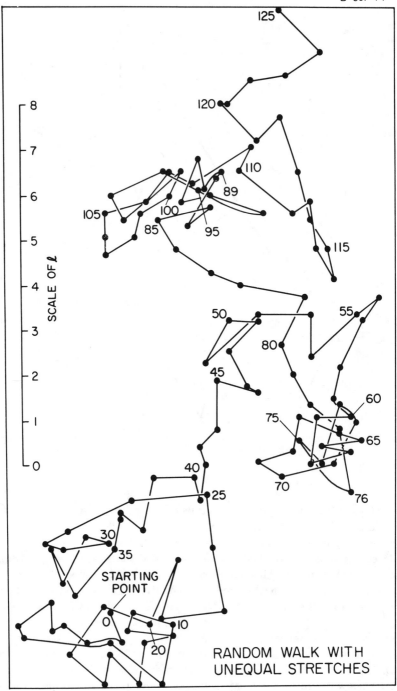

SCALE OF ℓ

125

120

110

89

105
100
95
85

115

50

55

80

45

60
75
65

40
25

76
70

30
35

STARTING
POINT

0

10

20

RANDOM WALK WITH
UNEQUAL STRETCHES

Fig. 38

from the origin and in the other it is the expectancy for the amplitudes of single deviations from the average vector. The latter expectancy is generally preferable when the average vector is so large that it can not be regarded as a sample from a population in which the amplitude of the average vector is zero.

From equations (54) and (56) we can write

$$W(K) = e^{-c^2/M^2} \tag{57}$$

from which $M = 0.262$ c for the data in Figure 36. The probability $W(K)$ that c exceeds any specified value can be obtained and thus the probability that c lies between specified limits. This latter probability times 378 gives the number of values of c to be expected in that range if the sample distribution of c values is governed by (57). Figure 39 (upper part) compares the observed number of c values (from Figure 36) in indicated ranges of c with the theoretical values from (57) shown by the smooth curve. The agreement between the two is satisfactory [1].

From equation (36), if $W(K) = 0.500$ then $K = 0.833$ or 0.833 M is the so-called probable radius since for the theoretical distribution half the points should lie inside and half outside a circle of this radius with the center at the origin if the mean vector in the harmonic dial does not differ significantly from zero. Otherwise the center of the circle is at the end point of the average vector. With $M = 0.262$ c the probable radius for Figure 36 is 0.216 c and this circle (third from center) contains very closely half the total of 378 points. It can be shown that the mean value of c for the theoretical distribution (53) is 0.886 M. The mean c for Figure 36 is 0.88 M which agrees well with the expected value. Thus, so far as these tests go, each of the 378 vectors in Figure 36 can be regarded as the result of a random walk of n stretches where the stretches vary at random about a mean value ℓ. The parameter M is prescribed by the observations ($M = 0.262$ c for Figure 36). This expectancy M for the equivalent random walk is given in terms of ℓ and the number of steps n by the relation $M = \ell\sqrt{n}$ (described previously) wherein n can be chosen arbitrarily. Thus $\ell = M/\sqrt{n}$.

Equations (52) to (57) also govern the distribution of the amplitudes of the average vector obtained from samples of n individual vectors with random directions. This follows from (48) and (49) since the distribution of means of a (or b) in samples of n from a normal population with σ_a (or σ_b) for the standard deviation of single values of a (or b) is governed by equations (48) and (49) with $\sigma_a/\sqrt{n}$ (or $\sigma_b/\sqrt{n}$) replacing σ_a (or σ_b). Thus for the distribution of means of such samples of n, equations (51) through (57) apply if M is replaced by m with

$$m = M/\sqrt{n} . \tag{58}$$

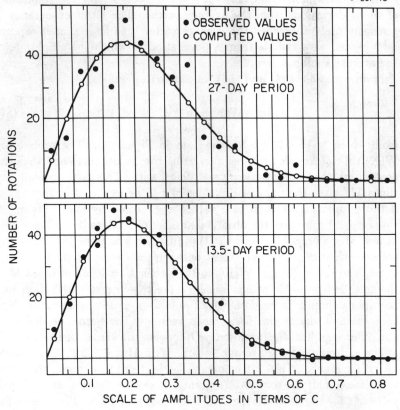

NUMBER OF ROTATIONS (INTERVALS OF 27 DAYS)
HAVING FOR INDICATED PERIODS IN INTERNATIONAL MAG-
NETIC CHARACTER - FIGURE C, 1906 - 1933, AMPLITUDES
BETWEEN 0 AND 0.036C, 0.036C AND 0.072C, ETC.

Fig. 39

For Figure 36, m = 0.262 c/$\sqrt{378}$ = 0.0135 c. In Figure 36 the aver-age vector has the amplitude 0.0336 c (c is the character figure unit) or 2.49 m. Using (56) with K = 2.49 we find W(K) = $e^{-2.49^2}$ = $e^{-6.2}$ = 0.002. Thus an amplitude exceeding that found in Figure 36 (i.e., 0.0135 c) should occur only about once in 500 times, i.e., in 500 samples of 378 with m = 2.49 one should expect to obtain an average vector as large or larger than that obtained in the single sample of Figure 36. This might be taken to indicate that the 27-day wave in Figure 36 is probably at least in part persistent or systematic and that the average 27-day wave is statistically significant (probability 500 to 1). However, it will be seen later why this interpretation, which is commonly applied to geophysical periodicities, is not generally justified. The understanding of the reasons for this is of paramount importance in the analysis of geophysical periodicities.

In order to understand the above statement, it is necessary to see how the expectancy depends on the length of period. Suppose that N sets of r ordinates are obtained by random sampling from a normal (Gaussian) population with standard deviation ζ and mean zero. Due to sampling fluctuations the mean ordinate in each set of r ordinates may differ from zero and also the sample standard deviation ζ^* may differ from ζ for the same reason. Let each set of r ordinates be harmonically analyzed and for each frequency ν let ϵ_ν, the expectancy, be obtained from the average of the squares of the large number of resulting amplitudes c_i, i.e., $\epsilon_\nu = \Sigma c_i^2/N$. Then it can be shown that

$$\epsilon_\nu = 2\zeta/\sqrt{r}; \quad \epsilon_{r/2} = \zeta/\sqrt{r} . \tag{59}$$

Thus for random ordinates (the distribution need not be Gaussian as here described but can vary within wide limits) the expectancy ϵ_ν is independent of frequency ν. This independence can be termed the law of the equipartition of variance (square of standard deviation). Because by taking r uneven we know from (44) that each amplitude c_ν contributes $c_\nu^2/2$ to the variance η_k^2 of the sum ϕ_k of sine waves. If (44) is written down for each set, summed, and divided by N we obtain

$$(\zeta^*)^2 = (c_1^2 + c_2^2 + \cdots c_{(r-1)/2}^2)/2 . \tag{60}$$

For equipartition $c_1^2 = c_2^2 = \cdots = c_{(r-1)/2}^2 = \epsilon^2$. Then it can be shown that $\epsilon^2 = 4\zeta^2/r$ as in (59). For the data used to obtain the results in Figure 36 the standard deviation ζ of the daily values of character figure C from their mean [1] for the interval 1906–1933 is $\zeta = 0.467$ C. Using (59) with $r = 27$, the number of ordinates in each set harmonically analyzed, the equipartition value of the expectancy ϵ (which would apply if, contrary to the actual case, the character figures C were randomly mixed) is $\epsilon = 2 \times 0.467$ C$/\sqrt{27} = 0.180$ C. The expectancy M determined from the mean of the squares of the 378 amplitudes in Figure 36 gives $M^2 = 0.0696$ C^2 or $M = 0.262$ C, which is greatly different than the value 0.180 C and shows that the expectancy derived from the amplitudes obtained by harmonic analysis from single rotations (one rotation here = 27 days) definitely depends on the length of the period. This result is a consequence of the fact that the character figures C for successive days are certainly not independent (random), a statement which characterizes many other geophysical variables. This nonindependence may be tested in the following way: Let $\zeta(1)$ be the standard deviation of C for single days from the mean of all available days, and let $\zeta(2)$, $\zeta(3)$, etc. be the standard deviations of the sums of C for 2, 3, etc. days, each (standard deviation) having been divided respectively by $\sqrt{2}$, $\sqrt{3}$, etc. Then for complete independence $\zeta(1) = \zeta(2) = \zeta(3)$, etc. (except for sampling fluctuations). Thus the ratios $\zeta(2)/\xi(1)$, $\zeta(3)/\zeta(1)$, etc. can be taken as measures of dependence. A similar test will be used later for testing quasi-persistent periods. To see how this expectancy affects the expectancy for sine waves, first suppose that the whole series of ordinates is arranged (for harmonic analysis) into sets of r successive ordinates. Then the arithmetic mean in each single set will generally differ more from the arithmetic mean of all ordinates than in the case of independence; or, in other words, the standard deviation of the means for sets of r ordinates will be greater than the random value $\zeta/\sqrt{r}$. For example [1], the standard deviation of single daily values of character figure C (1906–1933) is 0.467 C. The standard deviation for the means of C for 27-day intervals (378) is found to be 0.148 C. If the values of C for successive days were independent, this last value would be expected to be 0.467 C$/\sqrt{27} = 0.090$ C. On the other hand, if, in each single set of r ordinates, the deviations of each of the r ordinates from the mean for the set are formed, their standard deviation $\zeta(r)$ will be smaller than the standard deviation ζ of all ordinates, and the ratio $\zeta(r)/\zeta$ will increase to unity for increasing r. (In the example considered above, this ratio is, for r = 27, 0.444 C/0.467 C.) From (45) it follows that the expectancy for shorter periods (computed from sets of a few ordinates) will generally be less than that for longer periods.

In the harmonic dial, each vector (for which only the end point is plotted) may be considered as the sum of two vectors: one for any "persistent" wave that is present and another for the "accidental" wave. The vector for the "persistent" wave will be essentially the same for all points in the dial since it has essentially constant amplitude, C, and phase. The average vector derived from harmonic analysis of N sets of r ordinates is the sum of the average persistent vector with amplitude, C, and the average "accidental vector" which has amplitude of the order of $M/\sqrt{N}$ analogous to (58). Thus no matter how small C may be relative to M, in the average taken over a sufficient number N of periods, the persistent wave will eventually stick out above the "accidental" waves produced by the nonpersistent waves which mask the hidden periodicity in the original data. Figure 35 is an outstanding example of this $1/\sqrt{N}$ law.

4.8 Examples

Figure 40 shows the harmonic dials for the six-month waves in C from 28 years of data. The left dial was computed from 56 half years and the right one from 28 full years. The average wave with the amplitude c = 0.0675 C is represented by the vector OA, the same in both figures, with its maximum about March 22 and September 20 near the equinoxes. The

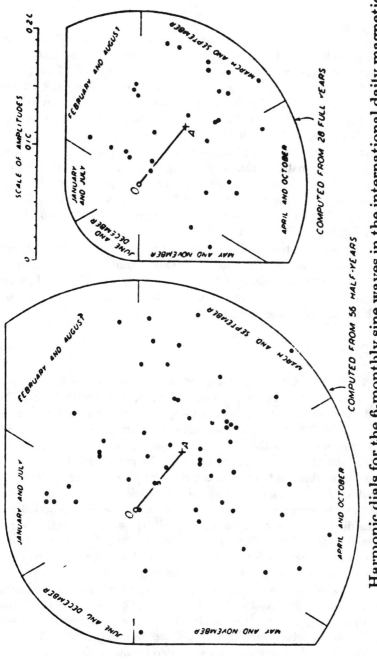

Harmonic dials for the 6-monthly sine waves in the international daily magnetic character-figures C

Fig. 40

75

expectancy for single vectors (reckoned from the origin) is M_0 obtained from

$$M_0^2 = (c_1^2 + c_2^2 + \cdots c_N^2)/N \qquad (61)$$

in which $c_1, c_2, \ldots$ etc. are the amplitudes (from the origin) of the individual vectors. For the left diagram of Figure 40, $M_0 = 0.111$ C and for the right $M_0 = 0.096$ C. For the average of 56 or 28 accidental waves we should expect the expectancy m_0 (see (58)) to be given by $m_0 = 0.111$ C$/\sqrt{56} = 0.0148$ C for the left diagram and $m_0 = 0.096$ C$/\sqrt{28} = 0.0181$ C for the right diagram. The average vector OA (amplitude $c = 0.0675$ C) is 4.6 and 3.6 times as large, i.e., K in (56) is 4.6 and 3.6 which gives W(K) from (56) about 10^{-9} and 10^{-6} for the probability that the average wave is accidental. The expectancy for the accidental vectors reckoned from A in Figure 40 (or as in equation (51)) is for single vectors 0.088 C and 0.068 C giving for the averages of 56 and 28 vectors the expectancies 0.0118 C and 0.0129 C and giving K = 5.7 and 5.2 and W(K) less than 10^{-12}.

Figure 41 shows the result of adding the vectors in the left diagram of Figure 40 for consecutive intervals. This "summation dial" bears no resemblance to the random walks shown in Figures 37 and 38, since the individual vectors in Figure 40 definitely prefer to "walk" toward March 22 (or September 20).

4.9 Periodogram

Figure 42 illustrates the periodogram for barometric pressure at Potsdam, Germany [1]. The expectancies for sine waves with 6-, 8-, 12-, and 24-hour periods, computed from single sets of r = 24 hourly values, are 0.11, 0.14, 0.30, and 1.11 mm Hg. These are plotted in Figure 42. Each curve represents the mean periodogram, i.e., the expectancy M as a function of period. The ordinates on the mean periodogram for N sets of r ordinates are the expectancies M for single vectors reduced by the factor $1/\sqrt{N}$. Persistent waves of amplitude c greater than $M/\sqrt{N}$ where M is the expectancy for single waves for that particular period will then be discovered, and the ratio $K = c/\sqrt{M}/N$ will indicate the degree of reliability [1]. Figure 42 (parts 13A to 13E, respectively) illustrates the mean periodograms for waves with 6- to 24-hour periods calculated from single days (13A) and the mean periodograms for waves computed from N = 5, 30, 365, and 22,000 days (60 years) obtained by reducing the curve for single days by the factor $1/\sqrt{N}$; for clarity part 13E has a ten times magnified ordinate scale. The persistent waves for periods 6, 8, 12, and 24 solar hours have amplitudes of 0.011, 0.026, 0.226, and 0.095 mm Hg and these have been indicated by the vertical lines in each periodogram as has the lunar tidal wave of period 12 hours 55 minutes

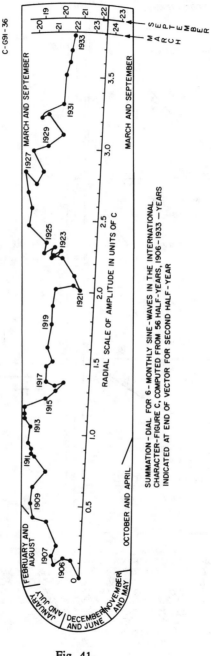

C-G9I-36

SUMMATION-DIAL FOR 6-MONTHLY SINE-WAVES IN THE INTERNATIONAL
CHARACTER-FIGURE C, COMPUTED FROM 56 HALF-YEARS, 1906-1933 — YEARS
INDICATED AT END OF VECTOR FOR SECOND HALF-YEAR

Fig. 41

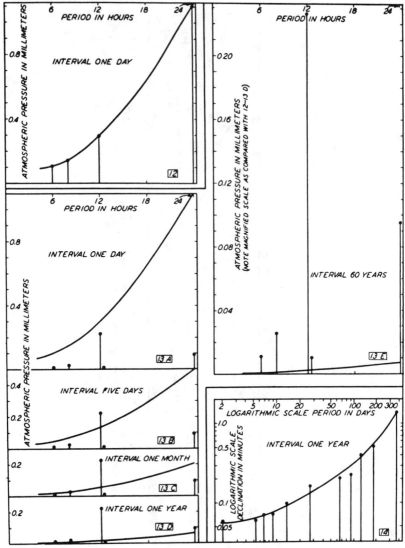

12 –MEAN PERIODOGRAM OF ATMOSPHERIC PRESSURE AT POTSDAM SHOW-
ING MEAN AMPLITUDES OF SINE-WAVES BASED ON HARMONIC ANALYSIS
FROM SINGLE-DAY INTERVALS FOR PERIODS 6, 8, 12, AND 24 SOLAR HOURS
13 –AMPLITUDES OF PERSISTENT SINE-WAVES FOR PERIODS 6, 8, 12, AND 24
SOLAR HOURS AND FOR 12 LUNAR (=12.4 SOLAR) HOURS CONTRASTED
WITH MEAN PERIODOGRAMS FOR INTERVALS AS INDICATED
14 -MEAN PERIODOGRAM OF MAGNETIC DECLINATION AT GREENWICH,
SHOWING MEAN AMPLITUDES OF SINE-WAVES BASED ON HARMONIC
ANALYSIS FROM SINGLE-YEAR INTERVALS (AFTER SCHUSTER)

Fig. 42

and amplitude 0.011 mm Hg. Figure 42 shows how persistent waves over an increasing number of days (N) gradually pierce through the mean periodogram, which represents the veil of the nonperiodic fluctuations hiding the persistent waves [1].

4.10   Quasi-Persistence and Effective Expectancy

Bartels [1] calls quasi-persistent such periodicities as are repeated with approximately the same amplitude and phase for a certain number of times forming what may be termed a sequence, each sequence ending more or less abruptly with no phase relation to other sequences. Such sequences in daily values of the character figure C were discussed in 3.1. This recurrence-phenomenon is expressed in quasi-persistence of the various sine waves with periods that are submultiples of 27 days. It is illustrated in Figure 43 which is the summation-dial for 27-day waves in character figure C for 378 solar rotations of 27 days each. The data are the same as in Figure 36. In Figure 43 the single vectors from Figure 36 are shown, summed for consecutive "rotations" 0 to 378. Figure 43 shows several sequences (for example from rotation 247 to 270) with the successive vectors in about the same direction. Other sequences appear more like a random walk. However, it is essential to obtain some numerical measure which clearly distinguishes between the random walk (Figure 38), quasi-persistence (Figure 43), and persistence (Figure 41). Suppose that we compute from the given N successive vectors the expectancy $M(1)$ for single vectors, $M(2)$ for the means of two vectors, and $M(h)$ for means of h successive (non-overlapping) vectors. Let $E(h) = M(h)\sqrt{h}$. For random vectors $M(h) = M(1)/\sqrt{h}$; consequently $E(h) = M(1)$ for random vectors. For persistent vectors $M(h) = M(1)$ and $E(h) = M(1)\sqrt{h}$. Figure 44 (22A) shows $E(h)$ as a function of h for random vectors. ($E(h)$ here is the same as Bartels c(h) with a bold face C as in Figure 44.) Diagram 22D of Figure 44 shows $E(h)$ for purely persistent waves, i.e., the linear increase with $\sqrt{h}$. For quasi-persistent waves $E(2) > E(1)$ and $E(3) > E(2)$, etc. but this increase is not proportional to $\sqrt{h}$, as for persistence, and in general $E(h)$ approaches asymptotically an upper limit $E(\infty)$ as k becomes large. If now we put $E(\infty)/E(1) = \sqrt{\sigma}$ then σ may be designated the equivalent length of the sequences. Equation (58) gave the expectancy m for the average of n random vectors in terms of the expectancy M for single vectors, i.e., $m = M/\sqrt{h}$; and following equation (58) the crucial ratio K was obtained by dividing the amplitude (0.0336 C) of the average vector in Figure 36 by $m = 0.262/\sqrt{378}$, giving K = 2.49 in order to test whether the amplitude of the average wave in Figure 36 could have been ascribed to chance. We found from (56) 0.002 for the probability that this average wave was "accidental". Now since the vectors

79

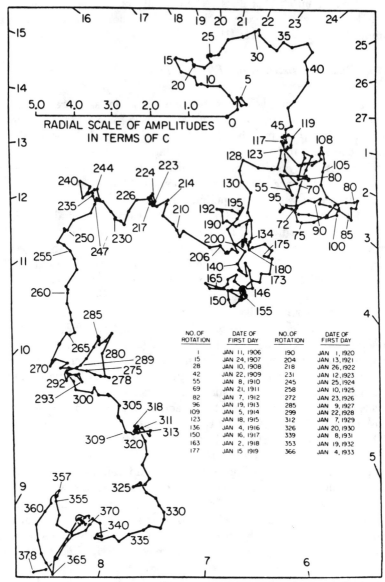

SCALE FOR TIME OF MAXIMUM (DAYS)

SUMMATION-DIAL FOR 27-DAY PERIOD IN INTERNATIONAL
MAGNETIC CHARACTER-FIGURE C, 1906-1933, COMPUTED FROM
378 ROTATIONS (INTERVALS OF 27 DAYS) WITH NUMBER OF
ROTATION INDICATED AT END OF CORRESPONDING VECTOR

Fig. 43

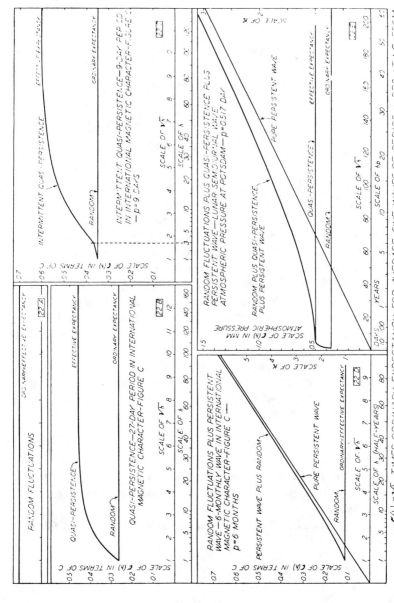

Fig. 44 HARMONIC ANALYSIS OF INTERVALS OF LENGTH h_p AS FUNCTIONS OF $\sqrt{h}$ — FIVE TYPICAL CASES

— $C(h) = \sqrt{h}$ TIMES ORDINARY EXPECTANCY FOR AVERAGE SINE-WAVES OF PERIOD p RESULTING FROM

in Figure 43 are not random but quasi-persistent and since Bartels [1] finds $E(\infty)/E1 = \sqrt{\sigma} = 1.74$ giving the equivalent length of sequences $\sigma = 1.74^2 = 3.0$ rotations, the *effective expectancy* m_e, which must be used to test the reality of the average wave in Figure 36, has to be computed with m_e given by:

$$m_e = M \sqrt{\sigma}/\sqrt{N} = M/\sqrt{N/3} \ . \tag{62}$$

Thus the effective number of independent random vectors is not 378 but 378/3 for Figure 36 (or 43). Since M was 0.262 C we have

$$m_e = 0.262 \ C/\sqrt{126} = 0.0234 \ C \ . \tag{63}$$

Since the amplitude of the average 27-day wave in Figure 36 is 0.0336 C we now find $K = 0.0336/0.0234 = 1.43$ and $W(K) = e^{-1.43^2} = 0.14$ instead of 0.002 as before. Thus, taking proper account of the effective expectancy there is no evidence for a persistent 27-day wave in international character figures. From this example as discussed in much greater detail in reference [1], it is evident that in testing geophysical periodicities the question of independence of vectors for successive periods of data should always be examined; otherwise false persistent periods may be claimed to be real. Figure 44 shows the "characteristic diagram" for various combinations of the kinds of persistence which can arise. These are fully discussed in reference [1].

Finally it should be emphasized that the distribution of points is not always circularly symmetric, i.e., σ_a is often not equal to σ_b (see below equation (51)). Many cases of elliptical distributions arise. Procedures for such cases are given in reference [3], with examples.

IV.   References

1. J. Bartels, Random Fluctuations, Persistence, and Quasi-Persistence in Geophysical and Cosmical Periodicities, *Terr. Mag.*, *40*, 1–60, 1935.
2. S. Chapman and J. Bartels, *Geomagnetism*, Oxford at the Clarendon Press (1940).
3. J. Bartels, Statistical Methods for Research on Diurnal Variations, *Terr. Mag.*, *37*, 291–302, 1932.
4. J. Bartels, Bemerkungen zur praktischen harmonischen Analyse, *Gerl. Beitr. z. Geoph.*, *28*, 1–10, 1930.

V. Time Variations of Cosmic-Ray Intensity

5.1 Introduction

The discovery of cosmic radiation is a good example of the way science progresses through experiments made to better understand some interesting or puzzling phenomenon. Such experiments may provide the desired answers but occasionally they reveal some new or wholly unexpected phenomenon leading investigators on to more experiments. In 1912 the Austrian physicist Dr. Victor Hess, now a professor of physics at Fordham University, ascended in a balloon to an altitude of 17,000 feet with an ionization chamber, in an experiment designed to test whether the small residual ionization always found in heavily shielded ionization chambers at ground level might be due to something other than radioactive contamination in the walls of the chamber. He found that the rate of ionization increased with altitude_after initially decreasing_. To explain this he postulated an ultra-penetrating radiation from outer space. For this discovery Dr. Hess was awarded the 1936 Nobel prize in physics jointly with Dr. Carl Anderson who discovered the positive electron during cosmic-ray experiments with cloud chambers.

Today it is known that about 90% of the primary cosmic rays which arrive at the top of the atmosphere are protons and that the balance is made up of alpha particles or helium nuclei and heavier nuclei. A very small fraction of the primary particles may have energies up to 10^{16} electron volts. From ingenious experiments to investigate the nature of the showers of secondary particles created by the interaction of primaries with atomic nuclei in the atmosphere, many important discoveries in nuclear physics have been made. Many new particles, some called "strange particles", have been discovered.

Figure 45 shows schematically some of the secondary particles and how these in turn give rise to further secondaries in complex chain reactions. The μ-meson or hard component is responsible for the ionization normally produced in ionization chambers at low altitudes. These very penetrating μ-mesons have a mass of about 210 electron masses, unit electronic charge + or −, and a lifetime (at rest) of about 2×10^{-6} seconds. They result from the decay of charged π mesons which have a lifetime (at rest) of about 10^{-8} seconds.

The primary particles also generate a cascade of nucleons (i.e., protons and neutrons). Even relatively low energy protons, with momenta only sufficient (a few BeV/c) to reach the top of the atmosphere at geomagnetic latitudes greater than about 50°, are effective in generating the nucleonic component.

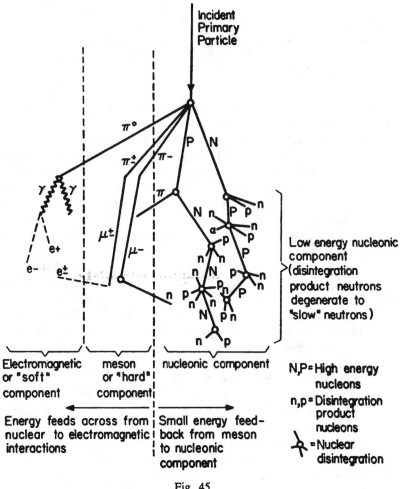

5.2   Variation of Cosmic-Ray Intensity with Latitude

Figure 46 shows the variation of neutron intensity with latitude (upper curve) and that of μ-meson intensity (lower curve) as obtained by Simpson from airplane flights in June 1948. At 30,000 feet altitude, the increase in neutron intensity between 0° and 60° geomagnetic latitude is about twice that of the meson component. The smaller latitude effect for the meson component is due to the fact that the lower energy primaries, which on account of the earth's magnetic field can reach the top of the atmosphere at higher latitudes but not at the equator, are less effective in generating μ-mesons than are the higher energy primaries which can reach the atmosphere at the equator (and at all other latitudes). At

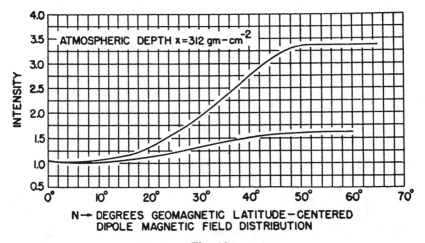

N → DEGREES GEOMAGNETIC LATITUDE − CENTERED
DIPOLE MAGNETIC FIELD DISTRIBUTION

Fig. 46

sea level the μ-meson intensity in shielded ionization chambers is only about 10% greater at geomagnetic latitude 50° (and higher) than at the equator. Thus at 50° all the additional primaries with energy less than that required to reach the equator and which can come in at latitude 50° produce only about 10% more μ-mesons (detectable at sea level) than are produced by all those primaries with energy equal to or exceeding that required to reach the geomagnetic equator. Since the number of primaries with energy greater than E is a rapidly decreasing function of E, it is evident that the μ-meson or hard component is a measure of the intensity of the higher energy primaries (above about 15 BeV for protons), whereas the neutron component as measured for example in IGY standardized neutron monitors [1] is much more sensitive to the primaries of lower energy (at least down to about 2 BeV for protons). Below this energy even the neutrons produced are not detected at sea level because of the blanketing effect of the atmosphere.

Because of their simplicity, ionization chambers have proved valuable for continuous registration over long periods of time. One of the longest series of observations is that made with the Carnegie Institution of Washington model C Compton-Bennett meter [2] which has operated continuously at the Instituto Geofísico de Huancayo since June 1936. Similar series of observations with similar instruments have been obtained at Godhavn (Greenland), Cheltenham (Maryland), now at Fredricksburg (Virginia), and at Christchurch (New Zealand). Simpson has operated neutron monitors at Chicago (Illinois), Climax (Colorado), Ciudad Universitaris (Mexico D.F.), and Huancayo (Peru) since about 1952. By the beginning of the IGY about 40 neutron monitors were operating in all parts of the world. There were also a large number of Geiger counter telescopes and several additional ionization chambers.

85

5.3 Magnetic Storm Effects

Figure 47 is an example of a large decrease in cosmic-ray intensity, associated with a period of severe magnetic disturbance. Daily means of cosmic-ray intensity from ionization chambers at three stations are shown in this figure, together with daily means of horizontal magnetic field, H, at Huancayo. The latter data show what may be described as a succession of three magnetic storms. It is evident that the decrease

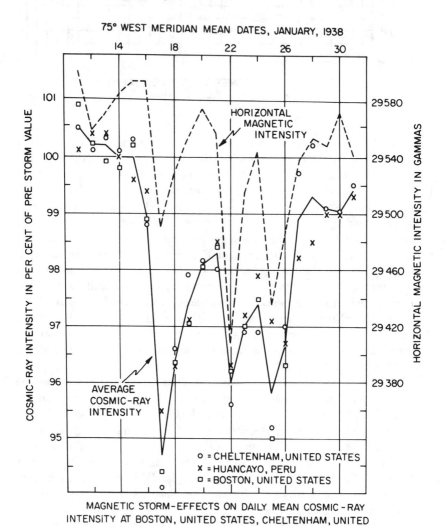

MAGNETIC STORM-EFFECTS ON DAILY MEAN COSMIC-RAY INTENSITY AT BOSTON, UNITED STATES, CHELTENHAM, UNITED STATES, AND HUANCAYO, PERU, AND ON DAILY MEAN MAGNETIC HORIZONTAL INTENSITY AT HUANCAYO, PERU

Fig. 47

of cosmic-ray intensity relative to that in H from January 16 to 17 was much greater than that from January 21 to 22, for example. For the comparable magnetic storm beginning August 21, 1937, Figure 48 shows no detectable change in cosmic-ray intensity.

Other magnetic storm effects are shown in Figure 49 in which are plotted the daily means of cosmic ray intensity for 1946 at Godhavn (g), Cheltenham (c), Christchurch (cc), and Huancayo (h) and also the daily mean horizontal intensity at Huancayo (H). One of the largest decreases occurred on July 25 after a large increase associated with a large solar flare. Despite the fact that not all magnetic storms are accompanied by a decrease in cosmic-ray intensity, Figure 50 shows that the average intensity for the five days of each month that are most disturbed magnetically tends to be definitely less than that for the five magnetically quietest days of each month. For the magnetic horizontal intensity at Huancayo the difference, ΔH, for disturbed minus quiet days is always negative, and Figure 50 shows that the corresponding differences ΔC for cosmic-ray intensity are preponderantly negative, although the correlation between ΔC and ΔH is not high. Theoretical attempts to ascribe the decrease of cosmic-ray intensity during magnetic storms to a ring current such as might account for the geomagnetic storm time variation give results to indicate that this is not the correct mechanism.

To obtain a better observational basis for understanding the cause for such cosmic-ray changes, Simpson [3] used aircraft-borne neutron detectors for latitude surveys at high altitude at times (in 1951) when the ground level neutron intensity was more or less normal and again a few days later when the ground level intensity was below normal. If, for example, the upper curve in Figure 46 were for a day with "normal" ground level intensity, then the curve for the day with subnormal intensity was found to be below that for the normal intensity at all latitudes between 40° and 65° N geomagnetic, to which range the survey was confined. If the depression in cosmic-ray intensity on a subnormal day were due, for example, to the effect of a magnetic storm caused by a current system in the lower ionosphere, then outside this current system the change in magnetic field, which would alter the trajectories of cosmic-ray particles, would be the same as if the magnetic moment of the earth had increased. In stating this conclusion the effect on cosmic-ray particles of the magnetic field of the hypothesized current system in the relatively short distance (say 100 km) between the earth's surface and the height of the current system is neglected. It can be shown from the Störmer-Vallarta theory for the motion of cosmic-ray particles in the dipole magnetic field of the earth that an increase in this dipole strength would only move the "knee" of the latitude curve toward the equator. The "knee" for the upper curve in Figure 46 is at about 50°. North of this point the intensity is about constant. However, Simpson's

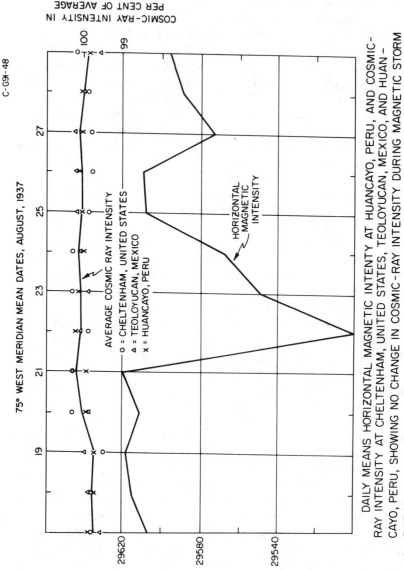

Fig. 48 DAILY MEANS HORIZONTAL MAGNETIC INTENTY AT HUANCAYO, PERU, AND COSMIC-RAY INTENSITY AT CHELTENHAM, UNITED STATES, TEOLOYUCAN, MEXICO, AND HUAN-CAYO, PERU, SHOWING NO CHANGE IN COSMIC-RAY INTENSITY DURING MAGNETIC STORM BEGINING AUGUST 21, 1937

88

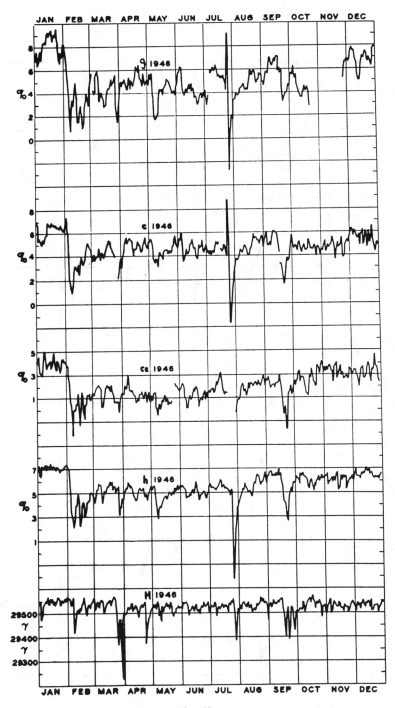

Fig. 49

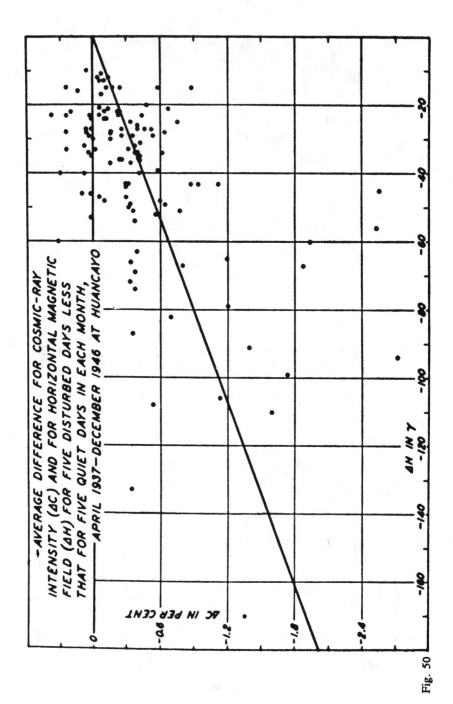

—AVERAGE DIFFERENCE FOR COSMIC-RAY INTENSITY (ΔC) AND FOR HORIZONTAL MAGNETIC FIELD (ΔH) FOR FIVE DISTURBED DAYS LESS THAT FOR FIVE QUIET DAYS IN EACH MONTH, APRIL 1937–DECEMBER 1946 AT HUANCAYO

Fig. 50

90

[3] results showed that on the days with subnormal intensity the knee was at the same latitude whereas the intensity was at all latitudes lower for the subnormal than for the normal day. This result showed that the depression of intensity on the subnormal day could not be ascribed to any magnetic field change equivalent to a change in the earth's dipole, i.e., to a storm-time variation current system close to the earth.

5.4 Twenty-Seven Day Variation

Figure 51 indicates the nature of the 27-day variation in cosmic-ray intensity by its correlation with the 27-day variation in international character figure C. It also illustrates a useful procedure first used by Bartels for testing the correlation between individual vectors (for common time intervals) in two harmonic dials. In each of the three harmonic dials of Figure 51 the phases (or times of maxima) for the cosmic-ray waves for intervals of 27-days are relative [4] to the phase or time of maximum of the 27-day wave from international magnetic character fig-

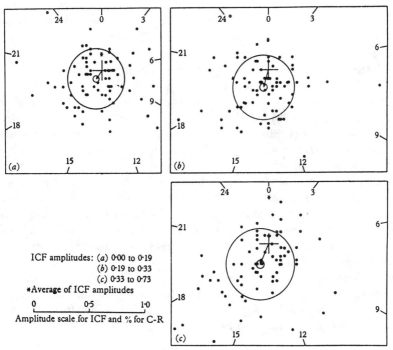

ICF amplitudes: (a) 0·00 to 0·19
(b) 0·19 to 0·33
(c) 0·33 to 0·73
∗Average of ICF amplitudes

0   0·5   1·0
Amplitude scale for ICF and % for C-R

Harmonic dials 27-day waves cosmic ray intensity (C–R), Huancayo (1936–54) phases relative to those for waves in international character figure (ICF) rotations with large C–R storm effects excluded.

Fig. 51

ures (ICF) for the same 27-day interval. In an harmonic dial for (ICF) such as in Figure 36 the points for the 27-day waves for each rotation (27-day interval) scatter around the origin. In the corresponding dial for 27-day waves in cosmic-ray intensity there is a similar scatter [5] around the origin. If in the ICF dial the vector for rotation number N_j has to be turned through an angle α_j to bring it vertical, then the vector, in the dial for cosmic-ray intensity, for the same rotation number N_j is turned through the same angle α_j.

Points in the dials of Figure 51 are for cosmic-ray intensity vectors (C-R) after each (C-R) vector has been turned through the angle α_j (different for each j). The average vector is shown and the large circle centered on its end point is the so-called probable error circle for single vector deviations from the average vector. The small circle is the probable error circle for the means [4] of 76, 79, and 75 such deviations, respectively, for (A), (B), and (C) of Figure 51. For (A), (B), and (C) the values of K (see 4.7) were 26, 4.4, and 4.5 and of W(K) 10^{-3}, 10^{-10}, and 10^{-9}, respectively, showing that the average vectors are not "accidental" and that the (C-R) vectors definitely tend to have their maxima about 15 days after the maxima of the ICF vectors. Or put another way, the minima of the 27-day waves in (C-R) tend to occur near (within two days or less) the times of maxima of the ICF vectors. In Section 4.10 the characteristic E(h) for quasi-persistent waves was discussed together with the equivalent length of sequences σ. The upper curve in Figure 52 shows E(h)/E(1) as ordinate (i.e., C(h)/C(1)) as a function of $\sqrt{h}$ as derived by Bartels for the international character figure C from 378 rotations (see Figure 36) for which $\sigma = 1.74$ or $\sigma^2 = 3.0$ for the equivalent length of sequences. Similarly (B) of Figure 52 shows the characteristic derived from 27-day waves in (C-R), not turned. For these $\sigma = 1.42$ or $\sigma^2 = 2.0$ for the equivalent number of sequences. Thus quasi persistence for 27-day waves in (C-R) prevails for about 2.0 rotations on the average for (C-R) and about 3.0 rotations [4] for ICF. These results are further evidence that the solar clouds responsible for magnetic storms and activity are somehow responsible for the changes or modulation of cosmic-ray intensity. Also it is noted that if a decrease in cosmic-ray intensity is associated with a magnetic storm such a decrease begins within an hour or so after the beginning of the main phase of the storm. This fact supports the belief that the mechanism for cosmic-ray modulation must not be far from the earth. More will be said on this point after discussing some recent results obtained by IGY investigators.

5.5 Variations with Solar Cycle

Before daily means (as in Figure 49) or monthly means (as in Figure 56) of cosmic-ray intensity from an ionization chamber can be com-

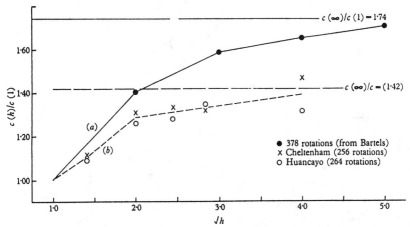

Quasipersistence in 27-day waves for international character figure (a), and for cosmic ray intensity (b).

Fig. 52

pared it is essential to correct the data for the seasonal variation shown in Figure 53. This variation arises from the fact that most μ-mesons are created at a height (about 16 km) in the atmosphere where the pressure is about 100 mb; when the height of the 100 mb level increases more μ-mesons decay into electrons and neutrinos before reaching the instruments and the cosmic-ray ionization decreases. This decrease in Figure 53 (B) is about 5.3% per km increase in height of the 100 mb

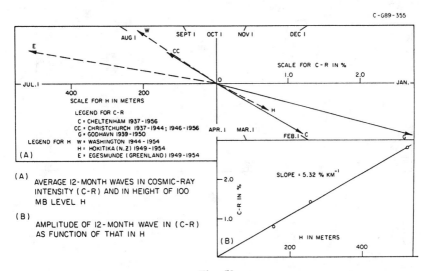

C-G89-355

(A) AVERAGE 12-MONTH WAVES IN COSMIC-RAY INTENSITY (C-R) AND IN HEIGHT OF 100 MB LEVEL H

(B) AMPLITUDE OF 12-MONTH WAVE IN (C-R) AS FUNCTION OF THAT IN H

Fig. 53

93

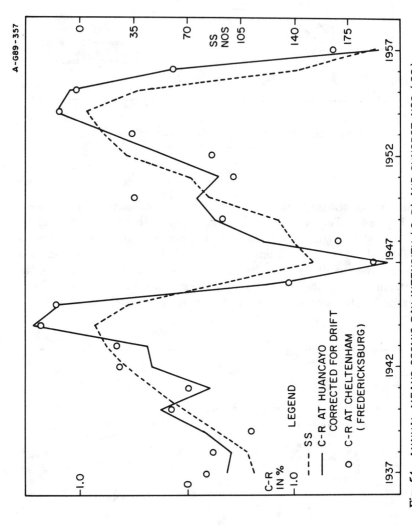

Fig. 54 ANNUAL MEANS COSMIC RAY INTENSITY (C-R) AND SUNSPOT NOS (SS)

LEGEND

--- SS

C-R AT HUANCAYO
CORRECTED FOR DRIFT

o C-R AT CHELTENHAM
(FREDERICKSBURG)

C-R
IN %

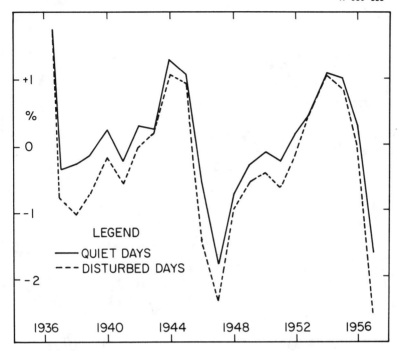

ANNUAL MEANS C-R INTENSITY AT HUANCAYO FOR
MAGNETICALLY QUIET AND DISTURBED DAYS,
ORDINATES ARE DEVIATIONS FROM 1937-1957 MEAN
FOR QUIET DAYS

Fig. 55

level. At Huancayo the seasonal variation in this height is evidently
quite small [5].

The variation of annual mean cosmic-ray intensity at Huancayo and
at Cheltenham (or Fredricksburg) is shown in Figure 54 from 1937 to
1957 together with the annual mean sunspot numbers. Figures 47 and 50
show that the cosmic-ray intensity is on the average less on magnetically
disturbed than on quiet days. Since there are more disturbed days near
sunspot maximum than near sunspot minimum one might expect an
11-year variation in the annual means of cosmic-ray intensity. However,
Figure 55 shows that this effect is small compared to the observed 11-year
variation which is only slightly less when derived from only the five quiet
days in each month than when derived from the five most disturbed days
in each month. Thus the 11-year variation is not due to the transient
decreases accompanying magnetic storms and disturbances. Other solar

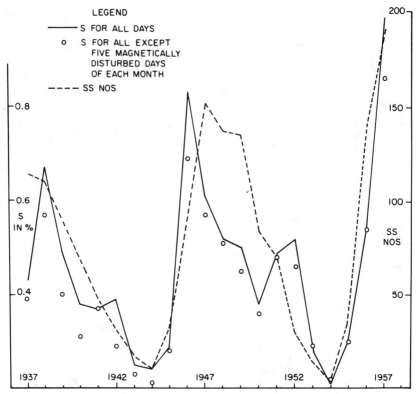

YEARLY POOLED STANDARD DEVIATION (S) OF DAILY MEANS OF COSMIC-RAY
INTENSITY AT HUANCAYO, AND YEARLY MEAN SUNSPOT NOS (SS)

Fig. 56

cycle effects on cosmic-ray intensity will be described before discussing possible explanations.

Figure 56 shows that the fluctuation of cosmic-ray intensity (i.e., activity) is greater near sunspot maximum than at sunspot minimum whereas the cosmic-ray intensity (Figures 54 and 55) is less near sunspot maximum than near sunspot minimum. For Figure 56 the cosmic-ray activity was measured by the standard deviation of daily means from their respective monthly means after pooling these standard deviations for each year to give S. Near sunspot minimum S is about 0.2% which is hardly more than the inevitable sampling fluctuations arising from the fact that only a finite number of μ-mesons traverse the ionization chamber in, say, one day.

Figure 57 compares the monthly mean cosmic-ray intensity from the ionization chamber at Huancayo with that from a neutron monitor at

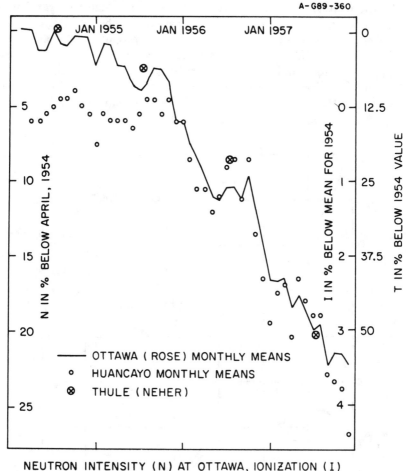

NEUTRON INTENSITY (N) AT OTTAWA, IONIZATION (I)
AT HUANCAYO, AND IONIZATION UNDER 15 GM CM⁻²
AT THULE, (T)

Fig. 57

Ottawa for the period January 1954 to December 1957. The figure
shows that during this period the intensity at Huancayo decreases by
about 4.5% while that at Ottawa decreases by about five times as much.
Neher's measurements at Thule of the ionization in high altitude balloon-
borne instruments are also shown. The latter show a decrease of about
50% between the summer of 1954 and that of 1957. The decrease of
intensity from 1954 through 1957 is that associated with the cycle of
increasing solar activity (Figure 55) and shows that the amplitude of
the 11-year variation of cosmic-ray intensity is much greater (by a factor
of five) for the lower energy part of the primary spectrum, as measured

by the neutron monitor at Ottawa, than for the higher energy part of the spectrum as measured by the ionization chamber at Huancayo. For Neher's measurements near the top of the atmosphere (where the effect of atmospheric absorption is less) the corresponding factor is at least 15. The decrease in neutron intensity at Ottawa (and in Neher's values) during 1954 and 1955 is not apparent at Huancayo. This fact indicates that as the solar cycle progresses from sunspot minimum towards maximum some "barrier" is set up which at first prevents the primaries in the

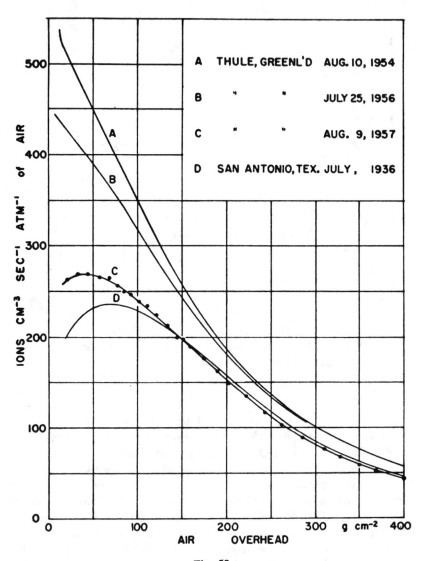

Fig. 58

lower energy part of the cosmic-ray spectrum from reaching the earth, and that as the cycle progresses the "barrier" becomes increasingly effective for increasing energy of primaries. This effect is also shown from Simpson's [1] latitude surveys with neutron detectors in aircraft, namely a southward shift of the latitude knee (Figure 46) as the solar cycle progresses through maximum.

Figure 58 shows Neher's results at Thule in greater detail for the summers of 1954, 1956, and 1957, together with results from similar flights in Texas in 1936. Comparison of curve (A) with (C) clearly shows that while particles down to very low energy (50 MeV or so if protons) were present at Thule in 1954, these were excluded in 1957 to the point where the energy spectrum of primaries over Thule in 1957 was not radically different from that over Texas in 1936.

V. References

1. J. A. Simpson, W. Fonger, and S. B. Treiman, Cosmic Radiation Intensity-Time Variations and Their Origin. I. Neutron Intensity Variation Method and Meteorological Factors, *Phys. Rev., 90*, 934–950, 1953.
2. A. H. Compton, E. O. Wollan, and R. D. Bennett, A Precision Recording Cosmic-Ray Meter, *Rev. Sci. Inst., 5*, 415–422, 1934.
3. J. A. Simpson, Cosmic-Radiation Intensity-Time Variations and Their Origin. III. The Origin of 27-day Variations, *Phys. Rev., 94*, 426–440, 1954.
4. S. E. Forbush, The 27-Day Variation in Cosmic Ray Intensity in Geomagnetic Activity, *Electromagnetic Phenomena in Cosmical Physics*, International Astronomical Union Symposium No. 6, pp. 332–344, Cambridge University Press (1958).
5. S. E. Forbush, World-Wide Cosmic-Ray Variations, 1937–1952, *J. Geophys. Res., 59*, 525–542, 1954.
6. H. V. Neher and H. Anderson, Cosmic-Ray Changes from 1954 to 1957, *Phys. Rev., 109*, 608, 1958.

VI. Cosmic-Ray Variations (Continued) and Some IGY Results

6.1 *Cosmic-Ray Equator and Intensity Variation Around It*

In 1956 Simpson [1] organized an aerial survey to determine the location of the cosmic-ray equator and the variation of intensity around it. The aircraft carried two neutron intensity recorders at a constant altitude of 18,000 feet. North-south (or S-N) flights at twelve different longitudes, spaced roughly 30° apart, were traversed about 12° of latitude on each side of the magnetic equator. For each traverse the curve of neutron intensity as a function of latitude showed a well defined minimum near the midpoint of the traverse. The locus of the twelve minima defined the location of the cosmic-ray equator at twelve different longitudes and the neutron intensity at these twelve points on the cosmic-ray equator defined the longitude variation along the cosmic-ray equator as shown in Figure 59. If the earth's magnetic field were that of a simple centered dipole with its axis parallel to the earth's rotational axis, the cosmic-ray equator would coincide with the geographic equator. Along this equator the cosmic-ray intensity would be constant. The observed variation of intensity with longitude is due in part to the fact that the eccentric dipole best approximating the earth's field has its center about 340 km from the earth's center (see Section 1.6). Simpson estimated from the data in Figure 59 that the magnetic center of the earth is displaced about 300 km from the earth's center. However, the maximum and minimum on the curve in Figure 59 are displaced about 40° to the west of where they would be predicted on the basis of the eccentric dipole, and in the direction of minimum and maximum surface magnetic fields.

The cosmic-ray equator is compared with the eccentric dipole equator and the dip equator (i.e., 0° inclination or vertical field zero) for the epoch 1945 in Figure 60. However, Rothwell and Quenby [2] pointed out that the dip equator for epoch 1955 follows the cosmic-ray equator about as closely as the dip equator follows the eccentric dipole field equator as

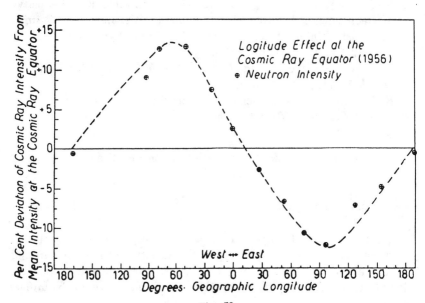

Fig. 59

shown in Figure 60. Quenby and Webber [3] estimated the location of the cosmic-ray equator that would result from considering the composite effect of the eccentric dipole and the nondipole parts of the earth's field on the motion of charged cosmic-ray particles. Their results are in quite good agreement with the observed cosmic-ray equator. Thus the discrepancies between Simpson's observations and the results predicted by the Lemaitre-Vallarta theory for the motion of charged particles in a dipole field arise from differences between the earth's actual field and the eccentric dipole approximation to it rather than to perturbing effects from fields external to the earth as had been suggested.

6.2 Solar Flare Effects

From continuous registration of cosmic-ray intensity at several stations over a period of two solar cycles (22 years) by the Department of Terrestrial Magnetism, Carnegie Institution of Washington, five large sudden increases of intensity have been observed [4]. All of these occurred within an hour or less after the start of a solar flare (or bright chromospheric eruption) or after a radio fade-out, which indicates the occurrence of a solar flare. Two of these are shown [5] in Figure 61 and one in Figure 62. The increase at Climax (Colorado) was about five times that at Cheltenham (Maryland). The latter station is at altitude 72 meters while Climax is at 3500 meters. Since the total ionization at Climax is about 2.5 times that at Cheltenham and the percentage (of the total) increase at Climax was 4.8 times that at Cheltenham, the

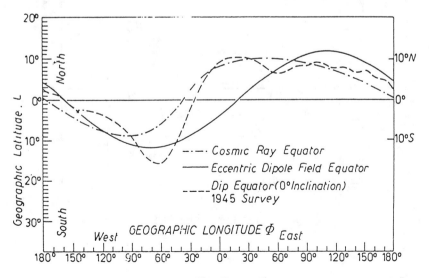

Fig. 60

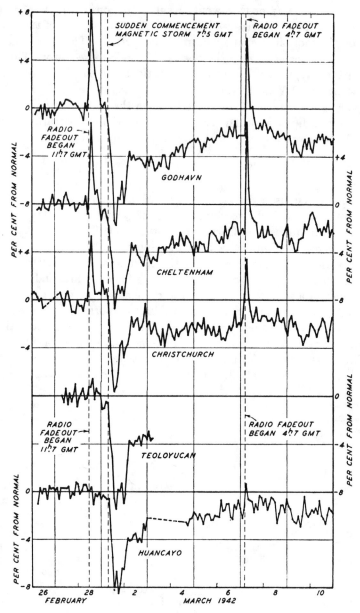

Increases of cosmic-ray intensity, February 28 and
March 7, 1942.

Fig. 61

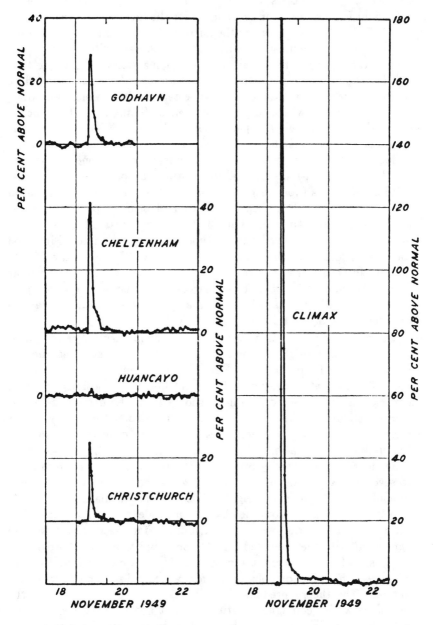

Increase of cosmic-ray intensity, November 19, 1949.

Fig. 62

actual magnitude of the increase at Climax was about twelve times that at Cheltenham. Since the difference in altitude between Climax and Cheltenham corresponds to 340 gm cm^{-2}, the exponential absorption coefficient for the radiation responsible for the increase is about 137 gm cm^{-2}. This is the rate at which the nucleonic component (responsible for star production in emulsions) was known to increase with altitude. The normal ratio (2.5) of the total cosmic-ray ionization at Climax to that at Cheltenham is mainly attributable to mesons. Thus the flare effect increased with altitude too rapidly to be ascribed to mesons. Moreover, the flare effect showed a large latitude effect (being zero at Huancayo on the geomagnetic equator) whereas the latitude effect for mesons is small. At Climax less than 10% of the total ionization is due normally to local radiation originating from the nucleonic component. Assuming that this radiation was produced entirely by particles in the same band of energy as those responsible for the 200% increase in ionization on November 19, 1949, then the number of primary particles, reaching there per unit time, in that band of energy, was estimated [5] to have increased to at least twenty times the normal value. Figure 63 shows a 24-fold increase of neutron intensity at Chicago and a 30-fold increase in Ottawa as observed with neutron monitors during the solar flare of February 23, 1956. This solar flare increase was recorded at a large number of stations, which fortunately has commenced operations in anticipation of the IGY program.

The earlier solar flare increases in cosmic-ray intensity were compared in some detail with results predicted on the basis that these charged particles came from the sun. Firor [6], using published results from numerical integrations and model experiments on the motion of charged particles in the earth's dipole field, showed that the magnitude of the increase should not be everywhere the same. Instead, he showed that there should be three "impact zones", one centered near 4:00 a.m., one centered near 9:00 a.m., and a third without any strong local time dependence. The intensity of the solar flare increases were from three to seven times greater in the morning zones than in the third or background zone. The magnetic rigidities used in the calculations were from 1 to 10 BV estimated from the observed latitude effect for the increases. Firor showed that the observed distributions for the increases were in agreement with the predicted impact zones and thus that the particles responsible for these increases must, therefore, have come from the sun. There was, however, one observed effect which did not accord with prediction. The theory predicts that there should be no increase in the region between geomagnetic latitude about 60° and the geomagnetic poles, whereas in this region large intensity increases have invariably been observed (e.g., at Godhavn as in Figures 61 and 62).

This aspect of the theory follows also directly from the equations of motion for auroral particles as discussed by Chapman and Bartels [7]

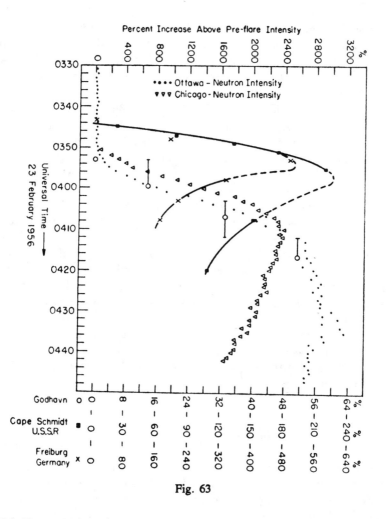

Fig. 63

(Vol. II, pp. 834–842). These show that for particles from the sun there is a region between the auroral zones and the geomagnetic poles inside which particles from the sun should not arrive. The angular distance from the poles to this inside boundary of the auroral zone depends of course on the lower limit for the particles' energy; for cosmic-ray primaries with energy sufficient to give effects at ground level this angular distance is about 30°. The observed increase at high latitudes requires a scattering or diffusing mechanism, as by ordered or disordered magnetic fields in highly conducting clouds, which came from the sun [8]. Simpson [9,10] and co-workers show that the model must be such that it does not "wipe out" the impact zones for the early part of the increase associated with the flare effect of February 23, 1956, and that it must also account for the fact that within about an hour after the start of the cosmic-ray

105

increase the impact zones are wiped out and thereafter the radiation is quite isotropic. The complete interpretation for these effects has important implications regarding the electrodynamic characteristics of the interplanetary medium. A better understanding of these characteristics will doubtless come from exploration with artificial satellites.

The cosmic-ray increase of February 23, 1956 also resulted in several effects on the ionosphere—effects on the dark side of the earth not due to wave radiation from the flare. One of these is the absorption of cosmic radio noise [11], due to increased ionization in the lower ionosphere. Leinbach and Reid [12] at the Geophysical Institute in Alaska have detected several increases in cosmic radio noise absorption associated with solar flares, but due to particles of energy too low to be detected by cosmic-ray recorders at ground level. Bailey [13] observed during the solar flare of February 23, 1956 a sudden increase in electron density at heights of about 90 km, coincident with the start of the cosmic-ray increase. These initial effects were observed at LF (low frequency) and VLF (very low frequency) as a sudden phase anomaly indicating a decrease in the height of reflection and as a decrease in field strength of certain radio signals received from distant stations. At the height of 90 km, the increase of electron density, during the early part of the cosmic-ray increase of February 23, 1946, was sufficient to transform the dark ionosphere at this height from nighttime characteristics to those of a sunlit hemisphere.

6.3 Alpha Particle Effects

The table in Figure 64 (from the work of the cosmic-ray group at the University of Minnesota) shows that the total flux of alpha particles in September 1957 was only about half that in October 1950 at Minneapolis. Also in Texas the flux decreased one-third between February 6, 1956 and October 19, 1957. Figure 65 shows the number distributions of α-particles according to track density in the emulsions flown on high altitude balloons. The track density is a measure of the ionization density from which the energy of the α-particles may be derived. Comparing these and additional similar results at other times Freier, Ney, and Fowler [14] find that the decrease in α-particle flux, at Minneapolis, for example, is associated with increasing activity of the solar cycle. They show that the decrease in the flux of α-particles was not brought about by a sharp cutoff which would exclude particles at the low energy end of the spectrum (i.e., those at the right in the diagrams of Figure 65) because low energy particles down to the energy limit imposed by the amount of air above are always observed. Thus the decrease in α-particle flux occurs throughout the spectrum; and even though both

106

DATE	PLACE	CUT-OFF ENERGY BEV/NUCLEON	FLUX α-PARTICLES PER M.² STER. SEC.	REFERENCE
4 OCT. 1950	MINNEAPOLIS MINNESOTA	$\leqq 0.15$	272 ± 20	WADDINGTON (1957)
18 JUNE 1954	SASKATOON CANADA	$\leqq 0.15$	290 ± 20	FREIER AND NEY
17 MAY 1956	WAUKON IOWA	$\leqq 0.15$	255 ± 20	FREIER AND NEY
17 MAY 1957	MINNEAPOLIS MINNESOTA	$\leqq 0.23$	157 ± 17	FREIER AND NEY
31 AUG.–1 SEPT. 1957	MINNEAPOLIS MINNESOTA	$\leqq 0.20$	138 ± 9	THIS WORK
6 FEB. 1956	SAN ANGELO TEXAS	~ 1.5	97 ± 8	FREIER AND NEY
19 OCT. 1957	BROWNWOOD TEXAS	~ 1.5	65^{+7}_{-5}	THIS WORK

Fig. 64

low and high energy particles are reduced in number some low energy particles still arrive. This is analogous to the fact that the solar cycle variation in cosmic-ray intensity is observed not only at latitudes north or south of, say, 35° as it would be if only those primary particles with energy less than that required to reach the atmosphere at 35° were excluded, but also at the equator. The mechanisms for these changes are not yet clearly understood, although it seems reasonably certain that large plasma clouds from the sun with "frozen in" magnetic fields play the important role. As solar activity increases, such clouds form some sort of barrier surface within which are both the sun and the earth. This barrier reduces the number of primary cosmic-ray particles which have sufficient energy to reach the equator (greater than 15 BeV if protons), and is even more effective for lower energy primaries, as shown by the fact that the solar cycle variation in intensity is greater for the neutron component at Ottawa (Figure 57) than for the μ-meson component at Huancayo. Nevertheless, the barrier still allows some low energy α-particles to pass through from outside the barrier. It is still not certain whether the magnetic fields are ordered or disordered. We have already seen that such a barrier seems necessary to explain the delay of arrival times for the particles from solar flare emission, by providing a reflecting or scattering boundary for such particles from the sun and for storage. Later some observations will be described which indicate that the low energy particles from the sun may be transferred to the earth in such plasma as also cause magnetic storms.

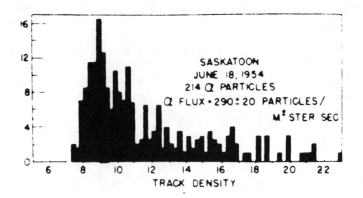

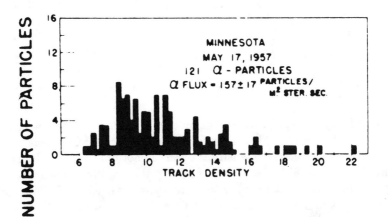

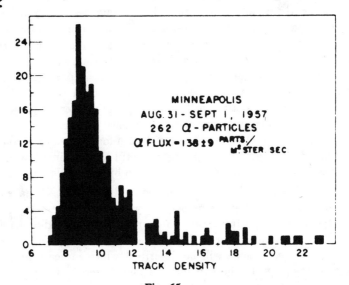

Fig. 65

During some high altitude balloon flights in the joint U.S.-Canadian IGY program, made on August 29–30, 1957, Anderson [15,16] discovered x rays during a magnetic storm with which was associated a Forbush decrease in cosmic-ray intensity. Anderson [16] estimated the x-ray energy to be in the region of 100 keV. The x rays were attributed to bremsstrahlung of electrons stopped in the high atmosphere above the balloons. The x rays were observed during the time the earth's horizontal magnetic field at Fredricksburg, Virginia was subnormal as in magnetic storms.

Figure 66 shows the results obtained by Winckler et al. [17] from a typical quiet day flight. The upper curve shows the counting rate for a single Geiger counter and the lower curve the rate of ionization in an ion chamber. Each of these rates is essentially constant after the balloon reached ceiling altitude at 1230 UT. Figure 67 shows the behavior of these counting rates on a similar flight by Winckler et al. [17] on June 30–July 1, 1957. The first burst of x rays occurred at 0330 UT, coincident with the development of strong auroral ray structure. The disturbances continued until 0645 UT. The upper curve of Figure 67 shows the ratio of the ionization rate (bottom curve) to that of the counter rate (middle curve). Prior to 0330 this ratio is characteristic of cosmic rays. The increase in the ratio at 0330 is due to x rays whose energy can be estimated on the basis of laboratory calibrations of the respective detectors. On other flights shielded photon counters showed conclusively that similar bursts arise from x rays. Figure 68, after Winckler et al. [18], shows another example of auroral x rays at Minneapolis (middle curve), cosmic noise absorption at Boulder (top curve), and horizontal magnetic intensity H at Fredricksburg. The H curve shows a strong magnetic bay at 0630 and another near 0900. These bays coincide with the periods of the x rays observed at Minneapolis and of strong absorption of cosmic noise at Boulder. During these same two intervals there passed over the zenith at Minneapolis a very large display of auroral luminosity. There was also a worldwide decrease in cosmic-ray intensity of about 5% at sea level and about 21% at balloon altitude. From the effects shown in Figure 68, the authors estimated that an electron flux of 0.6×10^6 electrons cm^{-2} sec^{-1} would be required and that the ionization which they produced in the D layer would account for the cosmic noise absorption. The x rays themselves carried 1000 times less energy than the electrons and could not provide sufficient ionization to account for the radio noise absorption. The increased ionization by these electrons is probably the cause of the increased conductivity which, by the dynamo theory, accounts for the magnetic bays or S_D-type disturbance. All these phenomena indicate the passage of the earth into an intense cloud which may have

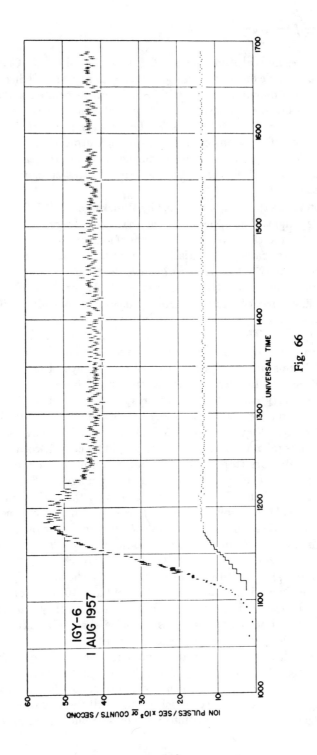

Fig. 66

IGY-6
1 AUG 1957

UNIVERSAL TIME

ION PULSES / SEC x10³ or COUNTS / SECOND

110

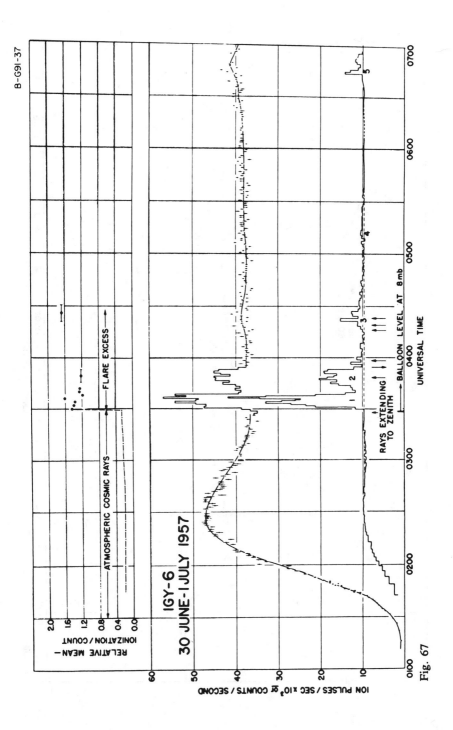

Fig. 67

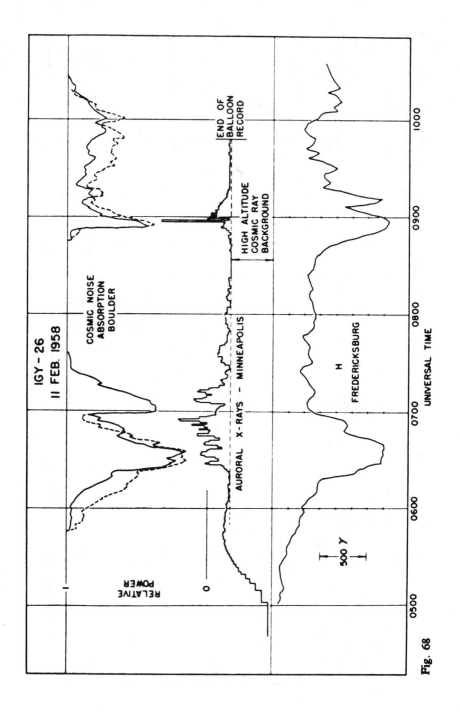

IGY – 26
11 FEB. 1958

COSMIC NOISE
ABSORPTION
BOULDER

END OF
BALLOON
RECORD

HIGH ALTITUDE
COSMIC RAY
BACKGROUND

AURORAL X-RAYS – MINNEAPOLIS

RELATIVE POWER

0

H
FREDERICKSBURG

500 γ

0500   0600   0700   0800   0900   1000

UNIVERSAL TIME

Fig. 68

112

been emitted from the sun during the solar flare at 2108 UT February 9. Other cases of x rays have also been observed by the Minnesota group with electron fluxes ten times as great as those on February 11, 1958. These estimated electron fluxes are about the same as the proton flux estimated by Chamberlain to be associated with the Doppler shifted Hα in the spectra of strong aurora at Yerkes Observatory. This approximate equality is consistent with electrical neutrality in the plasma from the sun as is required to account for the presence of particles much below the cutoff rigidity predicted by geomagnetic theory for the latitudes at which aurora are observed.

6.5 Protons from the Sun

On May 12, 1959, Ney et al. [19] in a balloon flight at 10 gm cm^{-2} over Minneapolis with ion chambers, Geiger and scintillation counters, and photographic emulsions in a period of unusual activity, found that the integral flux of particles at the top of the atmosphere increased by a factor of 1000. The composition of this incoming beam at 10 gm cm^{-2} was essentially pure hydrogen whereas the flux of α-particles was normal and not increased. Moreover, the particles in this event arrived at Minnesota at energies below the Störmer cutoff energy established for this latitude. In order that particles of such energy arrive at Minneapolis, the earth's magnetic field must be distorted by the incoming solar beam which followed the occurrence at 2000 UT, May 10, 1959, of the largest solar flare of the present solar cycle. The flare lasted more than three hours and was followed about one day later by a magnetic storm at 0300 UT May 12. A Forbush decrease in the neutron intensity on the ground at Minneapolis began at that time and reached a maximum of 15% at 0400 UT, May 12, 1959. Strong cosmic noise absorption (17 db) occurred from 0100 UT May 11 to at least 1700 UT May 12. The large increase of proton flux in the energy range between 110 MeV and 220 MeV has a spectrum similar to that for the solar flare effect of February 23, 1956. The fact that these particles arrived at Minneapolis probably indicates, according to Ney et al. [19], the arrival of a solar magnetic cloud which decreased the magnetic cutoffs and allowed these particles to be observed at Minneapolis where particles of such energies and less are normally forbidden or excluded by the earth's field. The arrival of this corpuscular beam at Minneapolis was accompanied by a very strong aurora. X rays were also found to accompany the aurora, and the flight instrumentation clearly separated these from the nucleonic component. This extremely important event was observed in data recovered from four of five balloons launched between 0200 UT and 1500 UT May 12. (From his recording of extra strong cosmic noise absorption Leinbach warned Ney et al. at Minnesota that something unusual was in progress, and the balloons were launched promptly by Ney and his co-workers.) They [19] state that this was the most spectacular nucleonic event that they had observed in two years' monitoring of cosmic rays during the IGY. Balloons were kept at the top of the atmosphere approximately 5% of that period of time. The authors observed that the energy spectrum for this event is much steeper than the energy spectrum of trapped protons in the Van Allen belt and that the integral fluxes are higher than those reported for the inner radiation belt.

Finally it seems quite likely that from the results of intensified cosmic-ray and other geophysical research during and following the IGY will come a coherent and quantitative explanation for many geophysical phenomena for which the sun is primarily responsible.

VI. References

1. L. Katz, P. Meyer, and J. A. Simpson, Further Experiments Concerning the Geomagnetic Field Effective for Cosmic Rays, Supplemento A1 Volume VIII, Serie X, *Del Nuovo Cimento*, 2° Trimestre, 277–282, 1958.

2. P. Rothwell and J. Quenby, Cosmic Rays in the Earth's Magnetic Field, Supplemento A1 Volume VIII, Serie X, *Del Nuovo Cimento*, No. 2 2° Trimestre, 249–256, 1958.

3. J. J. Quenby and W. R. Webber, Cosmic Ray Cut-Off Rigidities and the Earth's Magnetic Field, *Phil. Mag.*, *4*, 90–113, January 1959.

4. S. E. Forbush, Three Unusual Cosmic-Ray Increase Possibly Due to Charged Particles from the Sun, *Phys. Rev.*, *70*, 771–772, 1946.

5. S. E. Forbush, T. B. Stinchcomb, and M. Schein, The Extraordinary Increase of Cosmic-Ray Intensity on November 19, 1949, *Phys. Rev.*, *79*, 501–504, 1950.

6. J. Firor, Cosmic Radiation Intensity-Time Variations and Their Origin. IV. Increases Associated with Solar Flares, *Phys. Rev.*, *94*, 1017–1028, 1954.

7. S. Chapman and J. Bartels, *Geomagnetism*, Vol. II, p. 645, Oxford at the Clarendon Press (1940).

8. H. Alfvén, *Cosmical Electrodynamics*, Oxford at the Clarendon Press (1950).

9. R. Lüst and J. A. Simpson, Initial Stages in the Propagation of Cosmic Rays Produced by Solar Flares, *Phys. Rev.*, *108*, 1563–1576, 1957.

10. J. A. Simpson, Solar Flare Cosmic Rays and Their Propagation, Supplemento A1 Volume VIII, Serie X, *Del Nuovo Cimento*, 2° Trimestre, 133–160, 1958.

11. S. E. Forbush and B. F. Burke, Absorption of Cosmic Radio Noise at

22.2 Mc/Sec Following Solar Flare of February 23, 1956, *J. Geophys. Res.*, *61*, 573–575, 1956.

12. H. Leinbach and G. C. Reid, Ionization of the Upper Atmosphere by Low-Energy Charged Particles from a Solar Flare, *Phys. Rev. Letters*, *2*, 61, 1959.

13. D. K. Bailey, Disturbances in the Lower Ionosphere Observed at VHF Following the Solar Flare of 23 February 1956 with Particular Reference to the Auroral-Zone Absorption, *J. Geophys. Res.*, *62*, 431-463, 1957.

14. P. S. Freier, E. P. Ney, and P. H. Fowler, Primary α-Particle Intensity at Sunspot Maximum, *Nature*, *181*, 1319–1321, 1958.

15. K. A. Anderson, Cosmic-Ray Decrease of 29 August 1957, and Associated Soft Radiation, *Bull. Am. Phys. Soc.*, Ser. II, *2*, 349, 1957.

16. K. A. Anderson, Occurrence of Soft Radiation during the Magnetic Storm of 29 August 1957, *J. Geophys. Res.*, *62*, 641–644, 1957.

17. J. R. Winckler, L. Peterson, R. Arnoldy, and R. Hoffman, X-Rays from Visible Aurorae at Minneapolis, *Phys. Rev.*, *110*, 1221–1231, 1958.

18. J. R. Winckler, L. Peterson, R. Hoffman, and R. Arnoldy, Auroral X-Rays, Cosmic Rays, and Related Phenomena during the Storm of February 10–11, 1958, *J. Geophys. Res.*, *64*, 597–610, 1959.

19. E. P. Ney, J. R. Winckler, and P. S. Freier, Protons from the Sun on May 12, 1959, *Phys. Rev. Letters*, *3*, 183–185, 1959.

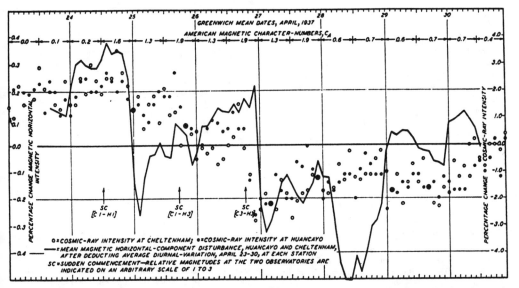

FIG. 1. Bi-hourly departures expressed in percentage of absolute values for cosmic-ray intensity and for disturbance of horizontal magnetic component April 23–30, 1937, Huancayo and Cheltenham magnetic observatories.

On the Effects in Cosmic-Ray Intensity Observed During the Recent Magnetic Storm

The purpose of this letter is to indicate some effects in cosmic-ray intensity which were observed simultaneously at two stations during the magnetic storm of April 25 to 30, 1937. The data were obtained with Compton-Bennett meters, one at the Cheltenham (Maryland) Magnetic Observatory of the United States Coast and Geodetic Survey and the other at the Huancayo (Peru) Magnetic Observatory of the Department of Terrestrial Magnetism of the Carnegie Institution of Washington. The accompanying figure summarizes the observed effects. On it are plotted for the two stations, in Greenwich mean time, the departures, in percent of the absolute value, of each bi-hourly mean of cosmic-ray intensity, after eliminating bursts, reduced to constant barometric pressure.

The striking similarity of the simultaneous changes in the bi-hourly means of cosmic-ray intensity at the two stations is obvious. The separation of the stations excludes the possibility that this is due to barometric changes.

The decrease of nearly four percent in cosmic-ray intensity from April 23 and 24 to April 27 is much greater than the largest changes which have been observed heretofore at either station over a similar period. The intensity of the magnetic storm was also the greatest since these meters have been operating.

The solid curve in the figure indicates the departures in percent of absolute value of bi-hourly means of the horizontal component of the earth's field derived from the magnetograms recorded at the two stations. It shows several storms following the usual course of a magnetic

disturbance, namely, an increase in the horizontal component, which often begins abruptly with a "sudden commencement" (SC); this increase usually continues for an hour or more and is followed by a rapid decrease and a slow recovery to normal after a day or two.

The primary field of a magnetic storm is known to arise from external causes and the major changes of that field can be ascribed to an external current-system.[1,2] This current system is such that, at great distances external to it, the changes are approximately those which would arise from a change in the earth's moment of the same sign as the observed change in horizontal intensity. Thus the decrease in cosmic-ray intensity which appears to follow the decrease in the horizontal component is to be expected on the basis of the theory of Lemaitre and Vallarta. The storm effect—if substantiated by future observations—when considered in the light of the theory of the allowed cones of Lemaitre and Vallarta, may provide further information on the energy distribution of cosmic rays.

Rigorous analysis based on the theory of Lemaitre and Vallarta has not yet been attempted pending receipt of additional magnetic data to determine the uniform part of its external field of the storm. Some discrepancies between the changes in cosmic-ray intensity and the horizontal component of the earth's field may be expected since cosmic rays are probably affected mostly by the equivalent dipole changes in the earth's field. Some discrepancy may be expected also when one considers the effects of the induced current system.

It may be remarked that there is a noticeable similarity between changes in the daily means of cosmic-ray intensity at these two stations. This would seem a necessary but not sufficient condition for the existence of the 27-day variation in cosmic-ray intensity noted by Hess.[3]

While the evidence here presented cannot be regarded in itself as conclusive proof that the observed changes in cosmic-ray intensity are due to the external field of the magnetic storm, this hypothesis seems to be the most reasonable one.

We are under obligation (1) to the staff of the United States Coast and Geodetic Survey at Cheltenham—in particular G. A. Hartnell who is in charge of the meter—and (2) to the staff at the Huancayo Magnetic Observatory.

S. E. FORBUSH

Department of Terrestrial Magnetism,
Carnegie Institution of Washington,
Washington, D. C.

[1] S. Chapman, Terr. Mag. 40, 349–370 (1935).
[2] L. Slaucitajs and A. G. McNish, Rep. and Comm., Edinburgh Assembly, Internat. Union Geod. Geophys., Ass. Terr. Mag. Electr Sept. 1936, 7 pp. (July 31, 1936).
[3] V. Hess, Terr. Mag. 41, 345–350 (1936).

LETTERS TO THE EDITOR

On Sidereal Diurnal Variation in Cosmic-Ray Intensity

In their investigation of the apparent effect of galactic rotation upon the intensity of cosmic rays, Compton and Getting[1] stressed the importance of testing the reality of the apparent 24-hour sidereal wave indicated by their discussion of one year's data obtained by Hess and Steinmaurer[2] on the Hafelekar, Austria. Illing[3] from additional data at this station, and Schonland, Delatizky, and Gaskell[4] from data at Capetown, found sidereal curves somewhat similar to the theoretical curve of Compton and Getting, which has an amplitude about 0.05 percent of the total intensity and maximum near 21^h. Comparison of their curves for separate years does not lend support for the reality of the average.

Although rigorous statistical methods are available[5, 6] for determining the reality of geophysical periodicities, these have not been applied heretofore to cosmic-ray data except in the case of the solar diurnal variation.[7] Analyzing data for 595 days obtained with a Compton-Bennett[8] precision recording cosmic-ray meter at the Magnetic Observatory of the United States Coast and Geodetic Survey near Cheltenham, Maryland, we find an apparent 24-hour sidereal wave with amplitude about 0.03 percent of the total intensity and with maximum near 22^h. To test whether this wave is statistically significant we utilize the amplitudes of the 24-hour *solar* wave which were obtained for each of 273 days in our analysis of the solar diurnal variation.[7] For each day the calculated amplitude of the 24-hour sidereal wave will be practically indentical with that for the solar wave—only the phase will differ. We assume that the distribution of amplitudes for the remaining 322 days (out of 595) is similar to that obtained for the 273 days. Following Bartels[5] we find that, *if the phases associated with the 595 single amplitudes are completely random*, the probability of obtaining an average 24-hour sidereal wave with amplitude at least as great as that which we actually found is about one in fifty.

Those experienced in applying these tests would not take this to indicate more than the slightest suspicion of statistical reality for the observed 24-hour sidereal wave which, even if statistically real, might be due to other causes[6] than an actual sidereal variation.

If a 24-hour sidereal wave actually exists in these data, the observations for 595 days are statistically inadequate to establish the fact. This conclusion is emphasized by analysis of data for 396 days from a Compton-Bennett meter at the Huancayo Magnetic Observatory of the Department of Terrestrial Magnetism. These data indicate an apparent 24-hour sidereal wave, the statistical significance of which has not been tested, with amplitude about 0.06 percent of the total intensity and with maximum near 5^h. Since this maximum differs by 7 hours, out of a possible 12, from that for the Cheltenham data, it is evident that the apparent sidereal wave at each station cannot be a true sidereal one.

Compton and Getting[1] found from the data of Hess and Steinmaurer[2] that the amplitude of the apparent 24-hour sidereal wave was nearly ten times their estimate of its probable error. Their exact procedure in obtaining this estimate was not given. Estimates of the probable errors in geophysical data are especially misleading. If based on the departures of observed points from a fitted wave, they are invariably too small when the departures for successive points are *statistically independent*. Tests on cosmic-ray data from Cheltenham indicate that such departures are *not* independent. Our conclusions regarding the reality of the 24-hour sidereal wave are based on a method of analysis[5] which takes account of this.

It is surely one of the most constructive recent developments in physics that such powerful tools have been evolved for evaluating the real or illusory nature of such interesting periodicities.

S. E. FORBUSH

Department of Terrestrial Magnetism,
Carnegie Institution of Washington,
Washington, D. C.,
November 30, 1937.

[1] A. H. Compton and I. A. Getting, Phys. Rev. 47, 817–821 (1935).
[2] V. Hess and R. Steinmaurer, Berlin, Sitz. Ber. Akad. Wiss. 521–542 (1933).
[3] W. Illing, Terr. Mag. 41, 185–191 (1936).
[4] B. F. J. Schonland, B. Delatizky, and J. P. Gaskell, Nature 138, 325 (1936).
[5] J. Bartels, Terr. Mag. 40, 1–60 (1935).
[6] J. Bartels, Sitz. Ber. Akad. Wiss. 504–522 (1935).
[7] S. E. Forbush, Terr. Mag. 42, 1–16 (1937).
[8] A. H. Compton, E. O. Wollan and R. D. Bennett, Rev. Sci. Inst. 5, 415–422 (1934).

Terrestrial Magnetism

and

Atmospheric Electricity

VOLUME 42   MARCH, 1937   No. 1

ON DIURNAL VARIATION IN COSMIC-RAY INTENSITY

By S. E. FORBUSH

Abstract—Cosmic-ray data from precision cosmic-ray meter records obtained at Cheltenham, Maryland, for 273 complete days during April 1935 to October 1936 are subjected to rigorous statistical analysis. The results indicate that the real barometric coefficient does not change from hour to hour or from month to month. It is shown that the barometric coefficient obtained at this station is in good agreement with certain results obtained from altitude-intensity curves. It is furthermore shown, in two independent ways, that there is no indication of any external air-temperature effect upon the recorded intensity. Finally, the data have been subjected to modern statistical methods which provide an objective measure for the probability that the observed diurnal-variation is real. The results indicate, for the period covered by this analysis, a physically significant 24-hour wave in apparent cosmic-ray intensity, with an amplitude of 0.17 per cent of the total intensity having its maximum at about 11^h, 75° west meridian mean time.

Instrument

The data used in this analysis were obtained from meter C-1, one of seven precision cosmic-ray meters which were constructed for the Carnegie Institution of Washington Committee on Coordination of Cosmic-Ray Investigations. This meter was designed by A. H. Compton, E. O. Wollan, and R. D. Bennett, and has been adequately described elsewhere[1]. The main ionization-chamber is that which was used in the meter designated as No. 3 in R. L. Doan's paper[2] "Fluctuations in cosmic-ray ionization as given by several recording meters located at the same station." Continuous records have been obtained since April 1935 by the staff of the United States Coast and Geodetic Survey Cheltenham Magnetic Observatory[3] cooperating with the Carnegie Institution's Department of Terrestrial Magnetism. At all times the meter has been fully shielded, the total shielding being equivalent to about 12 cm of pure solid lead[1].

For the total ionization produced within the chamber by cosmic rays alone we use the value 84 ions/cc/sec, which was kindly furnished us by Professor A. H. Compton. It was based on results at Chicago in connection with the determination of the residual ionization in each of the seven meters. This is essentially the same as the value 83 ions/cc/sec used by Doan[2].

Barometer-effect

For investigations of the diurnal variation in cosmic-ray intensity the necessity of determining the effect of changes in barometric pressure upon the recorded ionization is well known. This may ᴜe done by de-

---

[1] A. H. Compton, E. O. Wollan, and R. D. Bennett, Rev. Sci. Inst., 5, 415-422 (1934).
[2] R. L. Doan, Phys. Rev., 49, 107-122 (1936).
[3] Latitude 38° 44′ north, longitude 76° 50′ west, elevation above sea-level 72 meters.

termining the correlation between daily mean values of barometric pressure and of ionization, such as is shown graphically in Figure 1. All the

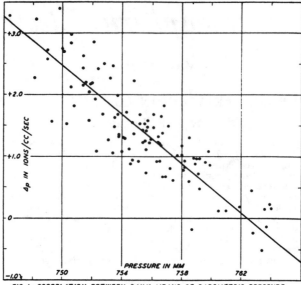

FIG. I—CORRELATION BETWEEN DAILY MEANS OF BAROMETRIC PRESSURE
AND OF DEPARTURE FROM BALANCE (Δ_P), 106 DAYS, JUNE I—SEPTEMBER 30, 1936

data used throughout this analysis are based upon the observed values of ionization exclusive of "bursts"[2]. We assume, for the ranges in pressure encountered at this station, a linear relation between the observed daily means of barometric pressure, P, and of departures from balance, Δ_P—the instrument automatically[1,2] measures the hourly mean values of the departures from a constant value of ionization—of the form

$$\Delta_{760} = \Delta_P - \beta(P - 760) \qquad (1)$$

in which Δ_{760} is the calculated value of Δ_P for $P = 760$ mm of mercury. If σ_Δ and σ_P be the standard deviations of Δ_P and P, and r the coefficient of correlation between the latter, then the value of β (usually called the regression-coefficient) given by

$$\beta = r(\sigma_\Delta/\sigma_P) \qquad (2)$$

is that which minimizes[4] the sums of the squares of the differences between the observed and calculated values of Δ_P. In this sense, the value β gives also the best estimate of Δ_{760}. We designate β as the barometric coefficient.

It should be emphasized that the value of β given by (2) is not in general the best value defining the actual relation between P and Δ_P. If, for example, we wish to obtain the best prediction of P, in the sense of least squares, from observed values of Δ_P, we would use[4] the regression-coefficient γ given by

$$\gamma = r(\sigma_P/\sigma_\Delta) \qquad (3)$$

[4]See, for example, Chapter IX, 9th edition, G. U. Yule, Introduction to the theory of statistics Charles Griffin and Co., London, 1929.

that is, the ratio of predicted changes in pressure to the observed changes in Δ_P should be γ. Thus $1/\gamma$ may be regarded as a second barometric coefficient which one would use to predict P from Δ_P. This is of interest because, in so far as our values of β and γ are correct, it sets an upper and lower limit for the coefficient, which we designate by β_0, expressing the actual relation between Δ_P and P. The determination of the best approximation to β_0 would involve a least-square adjustment in which the proper weights must first be assigned[5] to Δ_P and P. At any rate, the value β_0 must lie between β and $1/\gamma$. Values of r, β, and $(100\,\beta/84)$ are given in Table 1 for separate samples and their averages.

TABLE 1—*Correlation-coefficients* (r) *between daily means of barometric pressure and of cosmic-ray intensity at Cheltenham, Maryland; estimated standard errors* (S_r) *of* r, *barometric coefficient* (β) *expressed as change in rate of ionization in ion/cc/sec/mm, estimated probable error* (P_β), *and percentage-change in ionization per mm of pressure*[a]

Interval	No. days	r	S_r	β	P_β	$100\beta/84$
Apr. 20-June 30, 1935.........	58	−0.88	0.03	−0.209	0.010	−0.249
Aug. 1-Sep. 26, 1935..........	45	−0.68	0.08	−0.194	0.021	−0.231
Mar. 1-May 19, 1936.........	73	−0.84	0.04	−0.212	0.011	−0.252
June 1936...................	28	−0.83	0.06	−0.154	0.013	−0.183
July 1936..................	25	−0.86	0.05	−0.188	0.015	−0.224
August 1936................	31	−0.87	0.04	−0.185	0.013	−0.220
September 1936.............	22	−0.84	0.06	−0.141	0.013	−0.168
All.......................	282	−0.82	0.02	−0.198	0.006	−0.236
All except Aug. 1-Sep. 26, 1935.	237	−0.84	....	−0.198	.....	−0.236

[a]Total ionization due to cosmic rays is 84 ions/cc/sec.

For these r was obtained by measuring the deviations within each sample from the mean of that sample. From the data given in the last row of Table 1, $1/\gamma$ is found to be -0.333 per cent/mm of mercury. Consequently β_0, the actual percentage-change in ionization per mm of mercury, should lie between -0.333 and -0.236.

To compare these values with the changes which are actually observed when the same (effectively) instrument is taken to different altitudes we use Figure 4 of A. H. Compton's article[6] "Geographic study of cosmic rays." From the lower portion of the two upper curves (either of which may apply to our station, which is about 40° from the geomagnetic pole) we find values for the percentage-change in the total intensity per mm pressure which are roughly 0.28 and 0.24, respectively. Both values lie within the limits set above for β_0. However, in view of the differences in shielding (12 cm lead for our meter and 5.0 cm lead plus 2.5 cm bronze for the others) this agreement may be somewhat fortuitous.

According to Broxon, Merideth, and Strait[7] the value of barometric coefficient at sea-level, given by ionization-altitude curves obtained by Millikan and Cameron, using an instrument shielded by 7.6 cm of lead plus the 3-mm steel walls, is about 0.21 per cent change in ionization per mm change in pressure.

[5]W. E. Deming. On the application of least squares, Phil. Mag., 11, 146-158 (1931), and Phil. Mag., 17, 804-829 (1934).
[6]A. H. Compton, Phys. Rev., 43, 387-403 (1933).
[7]Phys. Rev., 43, 687-694 (1933).

Considering the various values of β given in Table 1 together with their estimated probable errors, P_β, there is no strong indication that the differences are real[8]. Indeed, these estimates of the probable errors are somewhat too small on account of the lack of independence between successive daily mean values of $\triangle_P$ (or of P).

Table 2 gives the values of r, β, $\sigma_\triangle$, σ_P calculated separately from each of the five indicated bihourly means for the same 58 days given in the first row of Table 1. Since each daily mean is the average of 12 such bihourly values, we should expect to find, if these were independent, the standard deviations of the *daily means* to be about $1/\sqrt{12}$ or 0.29 times the standard deviations given in the last two rows of Table 2. Actually $\sigma_\triangle$ and σ_P for the 58 daily means were found to be 0.82 and 3.45, respectively; these values differ little from the corresponding ones for the 58 bihourly means. This indicates that the 12 bihourly values are not independent; in fact, they are almost completely dependent. From this we may infer that the situation is not materially different with respect to the 24 hourly mean values on any given day.

The consequences of overlooking this lack of independence, and applying statistical formulas which are based on the hypothesis of independence, may be serious. We now examine the effect of overlooking this lack of independence upon the values of P_a given by Doan[2] in his Table 1. The differences between the various barometric coefficients a in that Table appear to be roughly 50 times larger than their probable errors. Accepting this, one would conclude that the barometric coefficient was definitely different for each meter, a situation which would be extremely disconcerting since the several meters are identical. Fortunately this is not the case. Using Dr. Doan's equation (6), and the values of P_r, and r given in the Table, we conclude that the value of n used was roughly 700 (the number of hours in the month). Using this value of n together with equations (5) and (6) and the values of r and a from the Table, we find values for P_a which are ten times larger than those in the Table.

After multiplying the values of P_a in the Table by 10, the values of a still appear to be different for the different instruments. Thus far, it is a matter of arithmetic; however, it is evident that the value of 700 (roughly) which was used for n in equations (6) and (7) ignored the fact that the 24 hourly mean values on each of the 31 days were not independent. As our earlier discussion showed, the 24 hourly mean values may for this purpose more safely be taken to be strictly dependent (that is, the same) so that instead of about 700 there are not more than 31 effectively independent values. Thus the values of P_a given in the Table should be roughly 0.2. Even the daily mean values are not strictly independent, so that the value 0.2 may still be somewhat underestimated. For the same reason the values of P_β in our Table 2 are also somewhat underestimated. Rough tests for independence of daily means indicate that our values of P_β (Table 1) might safely be nearly doubled.

Taking these things into account we conclude that the evidence in Table 1 of Dr. Doan's paper does not indicate that the barometric coefficient is different for different meters. From our results (Table 1) we conclude further that there is no good reason for suspecting that the barometric coefficient actually changes from time to time for the same meter. In addition the results indicated in our Table 2 indicate that the

[8]For general remarks on validity of probable errors, see p. 352 of reference footnote 4.

TABLE 2—*Correlation-coefficients (r) between alternate bihourly means of barometric pressure and of cosmic-ray intensity on 58 days during April 20 to June 30, 1935, Cheltenham, Maryland, with standard deviations (σ_Δ and σ_P) of Δ_P and of P*[a]

Bihourly interval	0-2	4-6	8-10	12-14	16-18	20-22
Correlation-coefficient (r)	−0.76	−0.81	−0.81	−0.80	−0.81	−0.78
Barometric coefficient (β) in ion/cc/sec/mm	−0.208	−0.183	−0.194	−0.181	−0.222	−0.195
σ_Δ in ion/cc/sec	0.99	0.86	0.91	0.84	0.98	0.88
σ_P in mm Hg	3.61	3.80	3.79	3.71	3.58	3.52

[a]Estimated standard error (S_r) is about 0.03 in each case and estimated probable error (P_β) is about 0.01 for each β.

barometric coefficient is not significantly different for different hours of the day; this was also indicated by Broxon, Merideth, and Strait.[7]

Although the value of β, based on 282 days of observations, is −0.198 ion/cc/sec/mm of mercury (which corresponds to the line drawn in Fig. 1), we had previously adopted, on the basis of the data then at hand, the value −0.206 ion/cc/sec/mm of mercury. This is the value of β which has been used in all our reductions.

Variations of the second kind and the external temperature-effect

On the basis of −0.206 ion/cc/sec/mm of mercury for β, the daily mean values of ionization corrected to pressure 760 mm of mercury are shown for 317 days in Figure 2. The variations in these daily means

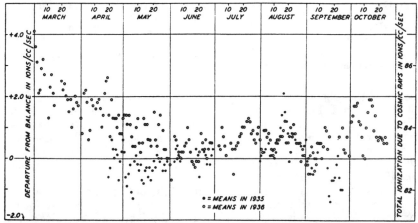

FIG. 2—DAILY MEANS OF APPARENT COSMIC-RAY INTENSITY AT BAROMETRIC PRESSURE 760 MM

have long been termed variations of the second kind. Our present concern is to investigate whether these may be due in part to variations in the temperature of the outside air, as has been reported by Hess and others.[9] As will be shown later, the question of the outside air-temperature effect has an extremely important bearing upon the results of the analysis of the diurnal variation. Indeed in this connection another independent test will be made to determine the magnitude of this effect. For this purpose we assume a linear relation of the form

[9]V. Hess and H. Graziadei, Terr. Mag., 41. 9-14 (1936).

$$\Delta_{P, T} = \Delta_{760} + \beta'(P - 760) + a(T - 20°.0) \qquad (4)$$

in which $\Delta_{P, T}$ is the observed daily mean departure from balance, and P and T are, respectively, the daily mean values of barometric pressure in mm and of outdoor air-temperature in degrees Centigrade. We determine the constants β' and a in this equation so that the observed values of $\Delta_{P, T}$ are in the best agreement in the sense of least squares, with the values predicted on the basis of the observed pressure and temperature; that is, we determine the constants by least squares assuming the errors of observation are confined solely to the observed values of $\Delta_{P, T}$. The reliability of β' and a is indicated by estimating the probable errors, following the procedure which is described by W. E. Deming in his paper[10] "On the significance of slopes and other parameters estimated by least squares"; that is, the probable errors for each sample are estimated on the assumption that the fit of equation (4) to the data is in each case an average fit.

The results are indicated in Table 3 for several samples. Here it will

TABLE 3—*Values of coefficients a and β' determined by least squares for formula* $\Delta_{P,T} = \Delta_0 + \beta'(P-760) + a(T-20.0)^a$ *and of their estimated probable errors*

Interval	No. days	β'	$P_{\beta'}$	a	P_a
		ion/cc/sec/mm		*ion/cc/sec/°C*	
Apr. 20-June 30, 1935........	58	−0.208	0.010	+0.012	0.009
June 1936.................	28	−0.181	0.016	−0.051	0.018
July 1936.................	25	−0.230	0.016	−0.114	0.025
Aug. 1936.................	31	−0.175	0.012	+0.051	0.013
Sep. 1936.................	22	−0.142	.....	−0.006	.....
June-Sep. 1936.............	106	−0.174	0.007	+0.016	0.009

$^a \Delta_0$ =departure from balance for P =760 mm and T =20°.0 C.

be noted that the values obtained for β' are in agreement with the values of β given in Table 1 for the same sample. Also the values of $P_{\beta'}$ are about the same as those for P_β in Table 1. This indicates that the barometric coefficient derived on the assumption that the observed ionization depends also on outdoor air-temperature is *no more reliable than when derived on the assumption that the observed ionization is independent of outside temperature.*

Finally, the scatter in the values of a, the outside-air temperature-coefficient, calculated from different samples is so great these can not be regarded as real. This is substantiated by the fact that the values of a in each case do not greatly exceed their probable errors. For the same reasons previously discussed in connection with the values of P_β in Table 1, the actual values of P_a (and of $P_{\beta'}$) in Table 3 may be nearly twice as large as those indicated. It is therefore unlikely that a real temperature-coefficient exists, which is greater in absolute value than about 0.02 ion cc/sec/°C, which corresponds to 0.024 for the percentage-change in total ionization per degree C. Hess[9] has used in some of his reductions a coefficient of −0.09 per cent per degree C (not −0.9 per cent as erroneously printed) when the apparatus was fully shielded but finds

[10]Phys. Rev., 49, 243-247 (1936)—see first paragraph of second column, p. 247.

that with partial (Halbpanzer) shielding the coefficient was so small that no correction for temperature was necessary.

The analysis of the diurnal variation

The adequate characterization of the diurnal variation in any geophysical phenomenon requires not simply a knowledge of its average value but also a full knowledge of its variability. The latter in general is made up of an irregular (or random) part and a systematic part such, for example, as a systematic variation with season in the amplitude (or phase) of the diurnal variation. These facts, together with the methods of analysis used in this discussion, have been set forth clearly in numerous papers by J. Bartels, who, as a research associate for the Department of Terrestrial Magnetism of the Carnegie Institution of Washington, has made important applications to problems in terrestrial magnetism. A great achievement of such methods was the determination of the lunar tide in the Earth's atmosphere at Potsdam. There the semi-diurnal lunar wave in atmospheric pressure [11,12] has an amplitude of only 0.01 mm of mercury, and this was determined with a precision of 0.001 mm of mercury from barograms which were read only to the nearest 0.1 mm.

Our raw material for the analysis of the diurnal variation consists for each day of 12 bihourly mean values of the departures from balance ("bursts" deducted). Each of these 12 values is corrected for the corresponding bihourly mean value of pressure in accordance with equation (1), using the value -0.206 ion/cc/sec/mm of mercury for β. The resulting 12 values of departures from balance (for $P = 760$ mm) are subjected to harmonic analysis, providing a set of harmonic constants a_1, b_1, a_2, b_2 for each day[13]. Following Bartels[14], the 24-hour wave for each day is completely characterized by a point, with coordinates a_1 and b_1, in a Cartesian diagram. This point may be considered the end-point of a clock-hand, which indicates the time of the wave-maximum on a suitable 24-hour dial, and the length of which indicates the amplitude of the wave. Such a diagram is called the 24-hour harmonic dial[14]. Similarly a 12-hour harmonic dial is constructed in which the time indicated by the hand (on the 12-hour dial) is that of the first maximum. For N days we have in each dial a cloud of N points and the analysis of the variability of the diurnal variation is now transformed into an analysis of the geometrical properties of these clouds[14].

The two-dimensional Gaussian frequency-distribution which fits the cloud best is in general elliptical, that is to say, the lines of equal frequency are ellipses. We need here only the expression for the axes of the *probable ellipse*, that is, the ellipse which contains $(N/2)$ points inside and $(N/2)$ points outside. The lengths P_1 and P_2 of the major and minor axes of the *probable ellipse* are given[14] by

$$P_1P_2 = 0.833\sqrt{(\sigma^2_{a_1} + \sigma^2_{b_1}) \mp \sqrt{(\sigma^2_{a_1} - \sigma^2_{b_1})^2 + 4r^2\sigma^2_{a_1}\sigma^2_{b_1}}} \qquad (5)$$

in which σ_{a_1} and σ_{b_1} are the standard deviations of a_1 and b_1 from their

[11]J. Bartels, Sci. Mon., 35, 110-130 (1932).
[12]J. Bartels, Berlin, Abh. Met. Inst., 8. No. 9, 1-51 (1927); Abstract in Naturw., 15, 860-865 (1927).
[13]a_2, b_2 were also determined but are not included in this discussion.
[14]J. Bartels, Terr. Mag., 37, 291-302. (1932).

respective means and r is the coefficient of correlation between the N values of a_1 and b_1. In case of circular symmetry $P_1 = P_2 = 0.833M$ in which $M = \sqrt{\sigma^2_{a_1} + \sigma^2_{b_1}}$. Similar expressions hold also for the cloud in the 12-hour dial.

The last line of Table 4 gives the constants for the 24-hour dial for

TABLE 4—*Constants for 24-hour harmonic dial of apparent cosmic-ray intensity in ionization-chamber, Cheltenham, Maryland, April 1935 to October 1936*

Sample	Interval	No. days	Average[a]			t_{max} 75°WMT	Standard deviation		$M_1 = (\sigma_{a_1}^2 + \sigma_{b_1}^2)^{1/2}$
			a_1	b_1	c_1		σ_{a_1}	σ_{b_1}	
			ion/cc/sec			*h*	*ion/cc/sec*		
A	Apr. 20-June 30, 1935..	58	−0.169	+0.075	0.185	9.4	0.188	0.196	0.271
B	Mar. 1-May 31, 1936..	82	−0.122	−0.046	0.130	12.4	0.210	0.204	0.292
C	June 1-July 31, 1936...	53	−0.163	+0.060	0.174	9.7	0.199	0.202	0.284
D	Aug. 1-Sep. 17, 1936...	48	−0.147	+0.007	0.147	10.8	0.203	0.202	0.286
E	Sep. 24-Oct. 27, 1936...	32	−0.107	−0.060	0.123	13.0	0.233	0.273	0.359
All samples		273	−0.142	+0.008	0.142	10.8	0.206[b] 0.206[c]	0.217[b] 0.213[c]	0.300[b] 0.297[c]
All samples without correction for non-cyclic change........ }		273	−0.145	+0.015	0.146	10.6	0.207[b]	0.248[b]	0.324[b]

[a] a_1 and b_1 positive upward and to right from origin, respectively, along axes 23^h and 5^h.
[b] When departures are from mean of all samples.
[c] When departures in each sample are from mean of that sample.

273 days when no correction for non-cyclic change was applied. For this case $\sigma_b > \sigma_a$ and the distribution is slightly elliptical; the axes of the ellipse were found to be in the ratio 1:1.26. After the coefficients a_1 and b_1 for each day were corrected for the apparent linear non-cyclic change[15] it will be seen from Table 4 that the value of σ_b was reduced by about 15 per cent, the ratio of the axes of the probable ellipse becoming 1.00. This is interpreted as justifying the correction for non-cyclic change, particularly since the presence of any variable linear non-cyclic change in the data would affect the coordinate b about four times[15] as much as a. The values in Table 4 refer then, except as otherwise noted, to values of a and b, which have been corrected for non-cyclic change, in which case the probable ellipse is a circle—called the probable-error circle. Incidentally, the coefficients have all been corrected for the use of bihourly means[16].

Figure 3 shows the points in the 24-hour harmonic dial for 273 single days which make up the five samples indicated in Table 4. The center of the cloud coincides with the center of the small circle. The large circle is the probable-error circle for single days. Its radius[14] is $0.833M = 0.250$ ion/cc/sec. The number of points inside the circle is 138, very closely half the total number of points. If now the points for successive days are independent, then the radius of the probable-error circle for the mean of 273 days should be $0.250/\sqrt{273} = 0.0152$. The actual amplitude,

[14] C. C. Ennis, Terr.Mag.. 32, 155-162 (1927).
[15] J. Bartels, Terr. Mag.. 40, 1-60 (1935).

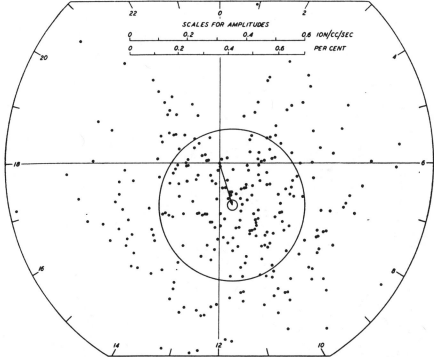

FIG. 3— 24-HOUR HARMONIC DIAL, APPARENT COSMIC-RAY INTENSITY, 273 SINGLE DAYS DURING APRIL 20, 1935 TO OCTOBER 27, 1936, CHELTENHAM, MARYLAND (TIMES OF MAXIMUM IN 75° WEST MERIDIAN MEAN HOURS)

C_1, from Table 4 is 0.142, which is about 9.3 times the radius of the probable-error circle. Therefore, the probability that the mean of another such sample of 273 days should by accident fall outside a circle of radius 0.142 with its center at the mean of the observed cloud is $(1/2)^{(9.3)^2}$ or only about 10^{-26}.

Another criterion for the physical significance of the observed amplitude is through equations (17.4), (17.6), and (19.1) of Bartels' discussion of the random walk[16]. For $M = 0.300$ (Table 4), $C_1 = 0.142$ (Table 4), $l = 0.332$ (Bartels' equation 19.1), and $n = 273$, we have $m = 0.0201$ and $\kappa^2 = 50.0$. From Bartels' equation (17.6) the probability that our observed amplitude 0.142 is the result of accident is $W(\kappa) = 10^{-21}$. Unless the 273 days are strictly independent, the estimate in the preceding paragraph for the radius of the probable-error circle is too small. The result of the test for independence is shown in Figure 5. The upper line, determined by the equation $M(h) = 0.300/\sqrt{h}$, shows how the value of M for means of h successive days should depend on h, *if the successive days were independent.* The calculated values of $M(h)$ for means of 4, 8, 12, 16, and 24 successive days are indicated by the circles; since their accuracy is only of the order $1/\sqrt{2h}$, there is no indication of lack of independence in the sample of 273 days. Thus it is practically certain that the 24-hour wave in *apparent* cosmic-ray intensity is real.

The 12-hour harmonic dial (constants in Table 5) is shown in Figure 2

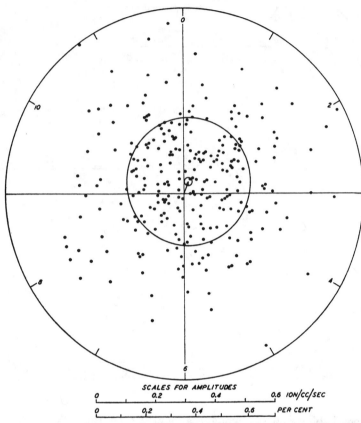

SCALES FOR AMPLITUDES

| 0 | 0.2 | 0.4 | 0.6 | ION/CC/SEC |

| 0 | 0.2 | 0.4 | 0.6 | PER CENT |

FIG. 4—12-HOUR HARMONIC DIAL, APPARENT COSMIC-RAY INTENSITY, 273 DAYS DURING APRIL 20, 1935, TO OCTOBER 27, 1936, CHELTENHAM, MARYLAND (TIMES OF FIRST MAXIMUM IN 75° WEST MERIDIAN MEAN HOURS)

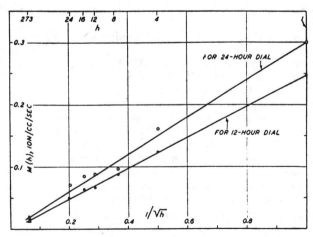

FIG. 5—TEST FOR INDEPENDENCE OF POINTS FOR SUCCESSIVE DAYS IN 24-HOUR AND 12-HOUR HARMONIC DIALS, APRIL 20, 1935, TO OCTOBER 27, 1936, CHELTENHAM, MARYLAND

128

4 together with the probable-error circles (the ratio of the axes of the ellipse, that is, P_1/P_2 was in this case actually 1.05) for single days and for the mean. As before, the number of points inside the former circle is not very different from the number outside. The lower part of Figure 5 shows the results of the test for independence of successive days in the 12-hour dial. As before, there is no indication of any lack of independence. Using the value of 0.246 given for M_2 in Table 5 the radius of the probable-error circle for the mean is $0.833 \times 0.246/\sqrt{273} = 0.0124$. The actual amplitude C_2 from Table 5 is 0.042, which is about 3.4 times the radius of the probable-error circle. The probability that the mean of another such sample of 273 points should fall by accident outside the circle which passes through the origin and the center of which is at the center of the observed cloud is $(1/2)^{3.4^2}$ or about one in 3000. Using the results of the random walk as above the probability for the observed amplitude being the result of accident is about one in 2000. Thus there may be a physically real 12-hour wave in the data considered, but more data are required to demonstrate it with a convincingly high degree of probability.

The points in the 12-hour dial obtained for the means of different

TABLE 5—*Constants for 12-hour harmonic dial of apparent cosmic-ray intensity in ionization-chamber, Cheltenham, Maryland, April 1935 to October 1936*

Sample	Interval	No. days	Average[a]			t_{max} 75°WMT	Standard deviation		$M_2 = (\sigma_{a_2}{}^2 + \sigma_{b_2}{}^2)^{1/2}$
			a_2	b_2	c_2		σ_{a_2}	σ_{b_2}	
			ion/cc/sec			*h*	*ion/cc/sec*		
A	Apr. 20-June 30, 1935..	58	+0.048	+0.018	0.051	11.7	0.168	0.162	0.234
B	Mar. 1-May 31, 1936..	82	+0.026	+0.031	0.040	0.7	0.179	0.192	0.262
C	June 1-July 31, 1936...	53	+0.055	+0.049	0.074	0.4	0.166	0.185	0.249
D	Aug. 1-Sep. 17, 1936...	48	−0.027	+0.015	0.031	4.0	0.162	0.162	0.224
E	Sep. 24-Oct. 27, 1936...	32	+0.038	+0.045	0.059	0.7	0.158	0.165	0.228
All samples		273	+0.028	+0.031	0.042	0.6	0.171[b]	0.176[b]	0.246[b]

[a] a_2 and b_2 positive upward and to right from origin, respectively, along axes 11^h and 2^h.
[b] Standard deviations of departures are measured from mean of 273 days.

samples in Table 5, together with the mean of all the samples and its probable-error circle are shown in Figure 8. The points for the 12-hour wave in pressure are also indicated.

The effect of the corrections for barometric pressure, and its uncertainty, upon the resultant 24-hour wave

Figure 6 shows the effect of the corrections for barometric pressure on the 24-hour wave in apparent cosmic-ray intensity. C is the point, for apparent cosmic-ray intensity, in the 24-hour dial for the mean of 273 single days. P is the point representing the average 24-hour wave in barometric pressure (without any correction for non-cyclic change) for the same 273 days. On account of the additive property[16] of harmonic coefficients (that is, the principle of superposition) the vector

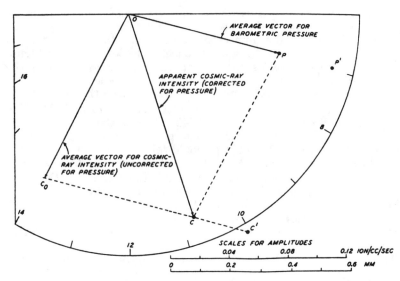

FIG. 6 – 24-HOUR HARMONIC DIAL SHOWING EFFECT OF PRESSURE-CORRECTIONS ON AVERAGE 24-HOUR WAVE IN APPARENT COSMIC-RAY INTENSITY, 273 DAYS DURING APRIL 20,1935 TO OCTOBER 27, 1936, CHELTENHAM, MARYLAND (TIMES OF MAXIMUM IN 75° WEST MERIDIAN MEAN HOURS)

OC may be conceived as the resultant of OP and another, OC_o. Thus C_o is the point in the harmonic dial which would have been found had no corrections for pressure been applied to the raw data. This shows that the actual 24-hour wave found in the data can not possibly be the result of the corrections applied for pressure. As far as its effect on OC is concerned, the effect of an error in the adopted value of the pressure-coefficient β amounts to an increase or decrease in the length OP. Thus, if the value 0.33 per cent per mm, which was the upper limit for β_o, had been adopted, the point C would have been shifted to C'. Therefore, the position of C, which would result from using the actual (unknown) value β_o for the barometric coefficient, is practically confined to the line segment CC' in Figure 6.

Corrections for barometric pressure were applied before making the harmonic analysis. As far as the average wave is concerned, the same result would have been found had the harmonic analysis been made first and the corrections for pressure then applied. However, the scatter of points in the dial for the 24-hour wave of apparent cosmic-ray intensity, uncorrected for pressure, would certainly have been greater and consequently the apparent probability for reality enormously smaller. This follows from the fact that the scatter in the 24-hour dial for barometric pressure derived from all days[17] is, relative to the average amplitude for all days, probably greater than the relative scatter found in the 24-hour dial for apparent cosmic-ray intensity corrected for pressure.

Comparison of results from different samples, including some tests for the effect of temperature on the instrument

Points in the 24-hour dial for different samples, designated in Table 4, are shown in Figure 7, together with their respective probable-error

[17]See, for example, Tables 6 and 19 of reference in footnote 12 for Potsdam results; also J. Bartels, Wien-Harms, Handbuch der Experimentalphysik, 25, I, 163-210 (1928).

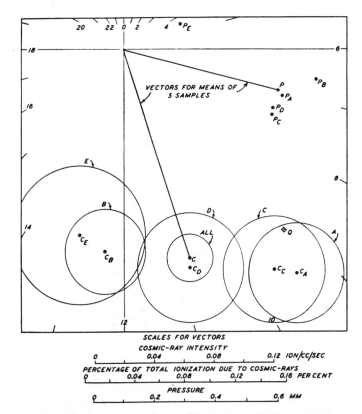

FIG. 7—24-HOUR HARMONIC DIAL, MEANS OF SEPARATE SAMPLES FOR APPARENT
COSMIC-RAY INTENSITY (C) AND FOR BAROMETRIC PRESSURE (P), APRIL 20, 1935
TO OCTOBER 27, 1936, CHELTENHAM, MARYLAND (TIMES OF MAXIMUM IN 75° WEST
MERIDIAN MEAN TIME)

circles. Samples A, C, and D were obtained with the instrument operating under nearly the same conditions of temperature. During these periods the instrument was subjected to a quite regular diurnal-variation of temperature with a maximum at 24^h and an amplitude of about $0°.5$ C (on account of the insulation of the room). For the sample B the room was, for 62 out of the 82 days, kept at constant temperature by thermostatic control. Sample E was secured with the door to the insulated room opened to an adjacent uninsulated room; under these circumstances the variations in temperature in the meter-room were somewhat irregular but the diurnal variation was, on the average, about $2°$ C with the maximum near 17^h. Now because the value of $M = \sqrt{\sigma^2_a + \sigma^2_b}$ (see Table 4), derived when the departures of a and b were from the mean of all the samples, was so little different from the value of M obtained[1x] when the departures within each sample were from the means for that sample, no very significant difference between the means of the different samples was to be expected. The probability that any two samples are separated a distance $K\sqrt{\rho^2_1 + \rho^2_2}$ by chance[14] is $(1/2)^{K^2}$ where

[1x]By application of equations (67) and (68) of W. E. Deming and Raymond T. Birge. Statistical theory of errors. Rev. Mod. Phys., 6, 119-161 (1934).

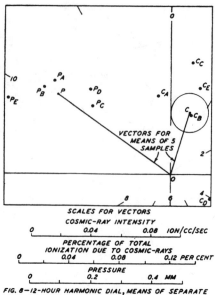

SCALES FOR VECTORS
COSMIC-RAY INTENSITY
0   0.04   0.08   ION/CC/SEC

PERCENTAGE OF TOTAL
IONIZATION DUE TO COSMIC-RAYS
0   0.04   0.08   0.12 PER CENT

PRESSURE
0   0.2   0.4 MM

FIG. 8—12-HOUR HARMONIC DIAL, MEANS OF SEPARATE
SAMPLES FOR APPARENT COSMIC-RAY INTENSITY (C) AND
FOR BAROMETRIC PRESSURE (P), APRIL 20, 1935 TO OCTOBER
27, 1936, CHELTENHAM, MARYLAND (TIMES OF MAXIMUM
IN 75° WEST MERIDIAN MEAN HOURS)

ρ_1 and ρ_2 are the radii of the two probable-error circles. In this way it is found that the probability that any sample is separated from the mean of all the samples by chance, is greater than one-eighth in all cases. Thus no sample differs significantly from the mean of all samples. The probability for separation by chance is least for the samples A and B and is for this case $1/500$ which is not small enough to be definitely significant. Therefore, this test indicates that the instrument is not affected by changes in temperature.

The point Q of Figure 7 is the result reported by Doan[2] referred to the same local-time origin as for the data at Cheltenham. Doan's paper does not indicate the value of barometric coefficient used in his reduction or the average diurnal-variation of barometric pressure on the ten days. It certainly may not be safely assumed that a series of observations with five instruments on ten days is equivalent to a series with one instrument for 50 days. This would depend on how much the scatter in the harmonic dial was due to external causes common to the five instruments. In any case Doan's results for Chicago are in accord with those for Cheltenham which is additional evidence that the Cheltenham meter has no temperature-effect since it is most unlikely that the five instruments at Chicago were subjected to the same diurnal-variation in temperature as the one at Cheltenham.

Additional tests for external temperature-effect

The results of additional tests for external temperature-effect are shown in Figure 9. C_1 and C_2 are the mean points in the 24-hour dial for apparent cosmic-ray intensity on two groups of days, with their probable-

132

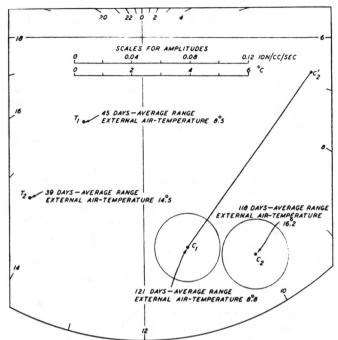

FIG. 9 — 24-HOUR HARMONIC DIAL, EFFECT OF TEMPERATURE EXTERNAL AIR ON APPARENT COSMIC-RAY INTENSITY, APRIL 20, 1935 TO SEPTEMBER 27, 1936, CHELTENHAM, MARYLAND (TIMES OF MAXIMUM IN 75° WEST MERIDIAN MEAN HOURS)

error circles. C_1 is for 121 days on which the average range in outdoor air-temperature was 8°.8C and C_2 is that for 118 days with an average range of 16°.2C. For group C_1 no daily range was greater than 12° C and for group C_2 no daily range was less than 12° C. T_1 and T_2 of Figure 9 represent the 24-hour wave in external air-temperature for averages of 45 and 39 days on which the average ranges in temperature were 8°.5 C and 14°.5 C, respectively. These days were taken from the two groups used for C_1 and C_2 and may be safely assumed to represent closely the diurnal variations in temperature for days to which C_1 and C_2 correspond. Were there any appreciable external temperature-effect, then it is to be expected that C_1 and C_2 would be definitely separated; the probability that the observed separation is due to chance is one in four. On the assumption of an external temperature-effect of −0.055 per cent of the total intensity per degree Centigrade, we should expect to find the point C_2 at C_2' and the probability of C_2 being so separated by accident from C_1 would be one in 10^6; thus were there a temperature-effect as great as that postulated, there certainly should be some indication of it. Hence this test supports the previous conclusions that there is no effect because of external temperature.

Comparison with other results and effect of local radiation

Hess and Graziadei[9] find, using an external temperature-coefficient of −0.09 per cent per degree Centigrade, a 24-hour wave which, on the basis of local time, agrees closely in phase and amplitude with that for Cheltenham for which no corrections for external temperature were applied. The former results were also in agreement with those obtained at Innsbruck by Steinmaurer[19] using the same external temperature-coefficient. However, when no correction for external air-temperature is applied to Steinmaurer's data, the maximum of the 24-hour wave occurs about 6^h L.M.T. This is probably also true for Hess's[9] data. Unless there is some definite reason why an external temperature-effect appears to exist at Innsbruck and on the Hafelekar but not at Cheltenham, it is evident that the results for the 24-hour wave at the last station are not in agreement, on the basis of local time, with those from the other stations.

Recently, at the Department of Terrestrial Magnetism, G. R. Wait has obtained results indicating the possibility of a diurnal variation of ionization in the air due to gamma rays alone; its amplitude may amount to as much as 0.3 ion/cc/sec in air, with the maximum about noon. In passing through 12 cm of lead the intensity of gamma rays may be expected to be reduced to about 0.25 per cent of the initial value (assuming the rays arise from radium C). Since the ionization produced by these within the chamber of the meter at Cheltenham should be about 67 times[1] greater than for normal air, a diurnal variation with amplitude as great as 0.05 ion/cc/sec in the chamber would be expected at Washington. This is about one-third the amplitude of the observed diurnal variation in apparent cosmic-ray intensity at Cheltemham. Any diurnal variation in gamma rays may vary with locality and thus distort determinations of the actual diurnal variation in cosmic-ray intensity.

To summarize, this analysis of the data for Cheltenham demonstrates the existence of a physically real 24-hour wave in apparent cosmic-ray intensity, which does not appear to be due to systematic instrumental effects but which may be due, in part at least, to variations in local radiation. It is hoped some statistical tests may be made to determine, if possible, whether the diurnal variation in the Earth's magnetic field may, as suggested by Gunn[20], be responsible for the apparent diurnal-variation.

Thanks are due the personnel of the Division of Terrestrial Magnetism and Seismology of the United States Coast and Geodetic Survey and particularly its staff at the Cheltenham Magnetic Observatory, whose generous cooperation makes possible the cosmic-ray records herein discussed. The writer wishes to acknowledge his indebtedness to Dr. J. A. Fleming, Director of the Department of Terrestrial Magnetism, for having made the investigation possible and for his continued interest in it.

[19]Beitr. Geophysik. **45**, 148-183 (1935).
[20]R. Gunn. Phys. Rev.. **41**, 683 (1932).

DEPARTMENT OF TERRESTRIAL MAGNETISM,
CARNEGIE INSTITUTION OF WASHINGTON,
Washington, D. C., December 8, 1936

Terrestrial Magnetism

and

Atmospheric Electricity

VOLUME 43   SEPTEMBER, 1938   No. 3

ON COSMIC-RAY EFFECTS ASSOCIATED WITH MAGNETIC STORMS

S. E. FORBUSH

Abstract—Evidence indicates that the storm-time field of some magnetic storms causes world-wide changes of several per cent in cosmic-ray intensity. That other magnetic storms of equal intensity at the Earth's surface occur with no appreciable cosmic-ray effects definitely indicates that the entire current-system for the storm-time field of both types of storms can not be located at the same distance above the Earth. In particular, the possibility that the current-systems responsible for the two types of storms both lie within the Earth's atmosphere appears remote. Assuming the current-system for the storm-time field of both types of storms to consist of a ring concentric with the Earth in the geomagnetic equatorial plane, magnetic data are analyzed to determine whether the radius of the assumed ring is, as would be expected, greater for magnetic storms which affect cosmic-ray intensity. Although the analysis is not conclusive on this point, the results satisfy a necessary condition for the existence of such a ring-current. The occurrence of aurora in temperate latitudes during most of the magnetic storms which affected cosmic-ray intensity is interpreted, after Störmer, to indicate the existence of such ring-currents. The percentage-changes in cosmic-ray intensity during magnetic storms is within the observational uncertainty the same at geomagnetic latitudes 50°.1 north and 0°.6 south. The significant correlation between changes in daily means of cosmic-ray intensity, for two stations separated 50° in latitude, probably results from the same mechanism responsible for the magnetic-storm effect.

Introduction

The world-wide decrease in cosmic-ray intensity which occurred during the magnetic storm commencing April 24, 1937, we ascribed [see reference 1 at end of paper] directly to the magnetic field of the storm. We assumed after S. Chapman [2] that the axially symmetric part of the field of the magnetic storm was due to a system of westward electric currents in the upper atmosphere. In the region outside such a current-system its magnetic field is approximately equivalent to that which would result from an increase in the Earth's dipole-moment. Provided the effect on cosmic rays of the field inside such a current-system is negligible, its effect on cosmic-ray intensity we concluded [1] would be similar to that resulting from an increase in the Earth's moment, namely, a decrease, according to the well-known theories of Störmer and of Lemaître and Vallarta.

S. Chapman [3] pointed out that the magnetic-storm effect on cosmic-ray intensity should provide determination of the height above the Earth at which flow the electric currents responsible for magnetic storms. He suggested, as a convenient model, a current-system, in the form of a spherical sheet concentric with the Earth, having an external field for which the existing theory of orbits of cosmic-ray particles would apply, but indicated that in the region inside this current-system the existing theory of orbits was not applicable. On the basis of the decrease in cosmic-ray intensity and in horizontal magnetic intensity which occurred during the magnetic storms near the end of April 1937, J. Clay and E. M. Bruins [4], neglecting the effects of the internal field of the current-system suggested by S. Chapman, estimated the radius of the assumed spherical sheet to be about three times that of the Earth. They concluded, however, that the field of the magnetic storm could not be due to the setting up of such a current-system since this should cause "an increase of cosmic-ray intensity in the same way as was found by Störmer [5] for the aurorae." To escape this difficulty Clay and Bruins assumed that the magnetic storm and the effect on cosmic-ray intensity could only be explained by a decrease of current-intensity in a normally existing current-system.

That aurora may appear in temperate latitudes coincident with definite decreases in cosmic-ray intensity is indicated by reports [6] of aurora in New England on April 25 and 26, 1937, during which period a significant world-wide decrease in cosmic-ray intensity was observed (see Fig. 5). V. F. Hess, R. Steinmaurer, and A. Demmelmair [7] noted a considerable decrease in cosmic-ray intensity, strikingly similar to that in our Figure 2, during the appearance, over most of Europe, of aurora on January 25, 1938.

An aurora was also seen on the evening of January 22, 1938, at the Watheroo Magnetic Observatory (latitude 30° south, longitude 116° east) of the Department of Terrestrial Magnetism. This corresponds to about 8^h, 75° west meridian mean time, January 22, for which Figure 2 indicates a considerable decrease in cosmic-ray intensity.

Thus a current-ring of several earth-radii, concentric with the Earth and in the geomagnetic equatorial plane of the type which Störmer [8] required to explain the incidence of aurorae farther southward during magnetic storms, may also explain the observed decrease in cosmic-ray intensity. Chapman [2, 3] in indicating objections to a ring of such large radius as was assigned by Störmer pointed out that such a ring would likely alter considerably the normal paths of cosmic-ray particles. He concluded [2] that if such a type of ring exists its radius is probably not more that a few times that of the Earth.

The cathode-ray experiments of Brüche [9] and of Birkeland [10] showed that a ring-current could be produced in the equatorial plane of a small magnetized model of the Earth. Brüche's experiments also showed that the presence of such a ring-current caused the incidence of aurora farther southward on the model of the Earth. A decrease in the current in this ring would then hardly be expected to cause the incidence of aurora farther southward as would be required on the basis of the hypothesis of Clay and Bruins.

T. H. Johnson [11], on the basis of observed changes in comic-ray intensity and in horizontal magnetic intensity, also estimated the radius of the assumed current-system proposed by Chapman [3] to be about four times that of the Earth for the magnetic storm near the end of April 1937. He cites the experiment of Brüche to indicate that a current-system of such a radius would be expected to cause an increase in cosmic-ray intensity. However, on three occasions a definite world-wide decrease

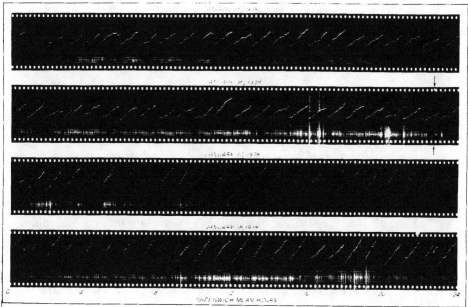

FIG. 1—COSMIC-RAY RECORDS, HUANCAYO, PERU, SHOWING EFFECT OF MAGNETIC STORM WHICH BEGAN 22ʰ7 GMT, JANUARY 16, 1935

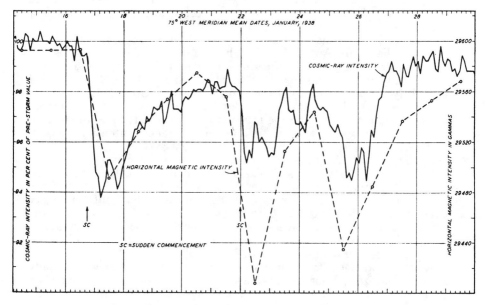

FIG. 2—MAGNETIC STORM-EFFECTS ON BIHOURLY MEAN COSMIC-RAY INTENSITY AVERAGED FOR BOSTON, UNITED STATES, CHELTENHAM, UNITED STATES, AND HUANCAYO, PERU, AND ON DAILY MEAN HORIZONTAL MAGNETIC INTENSITY, HUANCAYO, PERU

137

in cosmic-ray intensity was observed during the incidence of aurora in temperate latitudes. Whether both effects can be ascribed to a ring-current system can probably be decided only when the effects on cosmic-ray intensity for the field of such a ring have been calculated.

Results of observations on cosmic-ray effects during magnetic storms

From Compton-Bennett [12] precision recording cosmic-ray meters, operated, in a program for continuous registration, under the auspices of the Committee on Coordination of Cosmic-Ray Investigations of the Carnegie Institution of Washington, simultaneous cosmic-ray data have been obtained from two or more stations during three periods of intense magnetic storms. Cosmic-ray data used in this paper are corrected for barometric pressure [13] and were obtained from meters with a total shielding equivalent to 12 cm of lead.

Figure 1 is a reproduction of cosmic-ray records for four days obtained at the Huancayo Magnetic Observatory of the Department of Terrestrial Magnetism of the Carnegie Institution of Washington. Changes in barometric pressure during this period were small. The large increase in slope of the hourly electrometer-traces on the two lower records shows the effect of the magnetic storm which began at $22^h.7$, GMT, January 16, 1938.

The solid curve of Figure 2 indicates the changes in bihourly means of cosmic-ray intensity averaged for three stations. The dashed curve is drawn through the daily means of horizontal magnetic intensity for Huancayo. It will be noted that the ratio of changes in cosmic-ray intensity to those in horizontal magnetic intensity is greater for the interval January 17-21 than for the following period. This is also evident in Figure 3, which shows the agreement between changes in daily means of cosmic-ray intensity at three stations.

Figure 4 shows that the magnetic storm which began at $22^h.1$, 75° west meridian time, August 21, 1937, had no perceptible effect on cosmic-ray intensity at any one of the three stations indicated. The decrease in the daily mean of horizontal intensity from Agust 21-22 in Figure 4 is slightly greater than that from January 16-17 in Figure 3; otherwise, the changes in daily means of horizontal intensity for these two periods are very similar. If, during the interval January 16-21, 1938, the change in cosmic-ray intensity and that in horizontal magnetic intensity are both due to the magnetic field of the same current-system, then it follows that the current-system responsible for the changes in horizontal magnetic intensity from August 21-25, 1937, cannot be located at the same height above the Earth as that for January 16-21, 1938. That the type of current-system required to explain the decrease in horizontal intensity could certainly be similar in both cases will be shown later. From the greatly different effects on cosmic-ray intensity it appears impossible that the current-systems responsible for these two storms could both have been located within the Earth's atmosphere.

The proportionality, shown in Figure 6, between changes in daily means of horizontal magnetic intensity and in cosmic-ray intensity is an indication that one and the same current-system is responsible for both. That the ratios in Figure 6 of changes in cosmic-ray intensity to those in horizontal magnetic intensity are different also indicates that the

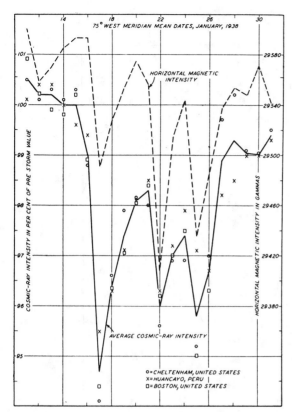

FIG. 3—MAGNETIC STORM-EFFECTS ON DAILY MEAN COSMIC-RAY INTENSITY AT BOSTON, UNITED STATES, CHELTENHAM, UNITED STATES, AND HUANCAYO, PERU, AND ON DAILY MEAN MAGNETIC HORIZONTAL INTENSITY AT HUANCAYO, PERU

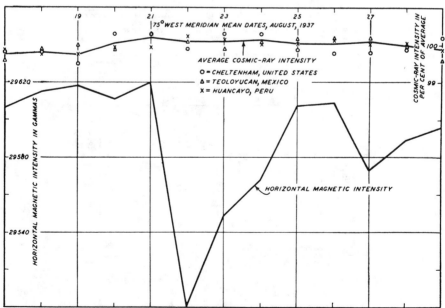

FIG. 4—DAILY MEANS HORIZONTAL MAGNETIC INTENSITY AT HUANCAYO, PERU, AND COSMIC-RAY INTENSITY AT CHELTENHAM, UNITED STATES, TEOLOYUCAN, MEXICO, AND HUANCAYO, PERU, SHOWING NO CHANGE IN COSMIC-RAY INTENSITY DURING MAGNETIC STORM BEGINNING AUGUST 21, 1937

139

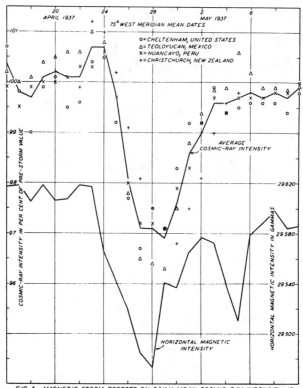

FIG. 5—MAGNETIC STORM-EFFECTS ON DAILY MEAN COSMIC-RAY INTENSITY AT
CHELTENHAM, UNITED STATES, TEOLOYUCAN, MEXICO, HUANCAYO, PERU, AND CHRIST-
CHURCH, NEW ZEALAND, AND ON MAGNETIC HORIZONTAL INTENSITY AT HUANCAYO,
PERU

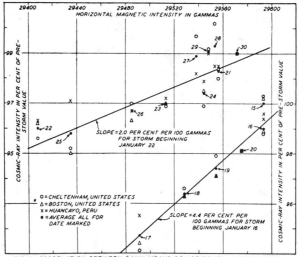

FIG. 6—CORRELATION BETWEEN DAILY MEANS OF HORIZONTAL MAGNETIC IN-
TENSITY AT HUANCAYO, PERU, AND OF COSMIC-RAY INTENSITY AT BOSTON, UNITED
STATES, CHELTENHAM, UNITED STATES, AND HUANCAYO, PERU, JANUARY 15-30, 1938

140

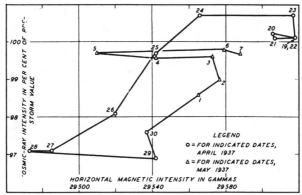

FIG. 7—RELATION BETWEEN DAILY MEANS OF HORIZONTAL MAG-
NETIC INTENSITY AT HU.ºNCAYO, AND OF COSMIC-RAY INTENSITY
AVERAGED FOR CHELTENHAM, TEOLOYUCAN, HUANCAYO, AND
CHRISTCHURCH (DATA PER FIG. 5), APRIL 19 TO MAY 7, 1937

current-system responsible for these two storms flowed at different heights
above the Earth.

Figure 5 shows the daily means of horizontal magnetic intensity at
Huancayo and of cosmic-ray intensity at four stations during April 16
to May 10, 1937. While the correspondence between changes in cosmic-
ray intensity at all the stations is evident, that between changes in cosmic-
ray intensity and in horizontal magnetic intensity is not close. This is
better shown in Figure 7, in which appear pronounced changes in horizontal magnetic intensity which are not accompanied by changes in
cosmic-ray intensity. Magnetic records from Huancayo and Cheltenham
showed sudden commencements on April 24, 25, 26, and May 4, 1937.
These indicate the onset of what probably constitutes distinct magnetic
storms. From Figures 5 and 7 it is evident that the first and last of these
had no apparent effect on cosmic-ray intensity. The shape of the diagram in Figure 7 could be partially explained on the basis that the current-
systems for the individual storms flowed at different distances above the
Earth. The complete explanation of Figure 7 may involve the question
of whether the current-systems for separate storms can simultaneously
exist more or less independently.

*Concerning the latitude-effect on changes in cosmic-ray intensity during
magnetic storms*

To determine whether a lattitude-effect is indicated in the changes
of cosmic-ray intensity associated with magnetic storms there are plotted
in Figure 8, for two periods of magnetic storms, daily means of cosmic-
ray intensity at Huancayo and at Cheltenham. If the percentage-
changes at Huancayo in geomagnetic latitude $0°.6$ south and at Cheltenham in geomagnetic latitude $50°.1$ north were equal, the points in Figure
8 should define the two lines with unit-slope. It is evident, however, that
the scatter of points does not exclude the possibility of a latitude-effect
of several per cent in the magnetic-storm effect.

This is also evident from Figure 9 in which is shown the high correlation ($r = 0.89$) between departures from the average for the interval

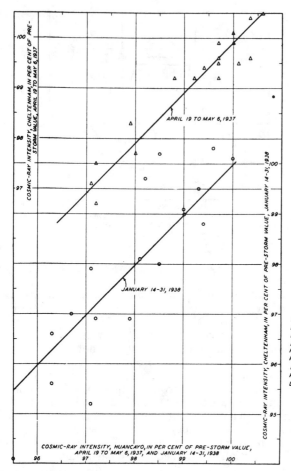

FIG. 8— CORRELATION BETWEEN DAILY MEANS OF COSMIC-RAY INTENSITY AT CHELTENHAM AND HUANCAYO FOR TWO PERIODS OF MAGNETIC STORMS, APRIL 19 TO MAY 6, 1937, AND JANUARY 14-31, 1938 (FOR NO LATITUDE-EFFECT POINTS WOULD DEFINE INDICATED LINES WITH UNIT-SLOPE)

April 23-30, 1937, of bihourly mean values of cosmic-ray intensity at Cheltenham and at Huancayo. The slopes of the two regression-lines are 1.11 and 0.88 The slope of the line defining the best value of the actual relation between departures at the two stations will lie between these two values, and will depend upon the relative weights assigned to the departures [14, 15]. If equal weight is assigned to both departures, the slope of the line is 0.98. On the basis of this series of data, which appears to be the better of the two in Figure 8, the possibility of the latitude-effect of several per cent is not excluded. The considerably greater scatter of points from the lower line in Figure 8 would allow a still larger latitude-effect. The difference in altitude between the two stations—72 meters at Cheltenham and 3350 meters at Huancayo—introduces further uncertainty concerning the latitude-effect for the same elevation. The effect of latitude on changes in cosmic-ray intensity during magnetic storms should be definitely answered when results are obtained from a Compton-Bennett meter now being installed at the Magnetic Observatory at Godhavn, Greenland, for the Committee on Coordination of Cosmic-Ray Investigations of the Carnegie Institution of Washington.

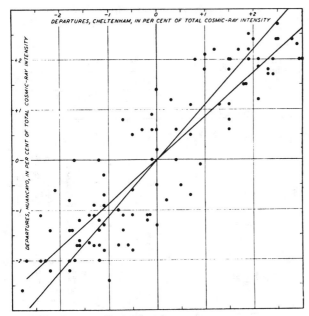

FIG. 9— CORRELATION BETWEEN
DEPARTURES, FROM AVERAGE FOR
THE INTERVAL, OF BIHOURLY
MEANS OF COSMIC-RAY INTEN-
SITY AT HUANCAYO AND CHEL-
TENHAM, APRIL 23-30, 1937 (RE-
GRESSION-LINES INDICATED)

Comparison of the fields of the magnetic storms of August 21, 1937, and January 16, 1938

Since no discernible cosmic-ray effects were observed during August 21-24, 1937, and since the largest effects were observed between January 16 and 19, 1938, these two storm-periods were selected for further investigation. Changes in daily means of horizontal intensity at Huancayo for these two periods were closely similar indicating that the fields, at the Earth, of the two storms were about equally intense.

Table 1 gives, for each of these storms, the value, at six observatories, for each of the three geomagnetic components of the storm-field. The components for the storm-field, for groups A and B, were obtained in the usual way [2, 16] by subtracting from the daily mean values of the components of total intensity for days during the storm, those for selected magnetically quiet days.

To obtain the components of the storm-field for group C, values of the total-intensity components were averaged for five periods of eight hours, each centered at local midnight on one of the five selected quiet days used in groups A and B. For each of three successive local midnights, the second of which occurred at each observatory near 54 hours after the sudden commencement, average values of the components of total intensity were obtained for the eight-hour interval centered at local midnight. These averages for each component at each station, when plotted against time, fell very nearly on a straight line. From these lines were read off values for the components of total intensity 54 hours after the sudden commencement. From these were subtracted the values of the components of total intensity averaged for the midnight-intervals preceding the storm, to give the storm-field components in group C of Table 1.

The world-wide features of the storm-field depend on universal or storm-time and constitute what is generally called [2] the storm-time

Observatory	Geomagnetic coordinates Φ	Λ	Ψ	Group (see notes)	August 1937 North ΔX'	East ΔY'	Down ΔZ	January 1938 North ΔX'	East ΔY'	Down ΔZ
	°	°	°		γ	γ	γ	γ	γ	γ
Cheltenham, Maryland	+50.1	350.5	+ 2.4	A	− 70	+17	+ 6	− 92	− 2	+15
				B	− 30	0	+12	− 54	+ 1	+22
				C	− 28	− 1	+ 7	− 36	+ 1	+17
Tucson, Arizona	+40.4	312.2	+10.1	A	− 88	+16	+15	− 78	−10	+11
				B	− 38	− 4	+ 8	− 57	0	+15
				C	− 37	− 1	+ 9	− 48	− 4	+14
San Juan, Puerto Rico	+29.8	3.1	− 0.6	A	− 86	+14	+10	−100	+10	+10
				B	− 37	+ 2	+10	− 64	+ 8	+18
				C	− 40	0	+ 8	− 53	0	+17
Honolulu, Hawaii	+21.0	266.5	+12.3	A	−112	+12	+14	− 80	− 1	+ 4
				B	− 42	+ 1	+ 2	− 60	+ 1	+ 7
				C	− 50	+ 3	+ 7	− 57	− 3	+ 6
Huancayo, Peru	− 0.6	353.8	+ 1.3	A	−110	+ 4	− 4	− 99	− 2	− 1
				B	− 48	− 3	+ 5	− 54	+ 8	−10
				C	− 46	0	+ 1	− 63	0	− 3
Watheroo, Western Australia	−41.8	185.6	+ 1.3	A	−102	+ 6	−39	− 76	−10	−19
				B	− 34	+ 3	−13	− 60	− 8	−14
				C	− 39	+ 4	−14	− 43	− 4	−12

Notes regarding groups:

A = mean for one day beginning six hours after sudden commencement of storm minus mean for five selected quiet days.

B for August 1937 = mean for two days beginning 30 hours after sudden commencement minus mean for five selected quiet days.

B for January 1938 = mean for two days beginning 22 and 54 hours, respectively, after sudden commencement minus mean for five selected quiet days.

C = value at 54 hours after sudden commencement obtained from linear adjustment through three eight-hour means centered at three successive local midnights minus average of eight-hour means centered at midnight on five selected quiet days.

field. The horizontal component of the storm-time field at each station is closely parallel to the geomagnetic meridians [16]. That the eastward components for group C in Table 1 are smaller than for the other groups suggests that the other components of the storm-time field in group C may more closely represent the actual components of the storm-time field than do those in groups A or B. It should be noted, however, that the components in group A pertain to the earlier and more intense part of the storm.

The storm-time field is very nearly symmetrical [16, 17] with respect to the geomagnetic equator so that if the eastward components in Table 1 are neglected a zonal harmonic series involving only harmonics of odd degree suffices to represent its magnetic potential. Because of the limited number of observatories for which data were immediately available for this analysis, the coefficients for harmonics of degree greater than three were disregarded. The expression of equation (1) was then assumed to represent the magnetic potential of the storm-time field.

$$V = [e_1 r + i_1(a^3/r^2)] P_1 + [e_3(r^3/a^2) + i_3(a^5/r^4)] P_3 \qquad (1)$$

in which a is the radius of the Earth and r is the distance from the Earth's center to the point at which the potential is to be evaluated. P_1 and P_3 are zonal-harmonic functions of the geomagnetic pole-distance; e and i

refer to the primary and induced systems, respectively. At the Earth's surface $(r=a)$ the northward geomagnetic and vertical components are as given by equations (2) and (3).

$$\triangle X' = [(i_1+e_1)\ P'_1 + (i_3+e_3)\ P'_3] \qquad (2)$$

$$\triangle Z = [(-2i_1+e_1)\ P_1 + (-4i_3+3e_3)\ P_3] \qquad (3)$$

The primes in (2) denote derivatives with respect to geomagetic pole-distance.

By a least-square adjustment of the data for each of the groups in Table 1 the coefficients in (2) and (3) were determined. These are tabulated in the columns headed intensity in Table 2, in which the unit is one gamma $= 10^{-5}$ gauss. Values of e_n and i_n are given in the columns headed potential.

TABLE 2—*Zonal-harmonic coefficients for differences in intensity for groups A, B, and C of Table 1 and for differences in intensity for international disturbed days minus international quiet days, 1927**

Storm	Group[a]	Degree n	Intensity		Potential		i_n/e_n
			North	Vertical	External e_n	Internal i_n	
			γ	γ	γ	γ	
Aug. 21, 1937	A	1	+113.7	+19.3	+82.2	+31.5	+0.38
		3	+ 1.9	−20.3	− 1.8	+ 3.7	−2.06
	B	1	+ 46.2	+16.7	+36.4	+ 9.8	+0.27
		3	0.0	+ 2.7	+ 0.4	− 0.4	−1.00
	C	1	+ 48.6	+11.4	+36.2	+12.4	+0.34
		3	− 0.2	− 8.4	− 1.3	+ 1.1	−0.85
Jan. 16, 1938	A	1	+102.0	+22.1	+75.4	+26.6	+0.35
		3	+ 6.8	+ 3.7	+ 4.4	+ 2.4	+0.55
	B	1	+ 66.8	+25.6	+53.1	+13.7	+0.26
		3	+ 6.8	− 0.9	+ 3.8	+ 3.0	+0.79
	C	1	+ 61.3	+20.3	+47.6	+13.7	+0.29
		3	− 1.1	− 4.7	− 1.2	+ 0.2	−0.17
Disturbed minus quiet days, 1927	D	1	+ 23.7	+ 3.8	+17.1	+ 6.6	+0.39
		3	− 0.8	− 3.0	− 0.9	+ 0.1	−0.11
	E	1	+ 24.7	+ 4.0	+17.8	+ 6.9	+0.39
		3	+ 0.1	− 5.4	− 0.7	+ 0.8	−1.12

*Using data of Table 1 of L. Slaucitajs and A. G. McNish [16].

[a]Notes regarding groups: D =from adjustment two coefficients to data for stations with $\Phi \leqq 52°.5$. E =from adjustment Slaucitajs and McNish based on all data using four coefficients.

A, B, and C of Figure 10 indicate, respectively, the degree with which the observed components of the storm-time field are approximated by the series (2) and (3) for the groups of data A, B, and C of Table 1. The poor fit of points to the curves in A of Figure 10 is doubtless due to the effect of irregular disturbances superposed upon the storm-time field. The deviations of the observed points from the curves are less in B of Figure 10 and still less for C of Figure 10.

While there exists an infinity of current-systems, which could give rise to the observed storm-time field [2], we assume that the external current-system is in the form of an anchor-ring, the cross-section of which for mathematical convenience is assumed infinitesimal, concentric with

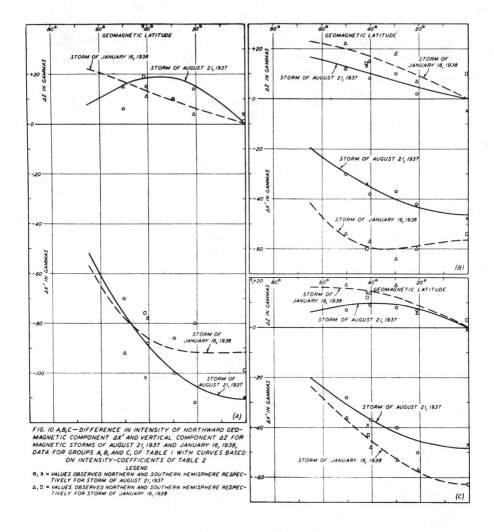

FIG. 10 A,B,C—DIFFERENCE IN INTENSITY OF NORTHWARD GEO-
MAGNETIC COMPONENT $\Delta X'$ AND VERTICAL COMPONENT ΔZ FOR
MAGNETIC STORMS OF AUGUST 21, 1937 AND JANUARY 16, 1938,
DATA FOR GROUPS A, B, AND C, OF TABLE I WITH CURVES BASED
ON INTENSITY-COEFFICIENTS OF TABLE 2
LEGEND
O, X = VALUES OBSERVED NORTHERN AND SOUTHERN HEMISPHERE RESPEC-
TIVELY FOR STORM OF AUGUST 21, 1937
△, ☐ = VALUES OBSERVED NORTHERN AND SOUTHERN HEMISPHERE RESPEC-
TIVELY FOR STORM OF JANUARY 16, 1938

the Earth and in the plane of the Earth's geomagnetic equator. For such
a ring, of radius R, in which flows unit westward current, the series for the
magnetic potential, on the coordinate-system used in (1), at the surface
of the Earth is given by equation (4), except for a constant.

$$V_R = 2\pi \left\{ (a/R)P_1 - (1/2)(a^3/R^3)P_3 + \ldots \right.$$
$$\left. + (-1)^n \left[\frac{1 \cdot 3 \ldots (2n-1)}{2 \cdot 4 \ldots 2n} \right] (a/R)^{2n+1} P_{2n+1} + \ldots \right\} \quad (4)$$

In (4) the ratio of the coefficient of P_1 to that of P_3 is $-2R^2/a^2$. In the
case of the storms under discussion, if the only source of potential were
the assumed ring-current, $\sqrt{-e_1/2e_3}$ should determine R/a, the radius
of the assumed ring in terms of the Earth's radius a.

146

However, the rather concentrated westward currents [2] in the auroral zones give rise to another source of potential. If the two zones, in each of which flows a westward current i, are considered as two circles of geomagnetic latitude on a sphere, concentric with the Earth and of radius c, and if the radius of each circle, one in the Northern and the other in the Southern Hemisphere, subtends an angle a at the Earth's center, then the potential of the system [18] for points on the Earth between the zones is, except for a constant, given by

$$\Omega = 4\pi i \sin^2 a \sum_{n=1}^{n=\infty} (1/n)(a^n/c^n)P'_n(a)P_n \quad \ldots \text{ for } n \text{ odd and } a < c \qquad (5)$$

P_n is the zonal-harmonic function of geomagnetic pole-distance and $P'_n(a)$ its derivative with respect to the cosine of pole-distance. When a and i in (5) are known, the coefficients for P_n in (5) can be determined. Subtracting these from the coefficients e_n given in Table 2, we have the coefficients for the potential of the assumed equatorial ring-current alone.

During the period from which the data in Table 1 for group C of the January storm were obtained, the character of the diurnal variations of vertical and horizontal magnetic intensity at Sitka indicated that the zonal currents flowed close to, but north of, that station. For the two-day interval centered 54 hours after the sudden commencement of the storm of January 16, 1938, daily means of horizontal intensity, H, and vertical intensity, Z, were computed for Sitka and from them the corresponding means for the five quiet days preceding the storm were subtracted. These vertical and horizontal components of the storm-time field fell respectively 30 gammas and 40 gammas below the values for $\triangle Z$ and $\triangle X'$ given by the curves at $\Phi = 60°$, for the storm of January 16, 1938, in C of Figure 10. The vertical and horizontal components of the field of the zonal current at Sitka may then be taken as 30 and 40 gammas, respectively. Assuming the zonal current to be linear at a height of 150 km, the horizontal distance S from Sitka to a point directly beneath the current is given by $S = 150 \times (40/30)$ or 200 km. The field of the current at a distance of 250 km is 50 gammas, which gives about 62,000 amperes for the zonal current. This is not very different from the value estimated by Chapman [2] for the total westward current in the auroral zone during moderate storms.

Thus at the time of the storm of January 16, 1938, to which group C of Table 2 applies, we estimate the westward zonal current in the northern and southern zones, in geomagnetic latitude 62°, to be 62,000 amperes. This estimate may also be safely used for group B for the storm of January 16, 1938. For group C of the August storm, the zone was too far north of Sitka to permit a reliable estimate of its location. We shall perhaps be not far wrong in assuming the geomagnetic latitude of the zones to be 70° and the current to be about 40,000 amperes, since the value of e_1 in Table 2 for group C for the storm of August, 1937, is about three-fourths of that in group C for the January storm. Similarly, for groups D and E, we adopt 70° for the latitude of the zones and 20,000 amperes for the current.

Using these data the coefficients of P_1 and P_3 in equation (5) are calculated. These are designated, with signs changed, in Table 3 as corrections, $\triangle e_n$, which are applied to the values of e_n in Table 2. E_n of

Table 3 thus gives the values of the coefficients which are ascribed to the field of the assumed equatorial ring-current.

In Table 3 it should be noted that all values of E_3 are negative, which according to equation (4) is a necessary condition for the existence of an equatorial ring carrying a westward current.

Unfortunately the values of (R/a) in Table 3 provide no convincing evidence concerning the important question of whether the radius of the assumed equatorial ring-current is, as would be expected, greater for the storm of January 16, 1938, which resulted in a large decrease in cosmic-ray intensity. However, it can be safely said that for none of the storms could (R/a) have been much less than two. In an unpublished manuscript, Dr. E. H. Vestine, of the Department of Terrestrial Magnetism, using the results of the analysis by L. Slaucitajs and A. G. McNish [16], finds, by a somewhat different procedure, a value between two and four for (R/a).

The method used here would of course be more effective the smaller the values of (R/a). The analysis is presented in the belief that the results may prove useful in checking future theories concerning the orbits of cosmic-ray particles in the field of a magnetic storm superposed on the Earth's permanent field.

TABLE 3—*Corrections $\triangle e_n$ to coefficients e_n for the potential in Table 2 on account of estimated currents i flowing westward in two zones in equal north and south geomagnetic colatitudes a and estimated radius (R/a) of assumed equatorial ring-current in earth-radii*

Storm	Group	Degree n	a	i	$\triangle e_n$	e_n	$E_n =$ $e_n + \triangle e_n$	$-(E_1/2E_3)$	(R/a)
			°	amp	γ	γ	γ		
Aug. 21, 1937	B	1	20	4.0×10^4	−0.9	+36.4	+35.5	16.2	4.0
		3			−1.5	+ 0.4	− 1.1		
	C	1	20	4.0×10^4	−0.9	+36.2	+35.3	6.3	2.5
		3			−1.5	− 1.3	− 2.8		
Jan. 16, 1938	B	1	28	6.2×10^4	−2.7	+53.1	+50.4	252.0	15.8
		3			−3.9	+ 3.8	− 0.1		
	C	1	28	6.2×10^4	−2.7	+47.6	+44.9	4.4	2.1
		3			−3.9	− 1.2	− 5.1		
Disturbed minus quiet days, 1937	D	1	20	2.0×10^4	−0.5	+17.1	+16.6	4.9	2.2
		3			−0.8	− 0.9	− 1.7		
	E	1	20	2.0×10^4	−0.5	+17.8	+17.3	5.8	2.4
		3			−0.8	− 0.7	− 1.5		

Evidence for world-wide effects on daily means of cosmic-ray intensity

That the daily means of cosmic-ray intensity at two widely separated stations are influenced in part by a common cause is indicated in Figure 11. To investigate the possible influence of solar rotation upon cosmic-ray intensity it was convenient to take averages of daily means of cosmic-ray intensity for alternate periods of four and five days. The correlation between the first differences of these (each average minus the preceding one) is shown in Figure 11. Data for days near the end of April 1937, on which the cosmic-ray intensity was definitely known to be affected by magnetic storms, were not included in deriving the correlation. After allowing for the fact that, at most only half of the 81 differences plotted

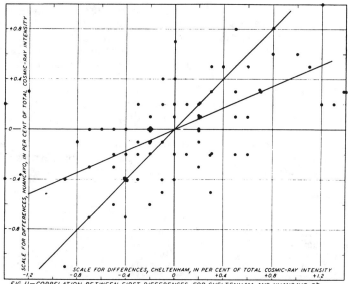

FIG. 11—CORRELATION BETWEEN FIRST DIFFERENCES, FOR CHELTENHAM AND HUANCAYO, OF
AVERAGE COSMIC-RAY INTENSITY FOR CORRESPONDING ALTERNATE INTERVALS OF 4 AND 5 DAYS,
JUNE 13 TO OCTOBER 2, 1936, AND APRIL 6 TO DECEMBER 31, 1937, EXCLUDING DAYS AFFECTED
BY MAGNETIC STORM OF APRIL 1937 (REGRESSION-LINES INDICATED)

in Figure 11 are statistically independent, the moderate correlation
($r = 0.64$) indicates a significant correspondence. That a line with unit-
slope, as for Figure 8, through the origin in Figure 11 lies between the
two indicated regression-lines is further indication that the correspondence is not accidental.

Whether the changes in daily means of cosmic-ray intensity at Huancayo and at Cheltenham are definitely associated with changes in the
Earth's magnetic field due to minor magnetic disturbances is not yet
certain.

Preliminary investigation of the 27-day period suggests the possibility of quasi-persistent [19] 27-day waves in cosmic-ray intensity having
similar phases at Cheltenham and Huancayo. This would indicate,
because of the 27-day quasi-persistent wave [19] in magnetic activity,
that the mechanism responsible for the correlation in Figure 11 is the
same as that for the world-wide changes observed during some magnetic
storms.

Acknowledgments

We are indebted to the Committee on Coordination of Cosmic-Ray
Investigations of the Carnegie Institution of Washington for providing
the assistance of W. R. Maltby whose aid in the reduction of the cosmic-
ray data and in the computations has been invaluable.

We are under obligation to the Director of the United States Coast
and Geodetic Survey for making available to us the magnetic data from
its five magnetic observatories and also to the staff of its Cheltenham
Magnetic Observatory for operation of a cosmic-ray meter. We are also
indebted to J. W. Beagley, of the staff of the magnetic observatory at
Christchurch, New Zealand, for operation of a meter and reduction of

records at that station, to Professor J. Gallo, Director of the National Astronomical Observatory at Tacubaya, Mexico, for supervising the operation of the cosmic-ray meter at Teoloyucan, to the staff of the Huancayo Magnetic Observatory of the Department of Terrestrial Magnetism of the Carnegie Institution of Washington for operation of a meter, to Professor R. D. Bennett of the Massachusetts Institute of Technology for making available to us the cosmic-ray records obtained there during the magnetic storm of January 1938, and to the National Carbon Company, Incorporated, for providing all batteries for operating the meters.

Finally, it is a pleasure to acknowledge the encouragement received from Dr. J. A. Fleming, Director of the Department of Terrestrial Magnetism of the Carnegie Institution of Washington.

References

[1] S. E. Forbush, Phys. Rev., **51**, 1108-1109 (1937).
[2] S. Chapman, Terr. Mag., **40**, 349-370 (1935).
[3] S. Chapman, Nature, **140**, 423-424 (1937).
[4] J. Clay and E. M. Bruins, Physica, **5**, 111-114 (1938).
[5] C. Störmer, Arch. Sci. Phys., Genève, **32**, 415-436 (1911).
[6] Terr. Mag., **42**, 211-214 (1937).
[7] V. F. Hess, R. Steinmaurer, and A. Demmelmair, Nature, **141**, 686-687 (1938).
[8] C. Störmer, Terr. Mag., **35**, 193-208 (1930).
[9] E. Brüche, Terr. Mag., **36**, 41-52 (1931).
[10] Kr. Birkeland, Norwegian Aurora Polaris Expedition, 1902-03, **1**, (1908-1913).
[11] T. H. Johnson, Terr. Mag., **43**, 1-6 (1938).
[12] A. H. Compton, E. O. Wollan, and R. D. Bennett, Rev. Sci. Instr., **5**, 415-422 (1934).
[13] S. E. Forbush, Terr. Mag., **42**, 1-16 (1937).
[14] H. S. Uhler, Optical Soc. Amer., **7**, 1043-1066 (1923).
[15] W. E. Deming, Phil. Mag., **11**, 146-158 (1931); **17**, 804-829 (1934).
[16] L. Slaucitajs and A. G. McNish, Trans. Edinburgh Meeting 1936, Internat Union Geod. Geophys., Ass. Terr. Mag. Electr., Bull. No. 10, 289-301 (1937).
[17] S. Chapman and A. T. Price, Phil. Trans. R. Soc., A, **229**, 427-460 (1930).
[18] S. Chapman and T. T. Whitehead, Proc. Internat. Math. Cong., Toronto, 1934, 313-337 (1928).
[19 J. Bartels, Terr. Mag., **40**, 1-60 (1935).

DEPARTMENT OF TERRESTRIAL MAGNETISM,
CARNEGIE INSTITUTION OF WASHINGTON,
Washington, D. C.

THE

PHYSICAL REVIEW

A Journal of Experimental and Theoretical Physics Established by E. L. Nichols in 1893

| VOL. 54, NO. 12 | DECEMBER 15, 1938 | SECOND SERIES |

On World-Wide Changes in Cosmic-Ray Intensity

S. E. FORBUSH

Department of Terrestrial Magnetism, Carnegie Institution of Washington, Washington, D. C.

(Received October 11, 1938)

Continuous cosmic-ray records for periods of 17 months or more obtained by Compton-Bennett meters at Cheltenham (United States), Teoloyucan (Mexico), Christchurch (New Zealand), and Huancayo (Peru) are reduced to a constant barometric pressure at each station. After deducting a 12-month wave from the data at each of these stations except at Huancayo, where none is found, a high correlation ($r \doteq 0.90$) is obtained between the means of cosmic-ray intensity for each one-third month for any two of the four stations. This high correlation, based on all available data for each pair of stations, is definite evidence that, except for the 12-month waves, the major changes in cosmic-ray intensity are world-wide. With published data from a Steinke instrument on the Hafelekar, near Innsbruck, t) e world-wide changes, when expressed in percent of the absolute intensity at each station, are found to increase rapidly with altitude for stations at the same latitude. A large increase in the world-wide effect for stations at high altitudes occurs between the equator and geomagnetic latitude 30° north, while no further very large increase occurs up to latitude 47° north. Results for Christchurch and Cheltenham indicate that the magnitude of the world-wide effect is symmetrical about the equator, and its connection with effects observed during certain magnetic storms is discussed. The amplitude of the 12-month wave, which in the Northern Hemisphere has its maximum near mid-January, increases from zero at the equator to about 1.0 percent at Teoloyucan, 1.6 percent at Cheltenham, and 1.9 percent on the Hafelekar. At Christchurch, its maximum occurs near the end of July, its amplitude being about 0.8 percent of the total intensity. Accordingly, it seems that the 12-month wave may not be ascribed to a solar-magnetic moment.

INTRODUCTION

THE decrease of several percent in cosmic-ray intensity which occurred[1-4] simultaneously at Cheltenham (United States), Huancayo (Peru), on the Hafelekar (Germany), Christchurch (New Zealand) and Teoloyucan (Mexico) during the magnetic storms which began April 24, 1937, and January 16, 1938, provided definite evidence that these changes were world-wide. The fact that the decrease in cosmic-ray intensity began within an hour or two after the sudden commencements of these magnetic storms led us to hypothesize[1,4] that the change in cosmic-ray intensity resulted from the magnetic field of the current-system responsible for magnetic storms. The nature of the field-changes which occur over the earth's surface during magnetic storms is such that the world-wide characteristics of the storm field can be ascribed either to a system of westward currents forming a spherical current-sheet, concentric with the earth, in the upper atmosphere or to a westward current-system approximated by a ring concentric with the earth and in the plane of the earth's geomagnetic equator.[5]

[1] S. E. Forbush, Phys. Rev. **51**, 1108–1109 (1937).
[2] V. F. Hess and A. Demmelmair, Nature **140**, 316–317 (1937).
[3] V. F. Hess, R. Steinmaurer and A. Demmelmair, Nature **141**, 686–687 (1938).
[4] S. E. Forbush, Terr. Mag. **43**, 203–218 (1938).

[5] S. Chapman, Terr. Mag. **40**, 349–370 (1935).

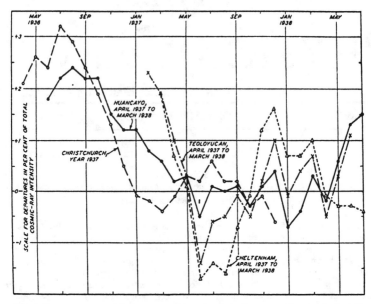

Fig. 1. Departures monthly means cosmic-ray intensity from mean for year indicated.

Further investigation[4] showed, however, two magnetic storms of equal intensity at the earth's surface, during one of which a world-wide decrease of about six percent in cosmic-ray intensity was observed, while no detectable change occurred in the other. Assuming the world-wide part of the storm field to arise from the above-mentioned ring-current system, we attempted[4] to calculate, using magnetic data from several observatories, the radius of the assumed ring for the two storms. Although this analysis provided no convincing evidence that the radius of the ring was significantly different for the two storms, it did indicate that a necessary condition for the existence of such a ring-current was satisfied by the magnetic data. We concluded that the radius for neither storm could be much less than about

twice that of the earth, although the possibility of an indefinitely larger radius for either of the storms was not excluded.

We also found that, while there was close correlation between changes in daily means of cosmic-ray intensity and of the horizontal magnetic component at the equator during individual magnetic storms, the ratio between the two was significantly different even for storms accompanied by world-wide changes in cosmic-ray intensity. We therefore concluded that, if the major current-system for magnetic storms were approximated by an equatorial ring, its radius must be different even for the different storms accompanied by changes in cosmic-ray intensity.

Evidence is presented in the present investigation that the world-wide character of changes in

TABLE I. *Location and elevation of cosmic-ray stations and magnitude of world-wide effect, relative to Huancayo, for each.*

STATION	GEOGRAPHIC		GEOMAG-NETIC LATITUDE	ELEVATION ABOVE SEA LEVEL	WORLD-WIDE EFFECT RELATIVE TO HUANCAYO
	LATITUDE	LONGITUDE			
Cheltenham, United States	38.7N	76.8W	50.1N	72	1.11
Teoloyucan, Mexico	19.2N	99.2W	29.7N	2285	1.58
Huancayo, Peru	12.0S	75.3W	0.6S	3350	1.00
Christchurch, New Zealand	43.5S	172.6E	48.0S	8	1.05
Hafelekar, Germany	47.3N	11.3E	48.4N	2300	1.59

cosmic-ray intensity is found throughout the data available to us, and hence is not restricted to periods of intense magnetic storms.

Table I indicates the location and elevation of the several cosmic-ray meters. Those at the first four locations are Compton-Bennett[6] precision recording instruments operated under the auspices of the Committee on Coordination of Cosmic-Ray Investigations of the Carnegie Institution of Washington. Data from these were obtained with a total shielding equivalent to 12 cm of lead. Data for the last station of Table I were obtained from publications by V. F. Hess[2, 3, 7] and his co-workers who used a Steinke apparatus at constant temperature, shielded on all sides with ten cm of lead and seven cm of iron. Observed values of ionization were corrected for bursts and for changes in barometric pressure, by applying to all data at each station a constant barometric coefficient.[8] The percentage changes in cosmic-ray intensity for each station were determined on the basis of the total ionization, corrected for residual, given by the meter at that station.

---

[6] A. H. Compton, E. O. Wollan and R. D. Bennett, Rev. Sci. Inst. 5, 415–422 (1934).
[7] A. Demmelmair, Ber. Akad. Wiss. Wien, 146, IIa, 643–659 (1937).
[8] S. E. Forbush, Terr. Mag. 42, 1–16 (1937).

ANALYSIS OF DATA

In Fig. 1 are plotted, for four stations, the departures of observed monthly means of cosmic-ray intensity from the indicated annual means. It is evident that, in general, no particularly close correspondence exists between the departures for the different stations. From the 13 monthly mean departures for Cheltenham, beginning with April, 1937, we obtained by harmonic analysis[9] the 12-month wave which best fits the departures in this interval corrected for noncyclic change, which was assumed linear and was taken as the departure for April, 1938 minus that for April, 1937. The smooth function plotted in Fig. 2 is the sum of this wave and the linear change. This function, indicated in Fig. 2 by a dashed line where extrapolated, departs considerably from the observed departures.

The dashed curve in Fig. 2 indicates the ordinates of the 12-month wave deducted from the observed departures for Cheltenham. This curve resembles closely that for the observed departures for Huancayo. The correlation between the two ($r = 0.90$) is shown in Fig. 10. This indicates that the variations in cosmic-ray intensity which occur at Huancayo also occur with about the same amplitude at Cheltenham, where they are superposed upon a 12-month wave.

---

[9] J. Bartels, Beitr. Geophysik 28, 1–10 (1930).

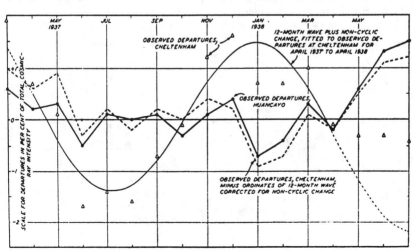

Fig. 2. Departures cosmic-ray intensity from Fig. 1, for Huancayo, and for Cheltenham before and after deducting 12-month wave.

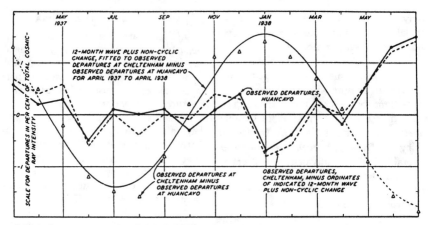

FIG. 3. Departures cosmic-ray intensity from Fig. 1, for Huancayo, and for Cheltenham after deducting departures for Huancayo.

When the departures at Huancayo for the interval April 1937 to April 1938 are corrected for noncyclic change and subjected to harmonic analysis, the amplitude of the resulting 12-month wave is practically zero. Thus, although the changes in cosmic-ray intensity at Huancayo are also found at Cheltenham superposed on a 12-month wave, our determination of the 12-month wave for Cheltenham is not materially affected by their presence. If, obscured by the changes which also occur at Cheltenham, a 12-month wave in cosmic-ray intensity existed at Huancayo, then the differences, taken month by month, between the two irregular curves of departures in Fig. 2 should determine the 12-month wave for Huancayo. In this way no significant amplitude for this wave is found.

If the indication is correct that changes in cosmic-ray intensity occur simultaneously at Huancayo and at Cheltenham superposed on a 12-month wave, then a 12-month wave should fit quite closely the differences obtained by subtracting the monthly mean departures of cosmic ray intensity for Huancayo from those at Cheltenham. The result is shown in Fig. 3 in which the better approximation to the 12-month wave is evident.

To obtain a better indication of the relation

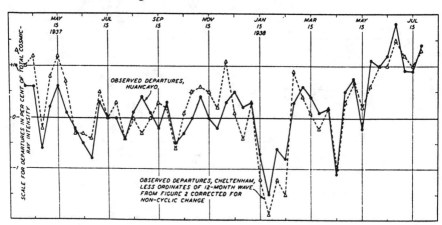

FIG. 4. Departures cosmic-ray intensity for each one-third month from mean April, 1937 to March, 1938 for Cheltenham after deducting 12-month wave, and for Huancayo.

between the world-wide changes at Huancayo and Cheltenham, the departures from the annual mean of those of cosmic-ray intensity for each one-third month were computed. From these means for Cheltenham were subtracted the ordinates of the 12-month wave of Fig. 2 and the result is shown in Fig. 4. The correlation between the two curves is 0.86 (see Fig. 12-*A*).

Figure 5 shows the close approximation of a 12-month wave to the result obtained by subtracting the monthly mean departures of cosmic-ray intensity at Huancayo from those at Christchurch. The dashed curve, which is remarkably similar to that for the observed departures at Huancayo, results from deducting the ordinates of this 12-month wave from the observed de-

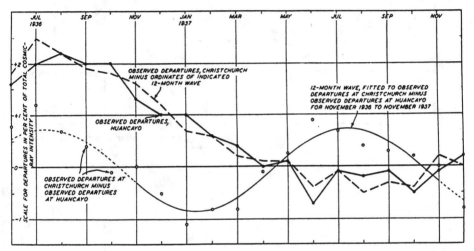

FIG. 5. Departures monthly means cosmic-ray intensity from mean for 1937 for Huancayo and for Christchurch after deducting departures for Huancayo.

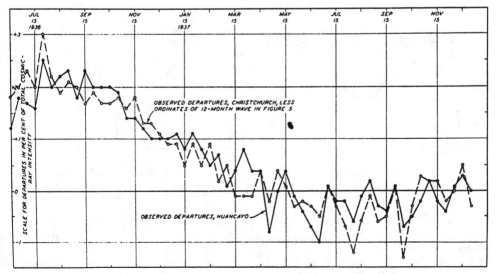

FIG. 6. Departures cosmic-ray intensity for each one-third month from mean for 1937, for Christchurch with 12-month wave deducted, and for Huancayo.

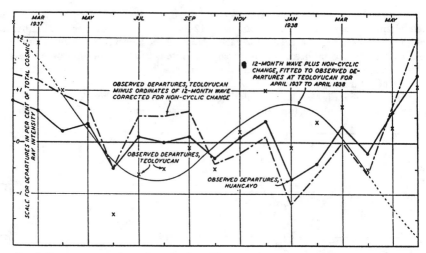

FIG. 7. Departures cosmic-ray intensity from Fig. 1, for Teoloyucan after deducting 12-month wave, and for Huancayo.

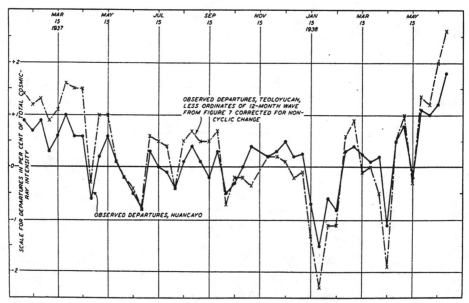

FIG. 8. Departures cosmic-ray intensity for each one-third month from mean for year beginning April 1937, for Teoloyucan after deducting 12-month wave, and for Huancayo.

partures (see Fig. 1) at Christchurch. The high correlation ($r = 0.97$, see Fig. 10) between these curves results from the similarity in their trends rather than from correspondence of individual changes. However, a parallelism between changes in the means of cosmic-ray intensity for each

one-third month at Christchurch, after deducting the 12-month wave, and those at Huancayo is evident in Fig. 6.

The similarity between the observed departures of monthly means of cosmic-ray intensity at Huancayo and those obtained by deducting

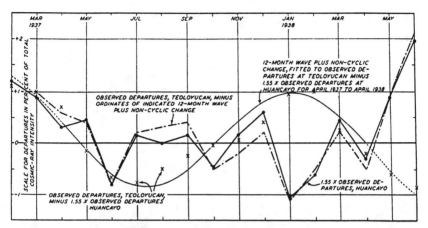

FIG. 9. Departures cosmic-ray intensity from Fig. 1, times 1.55 for Huancayo, and for Teoloyucan after deducting 1.55 times departures for Huancayo.

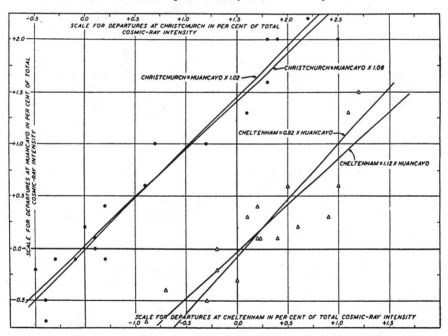

FIG. 10. Correlation between departures cosmic-ray intensity from Fig. 1, after deducting 12-month waves for Christchurch and Cheltenham.

the 12-month wave at Teoloyucan is shown in Fig. 7. The correlation between the two ($r = 0.92$) is shown in Fig. 11-*A*, which indicates that the world-wide changes at Teoloyucan are between 50 and 80 percent greater than at Huancayo. This is also evident from the curves in Fig. 8,

the correlation between which is 0.92 (see Fig. 12-*B*). It will later be shown that the world-wide changes in cosmic-ray intensity at Teoloyucan are about 55 percent greater than at Huancayo.

The close approximation of a 12-month wave, plus a linear change, to the result obtained by

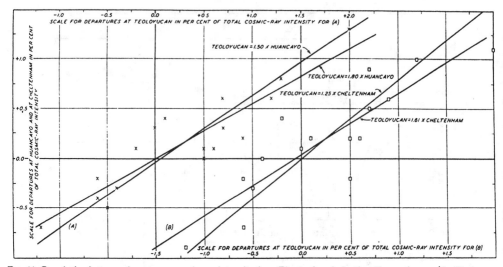

Fig. 11. Correlation between departures cosmic-ray intensity from Fig. 1, after deducting 12-month waves for Teoloyucan and Cheltenham.

subtracting 1.55 times the departures at Huancayo from those at Teoloyucan is shown in Fig. 9. The dashed curve, which so closely parallels the curve for 1.55 times the observed departures at Huancayo, results from deducting the ordinates of the indicated wave from the observed departures at Teoloyucan.

RELATIVE MAGNITUDE OF WORLD-WIDE
CHANGES IN COSMIC-RAY INTENSITY
AT DIFFERENT LATITUDES AND
ELEVATIONS

In Fig. 11-*B* is shown the correlation between monthly means of cosmic-ray intensity at Teoloyucan and Cheltenham after deducting the respective 12-month waves. If one arbitrarily assumes the best relation between the world-wide changes at any two stations to be the average of the two factors indicated for the regression lines in Figs. 10 and 11, it is found that (1) the world-wide changes at Cheltenham and at Christchurch are each a few percent greater than those at Huancayo, and (2) the world-wide changes at Teoloyucan are about 65 percent greater than at Huancayo and about 43 percent greater than at Cheltenham. However, these factors for the three relations, of which only two are independent, between Cheltenham, Teoloyucan, and Huancayo are mutually only roughly consistent. This is to be expected since the factors were selected somewhat arbitrarily.

To obtain more reliable values for these relations, assumed linear, between world-wide changes at different stations, use is made of the data in Fig. 12. *C* of Fig. 12 indicates the correlation between world-wide changes in the means for each one-third month at Teoloyucan and Cheltenham. For Figs. 12, 4, 6 and 8, only those days were used in each one-third month for which simultaneous data were available from both stations; there are, however, but few cases in which all days were not available for all stations.

The slopes of the lines indicated in Fig. 12 represent the most probable values, in the sense of least squares, consistent with the assignment of weights[10, 11] 1, 2 and 3, respectively, to the departures at Cheltenham, Teoloyucan, and Huancayo. Since these weights were assigned in an objective manner, there is nothing arbitrary, except the use of least squares, in the method used to obtain the slopes indicated.

The correlation coefficients for *A*, *B*, and *C* of Fig. 12 are 0.86, 0.92, and 0.86, respectively.

[10] W. E. Deming, Phil. Mag. **11**, 146–158 (1931); **17**, 804–829 (1934).
[11] H. S. Uhler, J. Opt. Soc. Am. **7**, 1043–1066 (1923).

The slopes of the regression lines, which result if infinite weight is given to one or the other of the two coordinates, are 1.01 and 1.36 for A, 1.42 and 1.68 for B, and 1.11 and 1.50 for C. Any slope between these limits may be obtained by least squares depending upon the relative weights assigned to the two coordinates. Suppose a slope midway between that for the two regression lines is arbitrarily adopted for A and B. With these slopes, values for the departures for Huancayo can be calculated separately from those at Cheltenham and at Teoloyucan. The variance (square of standard deviation) for single differences in each of the two sets of differences between the observed and calculated departures for Huancayo is computed. With the same arbitrary slopes, the variance is now determined for single differences between the two series of calculated departures for Huancayo. Since each of these three computed values of variance is the sum of two separate variances because of independent statistical variations at the two stations involved, one obtains three equations from which the variance, because of statistical fluctuations, at each of the stations is readily obtained. The variance thus obtained of statistical variations at each station

is measured in terms of the original unit for Huancayo, and thus in terms of units, which are larger (on account of the arbitrary factors used for calculating departures for Huancayo) for Cheltenham and Teoloyucan than the original units for those two stations. Correcting for this change of units, one obtains the variance for statistical variations at each station in terms of the original units used for that station. The weights are then assigned to each station in inverse proportion to the variance of statistical fluctuations at that station. The slopes indicated for the lines in Fig. 12 were obtained with these weights. These slopes were then used to obtain a closer approximation to the weights; these second approximations differed so little from the previous ones that the slopes calculated from them were not significantly different from those in Fig. 12.

The three slopes given in Fig. 12 are mutually quite consistent. Also the weights, thus objectively assigned, are not surprising since variations in daily means of barometric pressure are in general largest at Cheltenham and practically absent at Huancayo.

To determine the magnitude of the world-wide

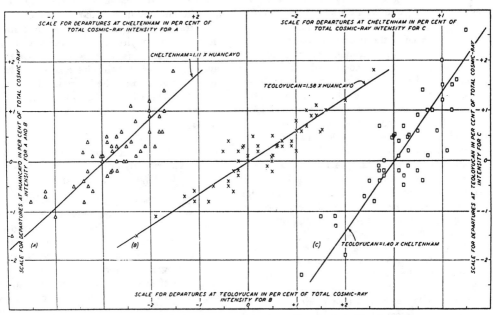

Fig. 12. Relation between departures cosmic-ray intensity from Figs. 4 and 8.

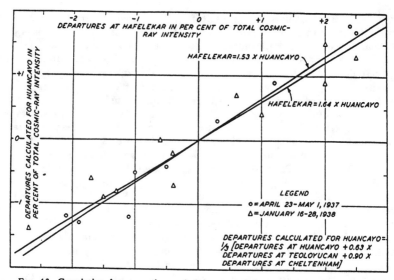

FIG. 13. Correlation between observed departures daily means cosmic-ray intensity from average during two magnetic storms for Hafelekar and calculated departures for Huancayo.

effect in cosmic-ray intensity for a station at great elevation and high latitude, data published by V. F. Hess[2, 3] and his co-workers for the Hafelekar during two intense magnetic storms were used. Departures of the scaled daily cosmic-ray intensity (75° west meridian day) from the means for each of the storms are plotted in Fig. 13. Daily (75° west meridian) means of cosmic-ray intensity were then computed separately for Huancayo, Teoloyucan, and Cheltenham. Their departures from the means for the respective storms were used in the equation of Fig. 13 to compute calculated departures for Huancayo; the high correlation of $r = 0.96$ between the two departures is evident. The indicated regression lines fit the separate data for the two periods of magnetic storms about equally well. The correlation in Fig. 13 is considerably higher than that found when the *observed* departures at Huancayo were used as ordinate. Since the factors used in the equation for calculated departures at Huancayo are those indicated in Fig. 12, we assumed that the ratio between the world-wide effects at different stations during individual magnetic storms is the same as that for means for each one-third month. However, this point was carefully tested for each station

in the same manner as for the Hafelekar in Fig 13. While the correlation in most cases is not so high as for Fig. 13, no evidence was found indicating that the relations in Fig. 12 do not also apply to changes in cosmic-ray intensity at the other stations during the same magnetic storms. Thus for the magnitude of the world-wide effect at different stations relative to that for Huancayo, the values of Table I are adopted. In the case of the station on the Hafelekar the average of the two slopes in Fig. 13 was taken since any reasonable weighting will not appreciably alter the result.

SUMMARY AND DISCUSSION OF TWELVE-MONTH WAVES IN COSMIC-RAY INTENSITY AT DIFFERENT STATIONS

In Fig. 14 are summarized the results for the 12-month waves in cosmic-ray intensity at different stations after deducting the world-wide effects. Each vector in the harmonic dial[12] of Fig. 14 points to the date of the wave maximum and its length represents the amplitude. The world-wide effects were determined, as previously described, from the data for Huancayo, where

[12] J. Bartels, Terr. Mag. 40, 1–60 (1935).

the 12-month wave is absent. The world-wide effects at each of the other stations, except Hafelekar as noted below, were computed from those at Huancayo with the factors given in Table I.

The 12-month wave at the Hafelekar was based on monthly means, corrected only for barometric pressure, for April 1936 to March 1937 obtained from Table III of A. Demmelmair's paper.[7] Since no data for Huancayo prior to June 1936 were available, the world-wide changes for April and May 1936 were determined from the data at Christchurch (see Fig. 15). Fig. 14 also indicates the 12-month waves in temperature of outside air based, except for Teoloyucan, on data for the same intervals used to obtain the 12-month waves in cosmic-ray intensity. For Teoloyucan the temperature-wave is the average for the five years,[13] 1917–1921.

The phase of the 12-month wave in temperature at each station differs, except for Teoloyucan, by about 180° from that of the 12-month wave in cosmic-ray intensity. The ratio of the amplitude of the 12-month wave in cosmic-ray intensity to that in temperature is about the same for Cheltenham and Christchurch, which have similar elevations and geomagnetic latitudes. However, this ratio is quite different for the other stations, excluding the possibility that the variation in cosmic-ray intensity is, in

general, closely connected with temperature. Whether or not the 12-month wave in cosmic-ray intensity is the result of the 12-month wave in temperature, the fact that the two are about opposite in phase would necessarily lead to a correlation between the two, which has been reported by Hess[14] and others. Possibly the cause of the apparent seasonal wave in cosmic-ray intensity may be associated with the seasonal variation[15] in the distribution of air-density with height in the earth's atmosphere, as was implied by Schonland.[16]

If we arbitrarily assume a seasonal wave in cosmic-ray intensity having the same amplitude but opposite phase at Cheltenham and Christchurch, the observed 12-month waves in cosmic-ray intensity at these two stations could then be attributed to an annual wave, with the same phase and amplitude at both stations, superposed upon the seasonal waves. It is readily seen from Fig. 14 that this annual wave would have its maximum in January. This is just opposite in phase to that of the annual wave which, according to the calculations of Vallarta,[17] would arise if the sun had a magnetic moment. Thus it seems impossible to ascribe the 12-month wave in cosmic-ray intensity to a solar magnetic moment,

[13] Bol. Obs. Astr., Tacubaya, No. 10, Table 11 (1928).

[14] V. F. Hess, Terr. Mag. 41, 345–350 (1936).
[15] W. J. Humphreys, *Physics of the Air*, second edition (1929).
[16] B. F. J. Schonland, B. Delatizky and J. Gaskell, Terr. Mag. 42, 137–152 (1937).
[17] M. S. Vallarta, Nature 139, 839 (1937).

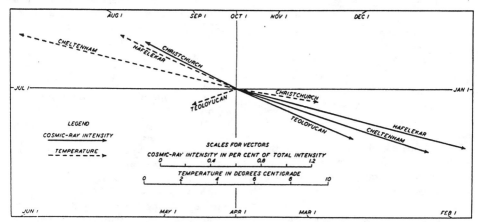

FIG. 14. Harmonic dial for 12-month waves in cosmic-ray intensity after deducting world-wide changes and in temperature showing dates and amplitudes of wave-maximum.

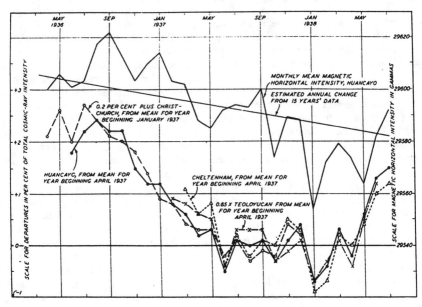

FIG. 15. Departures cosmic-ray intensity from Fig. 1 after deducting 12-month waves.

unless perhaps the annual variation in the angle between the earth's magnetic axis and the ecliptic introduces an unsuspected variation. This latter, however, would seem to be excluded on the basis that the diurnal variation of about 12° in the angle between the earth's magnetic axis and the ecliptic should then be expected to result in a diurnal variation in cosmic-ray intensity with amplitude many times that actually observed.[8]

Since the relative amplitudes of the 12-month waves are quite different from those of the world-wide changes at the different stations, the two effects must be ascribed to different causes. It would seem impossible to obtain the excellent agreement indicated in Figs. 4, 6, 8 and 12 unless the fundamental mechanism for the 12-month wave at each station varied almost perfectly sinusoidally.

It should be added that by varying the temperature of the Compton-Bennett meter at Cheltenham over a range of 15°C and comparing results over a period of many weeks with those from two Millikan-Neher electroscopes at constant temperature, the Compton-Bennett meter is found to be unaffected by temperature.

DISCUSSION OF THE WORLD-WIDE EFFECT IN COSMIC-RAY INTENSITY

The high correlation which exists between changes in cosmic-ray intensity at any two stations definitely establishes the fact that the major changes in cosmic-ray intensity, after eliminating the 12-month wave, are world-wide. The relative amplitude of the world-wide change at different stations derived from data extending over periods of more than a year is not significantly different from that derived from two short periods of intense magnetic storms. This suggests that the mechanism responsible for changes occurring over longer periods of time is similar to that which causes the decrease of intensity during certain individual magnetic storms.

The world-wide character of changes in monthly means of cosmic-ray intensity, after removing the 12-month wave, is summarized in Fig. 15. If the hypothesis that the decrease in intensity during certain magnetic storms is due to alteration of trajectories of cosmic-ray particles resulting from the magnetic field of an equatorial ring-current, or its equivalent, is correct, then the more or less continued world-wide changes in intensity evident in Fig. 15 would imply the continued existence of such a current-system with changing strength or radius. Störmer[18] concluded that such a current-system was necessary to account for the large diameter of the auroral zones.

For comparison, the monthly means of magnetic horizontal intensity are indicated in Fig. 15. The average rate of decrease, derived from fifteen years of data, of the horizontal magnetic intensity at this station is indicated by the slope of the straight line. Any correspondence between the trend in this line and that in cosmic-ray intensity must therefore be regarded as accidental. However, the correspondence between departures of monthly means in horizontal intensity from this line and those in cosmic-ray intensity is doubtless significant. That the correspondence in some instances is more marked than in others is not surprising in view of the fact[4] that some magnetic storms are not accompanied by significant changes in cosmic-ray intensity.

Since the period of minimum values for the departures in cosmic-ray intensity in Fig. 15 agrees roughly with that of maximum magnetic activity, and since we have also indicated[4] the existence of a 27-day wave, probably quasi-persistent,[12] in cosmic-ray intensity, it would not be unexpected to find, when adequate data are available, the 11-year cycle of sunspot-activity reflected in cosmic-ray intensity.

From values for the relative magnitudes of the world-wide effect in cosmic-ray intensity given in Table I, the effect at the Hafelekar, where the instrument had somewhat greater shielding than at the other stations, is about 40 percent greater than at Cheltenham. This indicates that the intensity of the cosmic-ray component which is affected by the mechanism causing the world-wide changes increases much faster with altitude than does the total intensity. This is consistent with the fact that Clay and Bruins[19] found, in an instrument shielded by 110 cm of iron, no measurable change in cosmic-ray intensity during the magnetic storm which began April 24, 1937. They were thus enabled to set an upper limit to the energy of particles affected by the field of the magnetic storm.

Table I also indicates for high altitude stations a rapid increase in the magnitude of the world-wide effect between the equator and geomagnetic latitude 30° north. While the value for the Hafelekar indicates no further increase up to latitude 47° north, the additional shielding may have somewhat reduced the ratio in Table I for that station. Since no data are available for a sea-level station at the equator, it is not possible to determine the latitude variation in the world-wide effect at sea level. However, the agreement between Cheltenham and Christchurch indicates that the variation with latitude is probably symmetrical about the equator. Whether the world-wide changes in cosmic-ray intensity are due to the magnetic field of an equatorial ring-current system, or its equivalent, can only be answered when the difficult problem of finding the trajectories of cosmic-ray particles in such a field superposed on that of the earth's doublet has been solved.

The world-wide effects indicated in Fig. 15 imply that special precautions are necessary to obtain accurate determinations of the longitude effect. Also, if the magnitude of the world-wide effect increases as rapidly with altitude and latitude as the results in Table I suggest, then the results for latitude variations at extreme altitudes may also be materially affected if the mechanism responsible for the world-wide effect operates somewhat continuously but with varying intensity. Although this suggests a possible explanation for the exclusion of low energy cosmic-ray particles from the top of the earth's atmosphere at high latitudes, it is open to the serious objection that it would also be expected to exclude the auroral particles.

Finally, it should be emphasized that the results of this investigation could not have been secured without remarkable stability in the several cosmic-ray meters by which the data were obtained. Thus the estimated probable error for each entry in the last column of Table I is roughly ± 0.03. Although an examination of barometric coefficients at the separate stations does not indicate the existence of appreciable instrumental differences, it may be desirable to obtain further

[18] C. Störmer, Terr. Mag. 35, 193–208 (1930).
[19] J. Clay and E. M. Bruins, Physica 5, 111–114 (1938).

confirmation of this through an intercomparison of instruments at some of the stations.

ACKNOWLEDGMENTS

We are indebted to the Committee on Coordination of Cosmic-Ray Investigations of the Carnegie Institution of Washington for providing the assistance of W. R. Maltby, without whose aid the reduction of the data here used would have been long delayed.

We are also under obligation to G. Hartnell and the staff of the Cheltenham Magnetic Observatory of the United States Coast and Geodetic Survey for operation of a cosmic-ray meter at that station, to J. W. Beagley of the Christchurch Magnetic Observatory for operation of a meter and reduction of records, to Professor J. Gallo, Director of the National Astronomical Observatory at Tacubaya, Mexico, for supervising the operation of the meter at Teoloyucan, and to the staff of the Huancayo Magnetic Observatory of the Department of Terrestrial Magnetism of the Carnegie Institution of Washington for operating the meter at that station. We are also especially indebted to Professor A. H. Compton and his assistants for making available to us the reduced data from Teoloyucan, and to the National Carbide Company, Incorporated, for providing all batteries for operating the Compton-Bennett meters.

Finally, it is a pleasure to acknowledge the encouragement received from Dr. J. A. Fleming, Director of the Department of Terrestrial Magnetism of the Carnegie Institution of Washington

JULY-OCTOBER, 1939   REVIEWS OF MODERN PHYSICS   VOLUME 11

World-Wide Changes in Cosmic-Ray Intensity

S. E. FORBUSH

Department of Terrestrial Magnetism, Carnegie Institution of Washington, Washington, D. C.

INTRODUCTION

IN a former investigation[1] variations in the average cosmic-ray intensity for intervals of one-third month, at each of four widely separated stations (see Table I) were shown to be well represented—over the entire period for which data were then available (minimum period 17 months)—as the sum of two components. One component consisted of a systematic 12-month seasonal wave, S, and the other of an irregular world-wide component, W.

The coefficient of correlation between the means of W, for intervals of one-third month, at any two of the stations was found to be about 0.90. This high correlation permitted a reliable determination of the ratio of W for each station to that for Huancayo. These ratios are given in Table I, in which are also indicated the amplitude and date of maximum of S originally obtained[1] for each station. (Procedure for separating S and W and for the derivation of the relative magnitudes of W are given in reference 1.)

WORLD-WIDE EFFECTS BASED ON RECENTLY EXTENDED DATA

The purpose of the present paper is to extend the determination of monthly means of W at each station to include all data now available. Fig. 1 indicates the variations in the observed monthly means of cosmic-ray intensity (after correction to constant barometric pressure) at each of the first four stations in Table I.

Figure 2 indicates the reliability of the original[1] determination of S and its constancy at Cheltenham and Christchurch. The wave in Fig. 2 is the original S for Christchurch minus that for Cheltenham. The points indicate, for each month, the difference of observed monthly means for Christchurch and for Cheltenham. The close approximation of the wave to the points indicates the excellent agreement between the monthly

means of W at these two stations (see Fig. 3). In Fig. 3, W for Huancayo, August, 1938, to March, 1939, was sometimes as much as two percent above W for the other stations. This discrepancy probably arises since, during this interval, the electrometer and insulators of the cosmic-ray meter at Huancayo were not kept properly dried, as they were at all other times. Fig. 3 shows that by May, 1939, when the original practice of renewing driers weekly—as is done regularly at Cheltenham and at Christchurch—was resumed, W for Huancayo was again in fair agreement with W for Cheltenham. The discrepancy at Huancayo is further indicated in Fig. 4, in which the wave is the original S for Christchurch. The circles result from deducting the observed monthly means (shown in Fig. 1) at Huancayo from those at Christchurch—the procedure used originally to determine S for Christchurch. These points are well fitted by the wave except from July to December, 1938—the interval during improper drying of electrometer and insulators at Huancayo. The open squares in Fig. 4 were obtained by subtracting from the observed monthly means at Christchurch the monthly means of W for Cheltenham, obtained by deducting S from the

TABLE I. *Location and elevation of cosmic-ray stations, magnitude of world-wide component relative to Huancayo, and amplitude and time of maximum for seasonal wave for each.*

STATION	LATI-TUDE	LONGI-TUDE	GEO-MAGNETIC LATITUDE	ELEVA-TION ABOVE SEA LEVEL	WORLD-WIDE EFFECT RELA-TIVE TO HUAN-CAYO	SEASONAL WAVE AMPLI-TUDE[a]	SEASONAL WAVE DATE OF MAXIMUM
Cheltenham, United States	38.7°N	76.8°W	50.1°N	72 m	1.11	1.6%	January 19
Teoloyucan, Mexico	19.2°N	99.2°W	29.7°N	2285	1.58	1.0	January 24
Huancayo, Peru	12 0°S	75.3°W	0.6°S	3350	1.00	0.0	——
Christchurch, New Zealand	43.5°S	172.6°E	48.0°S	8	1.05	0.8	July 28
Hafelekar, Germany	47.3°N	11.3°E	48.4°N	2300	1.59[b]	1.9	January 15

[a] In percent of total cosmic-ray intensity.
[b] From world-wide effects in two different magnetic storm· only.

[1] S. E. Forbush, Phys. Rev. 54, 975–988 (1938).

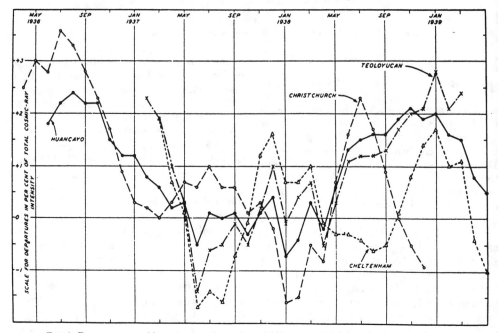

FIG. 1. Departures monthly means cosmic-ray intensity from mean for year beginning April, 1937.

observed monthly means at Cheltenham. The open squares closely fit the original S for Christchurch even in the interval when the circles depart from the wave.

The original S for Teoloyucan was derived from data for April, 1937, to April, 1938, and was obtained by deducting the monthly means of $W \times 1.58$ (see Table I) for Huancayo from the observed monthly means indicated in Fig. 1 for Teoloyucan.

In Fig. 5 is shown the result of deducting separately from the observed monthly means at Teoloyucan the monthly means of $W \times 1.42$ for Cheltenham and of $W \times 1.42$ for Christchurch (from Table I; 1.42 is taken for the ratio of W at Teoloyucan to W at Cheltenham or Christchurch). To fit satisfactorily the two sets of points in Fig. 5 it was necessary to assume, in addition to the original S for Teoloyucan, a linear increase with time in the observed monthly means. Such increase is quite artificial and is possibly an indication of inadequate insulation similar to that at Huancayo. For Teoloyucan W is now obtained by deducting the ordinates of

the smooth curve in Fig. 5 from the observed monthly means in Fig. 1. In Fig. 3 W for Teolo-

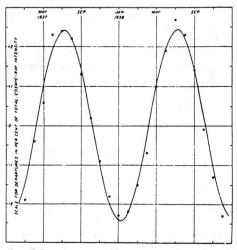

FIG. 2. Departures monthly means cosmic-ray intensity from mean for year beginning April, 1937, for Christchurch minus corresponding departures for Cheltenham; indicated wave derived from difference: adopted seasonal wave for Christchurch minus adopted seasonal wave for Cheltenham

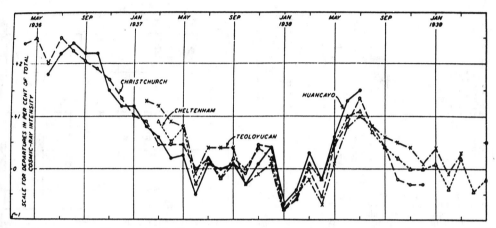

FIG. 3. Departures monthly means world-wide component cosmic-ray intensity from mean for year beginning April, 1937, at four stations referred to Huancayo.

yucan has been reduced to Huancayo (for comparison) using the factor 1/1.58 (see Table I).

It should be emphasized that excellent agreement in W for different stations, over long intervals of time, can be expected only in the complete absence of instrumental changes. If the indicated discrepancies result from such changes then the world-wide effect provides a reliable means of checking the stability of the different meters. Whether this is actually the case can doubtless be determined from further data.

In any case, these discrepancies cannot materially alter the high correlation[1] between changes in the means of W for each one-third month, of which one example is shown in Fig. 6.

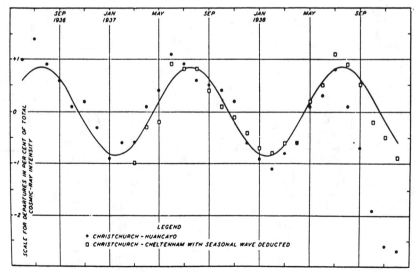

FIG. 4. Departures monthly means cosmic-ray intensity from mean for year beginning April, 1937, for Christchurch minus corresponding departures for Huancayo, and for Christchurch minus corresponding departures for Cheltenham with seasonal wave deducted; adopted seasonal wave for Christchurch indicated.

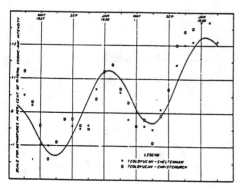

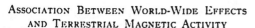

FIG. 5. Departures monthly means cosmic-ray intensity from mean for year beginning April, 1937, for Teoloyucan after deducting corresponding departures, world-wide component derived separately from data for Cheltenham and for Christchurch; adopted seasonal wave plus linear change for Teoloyucan indicated.

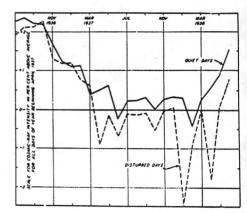

FIG. 7. Cosmic-ray intensity at Huancayo, Peru averaged for the five international magnetically quiet days and for the five international magnetically disturbed days of each month, July, 1936 to June, 1938.

ASSOCIATION BETWEEN WORLD-WIDE EFFECTS AND TERRESTRIAL MAGNETIC ACTIVITY

The largest world-wide effects in cosmic-ray intensity have occurred,[2] during magnetic storms, as decreases simultaneous with decreases in magnetic horizontal intensity at the equator.

[2] S. E. Forbush, Terr. Mag. **43**, 203–218 (1938).

While some magnetic storms[2] have occurred without detectable changes in cosmic-ray intensity there is a definite tendency for lower cosmic-ray intensity to occur on days of greatest magnetic disturbance. Designating each month's average value of cosmic-ray intensity for the five international magnetically disturbed days

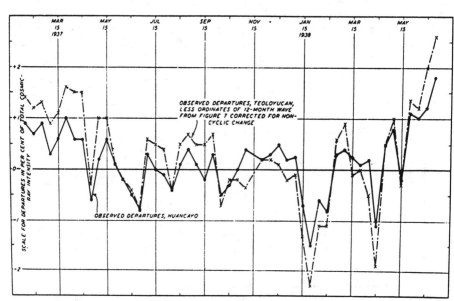

FIG. 6. Departures cosmic-ray intensity for each one-third month from mean for year beginning April, 1937, for Teoloyucan after deducting 12-month wave, and for Huancayo. (Fig. 7 in legend applies to Fig. 7 of reference 1.)

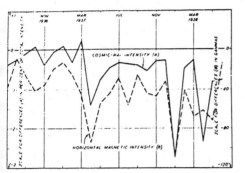

FIG. 8. Monthly difference, average for five international magnetically disturbed days minus average for five international magnetically quiet days, for cosmic-ray intensity and for horizontal magnetic intensity at Huancayo, Peru, July, 1936 to June, 1938.

by C_D and that for the five international magnetically quiet days by C_Q, then C_D is nearly always less than C_Q. In Fig. 7 C_D and C_Q are shown for July, 1936, to June, 1938, at Huancayo. In three cases C_D is slightly above C_Q. Fig. 8 indicates for each month the difference $(C_D - C_Q)$ and the corresponding difference $(H_D - H_Q)$ for magnetic horizontal intensity at Huancayo. Considerable correspondence is evident between the two curves of Fig. 8. This evidence, together with

that from magnetic-storm effects,[2] suggests that the world-wide changes in cosmic-ray intensity result from alteration of the trajectories of cosmic-ray particles in the external field superposed on that of the earth during magnetic disturbance. The main effect on cosmic-ray particles probably arises from the Störmer[3] equatorial ring-current. The curve for quiet days in Fig. 7 suggests that the Störmer ring may even be present on magnetically quiet days.

Dr. D. la Cour, Director of the Danish Meteorological Institute, which cooperates with the Carnegie Institution of Washington's Committee for Coordination of Cosmic-Ray Investigations by operating a Compton-Bennett cosmic-ray meter at its Godhavn Magnetic Observatory, has just advised that the cosmic-ray intensity was especially low at Godhavn during the magnetic storms which occurred near the end of April, 1939. During this period the cosmic-ray intensity at Cheltenham and Huancayo decreased about three percent. If the observation of a simultaneous decrease at Godhavn is confirmed, it will be of particular interest in view of the geomagnetic latitude (75° North) of that station.

[3] C. Störmer, Terr. Mag. 35, 193–208 (1930).

S. E. Forbush, *Department of Terrestrial Magnetism, Carnegie Institution of Washington:* M. S. Vallarta and O. Godart, in their paper on "Theory of Time Variations in Cosmic Rays," have calculated, for different latitudes, the theoretical amplitudes of the periodic 27-day variation in cosmic-ray intensity arising from a particular solar magnetic moment having a given inclination to the sun's axis of rotation. The results of our recent investigation* of the 27-day variation in cosmic-ray intensity at Huancayo, Peru (geomagnetic latitude 0.6° south) and at Christchurch, New Zealand (geomagnetic latitude 48.0° south) give no indication that the 27-day variation is significantly different at these two stations. The analysis also indicates that the 27-day variation is quasi-periodic in character, like the 27-day variation in magnetic activity.** The Vallarta-Godart theory of an inclined solar magnetic moment requires a periodic 27-day variation with amplitude several times larger at Christchurch than at Huancayo. The observed 27-day variations in cosmic-ray intensity at Huancayo and Christchurch are thus not adequately explained solely on the basis of an inclined solar magnetic moment.

Since world-wide changes in cosmic-ray intensity are associated with magnetic disturbance (see paper by writer on "World-Wide Changes in Cosmic-Ray Intensity"), which is known[2] to contain a 27-day quasi-periodic variation, a similar variation in cosmic-ray intensity is expected and probably arises from variations in the strength or size of the Störmer ring-current.

It has been shown† that the world-wide changes in cosmic-ray intensity at Christchurch are about 1.10 times greater than at Huancayo. From the calculations of Vallarta and Godart, any periodic 27-day variation arising from an inclined solar magnetic moment should have an amplitude which at Huancayo is negligible compared to that at Christchurch. Thus, if the amplitudes of the vectors in a harmonic dial[2] for the 27-day waves in cosmic-ray intensity at Huancayo are increased ten percent, and then subtracted from the vectors derived from corresponding intervals for the 27-day waves at Christchurch, the resulting vector-differences should be free of world-wide 27-day quasi-periodic effects. Statistical tests[1] on the harmonic dial for these vector differences should reveal any periodic 27-day wave at Christchurch. Using this procedure for 31 solar rotations we find the average amplitude of the periodic 27-day variation at Christchurch is not large enough to be regarded as statistically real. By this method a periodic 27-day variation at Christchurch with amplitude as great as 0.3 percent of the total cosmic-ray intensity should have been detected.

The absence of a periodic 27-day wave in cosmic-ray intensity at high latitudes would be interpreted solely, according to Vallarta and Godart, as indicating coincidence of the sun's magnetic moment and of its rotational axis. However, the search for a periodic 27-day variation in cosmic-ray intensity at high latitudes should be continued with more extensive data, for if such a variation is definitely established it would, on the basis of the Vallarta-Godart theory, confirm the much disputed existence of a permanent solar magnetic moment.

---

* To be published in Transactions, Washington Meeting, September, 1939, Int. Union Geod. Geophys., Ass. Terr. Mag. Elec.

** J. Bartels, Terr. Mag. 40, 1–60 (1935).

† S. E. Forbush, Phys. Rev. 54, 975–988 (1938).

Three Unusual Cosmic-Ray Increases Possibly Due to Charged Particles from the Sun

SCOTT E. FORBUSH
Department of Terrestrial Magnetism,
Carnegie Institution of Washington, Washington, D. C.
October 10, 1946

SEVERAL world-wide decreases in cosmic-ray intensity have been observed[1,2] during magnetic storms. These decreases have been ascribed[3] to ring currents, or their equivalents, required to account for the observed world-wide magnetic changes.

In about 10 years of continuous records of ionization in Compton-Bennett meters (shielded by 11-cm Pb) three obviously unusual increases in ionization have been noted. For Cheltenham, Maryland, geomagnetic latitude, $\Phi = 50°$ N, these are shown in Fig. 1, in which the bi-hourly means were corrected for barometric pressure. Curves very similar to the upper one in Fig. 1 obtain[2] simultaneously for Godhavn, Greenland, $\Phi = 78°$ N; and for Christchurch, New Zealand, $\Phi = 48°$ S. Except for the absence of significant increases on February 28, 1942; March 7, 1942; and July 25, 1946, the curves for Huancayo, Peru, $\Phi = 1°$ S, are otherwise quite similar to those for Cheltenham.

Figure 1 indicates each of the three unusual increases in cosmic-ray intensity began nearly simultaneously with a solar flare (bright chromospheric eruption) or radio fadeout (indicating a solar flare). Original records for February 28 and March 7, 1942, indicate increases in ionization which began within 0.3 hour after the commencement of the radio fadeout. The record for Cheltenham on July 25, 1946, first indicated an increase in ionization 1.0 hour after the reported commencement of the solar flare and fadeout. All three fadeouts were complete for about five hours and the flare was reportedly observed for three hours until clouds obscured it.

Magnetic records from several observatories indicate that the magnetic changes during these fadeouts or flares were probably caused by an augmentation of the diurnal variation.[4] Thus the magnetic changes do not appear ascribable to a ring current as first supposed.[3] Magnetic records on the night side of the Earth indicate no significant change in field at the time of the fadeouts yet the cosmic-ray records show the increases occurred simultaneously on the day and night side.

The known small diurnal variation[5] in cosmic-ray intensity excludes the possibility that these increases could have been caused by an augmentation in magnetic diurnal variation.

During a period of several hours on March 7, 1942, when the cosmic-ray ionization at Cheltenham increased to well above normal, that at Huancayo was sub-normal. Also the intensity at Cheltenham, Godhavn, and Christchurch was sub-normal before and after the unusual increase on March 7, 1942. Thus the mechanism responsible for the cosmic-ray increase on March 7, 1942, must have been distinct from that responsible for the otherwise sub-normal intensities during this period which have been ascribed[3] to ring currents.

These circumstances might suggest a change in the Sun's magnetic moment, arising perhaps from transient fields, as a possible cause[6] for the three unusual increases in cosmic-ray intensity. However, the effectiveness of such a mechanism should be independent of whether a particularly active area on the Sun was oriented toward the Earth. That each of the three fadeouts was followed, within about a day, by a magnetic storm, indicates that the three cosmic-ray increases did occur when a particularly active area on the Sun was preferentially oriented toward the Earth.

These considerations suggest the rather striking possibility that the three unusual increases in cosmic-ray intensity may have been caused by charged particles actually being emitted by the Sun with sufficient energy to reach the Earth at geomagnetic latitude 48° but not at the equator. It is recognized that particles of this energy should not escape from low latitudes on the Sun except in the absence of the much-disputed permanent solar magnetic field.

W. F. G. Swann[7] considered one mechanism for accelerating charged particles which involved changing magnetic fields of sunspots (near which observed flares occur) or stellar spots.

[1] E. B. Berry and V. F. Hess, Terr. Mag. 47, 251–256 (1942).
[2] S. E. Forbush and Isabelle Lange, Terr. Mag. 47, 331–334 (1942).
[3] S. E. Forbush, Terr. Mag. 43, 203–218 (1938).
[4] A. G. McNish, Terr. Mag. 42, 109–122 (1937).
[5] S. E. Forbush, Terr. Mag. 42, 1–16 (1937).
[6] M. S. Vallarta, Nature 139, 839 (1937).
[7] W. F. G. Swann, J. Frank. Inst. 215, 273–279 (1933).

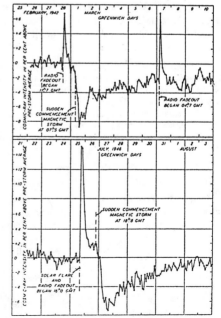

FIG. 1. Three unusual increases in cosmic-ray intensity at Cheltenham, Maryland, during solar flares and radio fadeouts.

REVIEWS OF MODERN PHYSICS   VOLUME 21, NUMBER 1   JANUARY, 1949

On the Mechanism of Sudden Increases of Cosmic Radiation Associated with Solar Flares

S. E. FORBUSH, P. S. GILL,* AND M. S. VALLARTA**

Department of Terrestrial Magnetism, Carnegie Institution of Washington, Washington, D. C.

A FEW sudden large increases of the intensity of cosmic radiation taking place soon after the appearance of a solar flare have been observed.[1] Most solar flares are not associated with such an increase. When the two phenomena appear together, the intensity of cosmic radiation begins to increase about an hour after the appearance of the solar flare, rises rapidly to its maximum value and then decreases rather slowly back to its normal value (Fig. 1). The whole process lasts for about a day. It appears only at high and intermediate geomagnetic latitudes. About a day after the appearance of the flare a magnetic-storm is felt all over the world, the intensity of cosmic radiation decreases and then slowly rises back to its normal value. We shall not be concerned in this paper with the magnetic storm effect.

1. In the following we shall attempt to give a preliminary theory of sudden increases of cosmic radiation associated with the appearance of solar flares. For the sake of clarity, we shall consider the following questions in turn:

(a) Is there any way that charged particles (protons and electrons) can be accelerated up to cosmic radiation energies (of the order of 10 Bev) by solar phenomena connected with the appearance of flares?

(b) If the answer to the previous question is in the affirmative, how do such particles escape from the sun?

(c) If such an escape is possible, can such particles reach the earth?

2. Question (a) has already been studied and answered affirmatively by W. F. G. Swann,[2] who solved, in relativistic mechanics, the problem of acquirement of cosmic-ray energies by charged particles, through the agency of variable mag-

netic fields associated with sunspots. It is well known that flares nearly always appear at the same location as sunspots. Swann assumes that the magnetic field of a sunspot varies linearly with time and inversely as the distance from the spot center. According to Cowling,[3] the field does not vary linearly with time except over an interval small compared with the time required for the field to reach its full value. Further, at sufficiently large distances, the field of a pair of spots of opposite polarities is that of a dipole.

The orbit of a charged particle being accelerated in the variable magnetic field of a sunspot is unstable both in the radial and the axial directions and pulls away from the plane and axis of the spot in a few seconds.

On the basis of Swann's results, and making the modifications suggested above, we compute that the maximum energy acquired by protons in the variable magnetic field of the pair of sunspots, which accompanied the solar flare of July 25, 1946, was 7.3 Bev, although probably most of the protons accelerated had an energy not exceeding 6 Bev. This pair had the following characteristics: spot radius, 28,000 km; time required to reach maximum value of the magnetic field, 6 days (0.518×10^6 sec.); maximum field, 380 gauss.

3. Regarding the second question (b), we must bear in mind that, if there is a permanent magnetic field of the sun of dipole moment 10 gauss-cm³, such as would account for a cut-off of the energy spectrum of primary cosmic radiation at about 2 Bev and correspondingly would place the knee of the latitude effect at about 50 geomagnetic latitude; then, because of the forbidden region of Störmer, particles having energies as computed in paragraph (3) can never leave the surface of the sun except at very high solar latitudes and then only over a small range

* Tata Institute of Fundamental Research, Bombay, India.
** Comisión Impulsora y Coordinadora de la Investigación Científica é Instituto de Física, Universidad de México, México, D. F.
[1] S. E. Forbush, Phys. Rev. 70, 771 (1946); H. V. Neher and W. C. Roesch, Rev. Mod. Phys. 20, 350 (1948).
[2] W. F. G. Swann, Phys. Rev. 43, 217 (1933).

[3] T. G. Cowling, Monthly Notices, Roy. Astr. Soc. 106, 218 (1946).

of latitude.[4] The sunspot pair of July 25, 1946, appeared at latitude 23° N, so that no particles of energy 7.3 Bev or less could escape from the sun, unless special means of exit are available.

Such means are provided by the combined action of the permanent field of the sun and the transient field of the pair of sunspots. As has already been pointed out, the field of the pair of sunspots drills a tunnel through the forbidden region of Störmer through which particles can then escape.[5] The existence of this tunnel depends on the relative strength and orientation of the

permanent and the transient dipoles, and on the ratio of their field strengths as a function of distance. It is assumed that the sun's permanent dipole coincides with the axis of rotation. Leaving out for the present the details of the calculation, it can be shown that the tunnel exists over a distance from the pair of sunspots such that the ratio of the permanent and transient magnetic field strengths is not constant. This condition seems to be necessary but not sufficient. Thus it is only seldom that the conditions for the existence of the tunnel are all satisfied, which explains

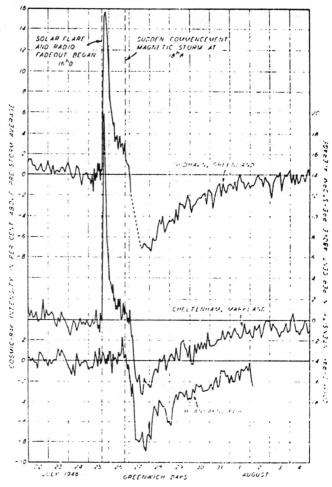

FIG. 1. Sudden increase of cosmic-ray intensity following solar flare of July 25, 1946.

[4] M. S. Vallarta, Nature 139, 839 (1937).
[5] M. S. Vallarta and O. Godart, Rev. Mod. Phys. 11, 180 (1939).

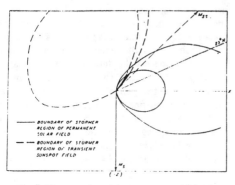

FIG. 2. The tunnel on July 25, 1946 at 17ʰ GMT
and at longitude 13° E.

why solar flares are seldom accompanied by an increase of cosmic-ray intensity.

The tunnel is usually very long and narrow. For the pair of sunspots of July 25, 1946, its length turns out to be 27 million km. Its aperture is between 2 and 3 degrees in latitude and not over 70 degrees in longitude. Conditions are illustrated in Figs. 2 and 3. The tunnel is shown in dark and the cross-hatched area corresponds to the region where the ratio of the variable to the steady field is constant and therefore the tunnel is closed.

A solar flare without a corresponding increase of cosmic radiation occurred on February 6, 1946. Figures 4 and 5 illustrate the conditions obtaining then. It is seen that there was no tunnel available anywhere. Therefore, the particles accelerated by the variable magnetic field

of the sunspots could not leave the surface of the sun.

The variation of permanent (H_s) and transient magnetic field (H_{ss}) as a function of distance from the sunspots at 16ʰ GMT on July 25, 1946, is shown in Fig. 6. It is seen that the end of the tunnel is just about, where $H_{ss}/H_s = $ const.

The data related to three flares we have studied are collected in Table I.

4. We now turn our attention to question (c). The tunnel through which particles must leave the sun is long, narrow, and extends over a limited range of solar longitude. Such particles can reach the earth only after traveling through the magnetic field of the sun, the earth, and the transient sunspot field. Therefore, they can arrive at the earth only along certain directions, and, by Liouville's theorem, with an intensity equal to what they had when they left the sun. During solar flares accompanied by an increase of cosmic radiation, one would therefore expect important departures from isotropy as observed at the earth. Such departures would occur only at certain points on the earth and only in certain directions.

Whether particles acquiring energy and leaving the sun by processes such as contemplated here will actually reach the earth cannot be decided without integrating the equations of motion. This laborious task will require the use of modern electronic computing machines. We intend to study this problem in the near future.

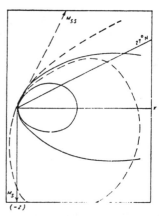

FIG. 4. No tunnel on February 6, 1946 at 16ʰ GMT
and at longitude 13° W.

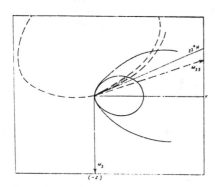

FIG. 3. No tunnel on July 25, 1946 at 17ʰ GMT
and at longitude 103° E.

174

TABLE I. Data on three flares.

No.	1	2	3
Date	Feb. 28, 1942	Feb. 6, 1946	July 25, 1946
Magnetic Moment gauss-cm³	7.8×10^{30}	3.5×10^{31}	3×10^{31}
Direction cosines:*			
α	0.01	0.11	−0.33
β	−0.90	0.97	0.90
γ	0.43	0.22	0.29
Spot diameter km	58000	52000	56000
Build-up time—days	~6	~6	~6
Maximum field gauss	4800	3800	3800
Solar latitude—degrees	7N	27N	23N
Solar longitude—degrees	7E	13W	13E
Time hours GMT	12	16	17
Energy Bev	8.2	6.8	7.3
Energy Störmers (protons)	0.39	0.35	0.37
Minimum geomagnetic latitude	22	34	30

* Note: Z axis along negative permanent dipole. X axis in central meridian.

5. The transit time for a particle of energy about 10 Bev traveling from the sun to the earth in the combined magnetic fields of the sun, sunspots, and earth would be of the order of magnitude of one hour, in agreement with observation.

The tunnel would shut off gradually as the relative magnitude and orientation of the permanent and transient dipoles changes. It seldom would stay open for longer than one day.

Particles can acquire energy so long as the sunspot field is changing with time. The duration of particle emission would in general be determined by the combination of two factors: time during which the sunspot field is changing, and time during which the tunnel stays open. It appears that seldom would it be greater than one day. These conditions are well illustrated by the solar flare of February 25, 1942. Unfortunately, the

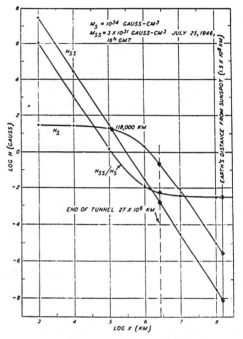

FIG. 6. The variation of permanent and transient fields as a function of distance from the sunspot.

orientation of the sunspot dipole could not be determined with good accuracy. However, from the information available to us we infer that a very narrow tunnel opened up for a short time late on February 27 or early on February 28,

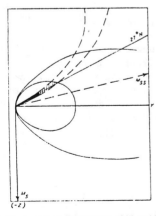

FIG. 5. No tunnel on February 6, 1946 at 16ʰ GMT and at longitude 77° E.

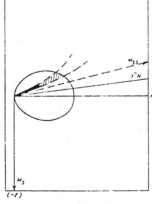

FIG. 7. No tunnel on February 25, 1942 at 18ʰ GMT and at longitude 43° E.

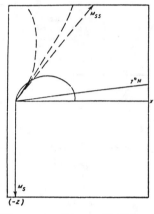

FIG. 8. The tunnel on February 27, 1942 at 21ʰ GMT and at longitude 15° E.

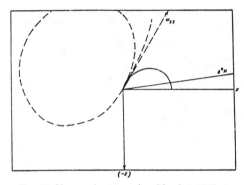

FIG. 10. Narrow short tunnel on March 1, 1942 at 22ʰ GMT and at longitude 11° W.

shut off late on February 28 and opened up again for a brief time on March 1 (Figs. 7 to 10). There was an increase of cosmic radiation on February 28 and again possibly a smaller one on March 1.

The energies imparted to particles by variable magnetic sunspot fields are in general of the order of magnitude of 0.3 to 0.4 Störmer. Hence, such particles can reach the earth only at intermediate and high latitudes. The sudden increases

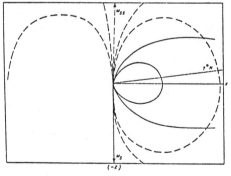

FIG. 9. No tunnel on February 28, 1942 at 12ʰ GMT and at longitude 7° E.

of intensity following solar flares should not be detected, in general, at equatorial latitudes in agreement with observation.

Particles of either positive or negative sign may be accelerated by variable magnetic fields of sunspots.

While it seems certain that particles of cosmic-ray energy can be occasionally ejected from the sun, it is not proved that all primary cosmic rays are generated by similar processes taking place in the stars.

The fact that very few flares are accompanied by sudden increases of cosmic radiation seems to provide a sound argument for the existence of a permanent solar magnetic field. Otherwise *all* flares would be expected to be associated with such increases. The role of the flare appears to be that of providing a supply of charged particles which are then accelerated by the variable field of the sunspots.

Finally, it should be emphasized again that the mechanism responsible for changes in the intensity of cosmic radiation during magnetic storms is clearly different from that which is connected with sudden increases of intensity which follows the occurrence of some solar flares and which is studied in this paper.[6]

[6] S. E. Forbush, see reference 1.

176

PHYSICAL REVIEW VOLUME 79, NUMBER 3 AUGUST 1, 1950

The Extraordinary Increase of Cosmic-Ray Intensity on November 19, 1949

Scott E. Forbush

Department of Terrestrial Magnetism, Carnegie Institution of Washington, Washington, D. C.

AND

Thomas B. Stinchcomb and Marcel Schein

Department of Physics, University of Chicago, Chicago, Illinois

(Received April 20, 1950)

Four sudden increases in cosmic-ray intensity associated with solar flares or chromospheric eruptions have so far been observed during more than a decade of continuous registration of cosmic-ray intensity. The last and largest of these increases occurred on November 19, 1949, when such an effect was recorded for the first time at a mountain station at Climax, Colorado. Here the intensity increased to about 200 percent above normal in half an hour. At the sea-level station at Cheltenham, Maryland, the increase was about 43 percent. No increase occurred at the equator. From the increase in the effect with altitude and latitude, it is concluded that the increase was due to the nucleonic component produced by relatively low energy primary charged particles probably accelerated by some solar mechanism.

THE sudden increase in cosmic-ray intensity which began at 10h45m GMT, November 19, 1949, was the largest yet recorded during more than a decade of continuous registration of cosmic-ray ionization at several stations (see Table I). Only three other unusual increases had been previously recorded.[1] These occurred on February 28, 1942, March 7, 1942, and July 25, 1946. All were registered with Compton-Bennett ionization chambers completely shielded with 12-cm Pb. Three of the increases began during intense chromospheric eruptions or solar flares.[1] While no solar flare was actually observed during the increase of cosmic-ray intensity on March 7, 1942, a radio fadeout occurred very near the time the increase in cosmic-ray intensity began. The fadeout, which occurred only on the day-

light side of the earth, quite definitely indicates the existence of a solar flare.

The terrestrial-magnetic effect of such solar flares is

TABLE I. Location and elevation of Compton-Bennett cosmic-ray meters.[a]

Station	Latitude (degrees)	Longitude (degrees)	Geomagnetic latitude (degrees)	Elevation (meters)	Operation began
Godhavn, Greenland	69.2 N	53.5 W	79.9 N	9	October, 1938
Cheltenham, Maryland	38.7 N	76.8 W	50.1 N	72	March, 1935
Climax, Colorado	39.4 N	106.2 W	48.1 N	3500	
Teoloyucan, Mexico	19.2 N	99.2 W	29.7 N	2285	February, 1937[b]
Huancayo, Peru	12.0 S	75.3 W	0.6 S	3350	June, 1936
Christchurch, New Zealand	43.5 S	172.6 E	48.6 S	8	June, 1936

[a] 12-cm Pb shield around all meters, except that the meter at Climax had, in addition, a slab of Fe 16.5 cm thick directly over the meter.
[b] Not operating after 1945.

[1] S. E. Forbush, Phys. Rev. **70**, 771 (1946).

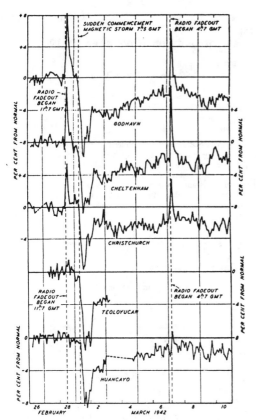

FIG. 1. Increases of cosmic-ray intensity, February 28 and March 7, 1942.

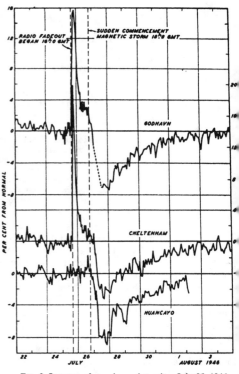

FIG. 2. Increase of cosmic-ray intensity, July 25, 1946.

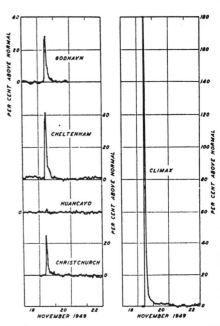

FIG. 3. Increase of cosmic-ray intensity, November 19, 1949

an increase, on the daylight side of the earth, in the normal diurnal variation in the earth's field.[2] The known small diurnal variation[3] in cosmic-ray intensity excludes the possibility that the increases were due to changes in the earth's external magnetic field resulting from an augmentation of the magnetic diurnal variation. The evidence thus indicates[1] that the four increases in cosmic-ray intensity were probably due to charged particles accelerated by some mechanism[4] on the sun. Unless the particles responsible for the increases were charged, it would be difficult to explain either the simultaneous occurrence of the increases on both the daylight and dark hemispheres or the absence of the increases at the equator.

The sudden increases in cosmic-ray intensity on February 28 and March 7, 1942, are shown in Fig. 1, in which the curves are drawn through the bihourly means, after correcting these to constant barometric pressure. It is evident that neither increase occurred at

[2] A. G. McNish, Terr. Mag. 42, 109 (1937).
[3] S. E. Forbush, Terr. Mag. 42, 1 (1937).
[4] Forbush, Gill, and Vallarta, Rev. Mod. Phys. 21, 44 (1949).

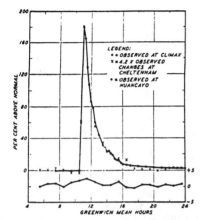

FIG. 4. Cosmic-ray intensity, November 19, 1949.

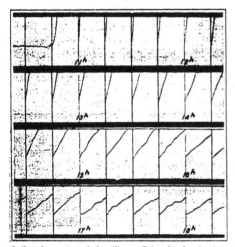

FIG. 5. Cosmic-ray records for Climax, Colorado, showing increase beginning at 10 hr. 45 min. GMT, November 19, 1949.

Huancayo and that the increase on February 28 did not occur at Teoloyucan. The decrease in cosmic-ray intensity during the magnetic storm following the sudden commencement on March 1 is evident at all the stations. The sudden increase in cosmic-ray intensity on July 25, 1946, is shown in Fig. 2. No data for this period were available from the other stations in Table I. Again, no increase occurred at Huancayo although the decrease during the subsequent magnetic storm is evident there.

The sudden increase in cosmic-ray intensity at the time of the solar flare on November 19, 1949, is shown in Fig. 3. This is the first instance when an increase in cosmic-ray intensity accompanying a solar flare has been recorded at a mountain station and at sea-level stations. The increase, in percent of the total cosmic-ray ionization, is obviously very much greater at Climax than at Cheltenham. In fact, if the ordinates on the

curve showing the increase at Cheltenham are multiplied by 4.2, the resulting points, shown in Fig. 4, lie on the curve for Climax. It may also be noted in Fig. 3 that the increase on November 19 was not followed by a decrease in cosmic-ray intensity during the magnetic storm which began about 18^h GMT on November 19.

The rapid increase on November 19, 1949, is evident beginning at 10^{h}45^m GMT in Fig. 5, which is a reproduction of part of the cosmic-ray record for Climax on that date. The electrometer is grounded every 15 min. The large increase in intensity caused the image of the electrometer needle to go off scale during the first few minutes in several of the 15-min. intervals. Departures of cosmic-ray intensity from the balance-value (electrometer trace horizontal) in percent of the total intensity are obtained by multiplying the slope by 9.8.

Figures 6 and 7 permit more detailed comparison of the solar-flare effects observed at several stations than

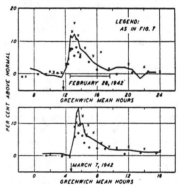

FIG. 6. Increases in cosmic-ray intensity during two solar flares, for different stations.

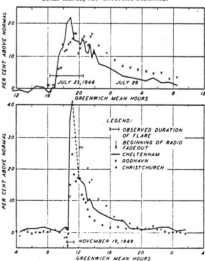

FIG. 7. Increases in cosmic-ray intensity during two solar flares, for different stations.

is possible in Figs. 1, 2, and 3. The observed durations of the solar flares are also indicated as well as the times of commencement of the radio fadeouts.

In addition to complete shielding by 12-cm Pb, the meter at Climax was under a rectangular iron shield 4 ft. long, 1 ft. wide, and 16.5 cm thick. The absorption mean free path for nucleons of medium energy in iron is approximately 240 g cm^{-2}. Taking the dimensions of the shield into account and figuring the zenith angle distribution for a radiation exponentially absorbed with an absorption coefficient of about 145 g cm^{-2}, it is estimated that the increase at Climax on November 19, 1949, would have been 15 percent greater without the iron shield. Thus it is estimated that the maximum of the increase on November 19 at Climax would have been about 207 percent without the iron shield, instead of the uncorrected 180 percent as shown in Figs. 3 and 4. This correction also results in a factor of 4.8 for the ratio of the percentage increase at Climax on November 19 relative to that at Cheltenham, instead of 4.2 as is indicated in Fig. 4.

Since the total ionization at Climax (under 12-cm Pb) is about 2.5 times that at Cheltenham, and since the percentage increase on November 19, 1949, was about 4.8 times that at Cheltenham, the actual magnitude of the increase on that date at Climax was about 12 times greater than at Cheltenham. Since the difference in the atmospheric layer is equivalent to 340 g cm^{-2}, the radiation responsible for the increase during the flare has an absorption coefficient of about 137 g cm^{-2}. This is just about the rate at which the nucleonic component, responsible for star production, increases with altitude.[5] The increase in total ionization under 12-cm Pb by a factor of 2.5 from Cheltenham to Climax is mainly due to mesons. It is thus evident that the magnitude of the

[5] J. J. Lord and Marcel Schein, Phys. Rev. 75, 1956 (1949).

flare effect increases too rapidly with altitude to be ascribed to ordinary mesons. The latitude effect in chambers under 12-cm Pb, due principally to mesons is small, whereas the flare effect exhibits a strong dependence on latitude (being zero at the equator). This also indicates that ordinary mesons contribute negligibly to the flare effect.

The results of Conversi[6] and those of Simpson[7] on the latitude variation of the proton and neutron intensity suggested that the cross section for nucleon production relative to that for meson production decreases rapidly with increasing energy of primary particles. This is in accord with the conclusion that the increase in intensity during the solar flare of November 19, 1949, was due principally to the nucleonic component or to local radiations originating from it, and not to ordinary mesons.

At Climax, under 12-cm Pb, probably not more than about 10 percent of the total ionization is normally due to local radiation originating from the nucleonic component. If we assume that this radiation is produced entirely by particles in the same band of energy as those responsible for the increase of 207 percent in ionization on November 19, 1949, then the number of primary particles, reaching there per unit time, in that band of energy, must have increased to at least 20 times the normal value.

Three sets of triple coincidence counters and one set of fourfold coincidence counters, located above the meter at Climax and arranged to record air showers were in continuous operation during the period of the increase in ionization on November 19, 1949. There was no evidence of any significant increase in the rate of air showers during this period.

[6] Marcello Conversi, Phys. Rev. 76, 444 (1949).
[7] J. A. Simpson, Jr., Phys. Rev. 76, 569 (1949).

HYSICAL REVIEW   VOLUME 87, NUMBER 5   SEPTEMBER 1, 1952

Correlation of Cosmic-Ray Ionization Measurements at High Altitudes, at Sea Level, and Neutron Intensities at Mountain Tops

H. V. NEHER*

California Institute of Technology, Pasadena, California

AND

S. E. FORBUSH

Carnegie Institution of Washington, Washington, D. C.

(Received May 19, 1952)

Although fluctuations in cosmic rays have been measured at sea level and high altitudes for a number of years, no serious attempt seems to have been made to correlate the two. As a result of a rather long series of balloon flights in the summer of 1951 it now becomes possible to correlate ionization measurements at high and low altitudes. Recent neutron intensity measurements during the same period by Simpson *et al.* permit a further comparison with this component. There seems to be a good correlation, during this period of observation, between the fluctuations as measured in (1) the ionization at 70,000 ft over North Dakota, (2) the meson component at Cheltenham, Maryland, (3) the meson component at the geomagnetic equator, and (4) the neutron component at mountain tops in Colorado and New Mexico.

IN a recent issue of this journal, Simpson *et al.*[1] reported a correlation of the fluctuations of neutron intensity, measured at three widely separated land stations. It is the purpose of this note to point out that correlations also exist between cosmic-ray intensity measurements at both high and low altitudes and the above neutron measurements.

In the summer of 1951 while these neutron measurements were being made, a group from the California Institute of Technology was making a series of balloon flights at Bismarck, North Dakota using ionization chambers. Also during this period, continuous measurements of cosmic-ray intensities were being made at Huancayo, Peru and at Cheltenham, Maryland by the Carnegie Institution of Washington. Plotted in Fig. 1 are the four sets of data. The neutron data were scaled from the curve given by Simpson *et al.*[1] for Climax, Colorado and are for the first 12-hour period of the respective days on which balloon flights were made. In three cases two flights were made on the same day at Bismarck, on July 27, August 6, and August 14. In such cases the trend of the neutron data at the particular times was taken, and what appeared to be a reasonable value was assigned. For the corresponding data at Cheltenham, the hourly values, corrected for bursts and barometric changes, for the 6 hours just before the balloon flights reached their maximum height and the 6 hours just afterward, are averaged. For the case of the data from Huancayo the daily mean values are plotted. Thus, the first point on July 18, is the mean for that day on 75° west meridian mean time. The points, for the respective times under consideration, are arbitrarily connected by straight lines in the figure. The balloon data at Bismarck are taken from the individual curves and are at 50 g cm⁻² pressure or about 70,000 ft.

In Fig. 2 are shown the data for Cheltenham and Huancayo compared with the neutron measurements of Simpson *et al.*, for the three months of July, August, and September of 1951. The ionization data are daily averages, while the neutron measurements are for 12-hour periods.

These data show that, during this period of observation, there was a very good correlation in time of the fluctuations measured (1) in the total ionization due to cosmic rays at 70,000 ft elevation over North Dakota; (2) in the neutron intensity at 11,000 ft elevation at Climax, Colarado; (3) in the sea level ionization, of particles that can penetrate 12 cm of lead, at Cheltenham, Maryland; and (4) in the ionization at 11,000 ft elevation at Huancayo, Peru as measured by an instrument similar to the one at Cheltenham.

The approximate ratios of the relative fluctuations

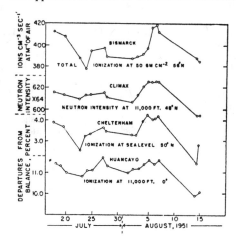

FIG. 1. Correlation between ionization measurements at high altitudes (about 70,000 ft), low altitudes, and neutron measurements at 11,000 ft.

* Assisted by the joint program of the ONR and AEC.
[1] Simpson, Fonger, and Wilcox, Phys. Rev. 85, 366 (1952).

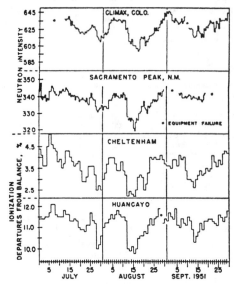

FIG. 2. Correlation between neutron measurements and ionization measurements at or near sea level.

as given in Fig. 1 are as follows: ionization at 70,000 ft to that at Cheltenham, 7:1; neutrons at 11,000 ft to ionization at Cheltenham, 3:1; ionization at Cheltenham to that at Huancayo, 1:1. A more detailed analysis[2] has shown that in the last case, the ratio between the fluctuations at Cheltenham and Huancayo is more nearly 1.1:1.

In addition to the above, two other pieces of experimental evidence are pertinent to the present discussion:

(1) A study of the differences in the balloon curves obtained at Bismarck during July and August, 1951 shows that there is a fairly wide distribution in the energy of the particles that fluctuate from day to day.

(2) In the summer of 1946 (data unpublished) an ionization chamber was sent up at Ft. Worth, Texas during the magnetic storm that followed the large flare of July 25 of that year. The percentage decrease in ionization compared with the normal value shown at 70,000 ft during this flight was approximately four

times the decrease found in an unshielded ionization chamber at Mt. Wilson, California during the same period.

The manner in which the radiation responsible for these fluctuations is absorbed in the atmosphere is consistent with the fact that the fluctuations are also present at Huancayo, Peru. Since these fluctuations are also present at Thule, Greenland,[3] it is evident that they are world-wide and hence represent real change in the total energy being brought into the earth by cosmic rays. The average energy of the particles must however, be somewhat less than that for the total cosmic-ray particles since the fluctuations in ionization increase with altitude, and are less pronounced at the equator.

These changes do not appear to be of the same type that have been measured during sudden increases in cosmic-rays, such as occurred on November 19, 1949 and on July 25, 1946. This conclusion is borne out by (a) the small changes in neutron intensity, relative to those measured with ionization chambers shown in Fig. 1, as compared with the much larger relative change measured[4] during the increase of November 19, 1949 (b) the fact that on no occasion has as appreciable increase in cosmic rays been measured at Huancayo when these increases occurred at intermediate and higher latitudes, while the changes shown in Fig. 1 do occur also at the equator. On the contrary, we believe that these fluctuations are similar to those that have been correlated with magnetic disturbances[3,5] both of the long period type and the short period type such as that which occurred one day following the solar flare of July 25, 1946.

In conclusion we wish to thank the U. S. Weather Bureau and particularly Mr. F. J. Bavendick for their cooperation at Bismarck. We also wish to thank Dr. Vincent Peterson, Mr. Edward Stern, and Mr. Alan Johnston for their help in preparing the instruments in carrying out the flights, and in reducing the data.

[2] S. E. Forbush, Phys. Rev. 54, 975 (1938).

[3] Neher, Peterson, and Stern, Phys. Rev. 85, 772 (1952).
[4] On this occasion the ratio of neutron increase to the increase in ionizing particles at Manchester, England was 60:1. See J. Adams, Phil. Mag. 41, 503 (1950).
[5] S. E. Forbush, Internatl. Assoc. of Terrest. Mag. and Elec. Washington Assembly (September, 1939); also, S. E. Forbush and I. Lange, Phys. Rev. 76, 1641 (1949).

WORLD-WIDE COSMIC-RAY VARIATIONS, 1937-1952

By Scott E. Forbush

Department of Terrestrial Magnetism, Carnegie Institution of Washington, Washington, D.C.

(Received September 28, 1954)

ABSTRACT

Annual means from continuous registration of cosmic-ray ionization at four stations from 1937 to 1952 show a variation of nearly four per cent, which is similar at all stations and which is negatively correlated with sunspot numbers. This variation in cosmic-ray intensity is quite similar for the annual means of all days, international magnetic quiet days, and international magnetic disturbed days, which indicates that it is not due to transient decreases accompanying some magnetic storms. Although the cosmic-ray intensity at some stations is affected by meteorological conditions, it is shown that on the average the cosmic-ray changes observed at Huancayo agree well with those at other stations. From an analysis of the variability of daily means at Huancayo and a sample comparison with Simpson's neutron data, it is concluded that the cosmic-ray ionization at Huancayo is very little affected by meteorological effects. Through a comparison with Neher's balloon observations, evidence is provided to indicate the reliability of cosmic-ray results at Huancayo over long periods of time. The relation between cosmic-ray decreases and some measures of geomagnetic activity is indicated, and it is shown that the major transient decreases in cosmic-ray intensity occur during magnetic disturbance. Graphs are included which depict the daily means of cosmic-ray intensity at Huancayo for all available data, 1937-1953.

Introduction

The uninterrupted cooperation of several organizations with the Department of Terrestrial Magnetism, Carnegie Institution of Washington, has resulted in continuous operation of Compton-Bennett cosmic-ray meters for the four stations listed in Table 1 from the dates indicated therein. The cooperating organizations are as follows: *Godhavn*—The Danish Meteorological Institute and the staff of its Godhavn Magnetic Observatory; *Cheltenham*—The United States Coast and Geodetic Survey and the staff of its Cheltenham Magnetic Observatory; *Huancayo* —The Government of Peru and the staff of its Instituto Geofísico de Huancayo; and *Christchurch*—The Department of Scientific and Industrial Research of New Zealand and the staff of its Christchurch Magnetic Observatory.

Additional details of the installations and of corrections for bursts and baro-

TABLE 1—*Location of Compton-Bennett meters*

Station	Lat.	Long.	Geomag. lat.	Elev.	Bar. coeff.	Continuous registration began
	°	°	°	meters	% $(cm\ Hg)^{-1}$	
Godhavn, Greenland	69.2 N	53.5 W	79.9 N	9	−0.18	Oct. 1938
Cheltenham, Maryland	38.7 N	76.8 W	50.1 N	72	−0.18	March 1937
Huancayo, Peru	12.0 S	75.3 W	0.6 S	3350	−0.30	June 1936
Christchurch, N.Z.	43.5 S	172.6 E	48.6 S	8	−0.18	April 1936

metric pressure are given in the publication [see 1 of "References" at end of paper]: "Cosmic-ray results from Huancayo Observatory, Peru, June 1936-December 1946, including summaries from observatories at Cheltenham, Christchurch, and Godhavn, through 1946." Tabulations for a similar publication extending through 1952 are now completed.

Seasonal wave

Early investigations [2] showed that, except for a seasonal wave, the major variations in cosmic-ray ionization were world-wide. Subsequently, four large [3, 4] increases of cosmic-ray intensity, associated with solar flares, were found at all stations except those at or near the equator. From the present long sequence of data, the average seasonal waves shown in Figure 1 were derived. No statistically

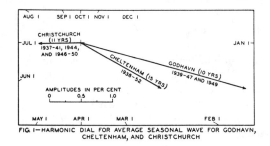

FIG. I—HARMONIC DIAL FOR AVERAGE SEASONAL WAVE FOR GODHAVN, CHELTENHAM, AND CHRISTCHURCH

significant seasonal wave was found for Huancayo. The seasonal waves are undoubtedly the consequence of seasonal variations in the vertical air-mass distribution, resulting in variations in the fraction of μ-mesons decaying before reaching the meters. These seasonal waves (corrected for non-cyclic change) have been deducted from all data used in this paper. This removes only the systematic seasonal variations. It does not remove variations arising from non-systematic, unpredictable, changes of vertical air-mass distribution, the magnitude of which for the different stations will be discussed later.

Sunspot-cycle variation in cosmic-ray intensity

When all available annual means of cosmic-ray ionization, corrected for bursts and barometric pressure, were examined, a large secular decrease was obvious in the results for Christchurch. Since there was no evidence in the results for Cheltenham of any significant secular change, other than the sunspot variation as shown in Figure 2A, no correction for drift was applied to the data for Cheltenham. By

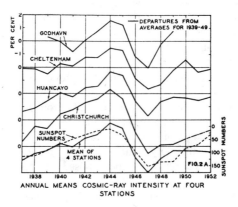

ANNUAL MEANS COSMIC-RAY INTENSITY AT FOUR
STATIONS

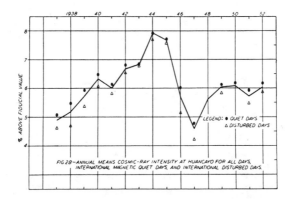

comparing results for the other stations with those for Cheltenham, the following linear changes were found: Christchurch, -1.40% yr^{-1}; Godhavn, -0.25% yr^{-1}; and Huancayo, $+0.40\%$ yr^{-1}. The annual means of Figure 2 have been corrected for the above linear changes, which are assumed to be instrumental and probably arise from decay of radioactive contamination in the main chamber or in the balance chamber of the meters. The agreement between the annual means of cosmic-ray intensity for the four stations, or their average, and that for annual mean sunspot numbers is evidence that the mechanism responsible for these changes in cosmic-ray intensity involves some phenomenon associated with solar activity.

It is known that some magnetic storms are accompanied by large decreases in cosmic-ray intensity, and it is shown later that most of the major decreases which occur during intervals of a few days are associated with magnetic storms or periods of magnetic disturbance. Thus, there arises the question of whether these decreases are mainly responsible for the variation of cosmic-ray intensity with sunspot numbers shown in Figure 2A. To answer this question, the variation of annual means of cosmic-ray intensity at Huancayo for all days (as used in Fig. 2A) is compared in Figure 2B with that for international magnetic quiet days and with that for international magnetic disturbed days. It is evident from Figure 2B that the variation of annual means for all days, which in Figure 2A was shown to follow the curve of sunspot numbers, is very little different from that for quiet days and not greatly different from that for disturbed days. Thus, the main features of the variation of cosmic-ray intensity with sunspot numbers persist for long periods (six months or more) and are not ascribable to transient decreases accompaning some magnetic storms.

Further evidence of effects that persist for long periods of time is indicated in Figure 16D, in which the curve (h) for daily means of cosmic-ray intensity at Huancayo shows a gradual increase of about 1.5% from January 1944 to September 1944 during a period in which there was no very large transient decrease in cosmic-ray intensity and no great magnetic storm. It thus appears that the transient decreases in cosmic-ray intensity which occur during some magnetic storms and magnetically disturbed periods are superimposed upon a variation with sunspot cycle.

Figure 3 shows the variation of the monthly means of cosmic-ray intensity

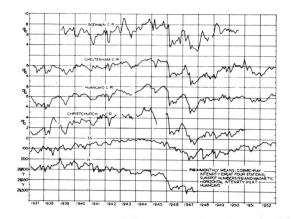

for four stations after removing the seasonal wave and linear trend. Also shown is the monthly mean horizontal magnetic component at Huancayo corrected for a linear estimate of secular change. It is evident that the horizontal intensity and cosmic-ray intensity were both markedly lower throughout 1946 and 1947 than in 1944. This fact suggests the possibility that the same mechanism may be responsible for both effects. In this connection, it should be mentioned that Vestine [5], from a long series of magnetic data from many observatories, found evidence for an 11-year variation in the horizontal component of the earth's field. In deriving the latitude distribution for the three geomagnetic components of the storm-time field, D_{st} (disturbed minus quiet days) he found [5] that the eastward geomagnetic component of D_{st} was zero, on the average. However, he points out that the yearly average of the east component for all days is not only not zero but varies during the sunspot cycle, indicating that the cause of this variation (and probably also that for horizontal intensity on all days) may be distinct from that for D_{st}. Thus, this unexplained variation in the earth's field is possibly connected with the sunspot variation in cosmic-ray intensity.

The variation of monthly means is further compared in Figure 4. To effect this comparison, the monthly means for Huancayo were categorized in six intervals of one per cent (3% to 9%). Monthly means for each of the other stations were averaged for each of these six categories. These group means are reasonably well fitted by the straight lines shown in Figure 4. These lines thus approximate the

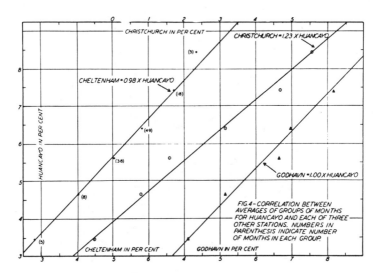

FIG 4 - CORRELATION BETWEEN AVERAGES OF GROUPS OF MONTHS FOR HUANCAYO AND EACH OF THREE OTHER STATIONS. NUMBERS IN PARENTHESIS INDICATE NUMBER OF MONTHS IN EACH GROUP.

regression lines obtained by assuming the monthly means at Huancayo are free of statistical errors, which are presumed present only in the means for the other stations. The factor of 1.23 for the ratio of changes at Christchurch to those at Huancayo is roughly 20% greater than that found earlier from a shorter series of data [2]. This factor is also greater than that derived in the following section. For Cheltenham and Godhavn, the factors shown on Figure 4 are more nearly consistent with those derived earlier [2] and with those derived in the following section.

Variation of daily means

Comparison of variations for average of ten selected 20-day intervals—From a plot of daily means for Huancayo (see Fig. 16), ten intervals of 20 days were selected with each interval exhibiting a variation similar to that shown for Huancayo in Figure 5. Figure 5 indicates the variation averaged for the same ten intervals for each of the other three stations. Figure 6 indicates the correlation between the averaged variation for Huancayo and that for each of the other three stations. The correlation coefficients and slopes of the regression lines are also indicated. The smaller of the two slopes results from the assumption of no statistical error in the means for Huancayo. Except for Christchurch, the factors are in fair agreement with those shown in Figure 4 and with those derived earlier [2, 6].

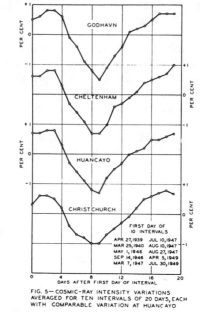

FIG. 5— COSMIC-RAY INTENSITY VARIATIONS
AVERAGED FOR TEN INTERVALS OF 20 DAYS, EACH
WITH COMPARABLE VARIATION AT HUANCAYO

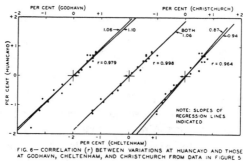

FIG. 6— CORRELATION (r) BETWEEN VARIATIONS AT HUANCAYO AND THOSE
AT GODHAVN, CHELTENHAM, AND CHRISTCHURCH FROM DATA IN FIGURE 5

Variability of daily means at Huancayo—Figure 7 indicates the standard deviation of daily means from monthly means at Huancayo for each year, 1937-1952. The curves show that standard deviations of departures from the monthly means are roughly four times smaller near sunspot minimum than near sunspot maximum. They are only slightly less when the five magnetically disturbed days of each month are excluded.

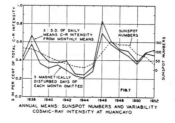

ANNUAL MEANS: SUNSPOT NUMBERS AND VARIABILITY
COSMIC-RAY INTENSITY AT HUANCAYO

For 1944, the standard deviation of daily means from monthly means is about 0.21% (excluding the five magnetically disturbed days). Since this figure includes the variability of the world-wide component, it is an upper limit for the combined effects of statistical fluctuations in the records and those from variations of μ-meson decay due to changes in vertical distribution of air-mass. It is thus evident that the latter effects are quite small at Huancayo, and that the daily means (relative to the mean of the month) at Huancayo are reliable to within at most 0.2% (that is, their s.d. $\leq$ 0.2%). From a previous investigation [2], it was found that the world-wide changes at Teoloyucan, Mexico, were about 1/0.63 times those at Huancayo. For 1937, daily means, with seasonal wave removed, were available from Teoloyucan for all months except January and November. These daily means for Teoloyucan were multiplied by 0.63 to reduce them to Huancayo. The difference between the daily mean at Huancayo and the reduced daily mean at Teoloyucan was found for each day of the ten months. The standard deviation of single differences about their average for the month was found to be 0.24% from pooling the ten samples of one month each. Assuming equal variance for statistical fluctuations at both stations, the standard deviation for the statistical fluctuations in single daily means is only 0.17%, which is slightly less than the figure of 0.2 derived from the fluctuations of daily means from the monthly means for Huancayo in 1944. The standard deviation of the ten monthly mean differences about their average is about three times greater than would be expected from the fluctuations in the differences in daily means. This may indicate some small systematic change which would arise if the seasonal variations at Teoloyucan deviated from a pure 12-month wave. It will be shown later that variations arising from non-world-wide changes are much greater at the other stations than at Huancayo and Teoloyucan and that the data from Huancayo and Teoloyucan provide more reliable measures of the world-wide component than do those from the other stations. The absence of any significant seasonal variation at Huancayo is further indication that the vertical distribution of air-mass there must vary little with season.

Comparison with neutron results—Figure 8 is a comparison of the variation of daily means for June 1951 from the Compton-Bennett meter at Huancayo and those published [7] by Simpson from neutron counters at Sacramento Peak, New Mexico. The standard deviation (s.d.) of the differences between daily means from the Compton-Bennett meter at Huancayo and those from the neutron counters (multiplied by 0.389) at Sacramento Peak is about 0.25%. The series is too short to determine whether there are systematic changes in background in either instrument involved. The occurrence of any such changes results in increasing the s.d. If the value of 0.17% is accepted for the s.d. of single daily means from the monthly average at Huancayo, then from the value of 0.25% for the s.d. of differences between the Huancayo daily means and those for neutrons in June 1951, the s.d. of daily means for the latter is found to be about 0.19%, in the reduced neutron units. Or, the s.d. of the neutron daily means would be $0.19/0.389 = 0.49\%$. Since this figure is many times greater than would be expected in view of Simpson's high neutron counting rate, it is evident that during the period of this comparison one of the two instruments was subject to variations, either real or instrumental, which did not affect the other. Since the Instituto Nacional de la Investigacion

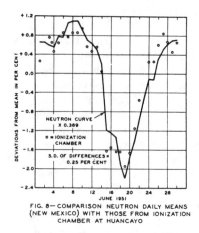

FIG. 8—COMPARISON NEUTRON DAILY MEANS
(NEW MEXICO) WITH THOSE FROM IONIZATION
CHAMBER AT HUANCAYO

Científica and the University of Mexico, Mexico, D.F., are now collaborating (since September 1, 1954) with the Department of Terrestrial Magnetism in the operation of a Compton-Bennett meter at the University of Mexico, Mexico, D.F., it will be possible in future comparison with neutron results to determine definitely whether there are variations in neutron values at New Mexico which do not occur at Huancayo or Mexico.

Comparison with results from balloon-borne ionization chambers—Figure 9

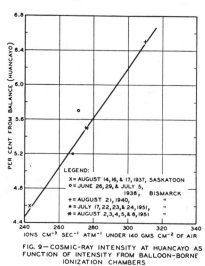

FIG. 9—COSMIC-RAY INTENSITY AT HUANCAYO AS
FUNCTION OF INTENSITY FROM BALLOON-BORNE
IONIZATION CHAMBERS

indicates a comparison between the changes in ionization at Huancayo and those obtained by Neher, *et al.* [8, 9], from balloon-borne ionization chambers under 140 gms cm^{-2} of air. The values for Huancayo are averages for those days in each of five different years on which the balloon flights were made. Comparing Neher's curve of ionization, as a function of the amount of air overhead, for June-July

190

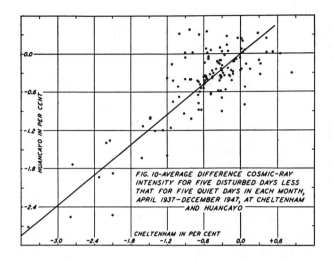

FIG. 10—AVERAGE DIFFERENCE COSMIC-RAY INTENSITY FOR FIVE DISTURBED DAYS LESS THAT FOR FIVE QUIET DAYS IN EACH MONTH, APRIL 1937–DECEMBER 1947, AT CHELTENHAM AND HUANCAYO

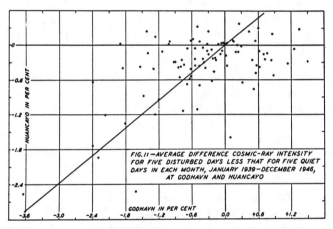

FIG. 11—AVERAGE DIFFERENCE COSMIC-RAY INTENSITY FOR FIVE DISTURBED DAYS LESS THAT FOR FIVE QUIET DAYS IN EACH MONTH, JANUARY 1939–DECEMBER 1946, AT GODHAVN AND HUANCAYO

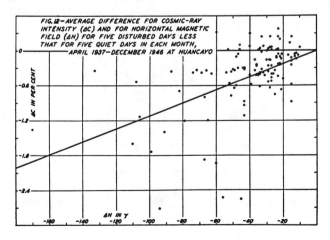

FIG. 12—AVERAGE DIFFERENCE FOR COSMIC-RAY INTENSITY (ΔC) AND FOR HORIZONTAL MAGNETIC FIELD (ΔH) FOR FIVE DISTURBED DAYS LESS THAT FOR FIVE QUIET DAYS IN EACH MONTH, APRIL 1937–DECEMBER 1946 AT HUANCAYO

1938 with that for August 1951 (Fig. 3 of reference 8), it is seen that the two agree for an amount of air overhead greater than about 140 gms cm^{-2}, but for less than 140 gms cm^{-2} the curves start to diverge considerably. For this reason, the ionization at 140 gms cm^{-2} was used in the comparisons with the Huancayo data. It would be of interest to obtain ionization-data from balloon flights which would involve a greater range in intensity at Huancayo than that of Figure 9. For example, with flights near sunspot minimum and maximum, a range of five per cent (if Fig. 3 is typical) could readily be realized. A further extension of range could be effected by flights during a period of large decrease which occurs during some magnetic storms. It would also be valuable to know whether decreases in intensity are observed from balloon flights during magnetic storms when no decrease is observed at Huancayo.

Cosmic-ray effects and geomagnetic activity

Cosmic-ray intensity for magnetically quiet and disturbed days—Figures 10 and 11 indicate, respectively, for Huancayo and Cheltenham, and Huancayo and Godhavn, the correlation between the average difference of cosmic-ray intensity for the five magnetically disturbed days of each month less that for the five quiet days. It is evident from Figure 10 that the frequency of positive values of the differences for Huancayo is only about one-fifth that for negative values, which indicates definitely that the cosmic-ray intensity tends to be less for the five magnetically disturbed days than for the five quiet days. It is evident from Figures 10 and 11 that the correlation between the differences for Huancayo and Godhavn is less than for Huancayo and Cheltenham. This, as will be shown later, is probably due to greater variations in vertical air-mass distribution at Godhavn as compared with those at Cheltenham.

Cosmic-ray intensity and magnetic horizontal intensity for disturbed minus quiet days—Figure 12 indicates the relation between the differences, disturbed minus quiet days, for cosmic-ray intensity and magnetic horizontal intensity at Huancayo. The differences are always negative for the horizontal intensity and preponderantly negative for cosmic-ray intensity at Huancayo. The correlation coefficient for data of Figure 12 would obviously be low. This is expected from the fact that the ratio between changes in cosmic-ray intensity to those in horizontal intensity is

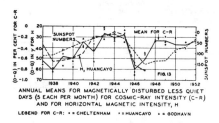

ANNUAL MEANS FOR MAGNETICALLY DISTURBED LESS QUIET DAYS (5 EACH PER MONTH) FOR COSMIC-RAY INTENSITY (C-R) AND FOR HORIZONTAL MAGNETIC INTENSITY, H

LEGEND FOR C-R: • = CHELTENHAM * = HUANCAYO ▪ = GODHAVN

known to vary from one storm to another [6]. Figure 13 indicates the variation in the annual means, for disturbed minus quiet days, in cosmic-ray intensity at three stations and in horizontal intensity, H, at Huancayo. Values for H were unavailable after 1947.

Twenty-seven day waves in cosmic-ray intensity, magnetic activity, and horizontal intensity—For completeness, Figures 14 and 15, published previously, are included here [10]. Figures 14 (A) and (B) are harmonic dials, indicating the departures from the average 27-day wave for American character-figure and for cosmic-ray

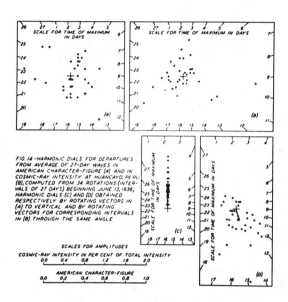

FIG. 14 –HARMONIC DIALS FOR DEPARTURES FROM AVERAGE OF 27-DAY WAVES IN AMERICAN CHARACTER-FIGURE (A) AND IN COSMIC-RAY INTENSITY AT HUANCAYO, PERU (B), COMPUTED FROM 34 ROTATIONS (INTERVALS OF 27 DAYS) BEGINNING JUNE 13, 1936; HARMONIC DIALS (C) AND (D) OBTAINED RESPECTIVELY BY ROTATING VECTORS IN (A) TO VERTICAL AND BY ROTATING VECTORS FOR CORRESPONDING INTERVALS IN (B) THROUGH THE SAME ANGLE

SCALES FOR AMPLITUDES

COSMIC-RAY INTENSITY IN PER CENT OF TOTAL INTENSITY
0.0   0.4   0.8   1.2   1.6   2.0

AMERICAN CHARACTER-FIGURE
0.0   0.2   0.4   0.6   0.8   1.0

intensity at Huancayo, respectively. Figures 14 (B) and (C) were obtained, respectively, by rotating the vectors in (A) to vertical and by rotating vectors for corresponding intervals (of 27 days) through the same angle. Statistical tests show that the probability, P, of obtaining an average vector as large or larger than that in Figure 14 (D) in a sample of 34 vectors from a population in which the components of the vectors are independent and random, with standard deviations estimated from the 34 vectors in (D), is only about 2×10^{-6}. This indicates that the vectors in (A) and (B) definitely tend to have similar phases. Furthermore, the phases in (C) and (D) of Figure 14 indicate that the maxima of the 27-day waves in cosmic-ray intensity tend to occur near the minima of the 27-day waves in magnetic activity, as measured by the American character-figure. Figure 15 indicates the results of a similar comparison between magnetic horizontal intensity and cosmic-ray intensity at Huancayo. For (C) of Figure 15, the probability, P, is 7×10^{-5}, indicating correlation between the phases of the vectors in (A) and (B) of Figure 15. Here it will be noted that the maxima of the waves in cosmic-ray intensity and those in horizontal intensity tend to be in phase, which is consistent with the results in Figure 14, since low values of horizontal intensity occur at times of high magnetic activity.

Variations in daily mean cosmic-ray intensity at Huancayo, 1937-1953, compared with variations in the earth's magnetic field and with cosmic-ray variations at other stations for selected years—Inspection of the graphs of cosmic-ray intensity for Huancayo in Figure 16 indicates a marked difference in the variability of daily

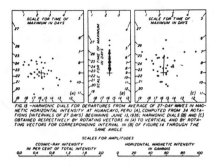

FIG. 15 —HARMONIC DIALS FOR DEPARTURES FROM AVERAGE OF 27-DAY WAVES IN MAG-
NETIC HORIZONTAL INTENSITY AT HUANCAYO, PERU (A), COMPUTED FROM 34 ROTA-
TIONS (INTERVALS OF 27 DAYS) BEGINNING JUNE 13, 1936; HARMONIC DIALS (B) AND (C)
OBTAINED RESPECTIVELY BY ROTATING VECTORS IN (A) TO VERTICAL AND BY ROTA-
TING VECTORS FOR CORRESPONDING INTERVAL IN (B) OF FIGURE 14 THROUGH THE
SAME ANGLE

SCALES FOR AMPLITUDES

means in different years, which is particularly evident if the curves for 1944 are
compared with those for 1946 and 1947; this variability was shown quantitatively
in Figure 7. During 1946 and 1947, there were large variations at Huancayo, which
in general follow those at Cheltenham, Godhavn (1946 only), and Christchurch
(1946 only). There is, of course, the large increase at Godhavn, and Cheltenham
on July 25, 1946, which occurred during a large solar flare on that date, and which
is absent at Huancayo, and at Christchurch where the meter was out of operation
for eight days starting July 23. On the other hand, a comparison of the graphs for
the four stations for 1944 shows that the variability of the daily means is decidedly
less at Huancayo than at the other three stations, and that the variability is greatest
at Godhavn. Moreover, the major variations at Godhavn, Cheltenham, and
Christchurch during 1944 were seen (by overlaying the original curves) to be
essentially uncorrelated. At Cheltenham, and to some extent at Godhavn, the
larger variations in 1944 (which were absent at Huancayo) occurred more often
in winter than in summer. At Godhavn and at Cheltenham, it was found that the
large variations in 1944 generally occurred during periods when the barometer
was changing rapidly. These large variations are thus probably due to changes of
the vertical air-mass distribution accompanying the movement of a front over the
station and the consequent effects arising from meson decay. Although smaller
variations occurring at Huancayo are often obscured at the other stations by this
meterorological effect, it has already been shown in Figures 5 and 6 that the
averages of a sample of such variations are very nearly the same at all four stations.

Figure 16 also shows the daily mean values of the horizontal magnetic component (H) at Huancayo, 1937-47, from which it can be seen whether decreases in
(H), which occur during magnetic storms, are accompanied by decreases in cosmic-
ray intensity. From these graphs, a tabulation showed 48 cases (1937-47) when
from one day to the next a decrease, in H, of 75 gammas or more occurred. In 36
of these cases, the change in cosmic-ray intensity at Huancayo was negative,
although in only 22 cases was the decrease in cosmic-ray intensity greater than

FIG. 16 (A to F)—Daily means cosmic-ray intensity: for Huancayo 1937-1953 (h), for Cheltenham
1944, 1946, and 1947 (c), for Godhavn 1944 and 1946 (g), and for Christchurch 1944 and 1946
(cc); and daily mean horizontal magnetic component at Huancayo 1937-1947 (H). (*Note: h, c,
H* on 75° WMT, *g* on 45° WMT, and *cc* on 172.°5 EMT; *h, c, g,* and *cc* in per cent above fiducial
values.)

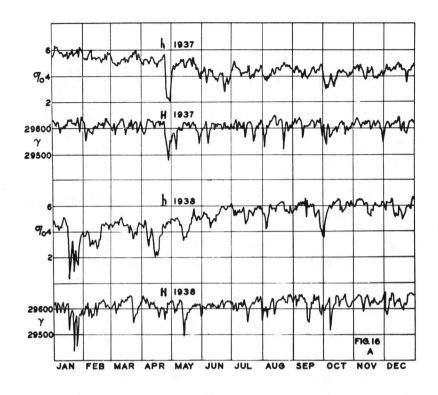

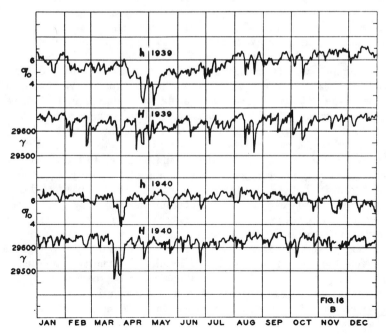

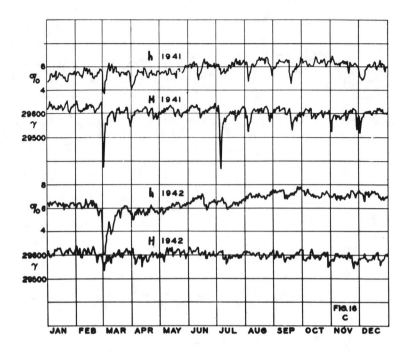

FIG. 16
C

0.4%. The graphs were also used to tabulate the dates between which the daily means of cosmic-ray intensity at Huancayo decreased continuously (successive days with no change were included) for a total decrease of 1.0% or more. There were 92 such intervals from 1937 to 1947. The change, ΔH, in daily mean horizontal magnetic intensity at Huancayo from the first to the last day of each of the above-selected intervals was also tabulated; in 71 (out of 92) of the intervals, ΔH was negative. Examination of magnetograms for Huancayo (Peru) and Watheroo (Australia) indicated magnetic disturbance in most of the 21 cases for which ΔH was either zero or positive. It thus seems evident that during most of the periods when the cosmic-ray intensity at Huancayo is decreasing there is evidence for magnetic disturbance, which suggests that the cause of the cosmic-ray decreases is quite probably connected with the mechanism giving rise to magnetic disturbance.

Finally, in this connection, attention should be called to the graphs of daily means for cosmic-ray intensity and magnetic horizontal intensity, H, for February 1946 in Figure 16. Between February 3 and 6, 1946, the five per cent decrease in in cosmic-ray intensity at Huancayo was accompanied by only a small decrease in H, while the large decrease in H after February 6 was accompanied by only a small further decrease in cosmic-ray intensity. There was at Huancayo and Watheroo a marked magnetic sudden commencement at 08^h42^m 75° WMT a few hours before the start of the decrease in cosmic-ray intensity. Attention should also be called to the fact (see also Fig. 3) that after February 6, 1946, both the cosmic-ray intensity and H at Huancayo remained low during the rest of the year. While it seems clear that most of the decreases in cosmic-ray intensity occur during

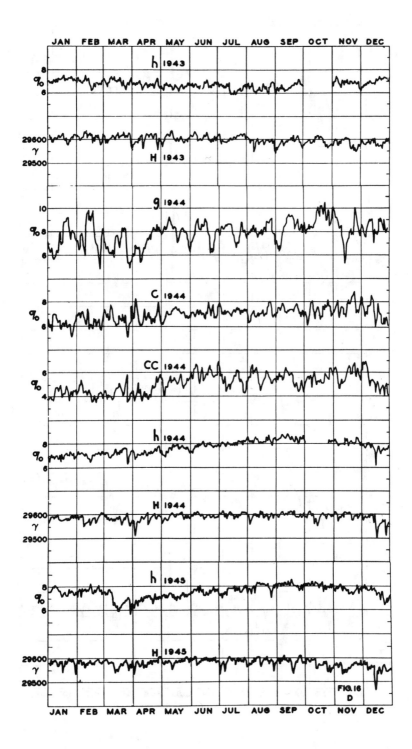

FIG.16
D

197

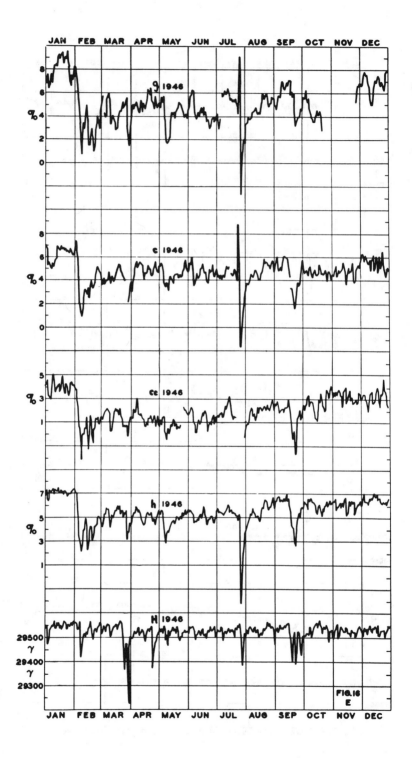

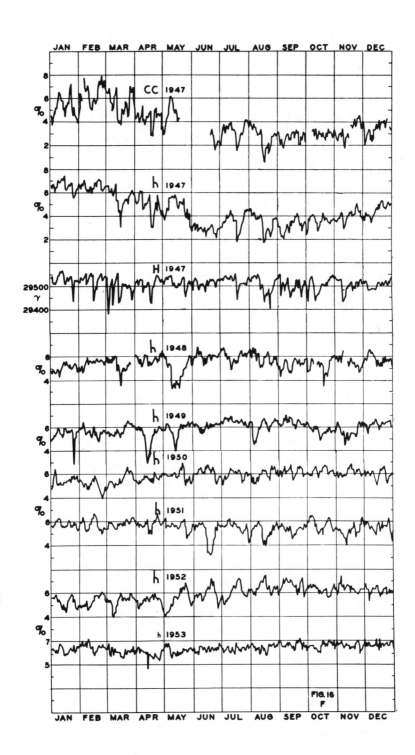

FIG. 16
F

periods of magnetic disturbance, no measurable characteristic of magnetic disturbance has yet been found which is quantitatively well correlated with changes in cosmic-ray intensity.

References

[1] I. Lange and S. E. Forbush, Cosmic-ray results from Huancayo Observatory, Peru, June 1936-December, 1946, including summaries from observatories at Cheltenham, Christchurch, and Godhavn, through 1946, Washington, D.C., Carnegie Inst. Pub. 175, Vol. 14 (1948).

[2] S. E. Forbush, Phys. Rev., 54, 975-988 (1938).

[3] S. E. Forbush, Phys. Rev., 70, 771 (1946).

[4] S. E. Forbush, T. B. Stinchcomb, and M. Schein, Phys. Rev., 79, 501-504 (1950).

[5] E. H. Vestine, L. Laporte, I. Lange, and W. E. Scott, The geomagnetic field, its description and analysis, Washington, D. C., Carnegie Inst. Pub. 580 (1947).

[6] S. E. Forbush, Terr. Mag., 43, 203-218 (1938).

[7] J. A. Simpson, Cosmic radiation intensity-time variations and their origin: III—On the origin of 27 day variations, Chicago, University of Chicago, Institute for Nuclear Studies (1954).

[8] H. V. Neher, V. Z. Peterson, and E. A. Stern, Phys. Rev., 90, 655-674 (1953).

[9] R. A. Millikan and H. V. Neher, Proc. Amer. Phil. Soc., 83, 409 (1940).

[10] S. E. Forbush, Trans. Washington Meeting, 1939; Internat. Union Geod. Geophys., Assoc. Terr. Mag. Electr., Bull. No. 11, 438-452 (1940).

DAYTIME ENHANCEMENT OF SIZE OF SUDDEN COMMENCEMENTS AND INITIAL PHASE OF MAGNETIC STORMS AT HUANCAYO

By S. E. Forbush and E. H. Vestine

Department of Terrestrial Magnetism, Carnegie Institution of Washington, Washington 15, D.C.

(Received May 19, 1955)

ABSTRACT

Applying statistical tests to 428 SC's, the frequency of occurrence is found to be independent of time of day. Statistical tests indicate that the average sizes of SC's and of IP are both significantly greater during the daylight hours at Huancayo. Also from 102 SC's occurring between 08^h and 14^h 75° WMT at Huancayo, we find that the average size of SC's is greater for those days with the larger diurnal variation (S_q) in H. This result is not only statistically significant but also the average size of SC's was about 50 per cent greater for the group of days with 50 per cent greater diurnal variation in H. The diurnal variation of SC's averaged on 75° WMT for San Juan and Honolulu is practically negligible. The augmentation of SC sizes at Huancayo with S_q in H at Huancayo was found to be the same whether the average size of the same SC's at San Juan and Honolulu was large or small. No significant diurnal variation was found in the frequency of occurrence of SC's observed both at Huancayo and Watheroo. A simple explanation is offered for the diurnal variation in the frequency of SC's found by Newton from Greenwich results.

The relation of daytime enhancement of the size of SC's to S_q in H at Huancayo indicates that the current system responsible is closely associated with the electrojet effect responsible for the large diurnal variation in H at Huancayo. The effects found are not predicted on the basis of the Chapman-Ferraro theory of magnetic storms in its present form. One possibility being examined is that the electric currents in the atmosphere near Huancayo are driven by electrojets of polar regions.

I. *Introduction*—Ever since Birkeland [see 1 of "References" at end of paper] first showed that the geomagnetic disturbance in auroral regions was best explained mainly by concentrated electric currents flowing linearly along the auroral zone, it has been clear that important sources of the field of magnetic storms were located within the atmosphere. He showed that magnetic disturbance near the auroral zone changed very rapidly with distance from the auroral zone, and found that the local field pattern on the ground was such that it closely resembled that due to an infinite linear current flowing at a height of about 100 km to perhaps 300 km above the ground. More refined estimates made since have shown that the current flow in high latitudes is probably closer to 100 km than to 300 km.

It is the purpose here to bring forward and discuss more critically new evidence for locating another concentration of current definitely within the atmosphere, and along the geomagnetic equator, where the current usually persists for an hour or so at the beginning of a magnetic storm. The effect of these currents is that of the local daytime enhancement of the sudden commencement (SC) and the initial phase (IP) of storms, seen most clearly in the horizontal magnetic component. This effect is shown to depend also upon the amplitude of the solar daily variation (S_q) on the day on which the SC or IP occurs. This suggests that substantial currents associated with the SC or IP flow in or near the E-region. The close association of the enhancements in IP and S_q may then be simply explained by supposing each to depend upon the locally enhanced conductivity in the E-region near the magnetic equator [2]. This result is of considerable interest, since it has long been known that SC's are larger in auroral regions than elsewhere. For instance, Chree [3] found those in the Antarctic to average about 4.5 times the values at Greenwich. Accordingly, in recent years it has been remarked that SC's must in part be due to atmospheric sources, because from potential theory it follows that local field patterns cannot readily arise from sources at a distance greater than the linear cross-section of the pattern [4,5,6]. In fact, it is quite clear that atmospheric sources of SC's and IP's are often dominant in the polar regions, and Nagata [4] has shown that the preliminary reverse impulse (SC*) of SC's also must be due to atmospheric sources. It thus appears likely that the sources of SC's and IP's are dominantly atmospheric, and consist of electric currents flowing in the ionosphere. These currents may be driven by emf's mainly originating in polar regions, as suggested for magnetic bays [7], and the so-called disturbance daily variation (D_S) [8].

The remarkable dependence of SC's and IP's upon the observed amplitude of S_q (as estimated here from the mean of the hour previous to an SC) makes possible the tentative removal of the very large part of SC's associated linearly with S_q amplitude at Huancayo. When this is done, the part remaining, whether due to currents flowing in solar streams or otherwise, is at least difficult to understand in terms of electric currents flowing beyond the atmosphere. It is found that the remainder of SC's (or IP's) at Huancayo is about twice the average for Honolulu and San Juan, which require no important addition of field values since the diurnal variation in SC's averaged on 75° WMT for San Juan and Honolulu is quite small. Hence, even this part is not readily assigned to current sources beyond the atmosphere, but is at least in fair accord with expectations based on atmospheric conductivity, assuming that all major currents actually flow within the atmosphere.

II. *Relation for Huancayo between diurnal variation (DV) in SC sizes*—In order to examine magnetic conditions near the magnetic equator during SC's and IP's, data for the years 1922-1946 were derived for Huancayo, Peru, with geographic position ($\phi = 12°.0$ south, $\lambda = 284°.7$ east), nearly directly on the magnetic equator, and the station Watheroo, Australia ($\phi = 30°.3$ south, $\lambda = 115°.9$ east).

Use was also made of data for the stations Honolulu ($\phi = 21°.3$ north, $\lambda = 201°.9$ east), San Juan ($\phi = 18°.4$ north, $\lambda = 293°.9$ east), and Cheltenham ($\phi = 38°.7$ north, $\lambda = 283°.2$ east). The only other established station near the magnetic equator is Kodaikanal ($\phi = 10°.2$ north, $\lambda = 77°.5$ east) in southern Asia, but results for this station have not been used here, though a similar investigation to that in this paper would be of particular interest.

The total number of SC's and IP's found at Huancayo and Watheroo was 428 after removing what were considered to be crochets, which often arise during solar flares, and have a field pattern resembling that of the quiet-day daily geomagnetic variation. The scalings of SC's at Huancayo and Watheroo were checked satisfactorily against those of Ferraro and Parkinson, whose earlier scalings of SC's at Honolulu, San Juan, and Cheltenham comprised the data of the latter stations. The convenient distinction by Ferraro, Parkinson, and Unthank of SC's into two classes, SC's and sudden impulses (SI's), depending on whether or not the initial sudden field departure was followed by a magnetic storm or otherwise, was not retained. This distinction may be somewhat artificial, since magnetic disturbance does not appear to change much in general type within wide ranges in the level of disturbances. It is also often found that the greater the intensity of a magnetic storm, the more rapidly does it go through its various phases; hence, very weak magnetic storms, barely distinguishable or indistinguishable from other fluctuations in amplitude during quiet days, may, at least at times, have SC's. Ferraro and Unthank [9] and Sugiura [6] called attention to the similarity of the DV in the average size of SC's in H at Huancayo to the DV of S_q in H there. This is illustrated by a new analysis here. The results are shown in Figure 1(A), in which the average S_q in H (solid curve) for those days on which 428 SC's occurred is compared with the median size (crosses) of SC's which occurred in each bihourly interval of the day. Figure 1(B) exhibits the correlation between the median sizes of SC's and the bihourly mean departures ΔH in S_q from the midnight value. The mean ΔH for each bihourly interval, for which the median size of SC was determined, was obtained in the following way. For the hour just preceding the hour of each SC, the departure of the hourly mean in H from the preceding midnight value was recorded as ΔH_t. For each of the 24 values of t, $(1/n) \sum \Delta H_t$ was computed with n the number of all SC's in the hour ($t + 1$). These averages are the circled points plotted in Figure 1(A). To obtain the averages of ΔH for the hours (and days) on which the SC's occurred, the amplitude of the 12-year average S_q curve [8] for ΔH at Huancayo was multiplied by 1.24 to fit the plotted points in Figure 1(A) for ΔH_t. From this curve, bihourly means of ΔH were obtained, for the same bihourly intervals for each of which the median size SC was determined. These bihourly means of ΔH are plotted in Figure 1(B) against the median size of SC which occurred in the corresponding bihourly intervals. The excellent fit of the curve to the points in Figure 1(A) indicates that the average ΔH is not noticeably affected by the lunar variation in H. If values of ΔH had been determined during or following the hour of the SC, these would certainly have been affected (and systematically) by the initial or main phase of the storms.

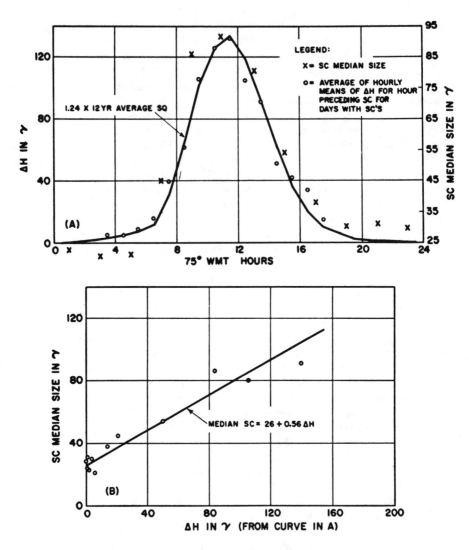

Fig. 1—(A) Comparison of diurnal variation in 428 SC's and S_q in H at Huancayo; (B) correlation between bihourly median sizes of SC's and mean bihourly departures ΔH in S_q from midnight value

III. *Normal distribution of log size of SC, and its utility for statistical tests*— In Figure 2 are plotted cumulative frequency distributions of SC sizes (A) for day and (B) for night at Huancayo. Since these cumulative distributions are well approximated by straight lines when plotted on logarithmic normal probability paper, it means the logarithm of the median size is the ordinate for 50 per cent on the abscissae scale. This size corresponds to the median of the frequency distribution of sizes (not their logs). The slope of the line determines the sample standard deviation of the logs of the sizes. Besides providing a simple representation of the

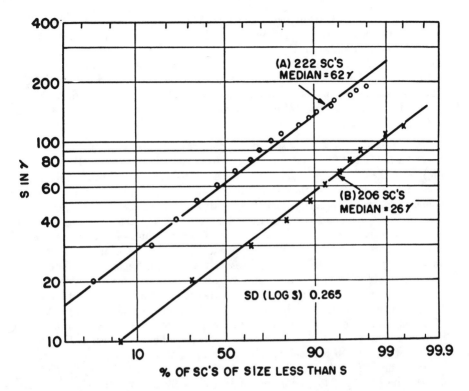

Fig. 2—Cumulative frequency distribution of SC sizes (S) at Huancayo: (A) for day (hours 0600-1800 75° WMT), (B) for night (hours 1800-0600 75° WMT)

frequency distributions, the fact that the distribution of log size is essentially normal makes it convenient for statistical tests of hypotheses concerning the data.

This convenience is illustrated by (A) and (B) of Figure 2. The sample standard deviations (SD) of log size are both approximately 0.265; the two sample means of log size are log 61.5 and log 25.5, and the difference in the two means of the logs is log $(61.5/25.5) = 0.384$. The SD for the difference between the two means is approximately $0.265 \sqrt{2}/\sqrt{200} = 0.0265$. Thus, the difference between the two means is about 14.5 times its SD and the difference is thus definitely significant. Hence, there is little doubt that the median sizes of SC's during the daytime at Huancayo are greater than for night.

There was scarcely need to test this point, but the method illustrates how the median size of SC for bihourly intervals was determined. Incidentally[10], if $m\{S\}$ is the mean of the sizes, and log ξ = mean log S, then

$$\log m\{S\} = \log \xi + 1.1513\ \sigma^2 \dots\dots\dots\dots\dots\dots (1)$$

$$m\{S\} = \xi \cdot k \dots\dots\dots\dots\dots\dots\dots\dots\dots (2)$$

in which log $k = 1.1513\ \sigma^2$ and σ^2 is the variance of log S. For example, from Figure 2 the sample estimate of $\sigma = 0.265$, so that $1.1513\ \sigma^2 \doteq 0.0809$ and $m\{S\} \doteq$

1.20 ξ; thus, for $\sigma \doteq 0.265$, the means are about 20 per cent greater than the medians. Consequently, the DV in mean SC at Huancayo is about 20 per cent greater than that shown in Figure 1(A).

Similarly, from Figure 6, the sample estimates $\sigma = 0.301$ for Watheroo; hence, the means at Watheroo are about 1.27 times the medians. For Honolulu and San Juan, the sample estimates of σ are, respectively, 0.248 and 0.261, and the factors by which the medians are multiplied to give the means are, respectively, 1.18 and 1.20.

IV. *Dichotomy of DV in size of SC, according to DV in H for Huancayo*— If the large DV in SC sizes is actually due to or connected with the DV in H, then the DV in the SC should be greater on days with large DV in H than on days with small DV in H. To test this point, the n values of ΔH_t corresponding to the n values of SC which occurred (on different days) in the hour ($t + 1$) were dichotomized. This was done in groups I and II, respectively, according to whether ΔH_t was greater than or less than the median of the n values of ΔH_t . Averages of ΔH_t were formed for each hour and for each group. These averages for group I and II are plotted, as circles and crosses, respectively, in Figure 3(A). The points are well fitted by the 12-year average curve for ΔH when its amplitude is multiplied by the indicated factor. Figure 3(A) indicates that the average amplitude of the DV in ΔH for days in group I is about 60 per cent greater than for days in group II. From each of these two curves in Figure 3(A), means of ΔH were obtained for each two-hour interval of the day. For the same bihourly intervals, the median size SC was obtained for each of the two groups.

In Figure 3(B), these median sizes of SC are plotted as ordinate; the abscissae are the means of ΔH for the corresponding bihourly interval. The line in Figure 3(B) is the same as that in Figure 1(B), and it is is seen to fit both sets of points in Figure 3(B). Moreover, the three uppermost points in (B) of Figure 2 are from group I, for which the DV in ΔH was greater than for group II. This result indicates that the median size of SC is greater on days (group I) with the larger DV in H, and that the median size SC increases linearly with ΔH.

The basis for a statistical test of whether the median size SC is significantly greater on days in group I (with the larger DV in ΔH) is indicated in Figure 4 by the frequency distributions of SC sizes. Here the difference in the logs of the two medians is log (81/60) = 0.130. Its SD is $\sqrt{2} \times 0.265 / \sqrt{60} = 0.047$. Thus, the ratio (0.130/0.047) of the difference to its SD is about 2.75, which indicates that in samples of size 60 the median for group (A) should exceed that in group (B) by as much or more than it does in this sample, with probability about 0.003. Thus, the excess of the median size for the group I over that for group II can be regarded as statistically significant.

V. *Comparison of DV in size of SC's for several observatories*—The median size SC for bihourly intervals for Huancayo and Watheroo (from the same 428 SC's) is shown in Figure 5(A). Median sizes of SC's are also shown for each four-hour interval of the day for Cheltenham, San Juan, and Honolulu, for which the numbers of SC's available for the samples were, respectively, 221, 225, and 271. All of these SC's occurred at Huancayo also.

206

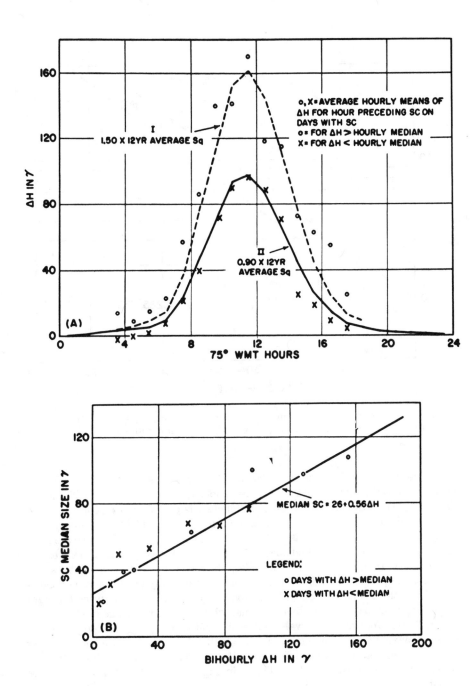

Fig. 3—(A) Diurnal variation of ΔH in S_q at Huancayo for days of large and small S_q ; (B) correlation between ΔH and median size of SC's showing increase in median size of SC on days with larger average diurnal variation S_q in H

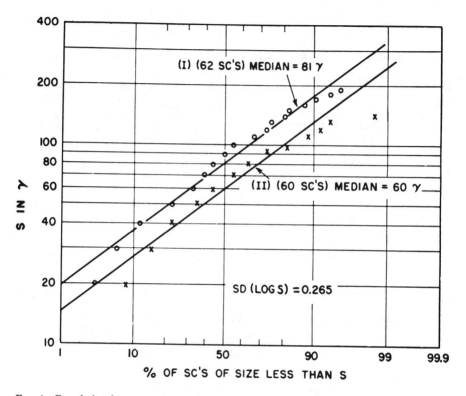

FIG. 4—Cumulative frequency distribution of SC sizes (*S*) at Huancayo for 75° WMT hours 0800 to 1500: (I) for days with large DV in *H*, (II) for days with small DV in *H*

Besides the large DV in median size of SC's at Huancayo (range 20 to 90 γ), a considerable DV in median size SC's is also evident for Cheltenham and Watheroo. The small DV in SC's at San Juan and Honolulu is probably not significant. The frequency distributions of SC's at Watheroo for day and night are compared in Figure 6. Testing the hypothesis that the samples (*A*) and (*B*) in Figure 6 are from the same population, it is found that differences between two medians, as great or greater than that for these two samples, should occur with probability about 0.002 for samples of the indicated size if these are actually from the same population. Thus, the median size SC at Watheroo is significantly greater during the night than during the day. It may be noted that while the daily average of the median size SC's (for four-hour intervals) is less at Watheroo than at Cheltenham, the ratio of maximum to minimum values is about 2.5 for each. Since the phase for the DV in SC's at Watheroo differs from that for Cheltenham by about 12 hours when both are plotted on 75° WMT, as in Figure 5, the two DV's would have about the same phase on LMT.

It is clear from Figure 5(*A*) that the DV in the ratio of sizes of SC's at Huancayo to that at Cheltenham, which was used by Sugiura [6], would differ quite considerably from the DV in SC's at Huancayo shown by the top graph in Figure 5.

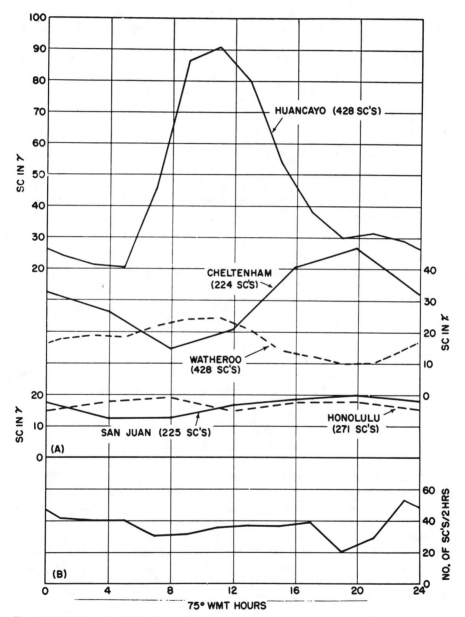

FIG. 5—(A) Diurnal variation in median size of SC's at several stations; (B) number of SC's for each two-hour interval of day for 428 SC's occurring both at Huancayo and Watheroo

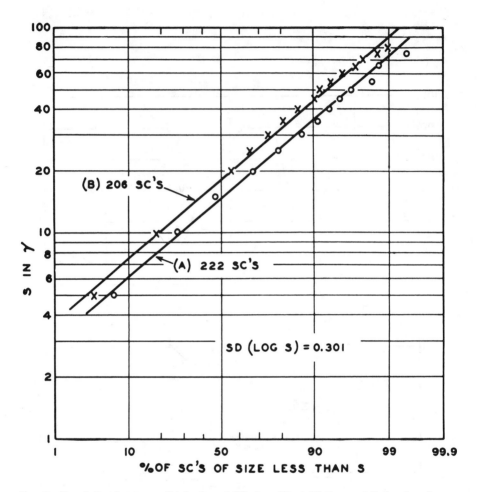

FIG. 6—Cumulative frequency distribution of SC sizes (S) at Watheroo: (A) for day (hours 0700-1900 120° EMT), (B) for night (hours 1900-0700 120° EMT)

The graphs for San Juan and Honolulu in Figure 5(A) indicate that SC sizes averaged for San Juan and Honolulu are nearly constant throughout the 24 hours of the day. This suggests that this average would be useful to normalize the sizes of SC's at other stations.

Since at Cheltenham there were some inverted (negative) SC's in H, the median size SC's were determined directly without the use of cumulated frequency distributions on log normal probability paper. In Figure 7, the DV in the mean size of SC's for four-hour intervals at Cheltenham and that for Newton's results [11] at Greenwich are seen to agree reasonably well.

VI. *Tests for DV in frequency of SC's and comparison with Newton's results*—Figure 5(B) shows for each bihourly interval of the day the number of SC's that

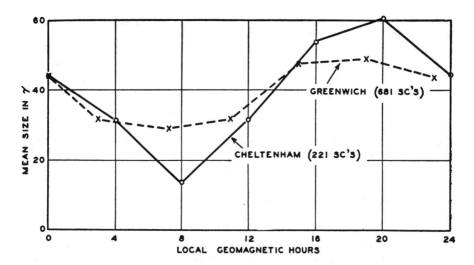

Fig. 7—Comparison of diurnal variation of mean size SC for four-hour intervals for Cheltenham and Greenwich

occurred both at Watheroo and Huancayo. The average number of SC's for each bihourly interval is about 36. A chi square test for homogeneity gives $P = 0.05$, which is scarcely small enough to reject the hypothesis of equal frequency (*viz.*, 36) in each of the 12 two-hour intervals. If the SC's occur randomly throughout the day, then a Poisson distribution is expected, for which the mean and variance are equal. For the sample in Figure 5(B), the mean and variance are, respectively, 36 and 66. However, this difference is not statistically significant ($P = 0.05$). The deviations for the hours 18-20 and 22-24 are the largest in the set.

The value in the interval 18-20 hours is undoubtedly somewhat low due to the loss of trace (or incomplete records of the SC's) during the interval when one or the other or both of the records were changed (*viz.*, *circa* 19ʰ 75° WMT). The largest number of SC's is for the interval 22-24 hours, which deviates by +17 from the mean. Using the normal approximation to the Poisson distribution, it is found (using variance = 36) that deviations greater than 17 in magnitude should occur with probability about 0.004, or once in about 20 sets of data like that in Figure 5(B), so that such a deviation is not too improbable in the sample under consideration.

Thus, there appears no evidence for any significant DV in the frequency of occurrence of SC's in the sample of 428 SC's which occurred simultaneously at Watheroo and Huancayo. This is quite different from the results found by Newton [11] from 681 SC's on the Greenwich records. Statistical tests for homogeneity based on the data in his Figure 3 definitely show that the DV indicated is real. Newton also indicated [11] that the SC's which occur near 08ʰ to 09ʰ GMT at Greenwich are often inverted (that is, negative in H). Effects from current systems which result in "inverting" the SC's during certain hours at one station and not at others, can doubtless also often "wipe out" SC's during those hours, with a resulting apparent decrease in frequency of occurrence during those hours. This would seem to explain why Newton's data indicate a DV in the frequency of SC's while the result in Figure 5(B) does not.

VII. *Dependence of the DV in SC sizes at Huancayo upon the size of SC's at San Juan and Honolulu*—Since the DV in median size SC's at Huancayo is greater on days with larger DV in H, it is of interest to determine whether this augmentation depends on the size of the SC's. For this purpose, data were available only 161 simultaneous SC's at Huancayo, San Juan, and Honolulu. The size for each of the 161 SC's was averaged for San Juan and Honolulu. Let $A_{i,j}$ indicate this average for the jth among the n_i SC's occurring in the ith four-hour interval $(i = 1, 2, \cdots 6; j = 1, 2, \cdots n_i)$ and let M_{i,n_i} represent the median of $A_{i,j}$. All $A_{i,j} > M_{i,n_i}$ were put in group I and the $A_{i,j} < M_{i,n_i}$ were put in group II. For each four-hour interval, the mean size SC at Huancayo was determined, and also $(1/n_i) \sum_{j=1}^{n_i} A_{i,j}$. The results are given in (a) and (b) of Table 1, together with ΔH_i from Figure 1(A) averaged for the indicated intervals, i. The results are plotted in Figure 8. The same values of ΔH_i were used in plotting the average size of SC's in groups I and II. It was assumed that the DV in ΔH for groups I and

TABLE 1—*Mean sizes of SC's in H at Huancayo for different intervals of the day, and for groups I and II dichotomized according to whether the size of SC averaged for San Juan and Honolulu was greater or less than the median of these averages for the interval*

75° WMT interval, i	No. of SC's in I	Mean SC in group I		No. of SC's in II	Mean SC in group II		ΔH_i at Huancayo*
		Huancayo	San Juan Honolulu		Huancayo	San Juan Honolulu	
hours		γ	γ		γ	γ	γ
(a)							
0000–0400	13	64	40	14	22	14	1
0400–0800	14	35	23	14	34	11	2
0800–1200	7	130	27	7	106	14	107
1200–1600	17	111	31	16	70	13	78
1600–2000	12	82	35	13	31	15	10
2000–2400	17	50	29	17	28	14	0
Mean	..	..	31	..	..	14	..
(b)							
0000–0400⎫ 2000–2400⎭	30	57	34	31	25	14	0
0040–0800⎫ 1600–2000⎭	26	58	29	27	32	14	6
0800–1200	7	130	27	7	106	14	107
1200–1600	17	111	35	16	70	13	78
Total	80	..	..	81	..	..	...

*Average for indicated interval from Figure 1(A).

II was that shown in Figure 1(A). Since this assumption may not be justified and since the number of cases in some intervals is small, the true slopes for groups I and II may differ considerably from those shown in Figure 8. If the augmentation of SC's at Huancayo depends on the electromotive driving force, a smaller slope would be expected for group II than for group I. A statistical test of this expectation is under way utilizing more extensive data.

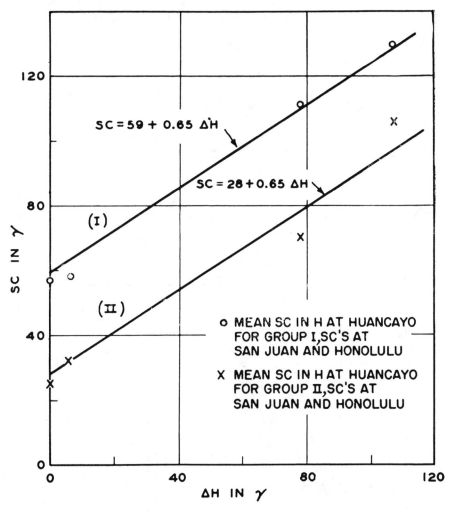

F$_{IG}$. 8—Correlation between mean size SC in H with departure ΔH in S_q from midnight value: (I) and (II), respectively, for large and small SC averages for San Juan and Honolulu

Figure 8 shows that the mean size, $\overline{SC_i}$, of SC's for a particular interval, i, of the day has the average approximation

$$\overline{SC_i} = A + 0.65\Delta H_i \dots\dots\dots\dots\dots\dots\dots(3)$$

in which ΔH_i is the departure of H, averaged for the same interval i, from midnight value. Only the value of A is greater for the group of SC's with the larger average for San Juan and Honolulu. The ratios of A to the mean size SC's at San Juan and Honolulu (Table 1) are, respectively, $59/31 = 1.9$ and $28/14 = 2.0$ for groups I and II. It may be noted that the slope of the lines in Figure 8 is somewhat greater than for those in Figures $1(B)$ and $2(B)$. However, the former applies to means and the latter to median size SC's. Since (at Huancayo) it was shown that the mean SC = median SC $\times$ 1.205, then a slope $= 0.56 \times 1.20 \doteq 0.67$ would be expected for Figure 8.

It was not surprising to find the value of A greater for group I than for group II, but it is surprising to find the value of A about twice the size of SC's averaged for San Juan and Honolulu for the same groups. This suggests a rapid change with latitude of what might be called the world-wide (analogous to D_{st}) component of SC's. Such a rapid latitude variation suggests that the current system for the world-wide component of SC's also flows in the atmosphere.

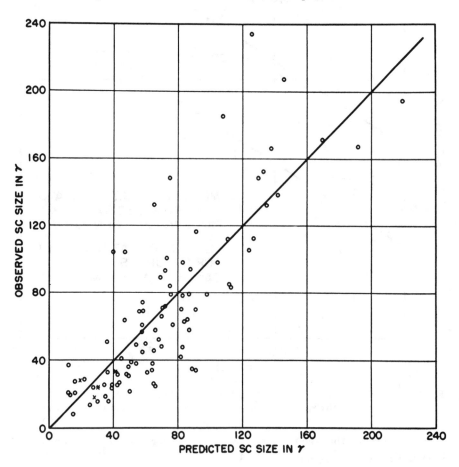

FIG. 9—Comparison of observed SC size at Huancayo with size predicted

In Figure 9, the observed size of 161 individual SC's at Huancayo is compared with the size predicted from the equation: SC (predicted) = 2 × size averaged for San Juan and Honolulu + 0.65 ΔH, in which ΔH is the estimated DV departure in H (from midnight) at the time of the SC. In view of the uncertainties in estimating ΔH (no correction for lunar variation, etc.), the correspondence is reasonably good.

VIII. *DV in magnitude of IP at Huancayo and its dependence on amplitude of the DV in H of S_q*—For each of the 428 SC's which occurred both at Huancayo and Watheroo, the IP averaged for one-half hour [IP_0 (1/2)] and for one hour [$IP_0(1)$] was measured, for Huancayo. $IP_0(1/2)$ is thus simply the difference between the mean value of H measured for the half-hour immediately following the SC and the value of H for the instant preceding the SC. $IP_0(1)$ is the measured mean of H for the whole hour immediately following the SC, less the value of H for the instant preceding the SC. These measured values, $IP_0(1/2)$ and $IP_0(1)$, are each somewhat affected (especially those between the hours 07-11 and 12-16, 75° WMT) by the change in H due to the large DV in H at Huancayo. The values of $IP(1/2)$ and $IP(1)$ are the measured values $IP_0(1/2)$ and $IP_0(1)$ corrected for the change in the mean of H, due to the DV in H, which occurs, respectively, during the half-hour and one-hour intervals following the SC. The individual IP's were dichotomized according to whether the DV in H was large or small, in the same manner as described for SC's in section IV. Since the same days apply for the IP's as for the SC's, the DV in ΔH shown by curves I and II in Figure 3(A) were used as a basis for the corrections in the IP's.

The curve in Figure 10(A) shows the DV in the size of $IP(1/2)$ for all days. The encircled points are the bihourly averages of $IP(1/2)$ for days in group I (larger DV in H) and the crosses the bihourly averages of $IP(1)$ for days in group II (small DV in H). The curve in Figure 9(A) is, except for amplitude, quite like that for SC's for Huancayo in Figure 5(A). Furthermore, the values of $IP(1/2)$, for the hours 08-14 75° WMT, are larger for the group with larger DV in H. This implies that the magnitude of the IP at Huancayo is augmented in the same way as is the size of SC's.

Figure 11 shows the cumulative frequency distribution of the sizes of $IP(1/2)$ for days with large (group I) and small (group II) DV in H. The difference between the mean log $IP(1/2)$ for the two groups is about 3.8 times the SD [log $IP(1/2)$], which leaves little doubt that the effect is real. This test, it should be observed, does not depend very critically upon whether the distribution of log $IP(1/2)$ is exactly normal. This follows from the central limit theorem of statistics, which states that the distribution of means of samples from non-normal distributions approaches normality. For samples of 30 (as in Fig. 11) the approximation is quite close.

In Figure 10(B), the average size of $IP(1/2)$ is seen to be about 0.87 times the median size SC's for the same bihourly intervals, or since the mean size of SC's at Huancayo is 20 per cent greater than the median, then the mean $IP(1/2)$ is about 0.72 times the mean size of the SC's.

In Figure 10(C), the means of $IP(1/2)$ are plotted against the means of $IP(1)$

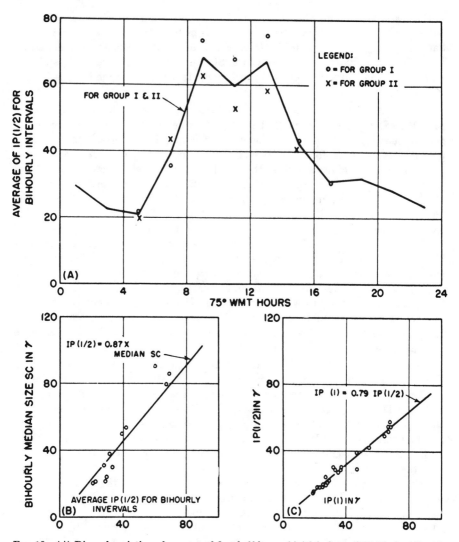

Fig. 10—(A) Diurnal variation of average of first half-hour of initial phase, IP(1/2), for bihourly intervals at Huancayo; (B) correlation between bihourly median size of SC's and average IP(1/2) for corresponding bihourly intervals; (C) correlation between average IP(1/2) and IP(1) for corresponding bihourly intervals

for corresponding bihourly intervals. The line indicates IP(1) = 0.79 IP(1/2), from which it follows that the IP for the interval 0.5 hour to 1.0 hour after the SC is, on the average, about 58 per cent as large as IP(1/2).

IX. *Discussion of results*—The results of the present paper clearly indicate the presence of a major and immediate atmospheric source of field in SC's and IP's. Although this conclusion accords with earlier tentative findings [4,6], the present results benefit by carefully made statistical tests. A new result found was the

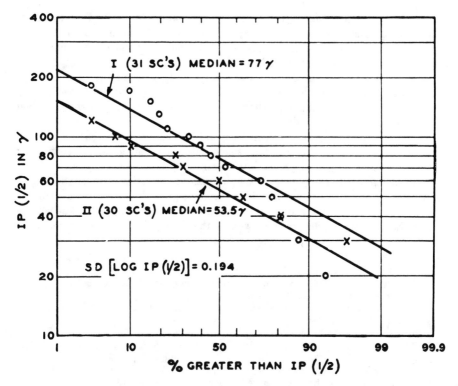

Fig. 11—Cumulative frequency distribution of size of initial phase (half-hour mean) IP(1/2), at Huancayo, hours 1000 to 1300 75° WMT: (I) for days with large diurnal variation in H, (II) for days with smaller diurnal variation in H

surprisingly good correlation between daytime SC and IP amplitude at Huancayo and the amplitude of S_q. This suggests that major currents responsible for SC and IP flow in or near the E-region above Huancayo. However, since the electromotive driving forces for S_q and SC and IP can presumably vary independently of one another, the linear relationship of Figure 9 is not explained in terms of electric conductivity alone. It may be that the charge accumulations which insure continuity of current flow are mutually dependent in some way, as might happen if the electromotive driving forces in middle and higher latitudes show some mutual resemblance in general type. For instance, electric fields originating in high latitudes, as in D_s, may provide the principal electromotive driving forces near Huancayo in the case of SC and IP. The simultaneous ionospheric motions at Huancayo due to possible polar electric doublets at times of SC and IP are in the course of examination and will be reported upon separately.

An interesting experimental program arises naturally from the present work. It was noted that the abnormal augmentation of SC, IP, and S_q at Huancayo is associated with a narrow belt of high electrical conductivity near the magnetic equator. In this event, a concentrated current should flow overhead at Huancayo, the height of which may be estimated using a north-south grouping of magnetic stations centered at Huancayo, measuring horizontal and vertical components (and space-gradients) of SC, IP, or S_q . In this way, the local currents may be estimated for SC and IP. If large enough, there would be no need to assign major sources of field to regions of space beyond the atmosphere.

For similar reasons, the current systems derived by Vestine [8] for the mean hourly field changes during the IP of magnetic storms should be revised, using instantaneous values, in the hope that this will reveal something of the electrojet at the magnetic equator by day.

It is also clear that it is established beyond reasonable doubt that IP tends to be enhanced by day at Huancayo, as suggested in a recent letter by Vestine [6], so that objections previously raised by Ferraro [12] seem now removed. It would also appear that extensions of existing rudimentary theories of magnetic storms must in future seek to explain the disturbance at ground level in terms of major sources located within the atmosphere, additional to those for D_S . These sources seem to be stronger in polar regions, whence they must ultimately arise from solar influences.

References

[1] Kr. Birkeland, Norwegian Aurora Polaris Expedition, 1902-1903, Christiania, 1, Pt. 1, 39-315 (1908), and Pt. 2, 319-551 (1913).

[2] M. Hirono, J. Geomag. Geoelectr., 4, 7-21 (1952); W. G. Baker and D. F. Martyn, Phil. Trans. R. Soc., 246, 282-320 (1953).

[3] C. Chree, Studies in terrestrial magnetism, Macmillan and Co., Ltd., London (1912).

[4] T. Nagata and N. Fukushima, Indian J. Met. Geophys., 5, Spl. Geomag. No. 75-88 (1954).

[5] T. Nagata, Nature, 169, 446 (1952); Rep. Ionosphere Res. Japan, 6, 13 (1952).

[6] M. Sugiura, J. Geophys. Res., 58, 558 (1953); E. H. Vestine, ibid., 539, 560 (1953); T. Yumura, Mem., Kakioka Magnetic Observatory, 7, 27-47 (1954).

[7] N. Fukushima, Polar magnetic storms and geomagnetic bays, J. Fac. Sci., Tokyo Univ., 8, 293-412 (1953).

[8] E. H. Vestine, Terr. Mag., 43, 261 (1938); E. H. Vestine, L. Laporte, I. Lange, and W. E. Scott, The geomagnetic field, its description and analysis, Carnegie Institution of Washington, Pub. No. 580 (1947).

[9] V. C. A. Ferraro and H. W. Unthank, Geofisica pura e appl., 20, 3-6 (1951); V. C. A. Ferraro, W. C. Parkinson, and H. W. Unthank, J. Geophys, Res., 56, 177-195 (1951).

[10] A. Hald, Statistical theory with engineering applications, John Wiley and Sons, New York (1952).

[11] H. W. Newton, Mon. Not. R. Astr. Soc., Geophys. Sup., 5, 159-185 (1948).

[12] V. C. A. Ferraro, J. Geophys. Res., 59, 309-311 (1954).

JOURNAL OF GEOPHYSICAL RESEARCH   VOLUME 61, No. 1   MARCH, 1956

VARIATIONS IN STRENGTH OF WIND SYSTEM, IN THE DYNAMO MECHANISM FOR THE MAGNETIC DIURNAL VARIATION, DEDUCED FROM SOLAR-FLARE EFFECTS AT HUANCAYO, PERU

BY SCOTT E. FORBUSH

*Department of Terrestrial Magnetism, Carnegie Institution
of Washington, Washington 15, D. C.*

(Received January 15, 1956)

ABSTRACT

Hourly means of solar-flare effects, or crochets, in magnetic horizontal intensity (H) at Huancayo exhibit, during daylight, a diurnal variation like that in H. This variation is expected, since McNish showed [see 1 of "References" at end of paper] that flares produce magnetic effects indistinguishable from those which would result from an increase in strength of the current system responsible for the normal magnetic diurnal variation, S_q . On the other hand, it is found that the average crochet size in H is greater by a statistically significant amount for groups of days with greater S_q amplitude. For example, the average crochet size is 80 per cent larger for a group of 49 days with average $\delta W_2 = 25.2$ (Bartels' measure of amplitude of S_q) than for a group of 48 days with average $\delta W_2 = 1.6$. Although δW_2 increases with sunspot number, the crochet size does not. Thus, the larger average size of crochets, at Huancayo, on days when S_q is larger, indicates that the strength of the wind system is, on the average, greater on days with larger S_q . From the change in crochet size, the strength of the wind system must vary by 50 per cent at least.

INTRODUCTION

The seasonal variation of the latitude of the northern and southern foci of the current system which causes the quiet-day magnetic diurnal variation, S_q , is well known [2]. Hasegawa [2] found two cases, in each of which from one day to the next the latitude of the focus of the northern current-system shifted about 15°. In addition, there were marked changes from one day to the next in the morphology of the current system. In neither of the two cases was there much change, on the successive days, in Bartels' δW_2 measure [3] for the amplitude of S_q at Huancayo. These facts indicate changes in the morphology of the air motions or wind patterns which, according to the dynamo theory, generate the S_q current-system. Since it has not yet been possible to measure directly the total conductivity in the ionospheric layer (below the E-region) where the S_q currents flow, the strength of the wind pattern or changes in its strength cannot be determined by this procedure. Suppose, however, we have a group (I) of crochets for days with large S_q and another group (II) for days with small S_q . Then, if the number of crochets in each group is large enough, the average increase in conductivity in the layer where the S_q currents flow should be the same for both groups. If we find that the average size of crochets is significantly greater in group I than in group II, this would indicate that the strength of the wind system for group I was greater than for group II and consequently that, in part, the larger S_q in group I is due to a stronger average wind-system. Thus far, we have assumed that the current system for crochets is at the same level (below the E-layer) as that for S_q . The fact that the magnetic effects from solar flares are indistinguishable from those which would

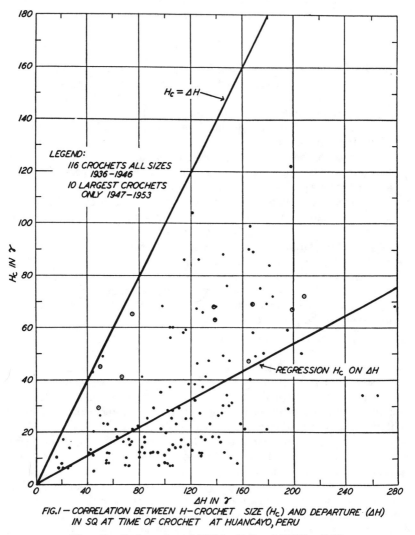

FIG.I — CORRELATION BETWEEN H—CROCHET SIZE (H_c) AND DEPARTURE (ΔH)
IN SQ AT TIME OF CROCHET AT HUANCAYO, PERU

(*Correction:* In Legend, read 1947 for 1946 and 1948 for 1947)

esult from an increased strength of the existing S_q current-system either indicates hat both are in the same layer or in layers in which the morphology of the wind pattern is the same.

INCREASE OF *H*-CROCHET SIZE
WITH AMPLITUDE OF DIURNAL VARIATION IN *H*

Figure 1 depicts, as ordinate, the size, H_c, of 126 crochets in horizontal intensity, *H*, at Huancayo (see Appendix *A* at end of paper). Each H_c is plotted against the departure, ΔH, of the value of *H* immediately preceding the crochet from the midnight bihourly mean value of *H*. The graph indicates, in spite of the large scatter, the tendency for larger crochet sizes to occur with larger values of ΔH.

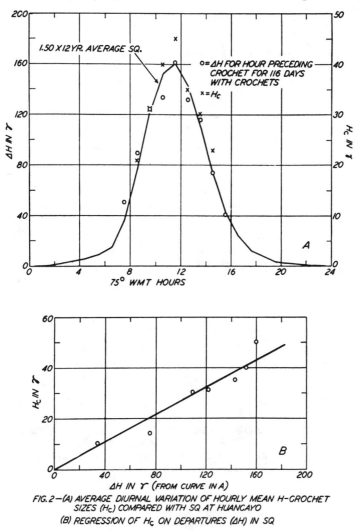

FIG. 2—(A) AVERAGE DIURNAL VARIATION OF HOURLY MEAN H-CROCHET SIZES (H_c) COMPARED WITH SQ AT HUANCAYO
(B) REGRESSION OF H_c ON DEPARTURES (ΔH) IN SQ

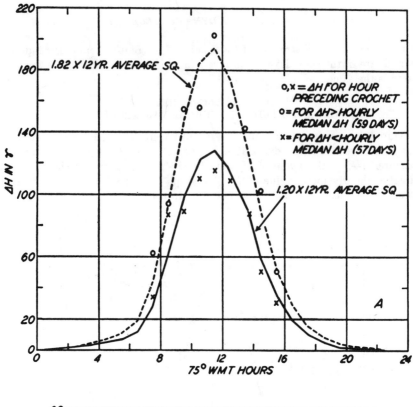

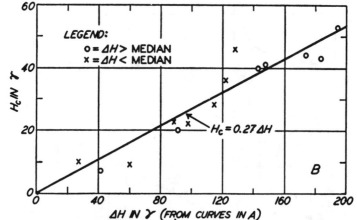

FIG. 3 — (A) AVERAGE DIURNAL VARIATION OF ΔH IN SQ AT HUANCAYO FOR
CROCHET DAYS WITH LARGE AND SMALL SQ

(B) REGRESSION OF H—CROCHET SIZE (Hc) ON DEPARTURES (ΔH)
IN SQ FROM MIDNIGHT VALUE

Since no point is above the line $H_c = \Delta H$, no crochet size is found to exceed ΔH. The lower line is the regression of H_c on ΔH from Figure 2. Figure 2(A) compares the hourly mean values of H_c with the average diurnal variation S_q estimated for the days on which the 116 crochets occurred. The solid curve is 1.50 times the 12-year average S_q in H. The values of ΔH (circles in Fig. 2A), to which the S_q curve is fitted, were obtained as follows: For those days on which crochets occurred in a particular hourly interval, t, the departure, δH, from the midnight bihourly mean H of the hourly mean for the interval $(t - 1)$ was computed; ΔH is the average of these departures δH for a given hour. This avoids basing the diurnal variation from days with crochets on hourly values which are augmented by the crochet. Since the open circles are well fitted by the solid curve, the mean hourly value of ΔH may be read from the curve for any particular hourly interval for which the mean H_c is shown. In Figure 2(B), the hourly mean crochet size H_c is plotted against the hourly mean ΔH for the same hour as read from the curve in Figure 2(A). Since the magnetic effect of solar flares is indistinguishable [1] from that which would result from an enhancement of the S_q current-system, the hourly averages of many H-crochets would be expected to exhibit a diurnal variation similar to that in H. The combined number of H-crochets in the two 75° WMT hourly intervals (0800–0900) and (1400–1500) was 19, which is about the average number observed in the single hourly intervals between 0900 and 1400. Since the average H-crochet size increases with ΔH, it is not surprising that fewer crochets are observed in the hours when ΔH is smaller. This occurs, not because the expected number of crochets is less, but because a larger fraction of them will be too small (less than about 7γ) to be seen in the background noise.

Figure 3(A) shows the estimated diurnal variation of ΔH, derived as described above, for two groups of days on which crochets were observed. Values of δH greater than the median δH for a given hour were averaged to give ΔH for the first group, as indicated by the circles in Figure 3(A). Values of δH less than the median for a given hour were averaged to obtain the values of ΔH for the second group, as indicated by the crosses. As before, the values of ΔH corresponding to any hour for which the average crochet size was obtained can be read from the curves in Figure 3(A). In Figure 3(B), the average for each hour of the H-crochet size, H_c, is plotted against the average ΔH for the same hour. The slope of the line in Figure 3(B) is the same as that in Figure 2(B). Figure 3(B) also indicates that H_c is proportional to ΔH and that the diurnal variation of the average H_c size is greater on days with larger S_q.

It remains to test the statistical reality of the difference between the average crochet size on days with large and small S_q in Figure 3. For each day with crochet, the value of Bartels' δW_2 measure [3] is used to determine the size of S_q in H at Huancayo. This measure is derived from the difference: average H for the 75° WMT interval, 0900 to 1400 hours, less the average H for the interval 0000 to 0500 hours. This difference is then corrected for the lunar variation in H at Huancayo, and for post-perturbation recovery. After normalizing for each month to eliminate seasonal influences, the final values, except for a scale-factor, give δW_2. For Figure 4, the crochet days with δW_2 greater than the median δW_2 (for 97 days with crochets) were placed in one group and those with δW_2 less than the median in another. Only crochets between 0900 and 1400 hours, 75° WMT, were used, since there were no statistically significant differences between the hourly frequencies of crochets within this interval. Also the δW_2 measure is based on these same hours. Before deriving the frequency distributions of crochet sizes, each observed crochet size H_c was corrected for diurnal variation between 0900

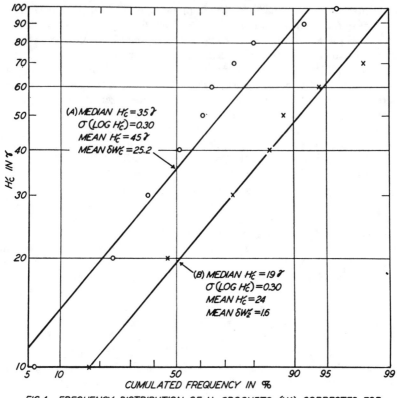

H_c IN γ

(A) MEDIAN $H_c'=35\,γ$
$\sigma\,(LOG\,H_c')=0.30$
MEAN $H_c'=45\,γ$
MEAN $\delta W_c'=25.2$

(B) MEDIAN $H_c'=19\,γ$
$\sigma\,(LOG\,H_c')=0.30$
MEAN $H_c'=24$
MEAN $\delta W_2'=1.6$

CUMULATED FREQUENCY IN %

FIG.4—FREQUENCY DISTRIBUTION OF H−CROCHETS (H_c') CORRECTED FOR
DIURNAL VARIATION 0900−1400, 75° WMT
(A)−FOR $\delta W_2 >12.5$ 49 CROCHETS (MEAN 49 γ)
(B)−FOR $\delta W_2 <12.5$ 48 CROCHETS (MEAN 25 γ)

and 1400 hours, 75° WMT. It was assumed, from the evidence in Figure 3(*B*), that the shape of the diurnal-variation curve for H_c is that of the 12-year average diurnal variation S_q in H. From this 12-year average, the hourly average for each of the five hourly intervals between 0900 and 1400 was taken, with the mean for the five hours 0000 to 0500 deducted. The ratio of these resulting five values of ΔH to the average ΔH for the five hours 0900 to 1400 was then found. These consecutive ratios, starting with the value for the interval 0900 to 1000 hours, were as follows: 1.13, 0.90, 0.86, 0.96, and 1.25. The observed size H_c of each crochet was multiplied by one of the above factors according to the hour in which the crochet occurred, to give the crochet size, H'_c , corrected for diurnal variation. This procedure provides a normalized crochet size, H'_c , which is independent of the time, between 0900 and 1400 hours, at which the crochet occurred but which does depend upon the amplitude of S_q in H. Finally, Figure 4 shows, on log normal probability paper, the frequency distribution of H'_c for 49 days with δW_2 greater

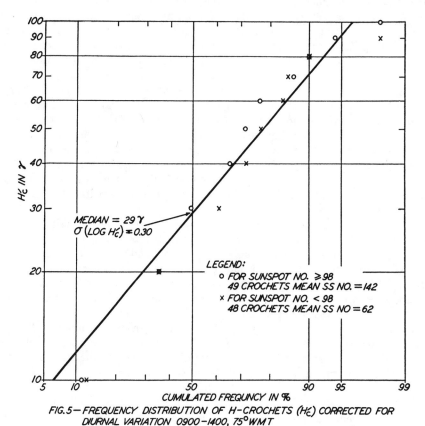

FIG. 5 — FREQUENCY DISTRIBUTION OF H-CROCHETS (H'_c) CORRECTED FOR
DIURNAL VARIATION 0900 –1400, 75° W M T

than the median δW_2 and for 48 days with δW_2 less than the median. Using the parameters indicated in Figure 4, the hypothesis that both samples came from the same population was tested. Differences as great or greater than those shown in Figure 4 between the samples should occur by chance only about once in about 10^5 such pairs of samples, thus leaving little doubt of the reality of the difference. Thus, the average crochet size is greater for the group of days with larger S_q . Although the frequency distributions in Figure 4 are approximately logarithmic normal, the means of samples of this size will, by the central limit theorem, be nearly normally distributed. The standard deviations of the means for the upper and lower distributions in Figure 4 are roughly 4γ and 2γ, respectively.

The seasonal variation in the amplitude of S_q at Huancayo is well known [3].

It is not due to a seasonal variation in solar radiation. It may arise from a seasonal variation in wind pattern or from the seasonal variation in conductivity of the S_q layer, at a given latitude, due to the seasonal variation in the sun's zenith distance. Either of the latter two causes would also result in a seasonal variation in the size of crochets. The frequency distributions of crochet sizes were plotted on log normal probability paper for the equinoxes and both solstices. The resulting estimates of mean crochet sizes, with estimated probable errors, were as follows: Equinoxes $42 \pm 4\gamma$, winter solstices $42 \pm 4\gamma$, and summer solstices $35 \pm 4\gamma$. The differences among these means are not statistically significant. From the sample of 97 days with crochets between 0900 and 1400 hours, 75° WMT, at Huancayo, the estimated mean values of ΔH for the interval 0900 to 1400 hours, 75° WMT, for the three seasons, were as follows: Equinoxes 149γ, winter solstices 129γ, and summer solstices 112γ. From Figure 2(B), the resulting mean crochet sizes would be 0.27 times the above values of ΔH, or as follows for the three seasons: Equinoxes 40γ, winter solstices 35γ, and summer solstices 30γ. The differences between these values and those obtained above for the mean crochet sizes are within the statistical uncertainties.

Since monthly and yearly averages of δW_2 were shown, by Bartels [3], to have a higher correlation with sunspot numbers than any other geophysical variable, it is expected that δW_2 in our sample will be correlated with sunspot number. If H_c' is also correlated with sunspot number, then the apparent correlation between H_c' and δW_2 might arise because both are influenced by causes related to sunspot numbers. Figure 5 shows the frequency distributions of H_c' for two groups with different mean sunspot number. There is no significant difference between the two distributions and thus no indication that H_c' varies with sunspot number. Thus, it seems reasonable to conclude that the reason for the larger average H-crochet on days with larger average S_q (or δW_2) is that on the average the strength of the wind system generating the S_q currents is greater on days with larger S_q.

CORRELATION ANALYSIS

The above conclusions may be verified from a study of the correlations between pairs of the three variables involved: Crochet size, δW_2, and sunspot number. While it has been indicated that the logarithms of crochet sizes are approximately normally distributed, the distribution of actual crochet sizes was not found to deviate sufficiently from a normal distribution to render a correlation analysis useless. Figure 6 shows the regression lines for the correlation berween H_c' and δW_2. From the slopes of the two regression lines, the correlation coefficient, r, can be obtained, as indicated on the graph. The value $P = 10^{-5}$ indicates the probability, based on Fisher's z transformation [4], of obtaining, from a sample of 97 pairs, a value of $r \geq 0.45$, if the sample came from a population with $r = 0$, and with the variates normally distributed. This result also leaves little doubt that H_c' depends on δW_2. Similarly, the correlation between δW_2 and daily mean sunspot numbers, for the sample of 97 days on which crochets were observed between 0900 and 1400 hours, 75° WMT, is shown in Figure 7. Here $r = 0.62$ and the probability of obtaining $r \geq 0.62$ in a sample of 97 from a population with $r = 0$ is less than 10^{-5}. Finally, Figure 8 indicates the correlation between sunspot

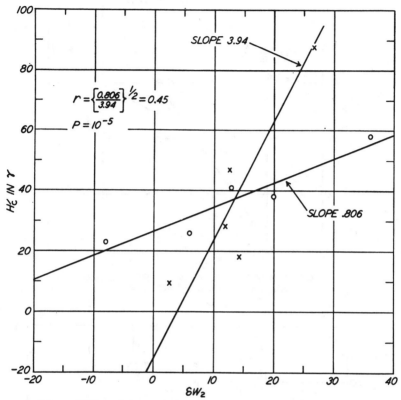

FIG. 6 — CORRELATION BETWEEN H_c' AND δW_2; REGRESSION LINES INDICATED

numbers and H_c'. Here $r = 0.12$ and the probability of reality is only 0.3, indicating that r does not differ significantly from zero. In general, if we have the three total correlation coefficients r_{12}, r_{13}, and r_{23} between three pairs of variables, then the partial correlation coefficient $r_{12.3}$, for example (to give the correlation between variables 1 and 2 that would obtain if variable 3 remained constant), is given by [4]

$$r_{12.3} = \frac{r_{12} - r_{13} \cdot r_{23}}{[(1 - r_{12}^2)(1 - r_{23}^2)]^{1/2}}$$

and similar formulas for the other two partial correlation coefficients. In the above formula, let variable 1 represent H_c', variable 2 represent δW_2, and variable 3 represent sunspot number. If we accept $r_{12} = 0.45$ (from Fig. 6) and $r_{23} = 0.62$ (from Fig. 7), it is readily found that $r_{12.3} = 0$ if $r_{13} = 0.73$; in other words, in order that the partial correlation between H_c' and δW_2 (with sunspot number constant) be zero (with $r_{12} = 0.45$ and $r_{23} = 0.62$), the total correlation between H_c' would need to be 0.73, a value which is most unlikely for the population value, since, from the sample in Figure 8, $r_{13} = 0.12$. Thus, the correlation between

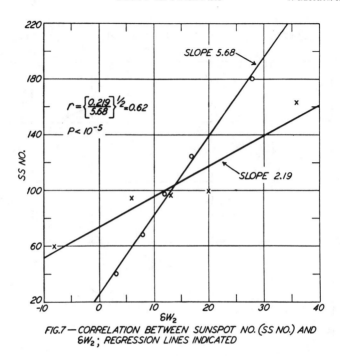

FIG.7 — CORRELATION BETWEEN SUNSPOT NO. (SS NO.) AND
δW_2; REGRESSION LINES INDICATED

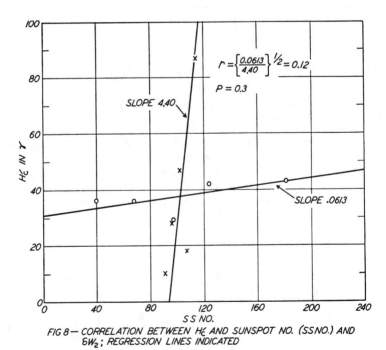

FIG 8 — CORRELATION BETWEEN H'_c AND SUNSPOT NO. (SS NO.) AND
δW_2; REGRESSION LINES INDICATED

H-crochet size and amplitude of S_q does not arise from a common effect on both due to a mechanism correlated with sunspot number.

A more direct test of this conclusion was made, as follows: As shown in Table 1,

TAFLE 1—*Arrangement for dichotomizing δW_2 to eliminate main effect of sunspot number on δW_2*

Sunspot numbers	No. of crochets	Group A		Group B	
		δW_2	No.	δW_2	No.
		γ		γ	
150–258	19	≥ 30	10	< 30	9
110–148	20	> 15	10	≤ 15	10
89–108	17	> 11	8	≤ 11	9
56– 87	21	≥ 8	11	< 8	10
0– 55	20	> 0	10	< 0	10
Totals	97		49		48

each day with crochet was placed in one of five groups according to the sunspot number for that day. There were about 20 such days in each group. Within each group, the median δW_2 was found; all days with $\delta W_2 >$ median were placed in group A and those with $\delta W_2 <$ median were placed in group B. The frequency distribution of crochet sizes H_c' in Table 2 was obtained for the 49 crochets in

TAFLE 2—*Cumulative frequency distribution of crochet sizes, H_c', for groups A and B*

H_c'	Group A		Group B	
	Cumulated number	Cumulated frequency	Cumulated number	Cumulated frequency
		per cent		*per cent*
0– 10	3	6	8	17
11– 20	14	29	20	42
21– 30	22	45	31	65
31– 40	27	55	39	81
41– 50	32	65	40	83
51– 60	33	67	44	92
61– 70	36	74	47	98
71– 80	39	80	48	100
81– 90	45	92	...	...
91–100	47	96	...	...
>100	49	100	...	...

group A and for the 48 crochets in group B. The mean sunspot numbers for groups A and B were, respectively, 106 and 98, a difference of only 8. However, the mean values of δW_2 for the two groups were, respectively, 23.4 and 3.4 for groups A and

B. The frequency distributions for groups A and B were sensibly logarithmically normal, with estimated standard deviation, s.d., of log size = 0.30. The median values of H'_c for groups A and B were, respectively, 34.0γ and 21.6γ, giving a difference in the mean logarithms of log $(34.0/21.6) = 0.197$. The s.d. for this difference is about $(0.30 \sqrt{2/7}) = 0.06$, or the difference is about 3.3 times its estimated s.d. A difference as large or larger than this should occur from samplng with probability about 10^{-3}. Thus, the difference between the means of the two distributions is statistically significant. The mean H'_c for group A is 43γ and for group B is 27γ, with corresponding δW_2 values 23.4 and 3.4, respectively. These two points lie quite close to the line with slope 0.806 in Figure 6. This result then shows directly that the correlation between H'_c and δW_2 is not due to a common effect of sunspot number on both.

CONCLUSION

The simplest explanation for the correlation between crochet size and amplitude of S_q is that it arises because of variations in the strength of the wind pattern, which by the dynamo theory generates the diurnal variation current-system. The stronger wind patterns result on the average in a proportionate increase in crochet size and in amplitude of S_q. The results obtained in Figure 4, for example, indicate that the strength of the wind pattern averages about 50 per cent greater for days in group A than for days in group B. On the other hand, the correlation between sunspot number and δW_2 probably is due to changes in solar radiation with sunspot number, and its effect on the ionospheric conductivity at the level where the S_q currents flow. The correlation coefficient $r = 0.62$ between δW_2 and sunspot number, as shown for the sample in Figure 7, can be used [5] to determine the magnitude of the part of δW_2 that is linearly correlated with sunspot numbers relative to the part which is not. This ratio is given [5] by $r/(1 - r^2)^{1/2}$. With $r = 0.62$, the ratio is about 0.8, which means that the standard deviation of the values of δW_2 predicted from daily sunspot numbers is for this sample somewhat smaller (0.8 as large) than the standard deviation of the residuals after making a linear prediction. Thus the variability of S_q due to variability in strength of wind patterns could be slightly larger than the variability due to changing sunspot number.

References

[1] A. G. McNish, Terr. Mag., **42**, 109-122 (1937).
[2] S. Chapman and J. Bartels, Geomagnetism, Oxford, Clarendon Press (1940).
[3] J. Bartels, Terr. Mag., **51**, 181-242 (1946).
[4] R. A. Fisher, Statistical methods for research workers, G. E. Stechert and Co., New York, 8th ed. (1941).
[5] J. Bartels, Terr. Mag., **37**, 1-52 (1932).

VARIATIONS OF WINDS IN DYNAMO MECHANISM FOR S_q

APPENDIX A—Date, GMT of beginning, and magnitude, H_c, of H-crochets at Huancayo, Peru, 1936-1947, with daily sunspot number, and other relevant data

Date	GMT	H_c	ΔH	ΔH^*	δW_2	Sunspot No.	Date	GMT	H_c	ΔH	ΔH^*	δW_2	Sunspot No.
	h m	γ	γ	γ				h m	γ	γ	γ		
1938 Apr. 26	12 50	+18	42	63	+23	92	1937 July 13	17 56	+12	158	127	+29	188
1936 Jun. 16	13 27	+10	27	70	+12	55	1937 Oct. 5	17 05	+68	289	278	+39	175
1936 July 15	13 26	+8	36	55	+10	67	1938 Jan. 15	17 08	+26	137	141	+25	118
1937 Jun. 24	13 27	+43	44	44	+19	133	1938 Feb. 3	17 30	+49	230	170	+21	68
1940 Feb. 8	13 45	+32	53	123	+13	64	1938 May 3	17 45	+41	164	131	+16	160
1946 Feb. 1	13 34	+22	56	79	-4	94	1938 Sep. 20	17 47	+84	199	181	+28	57
1946 Aug. 3	13 00	+12	40	56	+19	137	1939 Apr. 29	17 10	+75	163	174	+33	136
1947 Jun. 28	13 05	+21	73	97	+20	144	1940 Jan. 28	17 23	+10	171	151	+8	43
1936 Apr. 6	13 52	+25	37	86	+22	91	1940 Mar. 26	17 40	+34	256	263	+26p	108
1937 Apr. 21	13 58	+7	125	95	+11	127	1940 May 14	17 48	+28	100	102	0	93
1937 July 9	14 19	+86	80	126	+44	181	1941 Jun. 4	17 44	+60	106	106	+14	53
1941 Jun. 5	14 27	+17	52	72	+11	61	1942 Jan. 23	17 20	+12	136	128	-3	31
1941 Jun. 30	14 00	+20	32	45	-5	98	1942 Mar. 25	17 45	+31	152	152	+11	100
1942 Jan. 29	14 23	+35	72	97	-3	11	1943 Feb. 10	17 30	+38	114	118	-4	37
1943 Apr. 19	14 39	+7	53	72	-11	41	1945 Mar. 29	17 13	+10	106	80	-22	60
1943 Apr. 22	14 23	+12	40	80	-17	45	1946 Feb. 11	17 30	+37	106	104	+3	115
1943 Apr. 24	14 42	+25	52	105	+6	36	1947 Sep. 2	17 55	+29	227	196	+46	196
1946 Feb. 10	14 32	+29	30	91	+11	121	1946 Mar. 8	17 58	+122	198	198	+13	71
1946 Aug. 12	14 00	+14	98	108	+9	98	1936 Jun. 16	18 00	+12	100	90	+12	55
1947 Jan. 14	14 30	+58	118	168	+29	150	1936 Aug. 25	18 26	+68	151	139	+19	90
1947 Apr. 10	14 40	+30	152	150	+15	171	1937 Jun. 10	18 52	+11	97	44	+17	98
1947 July 3	14 50	+11	75	128	+17	148	1937 July 7	18 46	+17	166	131	+24	143
1947 Aug. 31	14 53	+88	110	145	+36p	213	1937 Oct. 28	18 24	+56	136	104	+16	104
1936 Feb. 14	15 26	+59	112	195	+6	77	1938 July 11	18 03	+17	120	113	+5	205
1936 Oct. 21	15 35	+23	140	159	+7	65	1938 July 25	18 00	+16	149	143	+38	202
1937 Jun. 14	15 32	+23	90	135	+42	185	1938 Oct. 15	18 59	+15	166	118	+10	103
1937 July 29	15 36	+8	70	102	+12	128	1938 Dec. 7	18 13	+12	12?	99	+18	117
1938 Jan. 13	15 30	+34	218	252	+28p	106	1939 July 15	18 30	+29	127	111	+11	87
1938 May 21	15 12	+15	79	106	+11	106	1939 Aug. 29	18 55	+13	152	121	+6	69
1938 Jun. 29	15 04	+12	69	94	+8	72	1939 Sep. 12	18 26	+41	111	82	+3	150
1938 Oct. 14	15 38	+21	133	179	+12	122	1940 Jan. 8	18 03	+23	94	53	+16	38
1939 Mar. 21	15 17	+31	86	120	-1p	89	1940 Feb. 17	18 34	+59	142	116	+14	51
1939 Sep. 3	15 25	+69	156	179	+10p	124	1940 May 25	18 29	+8	108	100	-9	76
1939 Sep. 13	15 36	+72	164	184	+12	127	1940 Sep. 20	18 59	+19	148	101	+11	98
1940 Feb. 24	15 40	+12	116	146	-3	40	1942 Jan. 12	18 34	+66	143	129	+18	48
1941 Feb. 27	15 44	+90	105	165	+16	50	1945 Apr. 27	18 23	+12	125	84	+11	56
1941 Feb. 28	15 27	+49	94	145	+4	55	1946 Feb. 24	18 42	+18	145	137	+11	87
1941 Apr. 3	15 43	+25	100	120	-13	21	1946 Feb. 28	18 05	+38	135	123	+18	90
1946 July 21	15 10	+86	115	115	-34p	110	1946 Mar. 21	18 15	+13	114	103	-8	101
1946 Aug. 7	15 35	+67	127	124	+15	116	1946 Sep. 13	18 20	+30	176	145	+25	92
1946 Oct. 13	15 30	+6	198	249	+25	92	1947 May 22	18 47	+34	181	148	+40	258
1947 Apr. 4	15 40	+50	211	206	+27	194	1947 July 31	18 48	+6	45	25	-27	113
1936 Apr. 8	16 46	+89	119	168	-3	76	1937 Jun. 2	19 40	+16	114	84	+18	89
1936 Apr. 25	16 53	+13	115	42	-5	72	1937 Aug. 28	19 25	+47	186	150	+32	130
1936 Aug. 5	16 05	+14	63	82	-4	87	1937 Sep. 29	19 55	+18	129	84	+19	126
1936 Nov. 6	16 10	+47	119	134	+14	151	1939 May 8	19 41	+18	152	128	+13p	105
1936 Nov. 27	16 50	+7	125	106	+10	145	1940 Feb. 15	19 17	+14	126	112	+1	78
1937 May 25	16 49	+99	166	166	+40p	171	1941 Feb. 26	19 11	+58	124	114	+9	46
1937 Sep. 29	16 31	+20	175	175	+19	126	1942 Nov. 6	19 27	+12	52	41	-12	33
1938 July 9	16 33	+18	134	142	+30	175	1944 Dec. 10	19 14	+21	75	75	+8	45
1938 July 29	16 10	+37	105	129	+6	151	1945 Apr. 25	19 47	+15	66	73	+6	52
1938 Sep. 23	16 54	+50	179	177	+19	86	1945 Sep. 3	19 59	+9	102	55	+11	26
1939 Apr. 21	16 56	+181	196	213	+36p	125	1945 Sep. 26	19 23	+11	129	100	+19	53
1939 July 16	16 40	+40	139	166	+22p	74	1945 Oct. 29	19 37	+16	122	68	+9	68
1939 Aug. 9	16 10	+8	122	156	+32	157	1936 May 8	20 19	+5	64	45	+15	46
1941 Mar. 7	16 30	+60	113	104	-1	47	1936 Jun. 10	20 53	+17	50	21	+1p	40
1942 Jan. 13	16 24	+76	129	148	+14	50	1937 May 1	20 27	+16	99	74	+19	89
1942 Mar. 31	16 23	+27	144	144	-4	55	1939 Sep. 6	20 00	+68	118	98	+29	136
1942 Apr. 29	16 23	+27	122	127	+2	65	1942 Mar. 6	20 40	+7	129	70	+5	20
1942 Oct. 26	16 18	+25	118	116	-9	13	1943 Apr. 19	20 30	+26	59	50	-11	41
1946 July 25	16 21	+104	121	121	+56p	117	1945 Apr. 27	20 50	+21	59	62	+11	56
1936 Aug. 4	17 26	+7	110	104	+1	65	1945 Apr. 28	20 44	+6	50	18	+6	55
1936 Sep. 4	17 11	+21	194	162	+25	57	1945 Aug. 17	20 20	+48	60	52	+3	34

*Used only in Figure 1.

LETTER TO EDITOR

LARGE INCREASE OF COSMIC-RAY INTENSITY FOLLOWING SOLAR FLARE ON FEBRUARY 23, 1956

The fifth and largest increase in cosmic-ray intensity yet recorded[1,2] at Cheltenham began at 03[h] 48[m] GMT, February 23, 1956. Mr. Roger Moore, of the North Atlantic Radio Warning Service, at Fort Belvoir, Virginia, has informed us that a large solar flare was reported first seen in Tokyo at 03[h] 34[m] GMT, February 23, 1956; and that at Kodaikanal, India, this solar flare was observed between 03[h]

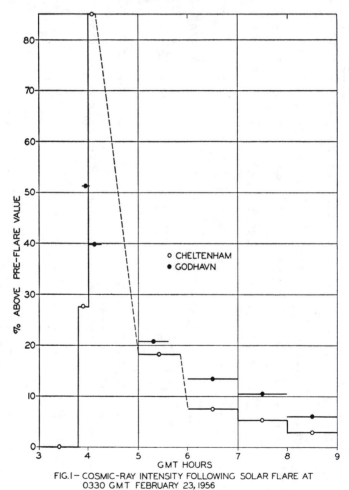

FIG. I — COSMIC-RAY INTENSITY FOLLOWING SOLAR FLARE AT 0330 GMT FEBRUARY 23, 1956

[1]S. E. Forbush, Phys. Rev., **70**, 771-772 (1946).
[2]S. E. Forbush, T. B. Stinchcomb, and M. Schein, Phys. Rev., **79**, 501-504 (1950).

30^m and 05^h 10^m GMT. Thus the increase in cosmic-ray intensity at Cheltenham began within about 18 minutes of the reported beginning of the flare. The Geophysical Observatory at Godhavn, Greenland, informed us that the increase there began at 03^h 53^m GMT.

The ionization in per cent above the pre-flare level is shown in Figure 1 for Cheltenham and for Godhavn, both derived from Compton-Bennett meters shielded by 11 cm Pb, for the period 03^h 00^m to 09^h 00^m GMT, February 23, 1956. The length of the horizontal bars attached to the points in the Figure indicates the duration of the interval in each hour for which a record was available. The maximum intensities recorded at Cheltenham and Godhavn were, respectively, about 85 and 50 per cent above the pre-flare values. At Cheltenham and at Godhavn, the record was off scale for the major part of the interval 04^h 00^m to 05^h 00^m. However, excellent averages were obtained for every six-minute interval from a large shielded ionization chamber at Derwood, Maryland (near Washington). These results, which will be reported later by others, indicate that the maximum intensity occurred between 04^h 00^m and 04^h 12^m GMT.

Dr. J. C. Barton, of the Physics Department of the University College of the West Indies, in Jamaica, B.W.I., wrote that a large increase was detected there, at 500 feet elevation, by Dr. J. H. Stockhausen, with a wide-angle telescope, shielded by 10 cm Pb. He indicated that the average counting rate for the interval 03^h 30^m to 04^h 00^m GMT, February 23, was about 33 per cent above the pre-flare value and that the rate for the next half-hour was about 27 per cent above normal; thereafter, the counting rate was about normal.

The results of Dr. Barton and Dr. Stockhausen at geomagnetic latitude 29° north show the largest increase yet reported for so low a geomagnetic latitude, thus indicating that the influx of particles on February 23, 1956, coming probably from the sun, contained a greater number of charged particles, above the cut-off momentum for this latitude, than has been observed in the four previously recorded large cosmic-ray increases.[1,2]

Note added in proof: Starting at 03^h 45^m GMT, February 23, 1956, the increase in cosmic-ray ionization from shielded Compton-Bennett meters averaged for the next 15 minutes was 18 per cent at Huancayo (geomagnetic latitude $\Phi = 0.6°S$) and about 36 per cent at Ciudad Universitaria, México, D.F. ($\Phi = 29.7°N$). Starting with either of these values, the increase averaged for subsequent 15-minute intervals decreased exponentially with half life 15 minutes. For the excellent records from Huancayo and from México, we are indebted to Mr. Alberto A. Giesecke, Jr., of the Instituto Geofísico de Huancayo, and to Dr. José y Coronado, of the Universidad Nacional de México.

Scott E. Forbush

Department of Terrestrial Magnetism,
  Carnegie Institution of Washington,
    Washington 15, D. C., February 28, 1956
    (Received February 28, 1956)

ABSORPTION OF COSMIC RADIO NOISE AT 22.2 MC/SEC
FOLLOWING SOLAR FLARE OF FEBRUARY 23, 1956

Reports of widespread ionospheric disturbances on the dark hemisphere of the earth, accompanying the great solar flare of February 23, 1956, have been noted by a number of groups. A. Shapley has reported an increase in ionospheric absorption at vertical sounding stations in high latitudes,[1] commencing a few minutes after the flare and lasting many hours, while Ellison and Reid,[2] and Gold and Palmer,[3] have reported a sudden decrease in 27-kc/sec atmospherics at the time of the flare, coming too early and much too suddenly to be due to sunrise effect. In the auroral zone, effects were noted by the Geophysical Institute at College, Alaska,[4] whose records of galactic radio noise at 30 Mc/sec showed an increase in absorption shortly after the flare, increasing gradually to a maximum of 1.8 db over the period $04^h 20^m$ to $09^h 00^m$ UT.

Absorption effects of a different nature have been found on radio-source records taken at the Seneca Radio Observatory of the Carnegie Institution, near Washington, D. C. The equipment consisted of a phase-switching interferometer[5] tuned to 22.2 Mc/sec, utilizing a pair of small antennas which were normally used for observations of the planet Jupiter, but which observe a number of other radio sources as well. Comparison of the record obtained near the time of the flare with a record taken the previous night showed a noticeable difference for a short period when Virgo A (IAU 12N1A) was being observed. The apparent intensity of the source on February 23 compared to the intensity of the previous night is shown in Figure 1. The horizontal lines give the relative intensity, averaged over the period indicated by the length of the line. Since the observation was made with a phase-switching interferometer which gives zero output when the source is near quadrature in the interference pattern, the averages were taken over the peaks, when the source is near a maximum or minimum in the interference pattern. From $04^h 02^m$ to $04^h 18^m$, the equipment was being calibrated, and hence no data are available over this period. The experimental uncertainty is determined by the background noise fluctuations, and since the source is not a strong one we estimate that the relative intensity may be subject to a probable error of the order of magnitude ±0.1.

For purposes of comparison, the cosmic-ray increase observed[6] by the large shielded Carnegie ion-chamber, also near Washington, is plotted in Figure 1, the

[1] A. Shapley, private communication.

[2] M. A. Ellison and J. H. Reid, J. Atmos. Terr. Phys., 8, 291 (1956).

[3] T. Gold and D. R. Palmer, *ibid.*, p. 287.

[4] Geophysical Institute, University of Alaska: Abstr. of terrestrial phenomena observed following solar flare 23 Feb. 1956, 0334.

[5] M. Ryle, Proc. R. Soc., A, 211, 351 (1952).

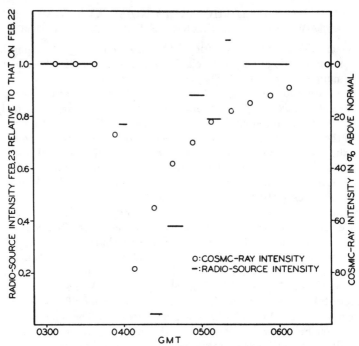

FIG. I—ABSORPTION ON 22 MC SEC⁻¹ FROM RADIO SOURCE VIRGO A AND
INCREASE IN COSMIC-RAY INTENSITY NEAR WASHINGTON, D.C.,
FOLLOWING SOLAR FLARE OF FEBRUARY 23,1956

similarity of the two curves showing clearly. The initial increase in absorption, at $04^h\ 00^m$, may not be real, however, since the uncertainty is large; the later points are unquestionably indicative of strong absorption, reaching a maximum of at least 6 db and probably greater. The commencement time is not well defined, but is certainly no later than $04^h\ 20^m$.

Ionospheric vertical-sounding records from the National Bureau of Standards station at Fort Belvoir, Virginia, showed no unusual increase in absorption at the time,[1] which would imply that the material responsible for the 22.2-Mc/sec absorption was not located below the $F2$ layer. Furthermore, the obscuration of a point source was observed, and it is not possible to say whether a localized absorbing cloud or a general absorbing blanket over the entire sky was responsible. Ellison and Reid[3] reported no observable absorption of the galactic background noise at 18 Mc/sec, an observation which might favor the hypothesis of a localized absorbing cloud, but since their observation was made at a different location on the earth, no definite conclusion can be reached.

S. E. Forbush
B. F. Burke

Department of Terrestrial Magnetism,
Carnegie Institution of Washington,
Washington 15, D. C., July 16, 1956
(Received July 17, 1956)

⁶H. E. Tatel, private communication.

SOLAR INFLUENCES ON COSMIC RAYS

By Scott E. Forbush

CARNEGIE INSTITUTION OF WASHINGTON, DEPARTMENT OF TERRESTRIAL MAGNETISM,
WASHINGTON D.C.

Introduction.—Probably all the established variations with time of cosmic-ray intensity are directly or indirectly due to solar influences. Ionization chambers, because they are relatively simple and reliable, have proved valuable for continuous registration of cosmic-ray intensity over long periods of time. The longest series of continuous observations with several identical instruments of this type is that which has been obtained with the Carnegie Institution of Washington model C Compton-Bennett meters.[1] With these meters, and with the unselfish co-operation of several organizations, continuous data have been obtained for nearly two decades at each of the following places:[2] Godhavn (Greenland), Cheltenham (Maryland), Huancayo (Peru), and Christchurch (New Zealand). Data for shorter periods have been obtained from Teoloyucan and Ciudad Universitaria, Mexico, D.F., and from Climax, Colorado. These ionization chambers are shielded by the equivalent of 12 cm. Pb to screen out local gamma rays, the intensity of which may vary with time.

Origin of μ-Meson Component Measured in Ionization Chambers and Its Seasonal Variation.—In ionization chambers the ionization is normally produced mainly by μ-mesons, which have a mass of about 210 electron masses, unit electronic charge (±), and a lifetime (at rest) of about 2×10^{-6} seconds. These very penetrating μ-mesons arise from the decay of short-lived charged π-mesons (lifetime at rest about 10^{-8} seconds), which in turn result from the interaction of high-energy protons of the primary cosmic-ray beam with atmospheric nuclei. There are other types of reaction with different decay products which need not concern us here. The maximum μ-meson intensity occurs in the region where the pressure is roughly 100 mb., which normally is near an altitude of 16 km. or so. An increase in the height of this pressure level lengthens the path for μ-mesons, so that more of them decay into not very penetrating electrons (and nutrinos) before reaching the instruments at the ground. Thus the seasonal variation in height of the 100-mb. level results in a seasonal variation in cosmic-ray intensity as measured by ionization chambers. The passage of meteorological "fronts" can also alter the height of the 100-mb. level and consequently the apparent cosmic-ray intensity as recorded in ionization chambers (or with Geiger counters). The seasonal waves and meteorological effects proved less interesting than troublesome, for they obscured for a while the more interesting remaining changes in cosmic-ray intensity which are world-wide[3] and result directly or indirectly from solar influences. Since ionization chambers are mainly sensitive to the μ-meson component, it is essential to remark that normally most of the ionization in such detectors arises from primaries of relatively high energy. For example, near sea level, the intensity in ionization chambers near the geomagnetic pole is ordinarily only about 10 per cent greater than at the equator. This means that about 90 per cent of the ionization results from primaries which have enough momentum to arrive at the earth's geomagnetic equator (about 15 Bev/c for protons). The primaries of lower momentum which are not prevented by the earth's magnetic field from reaching the earth, say at or north of geomagnetic latitude 50°, are ineffective in producing penetrating μ-mesons.

Origin of the Nucleonic Component and Solar-Flare Effects.—The primary cosmic-ray particles, protons or heavier nuclei, also generate a cascade of nucleons (protons and neutrons) in the atmosphere; even relatively low-energy protons, with momenta only great enough (a few Bev/c) to be allowed by the earth's field to reach the atmosphere at geomagnetic latitude 50°, are very effective generators of the nucleonic component. Most of the rare large increases in cosmic-ray intensity which have been accompanied by solar flares or chromospheric eruptions are due to relatively low-energy charged particles coming from the sun.[4] The momenta of most of these particles are too low, as shown by their large latitude effect, to result in production of μ-mesons (through the $\pi \rightarrow \mu$ decay). The large augmentation of ionization observed in ionization chambers during these solar-flare effects results from the nucleonic component, as was shown[4] by the fact that the magnitude of the augmentation increased with altitude at the same rate as that known for the nucleonic component. During the solar flare of November 19, 1949, the total ionization in a Compton-Bennett meter at Climax, Colorado, increased about 200 per cent. At Climax, under 12 cm. Pb, probably no more than 10 per cent of the total ionization is normally due to local radiation originating from the nucleonic component. Assuming that this normal radiation was produced by particles in the same energy band as that responsible for the increase of 200 per cent (at Climax) in ionization which accompanied the solar flare of November 19, 1949, then it was predicted[4] that the rate of arrival of primary particles in that energy band must have increased to at least 20 times the normal value. During the solar flare[5] of February 23, 1956, increases of 20-fold and greater have been reported from detectors sensitive only to the nucleonic component. Thus neutron detectors have many advantages over the ionization chamber for measuring effects which arise from variations in the intensity of the low-energy part of the cosmic-ray spectrum.

Magnetic-Storm Effects.—Figure 1 shows a decrease of daily means of cosmic-ray intensity similar at three stations during the magnetic storm of April 25–30, 1937. The bottom curve indicates the decrease and subsequent gradual recovery of daily means of horizontal magnetic intensity, H, at Huancayo. It will be noted that the rate of recovery of H toward normal is similar to that for cosmic-ray intensity. Figure 2 shows three examples of decreases in daily means of cosmic-ray intensity associated with decreases in horizontal magnetic intensity at Huancayo during the period January 11–31, 1938, which contains what may for convenience be called three separate magnetic storms. While the ratio of changes in cosmic-ray intensity to those in horizontal magnetic intensity is relatively constant throughout any one of these storms, it differs among the three storms. Not all storms are accompanied by decreases in cosmic-ray intensity. The decrease in horizontal magnetic intensity at Huancayo for the magnetic storm beginning August 21, 1937, was accompanied by no detectable decrease in cosmic-ray intensity in the Compton-Bennett ionization chambers. Figure 3 depicts the daily means of cosmic-ray intensity and of horizontal magnetic intensity, H, at Huancayo for 1946. The decrease of several per cent in cosmic-ray intensity between February 3 and 6 occurred before the major depression in H which took place on February 7. There was, however, on February 3 a sudden magnetic commencement, although this was not im-

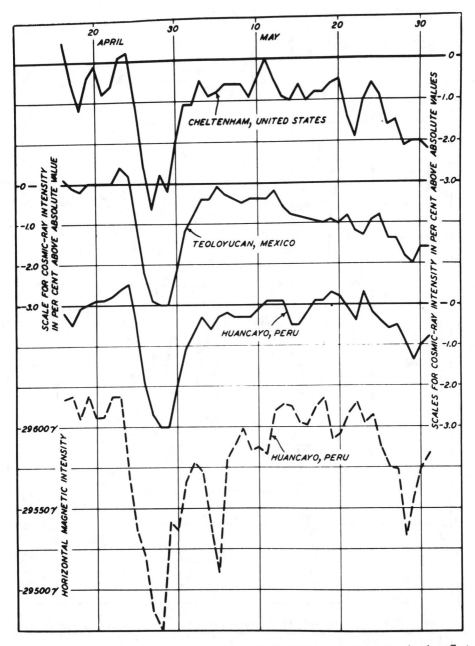

FIG. 1.—Daily means of cosmic-ray intensity and horizontal magnetic intensity, showing effect of magnetic storm of April 25–30, 1937, on cosmic-ray intensity.

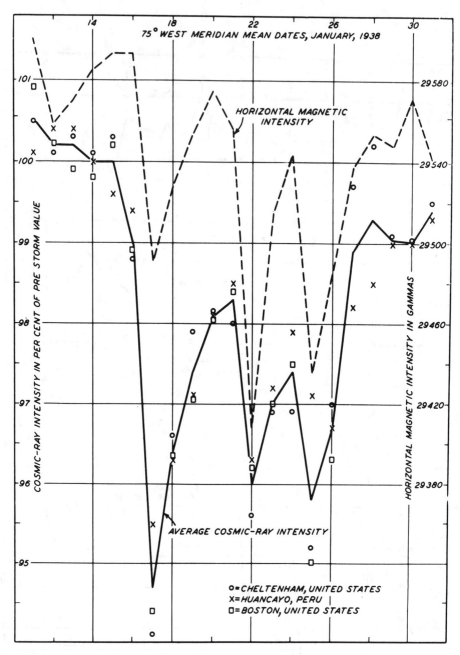

Fig. 2.—Magnetic-storm effects on daily mean cosmic-ray intensity at Boston, United States, Cheltenham, United States, and Huancayo, Peru, and on daily mean magnetic horizontal intensity at Huancayo, Peru.

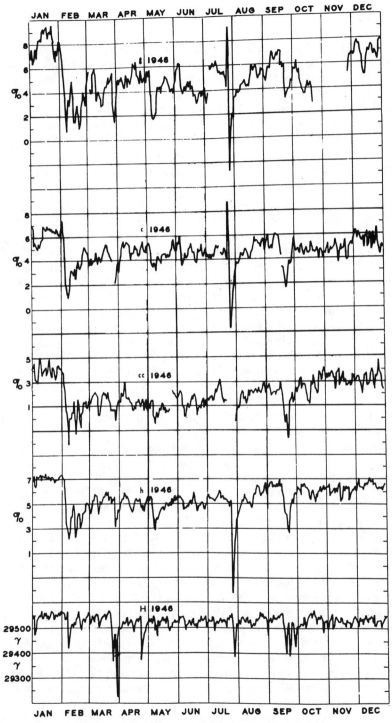

FIG. 3.—Daily means, cosmic-ray intensity, 1946. q = Godhavn, c = Cheltenham, cc = Christchurch, h = Huancayo; H = horizontal magnetic intensity at Huancayo.

mediately followed by a magnetic storm. This particular decrease in cosmic-ray intensity, which began February 3, 1946, is unusual in that the cosmic-ray intensity remained low for much of the remainder of the year. Incidentally, the large values for July 25 at Godhavn (*g*) and at Cheltenham (*c*) result from the large increase associated with a solar flare on that date. No flare effect was registered at Huancayo, since the charged particles from the sun had insufficient momenta for the earth's magnetic field to allow their arrival at the equator. At Christchurch (*cc*) the flare occurred during a period of a few days when the equipment was not operating. In Figure 3 it is evident that the major changes in daily means of cosmic-ray intensity are world-wide.[3]

Variability of Daily Means.—Figure 3 shows the variations in daily mean cosmic-ray intensity for several stations for a year near maximum sunspot activity. The curves for 1944 in Figure 4 indicate the variations of daily means for 1944 at sunspot minimum. The curve *h 1944* for daily means of cosmic-ray intensity at Huancayo exhibits much less variability than for 1946, and the same is true for the daily means of magnetic horizontal intensity (*H 1944*) at Huancayo. Each of the curves (*g 1944*, *c 1944*, and *cc 1944*) in Figure 8 exhibits greater variability, especially during winter, than does *h 1944*. Moreover, most of this variability is uncorrelated between different pairs of stations. Except at Huancayo, this variability is doubtless due to μ-meson decay effects resulting from changes in the vertical distribution of air mass with the passage of fronts. At Huancayo the variability of daily means in some months of 1944 is not greatly in excess of the inherent "noise level" of the instrument. Thus Huancayo is essentially free of the disturbing effects arising from the passage of meteorological "fronts" which do not occur there. The standard deviation of daily means from monthly means (pooled for each year) of cosmic-ray intensity at Huancayo is shown in Figure 5 for the period 1937–1955. This variability is least in the years of sunspot minima, 1944 and 1954, and increases near years of sunspot maxima. The variability is somewhat less when the five magnetically disturbed days of each month are excluded. This shows that in nearly every year significant cosmic-ray changes occur on at least some of the magnetically disturbed days. In some years much of the variability arises from the 27-day quasi-periodic variation.[2]

Disturbed minus Quiet-Day Difference.—For each month geomagneticians determine the five days when the earth's magnetic field is quietest and the five days when it is most disturbed. In Figure 6 the average difference in cosmic-ray intensity for the disturbed less that for the quiet days for each month at Huancayo is plotted against the corresponding difference for Cheltenham. These differences are preponderantly negative and correlated between the two stations, showing that the cosmic-ray intensity definitely tends to be less on magnetically disturbed than on magnetically quiet days. The annual means for magnetically disturbed (60 per year) less quiet days (60 per year) are shown in Figure 7 for cosmic-ray intensity at each of three stations (and their average) and for magnetic horizontal intensity. These differences are all negative and vary (algebraically) roughly with the sunspot cycle.

Twenty-seven-Day Variation.—Figure 8 (*A*) shows the variability of the 27-day waves in magnetic activity (American magnetic character figure) and in cosmic-ray intensity at Huancayo (*B*), for a sample of 34 solar rotations. The maxima of the

242

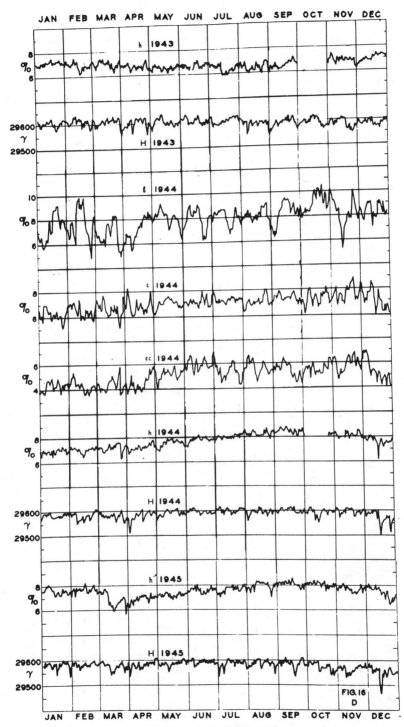

FIG. 4.—Daily means, cosmic-ray intensity, 1944 (see Fig. 3).

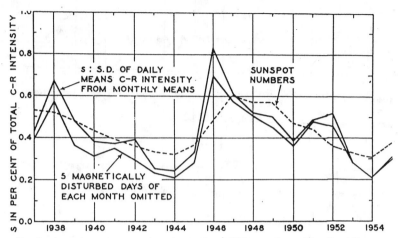

FIG. 5. –Annual means: sunspot numbers and variability, cosmic-ray intensity at Huancayo.

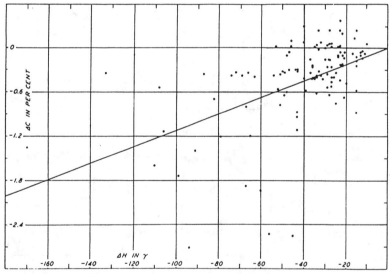

FIG. 6.—Average difference for cosmic-ray intensity (ΔC) and for horizontal magnetic field (ΔH) for five disturbed days less that for five quiet days in each month, April, 1937–December, 1946, at Huancayo.

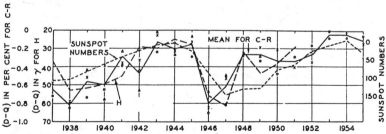

FIG. 7.—Annual means for magnetically disturbed less quiet days (5 each per month) for cosmic-ray intensity (C-R) and for horizontal magnetic intensity, H.
Legend for C-R: ° = Cheltenham, x = Huancayo, Δ = Godhavn.

244

27-day waves in cosmic-ray intensity have a statistically significant[2] tendency to occur near the times of the minima of the 27-day waves in character figure. This is shown in Figure 8 by the harmonic dials C and D obtained respectively by rotating vectors in A to the vertical and by rotating vectors (derived from the corresponding 27-day interval) in B through the same angle. If the phases of the 34 vectors in D were random, then the probability that the magnitude of the average vector

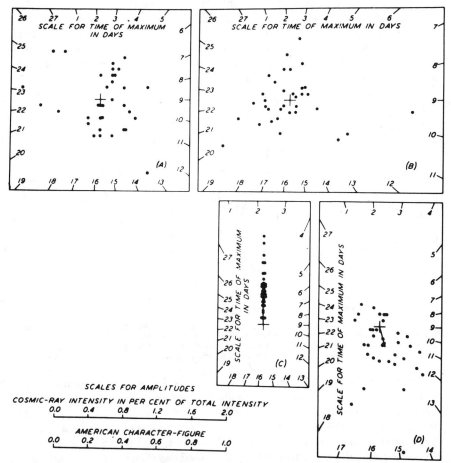

Fig. 8.—Harmonic dials for departures from average of 27-day waves in American character figure (A) and in cosmic-ray intensity at Huancayo, Peru (B), computed from 34 rotations (intervals of 27 days) beginning June 13, 1936; harmonic dials (C) and (D) obtained respectively by rotating vectors in (A) to vertical and by rotating vectors for corresponding intervals in (B) through the same angle.

would equal or exceed the magnitude of the average vector actually obtained in D would be about 2×10^{-6}. The results of a similar procedure show in Figure 9 that the maxima of the 27-day waves in cosmic-ray intensity have a statistically significant tendency to occur near the times of the maxima of the 27-day waves in magnetic horizontal, H, intensity at Huancayo. Since low values of H occur when magnetic activity is high, the results of Figures 8 and 9 are consistent. The magnetic-storm effects, the (magnetically) disturbed minus quiet-day differences, and the 27-day variations in cosmic-ray intensity and in the horizontal component, H, of the earth's magnetic field at the equator all indicate a significant tendency for decreased values of cosmic-ray intensity to occur with decreased values of H. The latter decreases are known to arise from magnetic fields with sources outside the earth. Thus the mechanism for most of these changes in cosmic-ray intensity is closely connected with the mechanism responsible for the magnetic changes. Several attempts have been made to calculate the expected cosmic-ray changes in magnetic storms, assuming that the external magnetic-storm field arises from a hypothetical ring current concentric with the earth and in the plane of the geomagnetic equator (this would explain one of the main features of magnetic storms). None of the results obtained has indicated that the decreases in cosmic-ray intensity arise from magnetic effects of the ring current.

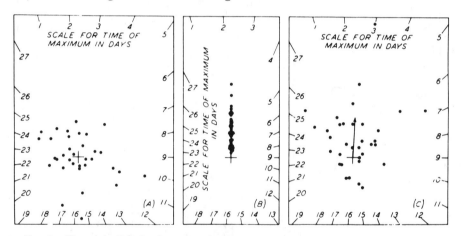

Fig. 9.—Harmonic dials for departures from average of 27-day waves in magnetic horizontal intensity at Huancayo, Peru (A), computed from 34 rotations (intervals of 27 days) beginning June 13, 1936; harmonic dials (B) and (C) obtained respectively by rotating vectors in (A) to vertical and by rotating vectors for corresponding interval in (B) of Fig. 8 through the same angle.

Variation with Sunspot Cycle.—Figure 10 shows the variation of annual means of cosmic-ray intensity at four stations and a comparison with sunspot numbers of the annual means averaged for all stations for the period 1937–1955. The latter curve indicates that cosmic-ray intensity is higher near the sunspot minima. This 11-year variation in cosmic-ray intensity at Huancayo, as shown in Figure 11, is about the same for all days, for magnetically disturbed days (5 per month), and for magnetically quiet days (5 per month). This indicates that the 11-year variation is not due directly to decreases in cosmic-ray intensity during magnetic storms, which are more frequent near times of sunspot maxima. It has been suggested that conducting solar streams (which give rise to magnetic storms) which carry "frozen-in" magnetic fields away from the sun during sunspot maxima may pervade the solar system to an extent which would reduce the flux of cosmic rays arriving at the earth from outside the solar system. In connection with the 11-year variation, it should be noted that Meyer and Simpson,[6] using nucleonic detectors in jet aircraft at 30,000 feet altitude, found that the knee of their latitude curve moved northward 3° between 1948 and 1951, showing that at the latter time additional low-rigidity

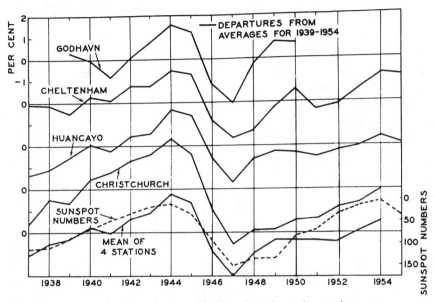

FIG. 10.—Annual means, cosmic-ray intensity, at four stations.

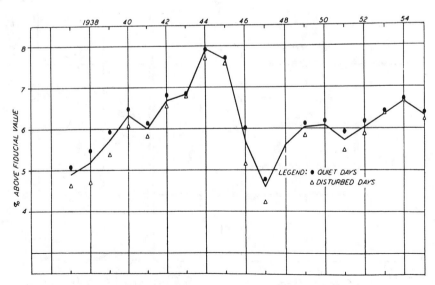

FIG. 11. Annual means, cosmic-ray intensity, at Huancayo for all days, international magnetic quiet days, and international disturbed days.

247

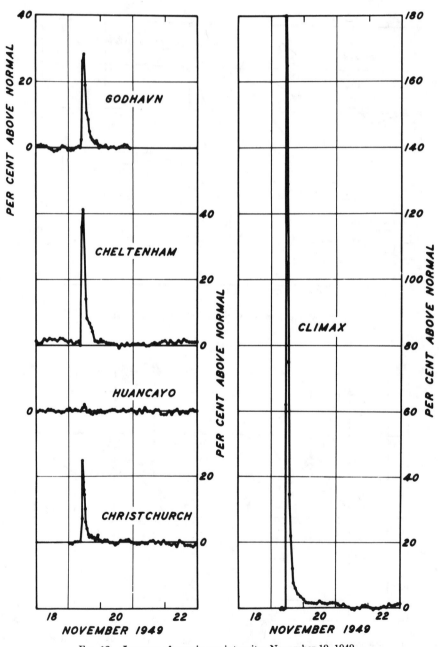

F_{IG}. 12.—Increase of cosmic-ray intensity, November 19, 1949.

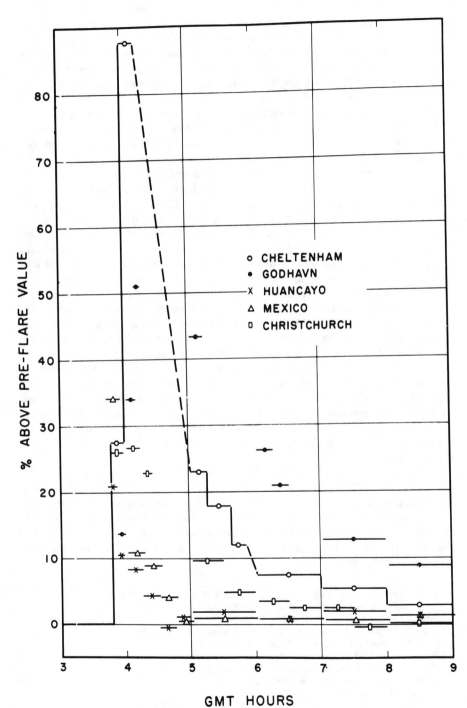

F_IG. 13.—Cosmic-ray intensity following solar flare at 0330 g.m.t., Feb. 23, 1956.

particles were reaching this level in the atmosphere. In addition, they found that the total cosmic-ray intensity was 13 per cent greater in 1951 than in 1948. Neher and Stern,[7] using high-altitude balloon-borne ionization chambers in the summers of 1951 and 1954, found, at the latter time, new low-energy particles arriving at geomagnetic latitudes north of 58°. Assuming that the new particles were protons, they calculated that there was no magnetic cutoff of primary protons above 150 Mev in 1954, whereas the cutoff for protons in 1951 was estimated at 800 Mev.

Solar-Flare Effects.—During nearly two decades of continuous operation of Compton-Bennett ionization chambers, five[4, 5] unusually large increases in cosmic-ray intensity have been recorded. Each of these followed within an hour (in all but one case, within a quarter-hour) the onset of a solar flare or radio fadeout. One of these increases is shown in Figure 3, another in Figure 12, and one in Figure 13. In four of the five increases no increase occurred at Huancayo (at the geomagnetic equator). The extra large increase at Climax relative to that at Cheltenham in Figure 12 was due mainly[4] to the high altitude of Climax relative to Cheltenham (3,428 meters). In the second section of this review ("Origin of the Nucleonic Component and the Solar-Flare Effect") the arguments were presented for the conclusion that the solar-flare increases were due to the nucleonic component generated by a great increase in the flux of charged particles in the lower-energy part of the cosmic-ray spectrum. Figure 13 shows the recent increase of February 23, 1956, as measured at several stations with Compton-Bennett ionization chambers. This is the first occasion on which a solar-flare increase has been observed at the equator and shows that charged particles with momenta of at least 15 Bev/c were involved. Schlüter[8] and Firor[9] have shown that if the particles responsible for the solar-flare increases come from the sun, then the observed intensities should be much greater in certain "impact zones" which depend on the geomagnetic latitude and local geomagnetic time. For example, in Figure 13 the large difference in the intensity at Cheltenham and Christchurch (both at about the same geomagnetic latitude, except for sign) is due to the location of these stations relative to the impact zones.

It thus seems reasonably certain that the solar-flare increases in cosmic-ray intensity are due to charged particles from the sun which are accelerated by some mechanism closely associated with the flare. The remaining variations of intensity are doubtless perturbations imposed by the magneto-hydrodynamical state of the solar system upon a steady influx of cosmic-ray particles coming from outside the solar system. The mechanism which effects these perturbations of intensity is not clearly understood.

[1] A. H. Compton, E. O. Wollan, and R. D. Bennett, *Rev. Sci. Instr.*, **5**, 415, 1934.

[2] Scott E. Forbush, *J. Geophys. Research*, **59**, 525, 1954.

[3] Scott E. Forbush, *Phys. Rev.*, **54**, 975, 1938.

[4] Scott E. Forbush, M. Schein, and T. B. Stinchcomb, *Phys. Rev.*, **79**, 501, 1950.

[5] Scott E. Forbush, *J. Geophys. Research*, **61**, 155, 1956.

[6] Peter Meyer and J. A. Simpson, *Phys. Rev.*, **99**, 1517, 1955.

[7] H. V. Neher and E. A. Stern, *Phys. Rev.*, **98**, 845, 1955.

[8] A. Schlüter, *Z. Naturforsch.*, **6a**, 613, 1951.

[9] J. Firor, *Phys. Rev.*, **94**, 1017, 1954.

Journal of

GEOPHYSICAL RESEARCH

The continuatiu: of

Terrestrial Magnetism and Atmospheric Electricity

VOLUME 63	DECEMBER, 1958	No. 4

COSMIC-RAY INTENSITY VARIATIONS DURING TWO SOLAR CYCLES

BY SCOTT E. FORBUSH

Department of Terrestrial Magnetism,
Carnegie Institution of Washington,
Washington 15, D. C.

(Received July 18, 1958)

ABSTRACT

To facilitate the use of cosmic-ray intensity data from ionization chambers which are being sent to International Geophysical Year World Data Centers, and of those which have appeared in two publications of the Carnegie Institution of Washington, an improved correction for instrumental drift at Huancayo is derived and the reliability of the corrections for seasonal variations is discussed.

From ionization chambers, the decrease of intensity from its maxima (near sunspot minima) is shown to lag a year or more behind the increase of solar activity following sunspot minima. This lag does not appear in the results obtained by Neher at high altitude and high latitude, nor in those obtained by Rose from a neutron monitor at Ottawa. The variability of daily means of cosmic-ray intensity (from monthly means) is in 1957 the largest observed during two complete solar cycles. Tables of monthly means corrected for seasonal wave are included, together with graphs of daily means at Huancayo for the period 1954–1957.

Introduction

In an earlier paper [see 1 of "References" at end of paper], the world-wide variation in cosmic-ray intensity and its negative correlation with the solar activity

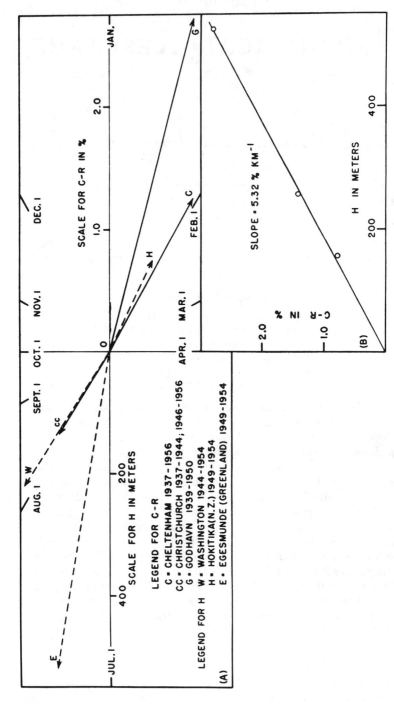

Fig. 1(A)—Average 12-month waves in cosmic-ray intensity (CR) and in height of 100-mb level H
Fig. 1(B)—Amplitude of 12-month wave in (C-R) as function of that in H

cycle was derived from continuous ionization-data using Compton-Bennett meters at several stations for the period 1937–1952. The occurrence of a maximum of solar activity, with the largest sunspot numbers on record in 1957, makes it of interest to extend the former analysis. In extending results to the interval 1937–1957, which includes the last two complete cycles of solar activity, some improvement in the reliability of the corrections for instrumental drift [1] for Huancayo has been effected.

In addition, the possibility of a seasonal wave in cosmic-ray intensity (corrected for barometric pressure) at Huancayo is examined, and measures for the reliability of the correction for seasonal variation at Chelthenham (Fredericksburg) and Christchurch are obtained. These results should be of interest to those who may wish to compare their own results with those which we are submitting to International Geophysical Year World Data Centers, and to those who may use the cosmic-ray data which have appeared in two publications [2,3] of the Carnegie Institution of Washington.

Seasonal Wave

Figure 1(A) summarizes the average 12-month waves in cosmic-ray intensity, C-R (after correcting for barometric pressure), at three stations for the years indicated. The 12-month waves in the height, H, of the 100-mb layer are also shown for approximately the same locations for the indicated intervals. In Figure 1(B), the amplitude of the 12-month wave in cosmic-ray intensity is plotted against that in H. Table 1 summarizes the result in Figure 1 and also indicates the change in cosmic-ray intensity in per cent for a change of one km in H. These ratios provide an approximate check on the over-all instrumental calibrations.

TABLE 1—*Parameters for 12-month waves in cosmic-ray intensity and height of 100-mb layer*

Cosmic-ray intensity				Height of 100-mb layer				
Station	Interval	Amplitude, %	Date of max.	Station	Interval	Amplitude, meters	Date of min.	(%)km^{-1}
Godhavn, Greenland	1939–1950	2.80	Jan. 14	Egesminde, Greenland	1949–1954	526	Jan. 10	5.3
Cheltenham, U. S.	1938–1956	1.42	Jan. 29	Washington, D. C.	1944–1954	262	Feb. 2	5.4
Christchurch, N. Z.	1937–1944 and 1945–1956	0.79	Aug. 2	Hokitika, N. Z.	1949–1954	156	July 27	5.1
Huancayo, Peru	1937–1956	....	...					

The harmonic dial in Figure 2 summarizes the 12-month waves in cosmic-ray intensity at Huancayo for the individual years, 1937–1956. The amplitude of

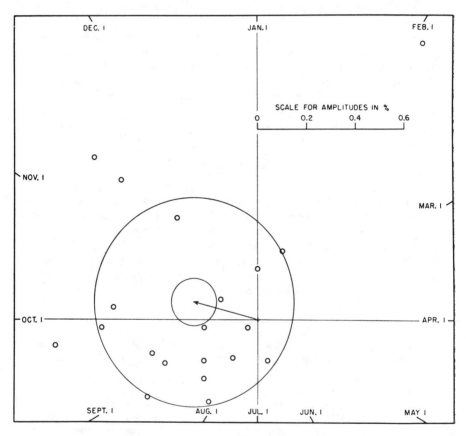

Fig. 2—Twelve-month waves in cosmic-ray intensity at Huancayo, 1937–1956

the average vector is 0.27 per cent. The expectancy [4] for single vectors is 0.50 per cent and that estimated for means of 20 is 0.11 per cent. The ratio, K, of the average to the expectancy for averages of 20 is $(0.27/0.11) = 2.45$. The probability of obtaining an average amplitude ≥ 0.27 per cent in samples of 20 random vectors having a population expectancy of 0.11 per cent for their average is $e^{-(2.45)^2} \doteq 0.003$. This indicates that the average in our sample is large enough to be statistically significant. However, the point (for 1947) in Figure 2, in the upper right corner, deviates from the mean by about 1.39 per cent. Using the above expectancy of 0.49 per cent, the probability of a deviation ≥ 1.39 per cent is about 3×10^{-4}. This result suggests that the statistical reality for the mean vector in Figure 2 should probably be accepted with some reservations. Also it is evident that the larger "probable error" circle in Figure 2 contains rather more than half the points.

As will be shown from Figure 3, the scatter of points in Figure 2 arises principally from world-wide changes in cosmic-ray intensity, either from those associated with the solar cycle or from other large transient decreases, and consequently these world-wide variations could, with probability 0.003 if they occurred randomly throughout the year (if not random this probability is increased), give rise to the average vector in Figure 2.

The radius of the smaller of the so-called [4] "probable error circles" in Figure 2 is 40 per cent of the length of the average vector, which indicates a considerable uncertainty in the average 12-month wave even though this average may be statistically significant. Because of these considerations, no correction for seasonal wave was applied to the data for Huancayo. The reality of the seasonal variation at Huancayo can probably be ascertained best from a comparison with neutron results there, or from reliable upper-air soundings if these become available.

The harmonic dial in Figure 3 is derived from that in Figure 2 (excluding

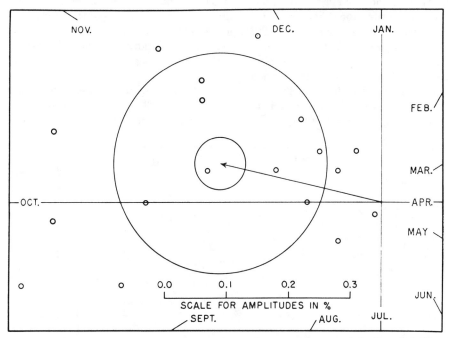

Fig. 3—Twelve-month waves in cosmic-ray intensity at Huancayo after deducting deviations from average 12-month wave for Cheltenham, 1938–1956

1937, for which data at Cheltenham were not complete). From each 12-month wave at Huancayo, there was deducted, for that year, the deviation of the 12-month wave at Cheltenham, from the average wave at Cheltenham for the years 1938–1956.

The expectancy [4] for single vectors in Figure 3 is about 0.21 per cent, or about half that in Figure 2. This reduction results from removing, in effect, the variability of world-wide changes, which are well correlated at Huancayo and Cheltenham. The average wave in Figure 3 is, of course, the same (0.27 per cent) as in Figure 2. However, the expectancy of 0.04 per cent for the means of 19 vectors (Fig. 3) cannot be used to test the statistical significance of this average which may arise from chance variations in the world-wide component; for this test, only the expectancy derived in Figure 2 is applicable. However, if this average does arise, principally from a real seasonal variation at Huancayo, then the "probable error circles" in Figure 3 provide upper limits (since the scatter of points in Fig. 3 includes contributions from Cheltenham and Huancayo) for the stability of the seasonal variation at Cheltenham and at Huancayo. The radius of the larger "probable error circle" in Figure 3 is about 0.17 per cent. Thus, if the average of Figures 2 and 3 results mainly from a seasonal variation at Huancayo, and if the 12-month wave for Cheltenham (see Fig. 1) is used to eliminate seasonal variations at Cheltenham, then the magnitude of the vector deviation of this average 12-month wave from the 12-month waves derived from individual years will be less than 0.17 per cent for about half the years. Only for about 10 per cent of the years will the deviation be expected to exceed 0.32 per cent. On the other hand, if the average wave in Figures 2 and 3 arises mainly from chance variations (though this appears improbable) in the world-wide component, then the seasonal variation at Cheltenham would be in error by this amount.

The same procedure illustrated in Figure 3 was also carried out using data from Huancayo and Christchurch for the 18 years, 1937–1944, and 1946–1956. In this case, the radius of the "probable error circle" for single 12-month vectors

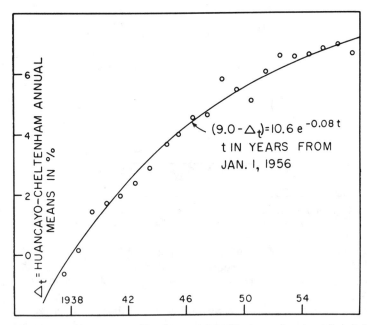

Fig. 4—Drift correction, Δ_t, to be subtracted from Huancayo departures from balance

was found to be about 0.21 per cent, which differs only slightly from the figure 0.17 per cent for Figure 3.

Corrections for Drift

As indicated previously [1], secular changes were evident in the curves for the annual means of cosmic-ray intensity at Huancayo, Christchurch, and Godhavn; no evidence for drift is indicated in the results for Cheltenham. Figure 4 indicates the results of redetermining the drift for Huancayo based on data for the period 1937–1957. This drift correction for Huancayo has been used to correct the results at Huancayo reported herewith, instead of the correction reported earlier [1].

The drift at Christchurch was also reexamined, and the values at the bottom of Table 4 were tentatively adopted in compiling the entries in Table 4 and the data for Christchurch for Figure 10. These drift corrections are regarded as somewhat arbitrary and unsatisfactory. In searching for reasons for the change in the rate of drift at the end of 1943, an apparent increase of about 10 per cent in barometric coefficient was found to occur between 1943 and 1944. This probably indicates either a change in the calibration of the voltmeter used to determine electrometer sensitivity or a change in the sensitivity of the meteorological baro-

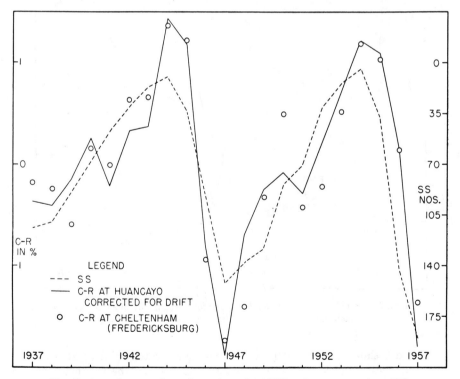

FIG. 5—Annual means of cosmic-ray intensity (C-R) and sunspot numbers (SS)

graph at Christchurch from which pressure values are furnished. These questions are being investigated at Christchurch. When the results are at hand, a further note on the drift corrections for Christchurch will be published, together with a revised table of monthly means if advisable. Incidentally, the barometric coefficient for Cheltenham was carefully redetermined from data for 1944 and 1954; these values did not differ significantly from 0.18 per cent/mm Hg, which was the value adopted earlier [2].

Variations with Solar-Activity Cycle

Figure 5 shows the variation of annual means of cosmic-ray intensity at Huancayo, after applying the drift correction shown in Figure 4, and at Cheltenham (Fredericksburg). It will be noted that the cosmic-ray intensities for 1947 and 1957 are less than the minimum in 1937, which is roughly in accord with the larger sunspot numbers which prevailed in 1947 and 1957 relative to that for 1937.

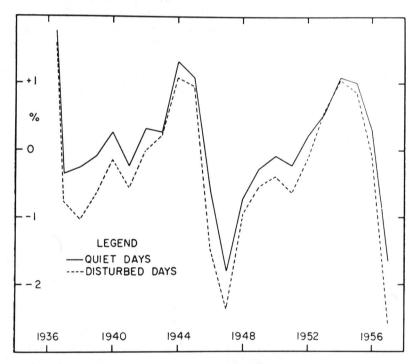

Fig. 6—Annual means of *C-R* intensity at Huancayo for magnetically quiet and disturbed days; ordinates are deviations from 1937–1957 mean for quiet days

Figure 6 shows the annual means of cosmic-ray intensity derived from the five magnetically quiet days of each month and that for the five magnetically disturbed days of each month. For 1936, only six months' data were available for these averages. As previously [1] pointed out, the variation with **sunspot** cycle

TABLE 2—*Monthly mean cosmic-ray intensity at Cheltenham (Fredericksburg), corrected for barometric pressure and seasonal wave, in units of 0.1 per cent from fiducial value*

Year	Jan.	Feb.	Mar.	Apr.	May	June	July	Aug.	Sep.	Oct.	Nov.	Dec.	Mean
1937	..	..	61	57	57	46	53	50	55	55	60	59	55
1938	45	46	52	48	54	60	61	59	58	57	56	56	54
1939	55	49	54	45	42	46	50	55	48	49	59	56	51
1940	63	59	60	58	64	61	60	58	58	56	53	50	58
1941	51	60	58	55	60	58	57	59	54	56	58	55	57
1942	62	66	51	58	64	66	62	64	64	65	65	68	63
1943	64	68	67	69	66	64	60	57	59	59	63	62	63
1944	62	65	67	69	71	71	73	70	74	73	76	72	70
1945	71	69	58	64	74	71	69	70	70	71	69	71	69
1946	62	38	44	48	46	48	42	47	44	47	47	55	47
1947	53	64	56	43	49	24	33	27	27	31	32	34	39
1948	39	42	41	47	40	50	43	43	37	41	42	46	43
1949	37	44	53	50	53	55	60	57	61	51	60	60	53
1950	49	54	61	66	61	63	61	64	58	60	66	76	62
1951	64	52	59	62	56	47	51	47	46	47	49	48	52
1952	43	47	47	52	55	56	60	58	56	59	60	60	54
1953	55	60	61	63	59	64	63	64	62	62	64	63	62
1954	63	67	69	64	70	68	72	70	69	69	72	70	69
1955	64	65	66	67	68	70	68	67	67	65	68	67	67
1956	64	57	58	60	55	57	63	61	56	63	57	45	58
1957	44	43	48	50	50	45	47	46	34	41	38	32	43

Mean 1937–1957 = 56.6

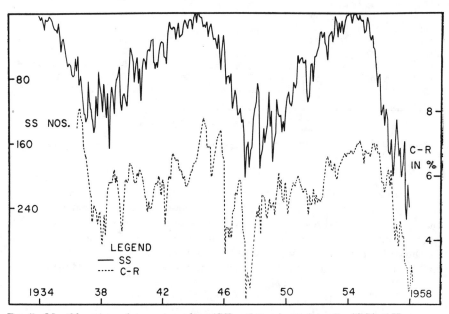

FIG. 7—Monthly means of sunspot numbers (*SS*) and cosmic-ray intensity (*C-R*) at Huancayo; ordinates from fiducial values

shown here in Figure 5 for all days is not due to magnetic storm effects on cosmic-ray intensity, since this variation is essentially the same for quiet and disturbed days (Fig. 6), and since the large transient decreases in cosmic-ray intensity almost never occur on magnetically quiet days. It will also be seen in Figure 6 that the yearly averages for the five disturbed days are almost always less than those for quiet days.

Table 3—*Monthly mean cosmic-ray intensity at Huancayo, corrected for barometric pressure and drift,[1] in units of 0.1 per cent from fiducial value*

Year	Jan.	Feb.	Mar.	Apr.	May	June	July	Aug.	Sep.	Oct.	Nov.	Dec.	Mean
1936	..	..	..	..	..	75	79	81	78	77	70	67	75
1937	66	62	59	54	54	46	51	50	49	45	48	49	53
1938	39	41	48	42	49	56	58	58	58	60	62	58	52
1939	60	55	53	46	43	49	54	59	58	59	64	63	55
1940	63	61	58	58	61	60	62	61	60	58	54	53	59
1941	49	52	50	50	52	56	57	58	57	60	58	54	54
1942	59	57	45	52	56	60	59	65	67	66	65	65	60
1943	65	62	62	61	59	60	57	57	58	..	61	62	60
1944	64	64	65	66	69	71	74	76	78	76	75	69	71
1945	69	69	60	62	65	68	70	72	73	74	70	66	68
1946	65	36	44	44	41	44	41	49	50	55	54	58	48
1947	59	58	49	42	40	22	30	28	25	30	34	38	38
1948	45	49	47	51	42	54	54	53	48	47	50	53	49
1949	50	51	52	48	54	56	61	56	59	53	52	58	54
1950	52	48	52	54	56	58	57	57	59	60	59	57	56
1951	56	55	56	56	54	48	54	51	52	56	51	54	54
1952	52	53	52	53	54	59	62	65	66	63	65	62	59
1953	62	64	60	59	62	63	64	65	66	67	66	67	64
1954	67	67	68	67	67	68	69	70	70	71	69	68	68
1955	64	68	67	67	67	66	68	70	70	68	70	67	68
1956	67	62	58	58	55	57	61	62	57	62	52	46	58
1957	40	44	45	38	46	43	41	41	33	32	31	24	38

Mean 1936–1957 = 57.3

[1]To correct for drift: Δ_t subtracted from observed values with $\Delta_t = 90 - 106e^{-0.08t}$; Δ_t in units of 0.1 per cent and t in years from January 1, 1936.

Figure 7 is a graph of monthly means of sunspot numbers and of cosmic-ray intensity at Huancayo (from Table 3). From these curves, it is evident that the maxima in the solar cycle variation of cosmic-ray intensity at Huancayo tend to occur as much as a year or so after the minima in the sunspot numbers. As will be shown later, this lag is not apparent in the results obtained by Neher [5] from balloon flights at Thule, nor in the neutron monitor results, at Ottawa, from a paper by H. G. Fenton [6], *et al.*

Figure 8 shows the monthly mean cosmic-ray intensity at Cheltenham and at Christchurch from Tables 2 and 4, respectively. These curves are generally

TABLE 4—*Monthly mean cosmic-ray intensity at Christchurch, corrected for barometric pressure, seasonal wave, and drift,[1] in units of 0.1 per cent from fiducial value*

Year	Jan.	Feb.	Mar.	Apr.	May	June	July	Aug.	Sep.	Oct.	Nov.	Dec.	Mean
1936	..	..	..	74	77	73	82	81	79	77	74	68	76
1937	62	58	53	52	53	51	55	53	56	56	59	57	55
1938	43	41	48	41	54	62	69	64	60	55	53	52	54
1939	50	40	38	33	33	44	52	58	55	52	60	57	48
1940	56	57	53	54	62	60	60	60	58	60	61	58	58
1941	51	51	49	56	57	66	62	70	64	72	70	65	61
1942	67	64	48	59	65	65	67	·72	71	..	..	..	64
1943	72	68	68	65	69	72	66	69	64	..	..	72	68
1944	69	70	70	73	79	86	82	82	81	82	86	81	78
1945	69	71	65	56	..	..	..	..	..	..	77	78	69
1946	69	39	43	44	38	42	49	48	45	56	62	61	50
1947	57	51	39	39	39	25	32	29	23	27	32	36	36
1948	38	38	37	43	38	42	49	48	41	45	52	48	43
1949	44	38	39	43	43	48	49	49	52	45	50	54	46
1950	48	43	55	48	48	55	53	58	60	57	60	57	54
1951	55	52	49	56	56	50	55	58	55	60	61	64	56
1952	58	52	55	55	60	64	64	64	65	67	73	69	62
1953	67	62	59	61	62	67	67	66	67	73	72	..	66
1954	70	65	66	71	70	71	80	76	75	74	74	77	72
1955	70	68	67	63	63	67	71	71	65	63	68	64	67
1956	59	53	47	46	48	51	57	59	51	58	53	43	52

Mean 1936–1956 = 58.8

[1]To correct for drift D (in units of 0.1 per cent), added to observed values, with: $D = +43$ from April 1, 1936, to October 31, 1938 (new ionization chamber used from November 11, 1938); $D_{t1} = +43 + 14t_1$ November 1, 1938, to December 31, 1943 (t_1 in years from November 1, 1938); $D_{t2} = +115 + 16t_2$ January 1, 1944, to December 31, 1957 (t_2 in years from January 1, 1944).

similar to those for cosmic-ray intensity at Huancayo in Figure 7, and the apparent maximum in 1936 on the latter curve is in agreement with that for Christchurch in Figure 8, although a minimum in sunspot numbers occurred near 1934.

Figure 9 shows the change, during two solar cycles, in the yearly pooled standard deviation of daily means from monthly means. These yearly pooled standard deviations [5] are slightly less when the five magnetically quiet days for each month are excluded; and it is noteworthy that near sunspot minima these values decreased to about 0.2 per cent. Since this figure includes any variability due to real variations in cosmic-ray intensity, meteorological causes, and that due to "noise level" in the meter, it indicates that the latter two together can at most account for 0.2 per cent. This figure then provides an upper limit to the uncertainty in the variation of daily means, within months, at Huancayo. For Cheltenham, Christchurch, and Godhavn, the corresponding is [1] considerably greater on account of transient deviations in the height of the 100-mb level from the smooth 12-month wave used to correct for seasonal variations.

In Figure 9, the largest value of S occurred in 1957, for which the sunspot number was also greatest.

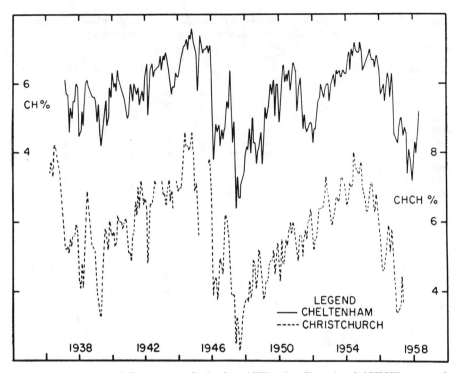

Fig. 8—Monthly mean *C-R* intensity at Cheltenham (*CH*) and at Christchurch (*CHCH*), corrected for seasonal variation and drift; ordinates from fiducial values

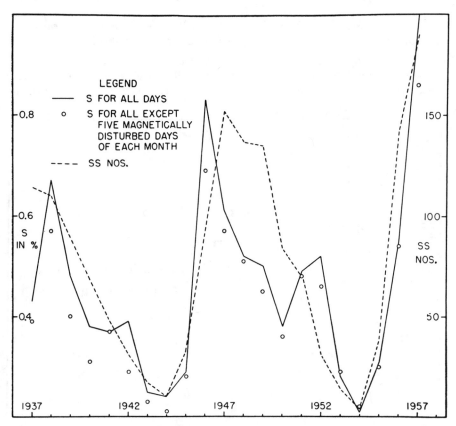

FIG. 9—Yearly pooled standard deviation (S) of daily means from monthly means of cosmic-ray intensity at Huancayo, and yearly mean sunspot numbers (SS)

263

Comparison of Solar Cycle Variations in Cosmic-Ray Ionization at Huancayo with Variations, at High Latitude, in Neutron Intensity and in High Altitude Ionization

In Figure 10, monthly means of ionization at Huancayo are compared with those in neutron intensity at Ottawa published [6] by Rose and co-workers. From about the beginning of 1956 through 1957, the percentage decrease in neutron

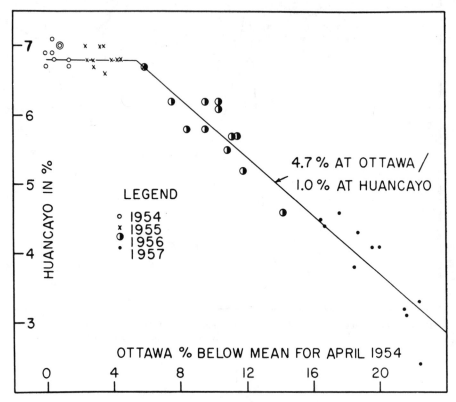

FIG. 10—Monthly means Huancayo (ionization chamber) *vs* monthly means neutron intensity at Ottawa (Rose), April 1954 to December 1957

intensity at Ottawa averaged about 4.7 times that at Huancayo. This is approximately the value found by Fenton, *et al.* [6], for the ratio of the decrease (about 22 per cent) in neutron intensity at Ottawa to that (about 5 per cent) from a meson telescope at Ottawa for the period 1954–1957. Figure 10, however, indicates that between 1954 and the end of 1955 the neutron intensity at Ottawa decreased about 5 per cent, while there was little, if any, decrease at Huancayo. This is also indicated in Figure 11, in which the ionization under 15 gm cm^{-2} at Thule is also plotted; these values were kindly provided by Neher [7] (see also reference [8]). From the summer of 1954 to the summer of 1955, the latter values indicate a decrease of about 9 per cent. Neher [9], Simpson [10], and others have shown that

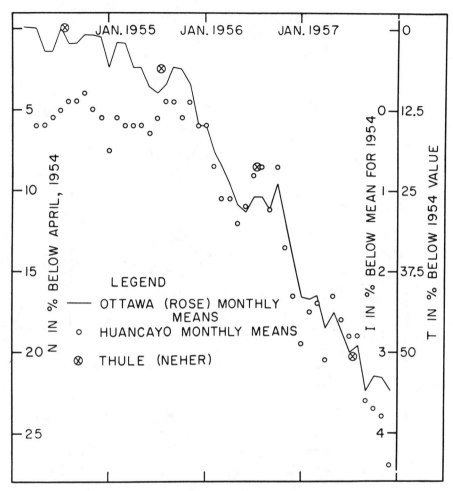

Fig. 11—Neutron intensity (N) at Ottawa, ionization (I) at Huancayo, and ionization under 15 gm cm^{-2} at Thule (T)

from its maximum at sunspot minimum, the decrease in primary cosmic radiation, which begins with the increase of solar activity following solar minimum, is initially due mainly to the exclusion of primary particles in the lower energy part of the spectrum. Figures 7, 10, and 11 show that the exclusion of primary particles with energy sufficient to reach the earth at the equator does not become evident until solar activity, as indicated by sunspot numbers, has increased to about 25–50 per cent of its maximum.

Daily Means at Huancayo, 1954–1957

Daily means of cosmic-ray intensity at Huancayo for the period 1954–1957 are plotted in Figures 12, 13, and 14. These daily means include the correction

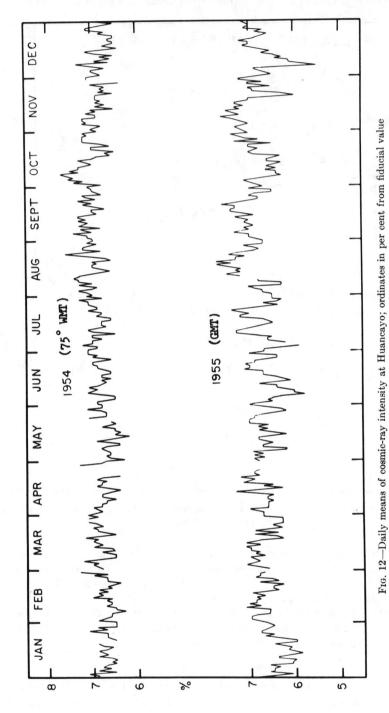

Fig. 12—Daily means of cosmic-ray intensity at Huancayo; ordinates in per cent from fiducial value

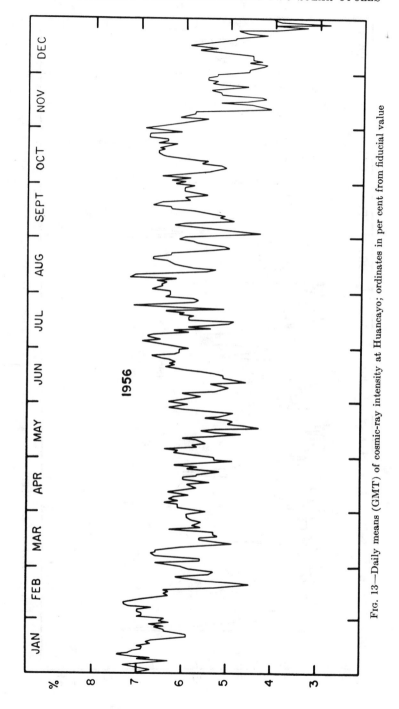

Fig. 13—Daily means (GMT) of cosmic-ray intensity at Huancayo; ordinates in per cent from fiducial value

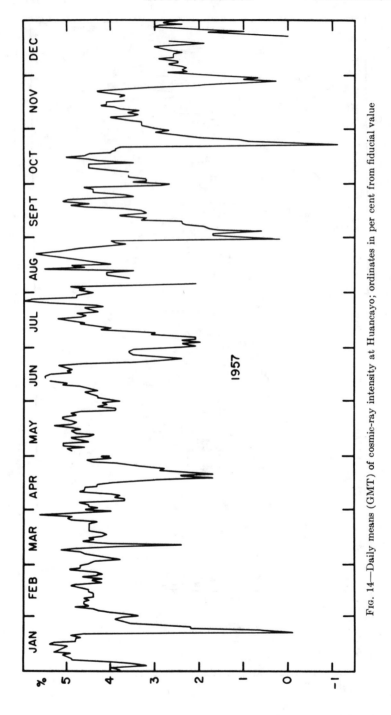

Fig. 14—Daily means (GMT) of cosmic-ray intensity at Huancayo; ordinates in per cent from fiducial value

for the drift indicated in Figure 4. Graphs of daily means at Huancayo for the period 1937–1953 are in a previous publication [1]; these daily means were corrected for the linear drift discussed in reference [1].

References

[1] S. E. Forbush, J. Geophys. Res., **59**, 525-542 (1954).

[2] I. Lange and S. E. Forbush, Cosmic-ray results from Huancayo Observatory, Peru, June 1936-December 1946, including summaries from observatories at Cheltenham, Christchurch, and Godhavn through 1946, Washington, D. C., Carnegie Inst. Pub. 175, Vol. 14 (1948).

[3] I. Lange and S. E. Forbush, Cosmic-ray results, Huancayo, Peru, January 1946-December 1955, Instituto Geofisico de Huancayo; Cheltenham, Maryland, March 1936-December 1955, U. S. Coast and Geodetic Survey, Cheltenham Magnetic Observatory; Christchurch, New Zealand, January 1947-December 1955; Godhavn, Greenland, January 1947-December 1950, Det Danske Meteorologiske Institut, Geofysik Observatorium, Washington, D. C., Carnegie Inst. Pub. 175, Vol. 20 (1957).

[4] J. Bartels, Terr. Mag., **40**, 1-60 (1935).

[5] H. V. Neher, in press.

[6] A. G. Fenton, K. B. Fenton, and D. C. Rose, in press.

]7] H. V. Neher, in press.

[8] H. V. Neher, V. Z. Peterson, and E. A. Stern, Phys. Rev., **90**, 655-674 (1953).

[9] H. V. Neher, Phys. Rev., **103**, 228 (1956).

[10] P. Meyer and J. A. Simpson, Phys. Rev., **106**, 568-571 (1957).

SUDDEN DECREASES IN COSMIC-RAY INTENSITY AT HUANCAYO, PERU, AND AT UPPSALA, SWEDEN

Fenton, Fenton, and Rose[1] showed that from 1954 to 1957 the sea-level neutron intensity at geomagnetic latitudes 57°N and 83°N decreased about 22 per cent. During the same period they found that the sea-level meson intensity at these

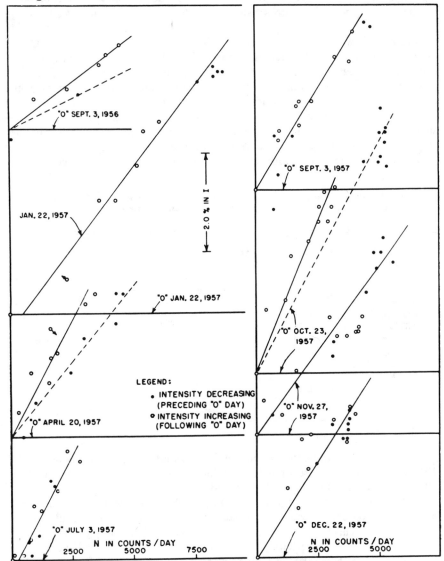

Fig. 1—Consecutive daily mean differences from minimum ("0" day) for eight rapid decreases and recoveries of ionization, *I*, at Huancayo and of neutron intensity, *N*, at Uppsala

[1] A. G. Fenton, K. B. Fenton, and D. C. Rose, Can. J. Phys., **36**, 824 (1958).

same latitudes decreased about 5 per cent. On the other hand, they found the ratio of the percentage decrease in neutron intensity changes at Ottawa, to that in meson intensity for several rapid decreases during 1957, varied from about 0.9 to 2.7. Since these ratios differed from the corresponding one (about 4.4) for the intensity changes from 1954 to 1957, this provided independent evidence supporting the conclusion of Forbush[2] that the transient decreases are only superposed on the intensity variation with sunspot cycle, but are not in themselves the cause of it.

Figure 1 shows the changes from the value on "0" day, in daily means (GMT) of neutron intensity at Uppsala relative to those in ionization at Huancayo during eight rapid decreases and recoveries between September 1956 and December 1957. The "0" day was that on which the minimum daily mean occurred. From the slopes of the lines in Figure 1, the ratios in Table 1 were obtained. These ratios seem definitely to vary for different transient decreases (and recoveries). Also, it is significant that the ratio for the recovery appears never to be less than that for the decrease, indicating for some cases that the energy spectrum of excluded particles changes differently with time during the decrease than during the recovery.

TABLE 1—*Ratio of change (ΔN) in neutron intensity at Uppsala to that in ionization (ΔI) at Huancayo during eight rapid decreases and recoveries*

Date at "0" day	$\Delta N/\Delta I$ (in 10^3 counts day^{-1} per cent^{-1})		$\Delta N/\Delta I$ (per cent)*(per cent)$^{-1}$	
	Decrease	Recovery	Decrease	Recovery
1956 Sep. 3	4.18	2.66	7.0	4.4
1957 Jan. 22	1.70	1.70	2.8	2.8
" Apr. 20	1.64	1.06	2.7	1.8
" July 3	1.03	1.03	1.7	1.7
" Sep. 3	1.24	1.24	2.1	2.1
" Oct. 23	1.07	0.83	1.8	1.4
" Nov. 27	1.47	1.47	2.4	2.4
" Dec. 22	1.28	1.28	2.1	2.1

*Based on 60,000 neutron counts per day.

In Table 1, the ratios for the sudden decrease, and recovery, for September 3, 1956, appear to be definitely greater than for any of the other cases. In September 1956, the neutron intensity at Ottawa[1] was about 10 per cent below that in 1954, while the average for 1957 was about 20 per cent below that for 1954; the corresponding figures for ionization at Huancayo[3] were, respectively, about one and three per cent. This indicates that the ratios in Table 1 probably decrease toward the maximum of solar activity; this possibility was indicated by Fenton, *et al.*[1] Such a change in these ratios would be expected, since with increasing solar activity there is an increase in the energy of primary particles excluded by the solar-

[2] S. E. Forbush, J. Geophys. Res., **59**, 525 (1954).

cycle modulation mechanism. These excluded particles would thus account for the smaller ratio for transient decreases at Uppsala relative to Huancayo.

Finally, it should be pointed out that Fonger[4] found that the ratios of the 27-day neutron intensity variations from July to October 1951 at Climax (geomagnetic latitude 48°N) were about five times those in ionization at Huancayo. At this time, the ionization at Huancayo[3] was about 1.5 per cent below the maximum in 1954. Thus, this ratio is in reasonable agreement with those in Table 1 for September 3, 1956, for a roughly similar level of intensity at Huancayo.

Arne E. Sandström*
Scott E. Forbush**

[3]S. E. Forbush, J. Geophys. Res., **63**, 651 (1958).
[4]W. H. Fonger, Phys. Rev., **91**, 351 (1953).

*Fysika Institutionen,
  Uppsala, Sweden, July 18, 1958
**Department of Terrestrial Magnetism,
  Carnegie Institution of Washington,
    Washington 15, D.C.
(Received October 15, 1958)

CORRELATION OF COSMIC-RAY INTENSITY AND SOLAR ACTIVITY

H. V. Neher

California Institute of Technology,
Pasadena, California

and

S. E. Forbush

Carnegie Institution of Washington,
Washington, D. C.
(Received August 4, 1958)

The present International Geophysical Year was chosen to include the most likely period of maximum activity of the sun. It is probably too early to tell whether or not the maximum of the current cycle has yet been reached, but it is already certain that the yearly average of the Zurich sunspot numbers for 1957 is much higher than ever before observed.[1] It is therefore of interest to see what has been the effect on cosmic rays.

In analyzing the data for long periods of time from the Carnegie Institution ionization chambers, Forbush[2] in 1954 found an inverse relationship between solar activity, as measured by Zurich sunspot numbers, and cosmic-ray intensity. Also Neher and Forbush[3] showed in 1952 that there was a good correlation for at least a few weeks between the ionization due to cosmic rays at balloon altitudes at geomagnetic latitude 56°N, the ionization at ground level at Cheltenham and Huancayo, and the neutron intensity at Sacramento Peak, New Mexico, and Climax, Colorado.

It is the purpose of this letter to point out the following relations: (a) The yearly averages of the ionization data at Huancayo correlate very well with the average value of the ionization measured at 90 000 ft, or 15 g cm^{-2} at Thule, Greenland. These latter values were made over about a 2-3 week period during the month of August of the particular year.[4] (b) There is also a very good anti-correlation with solar activity as measured by the yearly average of Zurich sunspot numbers for the same period. These relationships are shown in Fig. 1.

The large ratio of 19 to 1 for the percentage change near the north geomagnetic pole to that at the equator is due primarily to the large numbers of low-energy particles in the primary radiation which can get through the earth's magnetic field at Thule and penetrate 15 g cm^{-2} of air and which were present in some numbers during the solar minimum of 1954.

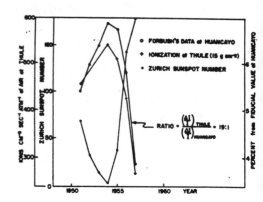

FIG. 1. Long-term correlation between ionization due to cosmic rays at high altitudes (15 g cm^{-2} pressure) near the north geomagnetic pole, the ionization at Huancayo, Peru, and the Zurich sunspot numbers.

The above therefore constitutes further evidence that for these long-time effects, (a) the changes are world wide, (b) the low energy particles are affected more than those of higher energy, (c) the average, yearly Zurich sunspot numbers are a good index of the long-term effect of the sun on the intensity of cosmic rays as measured on the earth.

[1]W. Waldmeier, J. Geophys. Research **63**, 411 (1958).

[2]S. E. Forbush, J. Geophys. Research **59**, 534 (1954).

[3]H. V. Neher and S. E. Forbush, Phys. Rev. **87**, 889 (1952).

[4]See Neher, Peterson, and Stern, Phys. Rev. **90**, 655 (1953); H. V. Neher, Phys. Rev. **103**, 228 (1956); **107**, 588 (1957); H. V. Neher and H. Anderson, Phys. Rev. **109**, 608 (1958).

Journal of

GEOPHYSICAL RESEARCH

VOLUME 65   AUGUST 1960   No. 8

Diurnal Variation in Cosmic-Ray Intensity, 1937–1959, at Cheltenham (Fredericksburg), Huancayo, and Christchurch

SCOTT E. FORBUSH AND D. VENKATESAN[1]

Department of Terrestrial Magnetism
Carnegie Institution of Washington
Washington, D. C.

Abstract. The 24-hour and 12-hour waves in cosmic-ray intensity at Cheltenham (Fredericksburg), Huancayo, and Christchurch and their variability are analyzed statistically, using data, corrected for pressure, for the period 1937–1959 from Compton-Bennett ionization chambers. The degree of correlation between the deviations of yearly mean 24-hour waves (from their 23-year means) at any two of the stations is almost as great as can be expected when account is taken of the noise level inherent in the instruments. The deviations of yearly means, from their 23-year averages, indicate large secular variations which may be due to a quasi-systematic 22-year variation. The phase difference between these yearly deviation vectors at Huancayo and Cheltenham (or Christchurch) is considerably less than that between the average vectors for 23 years. The statistical reality of the 12-hour wave is definitely established at all three stations, although, at least at Huancayo, the average 12-hour wave probably results entirely from systematic errors due to exceedingly small frictional effects in the barograph.

Introduction. The diurnal variation of cosmic-ray intensity as measured in ionization chambers is known to be of the same order of magnitude [*Forbush*, 1937] as the statistical uncertainty in its determination from data for individual days. The data for 23 years of cosmic-ray ionization at Cheltenham (Fredericksburg), Huancayo, and Christchurch provide material for a reliable statistical investigation of the year-to-year variability of the 24-hour wave in cosmic-ray intensity at three widely separated stations.

Barometric coefficients. Since at Cheltenham [*Forbush*, 1937], Huancayo, and Christchurch the 24-hour wave in pressure corrections is of the same order of magnitude as that for the 24-hour wave in cosmic-ray intensity corrected for pressure, it is essential to use reliable pres-

sure coefficients. For the 12-hour wave the consequence of errors in pressure coefficients is even more important.

Figure 1 shows the least-squares regression line and the barometric coefficient for Huan-

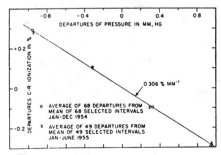

Fig. 1. Regression line and barometric coefficient for correcting cosmic-ray ionization for pressure at Huancayo.

[1] Visiting investigator, from Division of Pure Physics, National Research Council of Canada.

cayo determined from selected data in 1954 and 1955. Its value, −0.306%/mm Hg, differs insignificantly from the value −0.30%/mm Hg, determined from data in earlier years [*Lange and Forbush*, 1948], which has been used to correct the data at Huancayo from 1937 to 1959. It should be noted that the regression line in Figure 1 results from giving all weight to the pressure values. The data for Figure 1 were obtained as described in *Lange and Forbush* [1948]. Sequences of intervals of 4 days were selected with intervals characterized by monotonic increases (or decreases) of barometric pressure and by the absence of appreciable world-wide changes in cosmic-ray intensity. Generally the 4 selected days were consecutive, but occasionally they were alternate days. The selection was made to obtain the largest possible monotonic change in pressure, which nevertheless is quite small at Huancayo. For each interval the four departures of pressure and of uncorrected ionization from their means for that interval were computed, ranked according to the pressure departures, then averaged for all the selected intervals. These average departures for intervals with increasing pressure were then averaged with those for decreasing pressure after appropriately changing the sign of the latter departures. These departures are plotted in Figure 1. For Christchurch and Cheltenham the procedure was similar except that groups of 24 hourly intervals were selected with large monotonic pressure increases (or decreases) during these periods of 24 hours (starting at each of the 24 hours of the local day to eliminate diurnal variation effects). Pressure coefficients for Chel-

tenham and Christchurch differed insignificantly from the value −0.18%/mm Hg previously determined [*Lange and Forbush*, 1948] and used from 1937 to 1939.

At Cheltenham (Fredericksburg), pressure is recorded on the cosmic-ray record and the pressure at the middle of a 2-hour interval was used to correct the average ionization for that 2-hour interval. Figure 2 compares the average 24-hour and 12-hour waves, for 243 days in 1958, in pressure at Fredericksburg from the cosmic-ray barograph with those from a mercury barometer read hourly at the Washington airport on the same days. The phases of corresponding waves at the two places are practically identical. The small difference in amplitude may be real, since the stations are about 60 miles apart.

For Christchurch hourly mean pressures were scaled at Christchurch from a microbarograph. These are centered at 0.5, 1.5, 2.5, etc., hours LMT. To correct the bihourly mean ionization at Christchurch for the interval 0–2 hours LMT, for example, the average of the hourly mean pressures at 0.5 and 1.5 hours was used, and similarly for the other bi-hourly means.

For Huancayo instantaneous values of pressure at each local hour were scaled at Huancayo from an ordinary barograph. From 1936 to 1946 the bihourly values of ionization were corrected using instantaneous values of pressure read at the mid-point of these bihourly intervals. It was recently found that from January 1946 through December 1958 the procedure followed at Huancayo was changed, for reasons that will never be understood! This changed

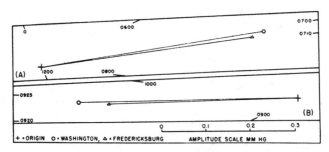

Fig. 2. Comparison of average of 24-hour (*A*) and 12-hour (*B*) waves in barometric pressure at Washington, D. C., from hourly mercury barometer readings and at Fredericksburg, Va., from cosmic-ray barograph. Times of maxima on 75° WMT.

TABLE 1. Coordinates in 24-Hour and 12-Hour Harmonic Dials for Yearly Mean Waves in Pressure and in Ionization Uncorrected for Pressure at Huancayo

Year	Barometric Pressure P $u = 0.01\%$ (1/30 mm Hg)				CR uncorrected for P $u = 0.01\%$			
	a_1	b_1	a_2	b_2	a_1	b_1	a_2	b_2
1937	+7	+31	+11	−29	−19	−22	−9	+31
1938	+6	+32	+13	−28	−21	−24	−11	+30
1939	+8	+29	+12	−28	−25	−27	−12	+31
1940	+8	+30	+11	−29	−25	−29	−11	+33
1941	+8	+32	+13	−29	−24	−30	−12	+33
1942	+8	+32	+14	−29	−23	−32	−14	+32
1943	+6	+30	+13	−28	−25	−31	−13	+32
1944	+5	+32	+15	−27	−15	−27	−13	+35
1945	+6	+32	+14	−27	−19	−28	−13	+33
1946	+6	+26	+10	−26	−24	−24	−10	+31
1947	+8	+28	+10	−27	−25	−30	−14	+33
1948	+6	+28	+13	−26	−20	−24	−13	+32
1949	+7	+29	+11	−27	−22	−21	−10	+29
1950	+6	+30	+11	−28	−24	−21	−12	+30
1951	+7	+31	+12	−28	−27	−18	−9	+32
1952	+7	+29	+12	−27	−20	−16	−10	+31
1953	+7	+31	+13	−27	−15	−13	−13	+32
1954	+4	+29	+15	−25	−4	−19	−15	+29
1955	+3	+29	+16	−24	−13	−15	−11	+34
1956	+5	+29	+14	−25	−23	−23	−12	+33
1957	+5	+29	+14	−25	−25	−26	−14	+33
1958	+4	+30	+14	−26	−24	−35	−12	+35
1959	+4	+28	+15	−24	−27	−34	−13	+37
Sum	+141	+686	+296	−619	−489	−569	−276	+741
Mean	+6.13	+29.83	+12.87	−26.91	−21.26	−24.74	−12.00	+32.22

erroneous procedure used the means of instantaneous values of pressure at hours 1 and 2, for example, to correct the bihourly value of ionization for the interval 0–2 hours. As a consequence the pressure waves used to correct the ionization have their phases one-half hour too late. Fortunately, the resulting error in published bihourly means for Huancayo never amounts to more than ±0.2 per cent, which is of no consequence except for effects on diurnal variation from a large number of days since the statistical uncertainty of bihourly values at Huancayo is about 0.4 per cent. The harmonic coefficients for the yearly mean 24-hour and 12-hour waves as listed in Tables 1 and 2 have all been corrected from January 1946 through December 1958 for this half-hour error in the phase of the pressure corrections. Since January 1959 the correct procedure has been reestablished. In addition, the barograms at Huancayo were examined for errors in time control. Except for 1937 these errors were never more than a few minutes of time. The data were also corrected for such errors.

Data from the ordinary barograph used at Huancayo were analyzed for detailed comparison with those from a microbarograph obtained by the Huancayo Observatory in 1938. Data from this microbarograph were never used since the phase of the 24-hour and 12-hour waves in pressure occurred 15 minutes late (owing to friction), and in addition the instrument was inadequately compensated for temperature. Before 1937 the ordinary barograph had been operated in some years in a meteorological shelter and in others in rooms where there was little diurnal variation in temperature. Under both circumstances the 24-hour and the 12-hour yearly mean waves were essentially the same, indicating no appreciable temperature coefficient. Af-

TABLE 2. Harmonic Coefficients, Origin Local Mean Time, for Yearly Mean
24-Hour and 12-Hour Waves in Pressure-Corrected Cosmic-Ray
Ionization in Units of 0.01 Per Cent of Total Intensity

Year	Cheltenham (Fredericksburg*)				Huancayo				Christchurch†			
	a_1	b_1	a_2	b_2	a_1	b_1	a_2	b_2	a_1	b_1	a_2	b_2
1937	−12	− 3	+2	+3	−12	+ 9	+2	+ 2	−11	− 3	0	+2
1938	−12	− 4	+4	+3	−15	+ 8	+2	+ 2	−15	− 4	+3	0
1939	−14	−11	+2	+5	−17	+ 2	0	+ 3	−14	−11	+3	−1
1940	−10	−11	0	+4	−17	+ 1	0	+ 4	−16	−14	0	+3
1941	−10	−10	+2	+4	−16	+ 2	+1	+ 4	−14	−13	+2	+3
1942	− 8	−11	+2	+2	−15	0	0	+ 3	− 9	−11	0	+1
1943	−12	−12	+2	+5	−19	− 1	0	+ 4	−13	−11	0	+3
1944	− 7	− 2	+2	+3	−10	+ 5	+2	+ 8	− 6	− 5	+3	+1
1945	− 8	− 7	+3	+4	−13	+ 4	+1	+ 6	− 8	− 9	+2	+2
1946	−10	−10	+4	+3	−18	+ 2	0	+ 5	−13	− 8	+3	+3
1947	− 5	−16	+2	0	−17	− 2	−4	+ 6	−10	−12	+1	+4
1948	−14	−11	+3	+2	−14	+ 4	0	+ 6	−12	− 5	+2	+1
1949	−14	− 9	+2	+1	−15	+ 8	+1	+ 2	−15	− 1	+3	−1
1950	−12	0	+4	+5	−18	+ 9	−1	+ 2	−14	− 2	+3	+1
1951	−15	− 2	+3	+4	−20	+13	+3	+ 4	−16	0	+4	+2
1952	−15	0	+3	+3	−13	+13	+2	+ 4	−13	+ 2	+2	+4
1953	−11	+ 8	+3	+1	− 8	+18	0	+ 5	− 4	+ 7	+2	0
1954	0	+ 6	+1	+3	0	+10	0	+ 4	− 1	+ 5	−1	0
1955	− 8	+14	+1	+3	−10	+14	+5	+10	− 7	+ 8	−1	0
1956	−17	− 3	0	+3	−18	+ 6	+2	+ 8	−13	+ 1	+2	+4
1957	−13	− 8	+2	+4	−20	+ 3	0	+ 8	−15	− 1	+3	+4
1958	−10	− 8	−2	+2	−20	− 5	+2	+ 9	−10	− 7	+2	+4
1959	− 9	−10	0	+3	−23	− 6	+2	+13	(−10)	− 9)	(+3	−1)
Sum	− 246	−130	+ 45	+ 70	− 348	+117	+ 20	+122	− 259	−103	+ 41	+ 39
Mean	−10.7	−5.7	+2.0	+3.0	−15.1	+5.1	+0.9	+5.3	−11.3	−4.5	+1.8	+1.7
C^1	12.1		3.6		15.9		5.4		12.1		2.5	
T^2	1348		0152		1046		0240		1326		0126	

C^1 = amplitude of mean; T^2 = time of maximum on LST.
* After October 5, 1956.
† 1942 9 months, 1943 10 months, 1945 6 months, 1959 extrapolated.

ter 1937 this instrument was operated in a room subject to little diurnal temperature change.

Recent tests were made to determine the phase lag in this instrument which might result from friction. Friction was found to retard the 12-hour and 24-hour waves by 11 ± 3 minutes. To determine the effect of friction the recording pen of the barograph was lifted (by a lever) from the barogram at 0700 LMT and the mercury barometer was read at this time. Recording was resumed after about an hour. Lifting the pen ensures that the barogram can be read at the exact time of the mercury reading. At Huancayo the pressure at 0700 LMT is increas-

ing. At 1300 LMT the pen was again lifted from the barogram, to interrupt recording for an hour, and the mercury barometer was read again. At 1300 the pressure at Huancayo is decreasing. The correction to the barogram at 0700 LMT less than at 1300 LMT was found to be +0.18 ± 0.04 mm Hg for the average of 20 determinations. Thus the barograph was assumed to read 0.09 mm Hg too low for increasing pressure and 0.09 mm Hg too high for decreasing pressure. Harmonic analysis of such differences between the average pressure recorded for a year and that corrected for the lag indicated a phase lag of 11.0 ± 2.5 minutes in the observed 12-hour and 24-hour waves in

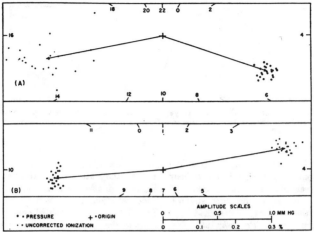

Fig. 3. Harmonic dials for yearly mean 24-hour (*A*) and 12-hour (*B*) waves in barometric pressure and uncorrected ionization at Huancayo, 1937–1959. Times of maxima in 75° WMT hours. Average vector for 23 years indicated.

pressure due to friction. In Tables 1 and 2 and those following no correction was made for this effect of friction.

Comparison of 12-hour and 24-hour waves in pressure and in uncorrected ionization at Huancayo, and effects of systematic errors in pressure on these. Figure 3 shows for Huancayo the yearly mean 24-hour (*A*) and 12-hour (*B*) waves in pressure and in uncorrected ionization for the years 1937–1959, as well as their averages. In (*B*) it is evident that the scatter of points for the yearly means of pressure is about the same as that for ionization uncorrected for pressure. Also, if the phase of the 12-hour wave for pressure is 11 minutes earlier (to correct for the estimated lag of pressure waves due to friction in the barograph), the phase of the pressure wave is then 180° from that in uncorrected ionization, indicating that the 12-hour wave in ionization may likely arise entirely from the 12-hour pressure wave. If this were so, the pressure coefficient would be the ratio of the amplitude of the 12-hour wave in uncorrected ionization to that in pressure. From the data of Table 1 this ratio is 0.344/0.983, or .35% mm Hg⁻¹, a value larger than that derived from Figure 1 (i.e., 0.306% mm Hg⁻¹).

The determination of the lag, due to friction, in the 12-hour pressure wave as described in

the section 'Barometric coefficients' reduces the amplitude of the 12-hour pressure wave only about 0.025 mm Hg, so that its corrected amplitude would be 1.008 (instead of 0.983), with a resulting pressure coefficient of 0.34% mm Hg, instead of the value 0.306 in Figure 1. In deriving the latter value the pressure values were assumed free of statistical error. If they are subject to even small statistical errors the value as derived in Figure 1 will be too small. In any event these considerations cast serious doubt on the physical reality of the 12-hour wave in pressure-corrected cosmic-ray intensity at least at Huancayo.

Twenty-four-hour wave in pressure-corrected ionization and its variability. Figure 4 shows the 24-hour harmonic dials for the yearly means, 1937–1959, of pressure-corrected ionization and the average wave for 23 years at Cheltenham, Huancayo, and Christchurch, from data in Table 2 in which are indicated the amplitudes and times of maximum for the average vectors. In Figure 5 are shown the harmonic dials for the vector summation from 1937 to 1959 of the departures of the yearly means from the mean for 1937–1959. These departures are given (without chronological summation) in the left half of Table 3.

The considerable similarity of the three sum-

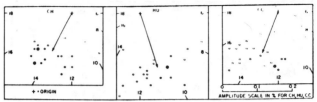

Fig. 4. Twenty-four-hour harmonic dials for yearly means, 1937–1959, and average vector for pressure-corrected cosmic-ray ionization at Cheltenham(CH), Huancayo (HU), and Christchurch (CC). Times of maxima on LST.

mation dials of Figure 5 indicates that the changes in amplitude and phase in the 24-hour waves, from year to year between 1937 and 1959, is rather similar at the three stations. The extent of this similarity, determined by a more objective procedure [*Forbush*, 1958] first used by Bartels, is shown in the upper three harmonic dials of Figure 6. In each of these dials, the vector departure (Table 3) of each yearly mean vector from the average for 23 years at one station is rotated to the vertical (0° on the dials of Fig. 6) and the vector departure (Table 3) for the second station for the corresponding year is rotated through the same angle. Only the average of the amplitudes for the vertical vectors is indicated, by crosses, in Figure 6. The amplitudes of the vectors rotated to the vertical and the coordinates for the rotated vectors at the second station are given in the right half of Table 3. The average vectors for the second station are also shown in Figure 6. It is quite evident from the upper three dials of that figure that there is a definite correspondence in phase between the changes from year to year in the 24-hour waves for any two of the three pairs of stations.

In Table 5, S_A and S_B indicate, respectively, the standard deviations of A and B (Table 3) from their means. If the amplitudes of the 23 vertical vectors used in deriving $(A-1)$, $(B-1)$, and $(C-1)$ of Figure 6 vary, and if these variations are correlated with the amplitudes of the rotated vectors for the second station, this variability will contribute to the scatter of points. For each pair of stations, the coefficient of correlation, r_1, between the 23 values of A_r and A (right half of Table 3) was determined and also the correlation coefficient r_2 between the 23 values of A_r and B (of Table 3). For each pair

of stations, r_1 and r_2 are given in Table 5. To remove the scatter, in the three upper dials of Figure 6, due to this correlation between amplitude variations of the vertical vectors for the first station and the rotated ones for the second station, the following equations were used:

$$\alpha = [A - \kappa(A_s - \bar{A}_s)] \quad (1)$$

$$\beta = [B - \lambda(A_s - \bar{A}_s)] \quad (2)$$

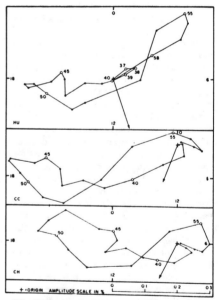

Fig. 5. Twenty-four-hour harmonic dials for summation of departures of yearly means, 1937–1959, from 23-year mean in pressure-corrected ionization at Huancayo (HU), Christchurch (CC), and Cheltenham (CH). Average vector also indicated. Times of maxima on LST.

TABLE 3. Coordinates in 24-Hour Harmonic Dial, Local Mean Time, for Departures Yearly Mean (Table 2) from Mean for 1937–1959 and Coordinates for Rotated Vector Departures Cheltenham (CH), Huancayo (HU), and Christchurch (CC) in Units of 0.001 Per Cent of Total Intensity*

Year	CH Δa_1	CH Δb_1	HU Δa_1	HU Δb_1	CC Δa_1	CC Δb_1	CH Vertical CH A_v†	CC A	CC B	HU Vertical HU A_v†	CH A	CH B	HU Vertical HU A_v†	CC A	CC B
1937	−13	+27	+31	+39	+3	+15	30	+12	−9	50	+13	+27	50	+14	+7
1938	−13	+17	+1	+29	−37	+5	21	+26	−26	29	+17	+13	29	+3	+37
1939	−33	−53	−19	−31	−27	−65	62	+70	−11	37	+63	0	37	+70	+11
1940	+7	−53	+19	−41	−47	−95	53	+89	−58	45	+45	+29	45	+106	+2
1941	+7	−43	+9	−31	−27	−85	44	+80	−40	32	+39	+19	32	+89	−1
1942	+27	−53	+1	−51	−23	−65	59	+69	−9	51	+54	+26	51	+66	+22
1943	−13	−63	+39	−61	+17	−65	64	+67	−3	73	+60	+23	73	+64	+21
1944	+37	+37	+51	−1	+53	−5	52	+34	−41	51	+36	+38	51	+53	+4
1945	+27	−13	+21	−11	−33	−45	30	+50	−26	24	+30	+1	24	+51	−24
1946	+7	−43	+29	−31	+17	−35	44	+32	−22	43	+26	+35	43	+37	+12
1947	+57	−103	+19	−71	+13	−75	118	+72	−25	74	+84	+82	74	+69	+32
1948	−33	−53	+11	−11	+7	−5	62	+8	−4	16	+16	+60	16	+1	+8
1949	−33	+33	+1	+29	−37	−35	47	+2	−51	29	+34	+32	29	+33	+39
1950	−13	+57	+29	+39	−27	+25	58	+31	−21	49	+54	+23	49	+37	+6
1951	−43	+37	+49	+79	−47	+45	57	+65	−3	93	+54	+17	93	+64	+16
1952	−43	+57	+21	+79	−17	+65	71	+62	−26	82	+44	+56	82	+59	+33
1953	+3	+137	+71	+129	+73	+115	137	+114	−76	148	+119	+69	148	+137	+8
1954	+107	+117	+151	+49	+103	+95	159	+140	−12	159	+138	+78	159	+127	+59
1955	+27	+97	+51	+89	+43	+125	101	+132	−8	103	+98	+25	103	+131	+25
1956	−63	+27	+29	+9	+17	+55	69	+37	−44	31	+68	+6	31	+34	+47
1957	−23	−23	+49	−21	−37	+35	32	+2	−51	53	+30	+12	53	+22	+47
1958	+7	−23	+49	−101	+13	−25	24	+28	+5	112	+17	+16	112	+18	+22
1959	+17	−43	−79	−111	+13	−45	46	+47	−1	137	+25	+39	137	+30	+36
Sum	+1	+11	−7	−3	+9	+5	1440	+1269	−444	1521	+1096	+548	1521	+1313	+237
Mean							63	+55	−19	66	+48	+24	66	+57	+10

* Coordinates used to 0.001 per cent only to avoid cumulative errors in coordinates for graphically rotated vectors.
† All $B_v = 0$, all A_v +; A, B coordinates of rotated vectors.

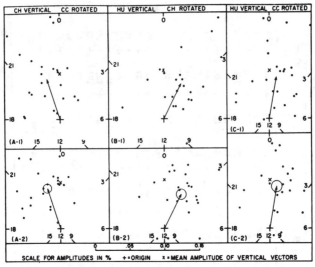

Fig. 6. (*A*-1), (*B*-1), (*C*-1), 24-hour harmonic dials after each deviation vector (yearly mean from mean for 1937–1959) at indicated station is rotated to 0^h LST, and corresponding deviation vector at second station is rotated through same angle. (*A*-2), (*B*-2), (*C*-2), and (*C*-1), respectively, after removing variability due to correlation between vertical and rotated vectors. Times of maxima relative.

In (1) and (2), α and β are the coordinates of the rotated vectors for the second station after removing the correlated variations. κ and λ are regression coefficients (all weight to A_s) determined from r_1 and r_2 in such a way as to minimize the standard deviations S_α and S_β of α and β.

Table 5 indicates the values of κ and λ and of S_α and S_β. The values of α and β using these values of κ and λ in equations 1 and 2 are given in Table 4. It will be seen that the standard deviations of α and β are in general somewhat less than those of A and B. Table 5 also indicates the expectancy [*Bartels*, 1935]

$$M = (S_\alpha{}^2 + S_\beta{}^2)^{1/2}$$

for single vectors, assuming statistical independence of vector deviations (from the mean) for successive years. There was no significant correlation between the values of α and β.

To determine whether there was indication for lack of statistical independence [*Bartels*, 1935] for vector deviations (from the average) for successive years, the expectancy for single years was also derived from the eleven differences, Δ_α and Δ_β, of α and of β (Table 4) between successive (nonoverlapping) years. Thus

$$M^* = \left[\left(\sum_1^{11} \Delta_\alpha{}^2 + \sum_1^{11} \Delta_\beta{}^2\right)/20\right]^{1/2}$$

gives the expectancy for single years if the deviations (of α and β) for successive years are statistically independent. The 20 degrees of freedom in the denominator for M^* arise from the fact that the means of Δ_α and Δ_β were assumed to be zero, which would give 22 degrees of freedom, but 2 degrees of freedom are deducted because of the two parameters κ and λ, leaving 20. Since M^* is about the same as M, there is no definite indication for lack of independence between the deviations of α and β (from their means) for successive years. Thus the expectancy, m, for means of 23 years may be estimated as $M/\sqrt{23}$.

The radii ρ_1 and ρ_{23} of the so-called probable-error circles for single years and for the mean of 23 years are, respectively, given [*Bartels*, 1935] by $\rho_1 = 0.833M$ and $\rho_{23} = 0.833m$. These probable-error circles are shown in (*A*-2), (*B*-2), and (*C*-2) of Figure 6 drawn with their centers

282

TABLE 4. Coordinates α, β in 24-Hour Harmonic Dial for Rotated Vectors;
Common Variations in Coordinates Arising from Correlation
between Amplitude of Vertical and Rotated Vectors Removed;
in Units of 0.001 Per Cent of Total Intensity

Year	CH Vertical CC Rotated		HU Vertical CH Rotated		HU Vertical CC Rotated	
	α	β	α	β	α	β
1937	+ 30	− 15	+ 22	+ 35	+ 22	+ 11
1938	+ 61	+ 19	+ 37	+ 31	+ 21	+ 47
1939	+ 71	+ 11	+ 79	+ 14	+ 84	+ 19
1940	+ 97	− 60	+ 57	+ 39	+ 116	+ 4
1941	+ 96	− 43	+ 58	+ 35	+ 106	+ 9
1942	+ 72	− 10	+ 62	+ 33	+ 74	+ 26
1943	+ 66	− 3	+ 56	+ 20	+ 60	+ 19
1944	+ 43	− 43	+ 44	+ 45	+ 61	0
1945	+ 76	− 32	+ 53	+ 21	+ 72	− 12
1946	+ 48	− 25	+ 39	+ 46	+ 49	+ 18
1947	+ 26	− 15	+ 80	+ 78	+ 65	+ 30
1948	+ 9	− 4	+ 43	− 36	+ 24	+ 6
1949	+ 15	− 54	− 14	+ 50	+ 51	+ 49
1950	+ 35	+ 20	+ 63	− 15	+ 45	+ 11
1951	+ 70	− 4	+ 39	+ 4	+ 50	+ 8
1952	+ 55	− 25	+ 35	+ 48	+ 51	+ 29
1953	+ 52	− 63	+ 74	+ 30	+ 96	− 31
1954	+ 60	+ 5	+ 87	+ 34	+ 81	+ 33
1955	+ 100	− 1	+ 78	+ 7	+ 113	+ 15
1956	+ 32	− 43	+ 87	+ 11	+ 52	− 37
1957	+ 28	− 56	+ 37	+ 18	+ 28	− 43
1958	+ 61	− 2	− 14	− 11	− 10	+ 6
1959	+ 61	− 2	− 14	+ 5	− 6	+ 16
Sum	+1264	−445	+1088	+542	+1305	+233
Mean	+ 55	− 19	+ 47	+ 24	+ 57	+ 10

C^1	58		53	58
T^2	2243		0146	0041
m^2	7.9		8.5	9.2
$\kappa = C/m$	7.3		6.2	6.3

C^1 = amplitude of mean; T^2 = time of maximum relative to vertical vector at 0 hr; m^2 = expectancy for mean.

at the end point of the average vector (coordinates $\bar{\alpha}$ and $\bar{\beta}$, Table 4). The probable-error circles for single years contain about half the points.

The amplitudes, C, of the average vectors (rotated for the second station) in the three dials in the lower half of Figure 6 are listed at the bottom of Table 4 together with m and $\kappa = C/m$. The probablity, P, of obtaining an average amplitude $\geq C$ in samples of 23 taken at random from a population with $C = 0$ and with the expectancy M for single vectors is given [Bartels, 1935] by $P = e^{-\kappa^2} (\kappa = C/m;$

$m = M/\sqrt{23})$. From the values of κ in Table 4, it is seen that P is less than e^{-38}, or less than about 10^{-15}, for any one of the three average vectors in the lower dials of Figure 6, leaving no doubt about the statistical reality of these average vectors.

Table 7 lists the probabilities for the average vectors (from further samples from the same population as those in the lower dials of Figure 6) to occur within various time intervals (note that the times are relative). For Figure 6(A-2), the probability 0.233 indicates that the chances are not too low that the phase for the average of the Christchurch rotated vectors (Chelten-

TABLE 5. Summary of Parameters in 24-Hour Harmonic Dials for Rotated Yearly Mean Vector Deviations from Mean for 1937–1959 for Cheltenham (CH), Christchurch (CC), and Huancayo (HU), in Units of 0.001 Per Cent of Total Intensity

	$\bar{A}_V$	$\bar{A}$	$\bar{B}$	C	C/A_V	θ°	S_A	S_B	r_1	κ	r_2	λ	S_α	S_β	M	M^*	m	ρ_1	ρ_{22}
CH vert. 63 CC rot.		+55	−19	58	0.92	−19	38	26	+0.76	+0.83	−0.23	−0.17	25	27	38	(42)	7.9	32	6.6
HU vert. 66 CH rot.		+48	+24	54	0.82	+27	37	31	+0.61	+0.55	+0.62	+0.48	32	26	41	(41)	8.5	34	7.1
HU vert. 66 CC rot.		+57	+10	58	0.88	+10	39	26	+0.52	+0.50	+0.44	+0.28	36	25	44	(35)	9.2	37	7.6

Legend

A_V, $\bar{A}$, $\bar{B}$ means from Table 3; $C^2 = \bar{A}^2 + \bar{B}^2$.
θ = angle in degrees clockwise from vertical to rotated vector.
S_A, S_B = standard deviations of A and B (Table 3) from their means; 22 degrees of freedom.
r_1 = correlation between A_V and A (Table 3).
r_2 = correlation between A_V and B (Table 3).
S_α, S_β = standard deviations of α and β (Table 4) with $\alpha = (A - \kappa A_V)$; $\beta = (B - \lambda A_V)$ and 20 degrees of freedom.
κ, λ = regression coefficients which minimize S_α and S_β (i.e., all weight to A_V).
M = expectancy for yearly vectors α and β (Table 4); $M^2 = S_\alpha^2 + S_\beta^2$.
M^* = expectancy for yearly vectors α and β (Table 4) derived from 11 differences from one year to next assuming statistical independence; 20 degrees of freedom.
$m = M/\sqrt{23}$ = expectancy for means of 23 years.
ρ_1, ρ_{22} = radii of 'probable error' circles for single years and for means of 23 years.

ham vertical) comes within 1 hour of the vertical. For Figure 6(C-2), HU vertical and CC rotated, the chances (0.399) are fairly good that the phase difference is not more than half an hour. For Figure 6(B-2), it seems rather improbable ($P = 0.03$) for the maxima to occur earlier than 0100. In any case, it seems rather improbable in Figure 6 that the times of maxima for Christchurch and Cheltenham occur as much later relative to Huancayo as do the average vectors for the total 24-hour wave (not deviations from the average) for which the times of maxima are given at the bottom of Table 2. Figure 7 shows the summation dial for the vectors in the three lower dials of Figure 6. These exhibit a remarkable consistency in phase.

Twelve-hour wave in pressure-corrected cosmic-ray ionization. The coordinates for the 12-hour waves in pressure-corrected cosmic-ray ionization are given in Table 2, and the harmonic dials in Figure 8. The statistical parameters for the dials are given in Table 6. Since the probability, P, of obtaining, from a population with expectancy m and average amplitude zero, waves of amplitude equal to or exceeding the average actually obtained is given by

$$P = e^{-\epsilon^2}$$

there is little doubt about the statistical reality of the 12-hour waves. Nevertheless, as discussed in the section 'Barometric coefficients,' it seems quite likely that for Huancayo, at least, the 12-hour wave may arise entirely as a consequence of small systematic errors in the barograph due to friction. It is of interest to note for Huancayo that M is slightly less (see Huancayo II, Table 6) when the average 12-hour wave in pressure is used to correct the 12-hour waves in ionization for each year (see HU$_2$ of Fig. 8) than when the latter are corrected by using the average 12-hour pressure wave for the same year (HU$_1$ of Fig. 8).

Comparison of expectancy for 24-hour and 12-hour harmonic dials with that estimated from statistical fluctuations in hourly values. For Cheltenham, the standard deviation of hourly values is about 0.7 per cent as determined from the hourly differences of two identical ionization chambers [*Lange and Forbush,* 1948]. For Huancayo, the value is about 0.6 per

TABLE 6. Summary of Parameters in 12-Hour Harmonic Dials for Pressure-Corrected Ionization, in Units of 0.001 Per Cent of Total Intensity

	a_2	b_2	C_2	T	S_{a_2}	S_{b_2}	M	M^*	m	ρ_1	ρ_{22}	$\kappa = C_2/m$
CH	+20	+30	36	1.9	15	13	20	18	4.2	17	3.5	10.3
HU I	+ 9	+53	54	2.7	17	29	33	17	6.9	28	5.8	7.8
II					17	19	25	17	5.2	21	4.3	10.4
CC	+18	+17	25	1.4	15	18	23	23	4.8	20	4.1	6.1

Legend

a_2, b_2 means from Table 2, $C_2{}^2 = a_2{}^2 + b_2{}^2$.
T = time of maximum on LMT hours.
S_{a_2}, S_{b_2}, = standard deviations of a_2 and b_2 from Table 2.
M = expectancy for single yearly means, $M^2 = S_{a_2}{}^2 + S_{b_2}{}^2$, 22 degrees of freedom.
M^* = expectancy for single yearly means derived from 11 differences one year to next assuming statistical independence; 22 degrees of freedom.
$m = M/\sqrt{23}$ = expectancy for means of 23 years.
In Huancayo I the pressure wave for each year was used to correct the ionization for that year. For Huancayo II parameters derived from using one single pressure wave, the mean for 23 years, to correct each yearly 12-hour wave in ionization for pressure.

cent, derived from differences one hour to the next. The expectancy e_1 for random fluctuations on single days is given [Bartels, 1935] by

$$e_1 = 2\xi/\sqrt{r}$$

with ξ = the standard deviation of ordinates (or bihourly values) and r = the number of ordinates (12) used in the harmonic analysis. Thus for Cheltenham

$$e_1 \doteq 1.4/\sqrt{12} \doteq 0.34\%$$

which compares with 0.36 per cent obtained from harmonic analysis of single days [Forbush, 1937]. For Huancayo,

$$e_1 = 1.2/\sqrt{12} \doteq 0.32.$$

Thus the expectancy, say e_2, for random fluctuations in the difference of 24-hour waves at Huancayo and Cheltenham with which Figure 6 is concerned is $e_2 \doteq 0.026$.

For Figure 6, the total number of complete days used was 7352 for Huancayo and 7414 for Cheltenham, or an average of 320 per year. The expected value of e_2 for differences of yearly means is thus about

$$[(0.32^2 + 0.26^2)/320]^{1/2} \doteq 0.026\%$$

which is, as might be anticipated, somewhat less than the value $M = 0.038$ per cent from the first row of Table 5. The larger value $M = 0.038$ arises in part from the fact that actual successive bihourly deviations from the true 24-hour wave are not strictly independent and that such deviations are not identical at the two stations. However, the fact that M actually obtained for differences in yearly means is only about 50 per cent greater than that expected from random statistical fluctuations alone means that the scatter in the dials of the lower part of Figure 6 is not greatly in excess of the expected lower limit.

In fact, with the observed $M = 0.038$ per cent for yearly differences in the 24-hour wave at Cheltenham and Huancayo and with $e_2 = 0.026$ per cent for the expectancy of such differences due only to purely statistical sampling errors (counting rate) of hourly values, one finds the value 0.028 per cent for the expectancy of yearly differences arising from all other causes, or about 0.020 per cent at each station. Thus the magnitude of any variations in the yearly mean 24-hour waves which are not common to any two of the three stations must be small indeed. The values of M for the yearly values of the 12-hour wave are only slightly larger than those for the 24-hour wave, although the expectancy arising from statistical fluctuations only in hourly values is the same as for the 24-hour wave [Bartels, 1935].

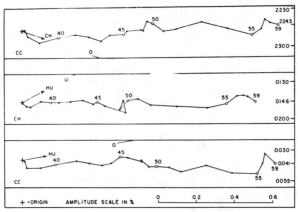

Fig. 7. Twenty-four-hour harmonic dials for summation of departures of yearly means of pressure-corrected ionization from mean 1937–1939 after vector departures at one station are rotated to 0^h LST, for which only the average vector is shown, and corresponding vector departures at second station are rotated through same angle. Times of maxima relative.

Test for spurious effects which might give rise to similar apparent changes in the 24-hour wave at different stations. Bartels [1935] was the first to show that 'curvature' effects may sometimes give rise to apparent diurnal variations which are not real. For example, the cosmic-ray ionization reaches a maximum near sunspot minimum, and at Huancayo the curve through the monthly means of cosmic-ray intensity as a function of time from January 1952 to December 1957 can be approximated by the parabola $\Delta = -t^2/500$, in which t is the time in months from the maximum (about November 1954) and Δ is the monthly mean cosmic-ray intensity in per cent from its value for November 1954. If any two points 12 months apart on this parabola are connected by a straight line, the parabola between these points, of course, lies entirely above the line with the maximum distance from the line to the parabola halfway between the points. Such a curvature effect gives rise to a small false annual variation with maximum in July (amplitude $\doteq 0.07$ per cent) if the year begins in January, or to a maximum in January if the year begins in July. In principle, this same parabolic curvature would lead to a maximum at the center of the day (for whatever hour the day starts), but, from the above equa-

tions, this effect is readily found negligibly small for the diurnal variation.

In addition to the above, a test for curvature effects on the 24-hour wave was made for Fredericksburg (Cheltenham), for 1957, during which year there were several large decreases in intensity. The average diurnal variation was computed for 328 complete days for the interval from 0.0 to 24.0 hours GMT (group 1). The average diurnal variation was also computed for 325 complete days with the 24-hour intervals starting at 12.0 hours GMT (group 2). The 24-hour harmonic coefficients, in percentage, for the two groups were, for group 1: $a_1 = -0.129$ and $b_1 = -0.081$; and for group 2: $a_1 = -0.134$, and $b_1 = -0.087$. These two sets of coefficients are each referred to 75° WMT (as in Table 2).

For the two groups, the times of maximum are practically identical and the amplitudes differ only by about 0.007 per cent, which indicates that the effect of curvature is indeed small. Actually, if the curvature is systematic, its effect in group 1 would add (or subtract, according to the sign of the curvature) a 24-hour wave with maximum (or minimum) at 12 hours GMT or 7 hours 75° WMT, whereas in group 2

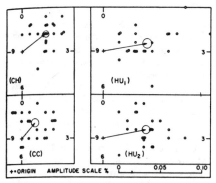

Fig. 8. Twelve-hour harmonic dials for yearly means and average vector for pressure-corrected ionization for Cheltenham (CH), Christchurch (CC), and Huancayo (HU). In (HU_2), pressure correction for each year made from 23-year average 12-hour pressure wave. Times of maxima on LST.

the effect of the same curvature would be to add (or subtract) a 24-hour wave with maximum (or minimum) at 0 hours GMT or 19 hours 75° WMT. Thus, since the maxima for groups 1 and 2 (on 75° WMT) occur near 1400 hours 75° WMT, the effect of curvature would be mainly to produce a difference in phase be-

tween the 24-hour waves for groups 1 and 2. Since the phase difference is only about 2 minutes of time, there is no evident difference that can be ascribed to curvature.

World-wide changes in intensity, which occur simultaneously on GMT, might contribute a spurious 24-hour wave to the yearly mean 24-hour waves (unless such effects averaged to zero). Although at Cheltenham and Huancayo any such spurious contributions to the 24-hour waves would be the same on LMT, such spurious waves would have phases at Christchurch (on LMT) which would differ by 7.5 hours from those at Cheltenham and Huancayo for the same years. If these effects were appreciable, they would increase the values of M in Table 5 for the cases where Christchurch is involved. Since this is not so, such effects cannot be large. Nevertheless, such effects may possibly explain the phase differences between Christchurch and Cheltenham in Figure 6(A-1) and (A-2) and account also for the differences in phases of Christchurch and Cheltenham, relative to Huancayo, which are evident in Figure 6(B-1) and (B-2).

The secular trend in the 24-hour variation from 1937 to 1959. Figure 9 shows the summation 24-hour harmonic dial for the depar-

TABLE 7. Probability P_1 for Maxima in Figure 6, A-2, B-2, and C-2 to Occur (in further samples) Later Than the Relative Time T_1 and Probability P_2 for the Maxima to Occur between Times T_1 and T_2

Vertical	Rotated	Fig. No.	T_1		P_1	T_1		T_2		P_2
			h	m		h	m	h	m	
CH	CC	6(A-2)	00	00	0.0003	23	00	01	00	0.233
			23	00	0.233	22	30	23	00	0.499
			22	43	0.500	22	00	22	30	0.246
			22	30	0.732					
			22	00	0.978					
HU	CH	6(B-2)	02	30	0.063	02	00	02	30	0.258
			02	00	0.321	01	30	02	00	0.437
			01	46	0.500	01	00	01	30	0.209
			01	30	0.758	23	00	01	00	0.033
			01	00	0.967					
HU	CC	6(C-2)	02	00	0.002	23	30	00	30	0.399
			01	30	0.031	00	30	01	00	0.381
			01	00	0.214	01	00	01	30	0.183
			00	41	0.500	01	30	02	00	0.029
			00	30	0.595					
			00	00	0.930					
			23	30	0.994					

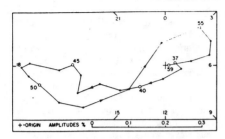

Fig. 9. Twenty-four-hour harmonic dial for summation of departures, averaged for Cheltenham, Christchurch, and Huancayo, of yearly means from mean for 1937–1959 of pressure-corrected ionisation. Times of maxima on LST.

tures of yearly means (from the mean for 1937–1959) averaged for Cheltenham, Huancayo, and Christchurch. The coordinates for Figure 9 are derived from those in the left half of Table 3. Since the summation 1937–1959 of the departures adds to zero, the resulting diagram is of necessity a closed figure. Nevertheless, from 1937 to 1948, the 'steps' tend generally toward about 1800 hours, whereas, from 1948 to 1955 or so, the steps tend toward about 0600. This suggests the possibility that the 24-hour wave varies quasi-systematically over a period of 22 years, although additional data are needed for a more definite indication. Nevertheless, the 24-hour wave certainly undergoes large changes, similar at all three stations. Moreover, these

changes in the 24-hour wave cannot be attributed to meteorological influences, since such effects would be expected to be about the same in each year, and the deviation of yearly means from the means for 1937–1959 should be essentially free of such influences. Moreover, it would scarcely be expected that any year-to-year variations due to meteorological causes would be similar at any two of the three stations.

On the other hand, the average 24-hour wave at any one of the stations will, of course, contain the 24-hour wave that might arise from systematic 24-hour variations in the height of the 100-millibar layer, for example.

Acknowledgment. This work was carried out under a National Science Foundation Grant for the International Geophysical Year Cosmic Ray Program, which support is gratefully acknowledged.

REFERENCES

Bartels, J., *Terrest. Magnetism and Atmospheric Elec., 40,* 1–60, 1935.
Forbush, S. E., *Terrest. Magnetism and Atmospheric Elec., 42,* 1–16, 1937.
Forbush, S. E., Electromagnetic phenomena in cosmical physics, *Intern. Astron. Symposium, 6,* 332–344, 1958.
Lange, Isabelle, and S. E. Forbush, Cosmic-ray results from Huancayo Observatory, Peru, June 1936–December 1946, *Carnegie Inst. Wash. Publ. 175,* vol. xiv, Washington, D. C., 1948.

(Manuscript received May 31, 1960.)

Journal of

GEOPHYSICAL RESEARCH

VOLUME 66   AUGUST 1961   No. 8

Intensity Variations in Outer Van Allen Radiation Belt

S. E. FORBUSH,[1] D. VENKATESAN, AND C. E. McILWAIN

Department of Physics and Astronomy, State University of Iowa
Iowa City, Iowa

Abstract. Using data from Explorer VII, the changes in the intensity of the outer Van Allen radiation belt were investigated in detail over the period October 26, 1959 to December 9, 1959. To relate the intensities to location in the belt, the parameter L was used. L is defined as a function of the integral invariant I and scalar magnetic field B, such that everywhere on the shell described by the motion of a trapped particle in the earth's magnetic field, L closely approximates the equatorial radius of the shell. For the period under study, the intensity variations at selected values of L were negatively correlated with geomagnetic activity. However, an analysis of data over a much longer period is required to determine whether this correlation is statistically significant.

For the period studied, the decrease in measured intensity during periods of magnetic storms (large a_p) is consistent with the idea that some of the particles are 'dumped' into the auroral and subauroral zones where they may partly contribute to the auroral zone currents that account for a_p. In addition, the reduction in intensity generally occurred in the outer part of the zone. The low values of intensity, for $L = 3.5$, 4.1, and 4.7, occurred in the interval November 22 to December 9, at times when the equatorial ring current field, U (southward on the earth at the equator), was large. Thus, U cannot be due to the westward current due to the longitudinal drift of measured trapped particles in the region for which L is greater than 3.5, since a decrease in particle density therein would diminish the westward drift current with a consequent decrease in U, which is opposite to what is observed. This result indicates that the site of the ring current is either elsewhere, or that the principal contribution to it comes from drifting particles having energies below the detection thresholds of Explorer VII equipment. The phenomena were investigated over a longer period, using daily averages of the maximum counting rates. This procedure had serious limitations.

Introduction. A study of the temporal fluctuations of the intensity of the outer Van Allen radiation belt and its association with magnetic storms is of importance [*Rothwell and McIlwain*, 1960; *Arnoldy, Hoffman and Winckler*, 1960; *Van Allen and Lin*, 1960]. The present study uses data from Explorer VII and covers in detail the period from October 26 to December 9, 1959.

Counting rates are available as a function of time since October 1959 from passes of Explorer

VII through the outer Van Allen radiation belt. The equipment was prepared by the Department of Physics and Astronomy of the State University of Iowa. The details of the equipment and the characteristics of the counters have been described earlier [*Ludwig and Whelpley*, 1960]. The approximate detection thresholds for the 112 Geiger counter are 30 Mev for protons, 2.5 Mev for electrons, and 80 kev for X rays (5 per cent transmission), and for the 302 Geiger counter the corresponding values are respectively 18 Mev, 1.1 Mev, and 30 kev. The counting-rate data for the 302 and 112 counters, corrected for dead time, are available for each pass as a func-

[1] Visiting Professor from the Department of Terrestrial Magnetism, Carnegie Institution, Washington, D. C.

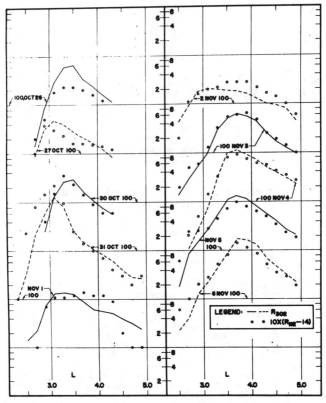

Fig. 1. Average counting rates vs. L from passes for each 'day,' October 26–November 6, 1959.

tion of time. The ratio of counting rates provides a measure of hardness of the radiation. It is believed that in the outer zone the counters are responding primarily to the bremsstrahlung generated by the impact of electrons in the energy range from a few tens to a few hundreds kiloelectronvolts (Van Allen, private communication).

In a typical transit with complete registration two maxima of intensity are observed as the satellite crosses the belt in its northbound and southbound passes. All the data used in the analysis were principally recorded by receivers at SUI, Iowa City, Iowa, and Blossom Point, Maryland, with occasional additional recordings from San Diego, California, and Ottawa, Canada. Data were used only for those passes that

clearly crossed the region of maximum intensity in the belt.

Method of analysis. In order to relate the measured intensities to location within the belt, it is necessary either to use the three geographic coordinates, latitude, longitude, and altitude, or to replace them with some other suitable coordinates such as L and B. McIlwain (unpublished communication, 1960) has pointed out that all the desirable characteristics of the integral invariant, I, can be retained by a parameter L which is nearly constant along a line of force in the earth's magnetic field. L is defined as a function of the integral invariant I [*Northrup and Teller*, 1960; *Vestine and Sibley*, 1960] and scalar magnetic field B, such that everywhere on the shell described by the motion of a trapped

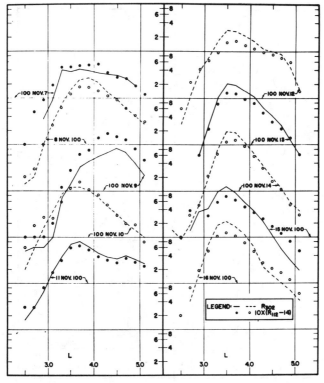

Fig. 2. Average counting rates vs. L from passes for each 'day,' November 7–16, 1959.

harged particle in the earth's magnetic field, L losely approximates the equatorial radius of the hell. The computation of L was supervised by lcIlwain; L is derived on the basis that the arth's field for epoch 1955 is represented by the x spherical harmonics obtained by *Finch and eaton* [1957]. Because of the strong intensity ariations with time, it has not as yet been posble to determine the dependence of intensity pon B. As a result, it has been necessary to igore this important effect.

For a constant L, the intensity should depend nly on B, which in turn will depend on the altide and longtitude at which the satellite crosses le zone. In addition there may be absorption fects resulting in additional intensity variations ith altitude. The motion of the satellite is such lat the altitude, B, and the longitude for the me L, all vary systematically rather than ran-

domly and independently. As a consequence, to determine the effect of each of these parameters upon the intensity, a much longer sequence of observations is required than would be the case if the consecutive values of these variables were both random and independent. These effects, together with relatively large and rapid time variations of intensity associated with magnetic disturbances, indicate that a systematic analysis of data over a long period of time is required to reveal quantitatively the effects of the individual parameter with a reasonable degree of certainty.

In general, all passes for which the data were received by the tracking stations listed above occurred within successive intervals of about 10 hours or less, which were interspersed with intervals of 14 hours or so, when no data could be recorded. The intensities for the same values of L were averaged for the passes in the 10-hour

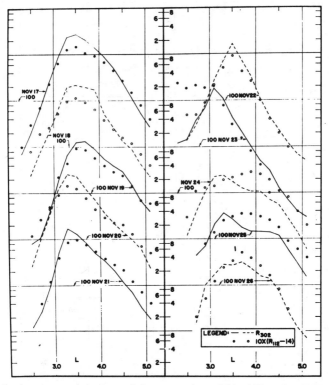

Fig. 3. Average counting rates vs. L from passes for each 'day,' November 17-26, 1959.

interval for each 'day.' Occasionally the 'day' extended from about 20 hours on one GMT day to the early hours of the next GMT day, in which case the 'day' was taken as the latter GMT day, since the center of the interval always was on that day. An indication of the variability of the 302 counting rate within 'days' is shown in the second graph from the bottom of Figure 9, in which for $L = 4.1$, R_{302} (the 302 counting rate) is plotted for each pass. The large scatter (even on some magnetically quiet days) is evident. This scatter is more or less typically similar to that at other values of L.

Average intensity for each 'day' as function of L. Figures 1 to 5 show for each 'day' the average intensity as a function of L. The counting-rate scale is logarithmic. The curve for each day is indicated together with the horizontal line that shows where, on the curve for that day,

the counting rate is 100 counts per second. To facilitate comparison with 302 rates (R_{302}), the rates for the 112 counter are multiplied by 10 after deducting 14 counts ($R_{112}-14$) per second to eliminate the cosmic-ray background. For the 302 counter, the normal cosmic-ray background count is negligible. Thus when the points for the 302 and 112 counters coincide, the ratio $R_{302}/(R_{112}-14)$ is 10. Large differences in the curves for different days are evident. Even on the magnetically quiet days, November 9, 11, and 12, 1959, there were considerable differences in the curves for the individual passes.

Figures 6 and 7 show the counting rates for individual passes as a function of L for November 23 and 28, 1959, which were days of large magnetic disturbance. Shown also for comparison are the average curves for each day preceding and following those dates. Within each of

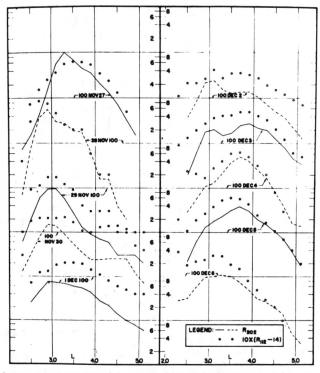

Fig. 4. Average counting rates vs. L from passes for each 'day,' November 27–December 6, 1959.

these 4 days the curves for the individual passes were quite similar. The curves for 3.6 hours UT and 3.8 hours UT on November 28, 1959, correspond to those in the center part of Figure 1, discussed by *O'Brien, Van Allen, Roach, and Gartlein* [1960], who showed that low-latitude aurora occurred at a location that corresponds to the L for the maximum rate near these times. The curves on November 28 at 3.6 and 3.8 hours UT show that for $L > 3.6$ the counting rates are essentially zero, although the maximum counting rates for these two passes are at least as high as on November 27, before the start of the storm. Incidentally, the curve in Figure 7 for 1.9 hours UT on November 28, 1959, was obtained during the initial phase of the storm. (The sudden commencement was at 2351 UT, November 27.) This curve is not very different (except perhaps for $L > 4.4$) from the average curve for November 27; the curve for 3.6 hours UT, No-

vember 28, 1959, however, is markedly different. It should be noted that the value of 8o for K_p (corresponding to $a_p = 414$ gammas) for t! ʲ interval 3 to 6 hours UT was one of the two highest for any 3-hour interval during the entire period October 26 to December 9, 1959. The other case was 18 to 21 hours UT, December 5, 1959, for which K_p was 8- (corresponding to $a_p = 358$ gammas). The values for U for the two corresponding cases were respectively 149 and 194 gammas.

The curves for the individual passes on November 23 differ less from the average for the previous day than do those for November 28. On November 23 the maximum value of K_p for any 3-hour interval in which one or more passes occurred was 6- ($a_p = 134$); also, it may be noted that the beginning of the storm of November 28 was marked by a sudden commencement, whereas that of November 23 was not.

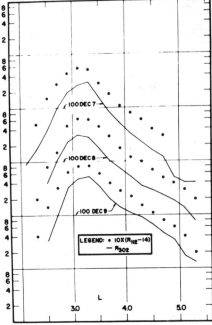

Fig. 5. Average counting rates vs. L during December 7-9, 1959.

storms on November 30 and December 3 with 3-hour values of K_p of 6o. This complicates the study of storm of December 5. The individual passes on that day, which occur just before the sudden commencement, are similar to the curves on December 4. Unfortunately, the next available pass is 16 hours after sudden commencement. However, this pass reveals a reduction in intensity, especially for $L > 3.4$, which is observed in the early hours of the next day also. A recovery to higher intensities is observed starting from the pass at $\sim$ 2230 UT on December 6.

Another storm in the beginning of the period —namely, October 29—with sudden commencement at 2347 UT initiates a disturbance lasting up to November 4, with the peak disturbances being around November 1-2. It can definitely be stated that the intensity on these 2 days was the lowest, and that recovery was observed from November 3.

Further analysis of data over long periods is essential before one can answer whether the counting rate always decreases with storms, and, if so, whether the decrease occurs only during the main phase, and whether a quantitative relation between the decrease in intensity and some characteristic parameter of the geomagnetic disturbance can be derived.

To get an average picture of intensity changes during magnetic disturbances, the four storms were considered together. The days—namely November 1, 23, and 28, and December 6, when U was the highest—were chosen as the 'O' days. The average variation of the intensity of the 302 counter (R_{302}) plotted as a function of L is shown in Figure 8 from (-4) day to ($+5$) day. The values are also given in Table 1.

Which, if any, of these factors was connected with the difference in the counting-rate curves for November 23 and 28 is uncertain.

The magnetic storm on December 5, 1959, with a 3-hour value for U and K_p as high as 194 gammas and 8- (corresponding $a_p = 358$ gammas), and with sudden commencement at 0659 UT, was preceded closely by two other moderate

TABLE 1. Average Counting Rate R_{302} Vs. L for Chosen Epoch Days. 'O' Days are November 1, 23, and 28, and December 6, 1959.

Designation of Day	$L \to$ 2.5	2.7	2.9	3.1	3.3	3.5	3.7	3.9	4.1	4.3	4.5	4.7	4.9	5.1	5.3
-4	8	11	41	182	670	1281	1305	910	560	392	291	152	70	40	17
-3	23	69	197	544	1028	919	563	373	245	185	130	75	45	24	15
-2	22	32	105	275	649	796	707	497	303	215	121	61	32	18	11
-1	22	86	235	563	664	640	388	258	142	96	59	35	22	12	8
0	64	202	352	597	538	287	140	70	39	32	19	14	9	3	2
$+1$	29	72	143	197	200	137	104	84	70	51	42	31	17	4	4
$+2$	22	65	128	194	266	252	242	204	169	117	86	59	40	12	7
$+3$	15	44	133	200	296	418	426	337	254	190	138	98	71	49	8
$+4$	25	76	184	242	287	366	566	519	373	274	192	126	92	24	14
$+5$	22	52	155	216	308	536	1010	967	768	444	298	200	138		

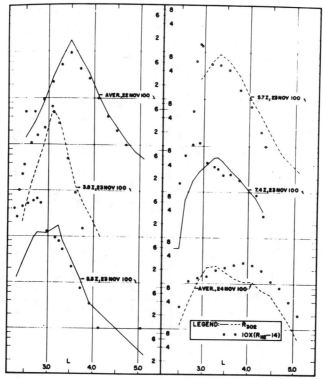

Fig. 6. Counting rates vs. L for individual passes, November 23, 1959. Average counting rates vs. L from passes on November 22 and 24, 1959.

The number of storms considered, namely, four, is not sufficient for drawing any quantitative conclusions. Nevertheless it can be seen from the figure that the intensity decreases to a minimum value on day (+1) and gradually recovers from day (+2). It should also be noted that the maximum intensity apparently shifts to a lower value of L, and subsequently recovers gradually to its original value of $L \sim 3.7$. Particularly noticeable are the large changes in intensity, especially at the higher L values.

Time variations of intensity at fixed values of L. During the period October 26 to December 9, 1959, the average counting rates for each 'day' for the 302 and 112 counters, at $L = 2.9$, 3.5, 4.1, and 4.7, are given in Table 2 and plotted in Figure 9. As pointed out earlier, the rate for the 112 counter, after subtracting 14

for cosmic-ray background, is multiplied by 10 before plotting. The table also gives ρ, the ratio $R_{302}/(R_{112}-14)$. The GMT time in hours refers to the middle of the pass. The U and a_p values are computed for the 3-hour intervals in which the passes occur.

At $L = 2.9$, the counting rates show large fluctuations from day to day. Indeed, within days, at $L = 2.9$, the fluctuations were often larger than at larger L values. The curve at $L = 2.9$ for the whole period of study shows no correspondence with those for larger values of L. However, the curves at $L = 3.5$, 4.1, and 4.7 exhibit a similar variation during this period. Figures 10 and 11 show that for the period October 26 to December 9, the average counting rates for each 'day' at $L = 3.5$ are about 2.0 times those at $L = 4.1$, and the average count-

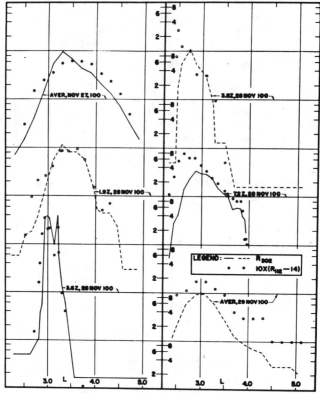

Fig. 7. Counting rates vs. *L* for individual passes November 28, 1959. Average counting rates vs. *L* from passes on November 27 and 29, 1959.

ing rates at $L = 4.1$ are about 3.8 times those at 4.7. Thus the percentage changes in the counting rates during this period are about the same at $L = 3.5, 4.1,$ and 4.7.

Relation between changes of counting rate and magnetic activity, October 26–December 9, 1959. In Figure 12, the average 302 counting rate (R_{302}) for each 'day' at $L = 4.1$ is shown (solid curve) for the period October 26 to December 9, 1959, along with the averages of the values of U (dashed curve) and of the a_p (crosses) values for the 3-hour intervals in which the individual passes occurred. U is the southward geomagnetic component, in gammas, of the equatorial ring current, ERC (primary plus induced). U is thus a measure of the storm-time

field, D_{st}, and thus it should not be confused with Bartels' U measure for monthly mean magnetic activity. The derivation of U was first described by Kertz [1958]. U is derived for each 3-hour Greenwich interval from three 3-hourly mean values of H from each of four observatories separated about 90° in longitude and near the equator. At each observatory these three 3-hour means of H cover a 9-hour interval centered at midnight, and are consequently unaffected by the ordinary diurnal variation, S_q, and it variability. By taking advantage of the antisymmetry, about local midnight, of the disturbance diurnal variation, S_D in H, the effect of S_D can be effectively eliminated. In addition, the secular variation is removed to permit compari-

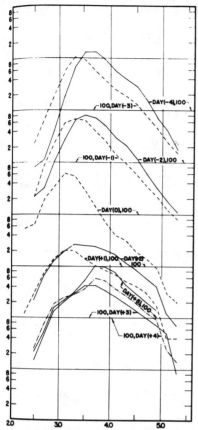

Fig. 8. Average counting rates (R_{303}) vs. L during four magnetic storms in period October 26–December 9, 1959. 'O' days: November 1, 23, 28, and December 6, 1959.

son of U values separated by long periods of time.

The correspondence between the changes in R_{303} and those in U and a_p [*Annals IGY*, 1957; *CRPL*, 1960] is shown in Figure 12. During this period the higher counting rates at $L = 4.1$ occur for the periods when U and a_p are smallest, that is, during magnetically quiet periods. The decrease in the counting rate R_{302} and R_{112} during the storms on November 23, November

28, and December 6 is notable as is the similarity in the 'recovery' of the counting rate and of U and of a_p during the post-perturbation period of the storms. The similarity in the 'recovery' part of the earlier storm on November 1 can also be seen. It is not possible from the study to say which of the two parameters U and a_p is to be preferred.

For a geophysical phenomenon that is affected through some link with solar activity, experience dictates that it would be thoroughly unjustified on any statistical basis (or otherwise) to conclude from such a very limited sample of data that the same relationships shown in Figure 12 would apply to subsequent data. It must be emphasized that the similarity between counting rate, U, and a_p shown in Figure 12, has been maximized by the choice of base values and of scales. Although these results are suggestive, it is essential to apply the analysis to additional data over a much longer period before accepting the conclusions suggested by this analysis. For this period, the decrease in measured intensity during periods of magnetic storms (large a_p) is consistent with the idea that some of the particles are 'dumped' into the auroral zone where they may partly contribute to the auroral zone currents that account for a_p. In addition, the reduction in intensity generally occurred in the outer part of the zone. The low values of intensity, for $L = 3.5$, 4.1, and 4.7, occurred in the interval November 22 to December 9, at times when the equatorial ring current field, U (southward on the earth at the equator), was large. Thus, U cannot be due to the westward current due to the longitudinal drift of the trapped particles measured by the equipment of Explorer VII, in the region for which L is greater than 3.5, since a decrease in particle density therein would diminish the westward drift current with a consequent decrease in U, which is opposite to what is observed. This result indicates that the site of the ring current is either elsewhere, or that the principal contribution to it comes from drifting particles having energies below the detection thresholds of Explorer VII equipment.

Analysis of the maximum counting rates for passes. Since values of L were not yet computed for the period after December 9, 1959, an alternative procedure was adopted on which the average of the maximum counting rates for all the passes in the day was derived. For the period

TABLE 2. Average Counting Rates R_{202} and $(R_{112}-14)$ at $L = 2.9, 3.5, 4.1,$ and 4.7 for the Period Oct. 26 to Dec. 9, 1959: ρ is the Ratio $R_{202}/(R_{112}-14)$

Date, month, day	Time GMT, hours	U	A_p gammas	$L = 2.9$ R_{202}	$R_{112}-14$	ρ	$L = 3.5$ R_{202}	$R_{112}-14$	ρ	$L = 4.1$ R_{202}	$R_{112}-14$	ρ	$L = 4.7$ R_{202}	$R_{112}-14$	ρ
Oct. 26	12.6	72	112	82	5	16.4	650	33	20.3	165	12	13.8			
27	12.2, 17.5	59	18	315	37	8.5	285	16	17.8	126	14	9.0			
30	11.1	27	18	(25)	4	6.2	295	23	12.8	76	6	12.6			14.5
31	12.4	97	64	590	146	4.0	265	25	10.6	85	7	12.1	29	2	32.0
Nov. 1	12.1	116	112	75	7	10.7	130	12	10.8	56	12	4.8	32	1	8.2
2	10.0, 11.7, 13.5	95	109	181	16	11.4	190	28	6.8	132	22	6.0	82	10	9.3
3	9.6, 11.3	86	96	56	6	9.3	520	52	10.0	455	48	9.5	130	14	9.1
4	11.0, 14.4	74	80	27	5	5.4	1000	80	12.5	725	60	12.1	310	34	11.7
5	8.8, 10.6, 12.3, 14.1	84	64	158	25	6.3	1102	71	15.6	805	63	12.8	270	23	11.4
6	8.4, 10.2	70	54	132	20	6.6	905	82	11.0	1325	85	15.6	330	29	10.2
7	9.8, 16.6	56	13	35	9	3.9	370	46	8.0	348	53	6.6	265	26	10.0
8	7.6, 9.4, 11.2	55	32	98	10	9.8	1500	121	12.4	2200	141	15.7	600	60	10.0
9	9.0, 10.8, 12.6	34	8	6	1	6.0	139	28	5.0	556	140	4.0	687	128	5.3
10	6.9, 8.7, 10.5	41	15	160	27	5.9	1214	115	10.5	740	82	9.0	210	20	10.5
11	8.3, 11.8, 10.1	44	10	68	8	8.5	727	59	12.3	460	38	12.1	407	35	11.7
12	6.2, 8.0, 9.7, 11.5	39	8	400	42	9.5	2837	156	18.2	1613	115	14.0	894	73	12.2
13	5.9, 9.4, 11.1	41	13	60	6	10.0	2090	131	15.9	1080	78	13.8	260	22	11.8
14	5.5, 7.2, 9.0, 10.8	55	58	52	5	10.4	1958	115	17.0	665	56	11.9	119	11	10.8
Nov. 15	6.9, 8.5, 10.3	31	11	360	46	7.8	1245	77	16.2	369	34	10.9	56	12	4.7
16	4.7, 6.9, 8.2	31	9	161	19	8.5	2233	126	18.6	605	49	12.3	106	17	6.2
17	6.1, 7.9, 9.5	52	24	252	23	11.0	2727	147	18.6	913	67	13.6	117	15	7.8
18	5.7, 7.4, 9.2	63	18	214	26	8.2	2132	112	19.2	515	38	13.6	90	12	7.5
19	5.2, 7.1, 8.8	54	37	41	5	8.2	1281	92	14.0	560	39	14.4	152	14	10.7
20	4.9, 6.6, 8.4	48	11	343	46	7.5	2313	124	18.7	378	30	12.3	98	13	7.5
21	4.5, 6.3, 8.1	43	15	104	12	8.7	1739	97	17.9	496	38	13.1	116	12	9.7
22	2.5, 4.2, 6.0	62	48	67	9	7.4	1282	91	14.1	76	10	7.6	10	1	10.0
23	3.8, 5.6, 7.4	117	127	672	187	3.6	440	30	14.7	50	4	12.2	7	0	
24	3.4, 5.2, 6.9	60	13	106	13	8.2	174	22	7.9	107	25	4.3	28	8	3.5
25	3.1, 3.9, 6.6	45	11	71	8	8.9	276	31	9.0	145	31	4.7	58	9	6.4
26	2.7, 4.5, 6.2	47	23	102	5	20.0	272	33	8.2	158	26	6.1	26	3	8.7
27	0.5, 2.3, 4.0, 5.8	54	31	155	26	6.0	720	63	11.4	240	32	7.5	32	7	4.7
28	1.9, 3.7, 5.4, 7.2	119	324	563	75	7.5	189	20	9.4	16	2	8.0	2	0	
29	1.6, 3.3, 5.1, 6.9	102	26	91	17	5.3	28	9	3.1	7	3	2.3	3	1	3.0
30	1.2, 2.9, 4.8	76	22	150	21	7.1	55	17	3.2	24	13	1.9	25	12	2.1
Dec. 1	2.6, 4.3	83	80	76	13	5.8	60	20	3.0	28	11	2.5	10	5	2.0
2	2.2	70	36	215	26	8.3	128	34	3.8	104	27	4.2	38	12	3.2
3	-0.1, 1.8, 3.6, 5.3	77	87	178	27	6.6	168	41	4.1	212	39	5.4	70	12	5.8
4	-0.3, 1.4, 3.2, 4.9	59	64	108	17	6.4	378	59	6.4	256	32	8.0	41	3	13.7
5	-0.7, 1.1, 2.8, 4.5	56	5	131	27	4.9	325	61	5.3	200	25	8.0	72	7	10.3
6	0.7, 2.5, 4.2	118	78	95	24	4.0	101	19	5.3	32	5	6.4	14	1	14.0
7	-1.1, 0.3, 2.0, 3.8	80	28	194	44	4.4	156	29	5.4	33	12	4.1	12	3	4.0
8	-1.4, -0.1, 1.7, 3.4	58	11	236	47	5.0	180	45	4.0	53	12	4.4	24	5	4.8
9	-2.2, 0.5, 1.3, 3.1	54	11	325	62	5.2	340	59	5.8	108	22	4.9	45	8	5.6

298

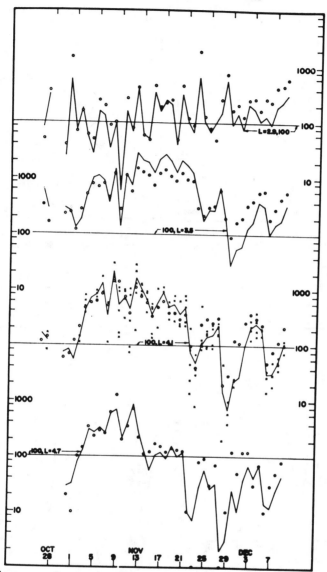

Fig. 9. Average counting rates from passes for each 'day,' October 26–December 9, 1959, at $L = 2.9, 3.5, 4.1,$ and 4.7, and rates from individual passes at $L = 4.1$.

October 26 to December 9, 1959, these daily averages of maximum counting rates fell on a curve quite similar to that for $L = 4.1$ in Figure 12. This indicated that such averages of 'intensity maxima" for the passes of the day might be used to determine the effect of magnetic disturbance. However, it was anticipated from the graph for $L = 2.9$, in Figure 9, that if the in-

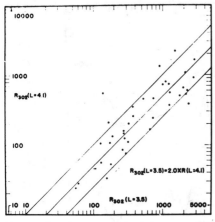

Fig. 10. R_{302} ($L = 4.1$) vs. R_{302} ($L = 3.5$) from daily averages, October 26–December 9, 1959.

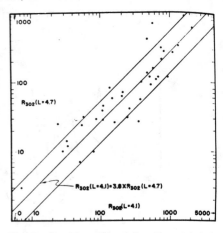

Fig. 11. R_{302} ($L = 4.7$) vs. R_{302} ($L = 4.1$) from daily averages, October 26–December 9, 1959.

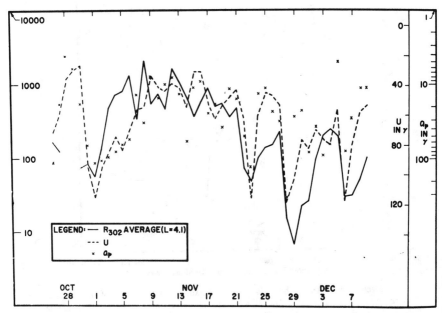

Fig. 12. Average counting rate, R_{302} at $L = 4.1$ from passes for each 'day,' equatorial ring current measure U, and magnetic activity a_p, October 26–December 9, 1959.

300

tensity maxima occurred at values of L of about 2.9 or so, then these intensity maxima would not be likely to show a consistent effect of magnetic disturbance. Except during April and May 1960, the departures from their monthly means of these averages of 'intensity maxima' for the daily passes indicated some negative correlation with the corresponding departures for magnetic activity. During April and May 1960, a rough evaluation reveals many cases of maxima of intensity occurred at values of L less than about 3.0, which possibly may account for the absence of any consistent correspondence btween the 'intensity maxima' and magnetic activity; if so, the use of these average maximum counting rates for long-term investigations is of dubious value.

Between the departures of the average maximum counting rates for the day, from their mean over the interval October 26, 1959, to July 7, 1960, and the corresponding departures in U, the correlation is of doubtful significance. There are indications, however, that this may be due to an effect of altitude on counting rate, although this cannot be reliably determined before all the data are examined.

Ratio of counting rate for the 302 counter to that for the 112 counter. From Figure 9, for October 26 to December 9 (for $L = 3.5$, 4.1, and 4.7) the ratio $\rho = R_{302}/(R_{112}-14)$ is generally seen to be greater for periods when R_{302} is high, although the rates for both counters vary somewhat similarly. This shows that the radiation is softer (i.e., more easily absorbed by the lead shield of the 112) at times of high intensity.

Acknowledgment. Work for this paper was assisted by the joint program of the Office of Naval Research and Atomic Energy Commission under contract N9onr—93803.

REFERENCES

Annals of the International Geophysical Year, 4, Parts IV–VII, Pergamon Press, 1957.

Arnoldy, R., R. Hoffman, and J. R. Winckler, Observation of the Van Allen Radiation Regions during August and September 1959, Part I, *J. Geophys. Research, 65,* 1361–1375, 1960.

CRPL—F, Part B, January, February, 1960. Publication of the U. S. Department of Commerce National Bureau of Standards Central Radio Propagation Laboratory, Boulder, Colorado, 1960.

Finch, H. F., and B. R. Leaton, The earth's main magnetic field—Epoch 1955·0, *Mon. Not. Roy. Astronom. Soc.* (Geophys. Suppl.) 7 (6), 314–317, 1957.

Kertz, Walter, Ein Neues Mass für die Feldstärke des Erdmagnetischen Äquatorialen Ringstroms Abhandlungen. der Akademie der Wissenschaften in Göttingen Mathematisch Physikalische Klasse Beiträge zum Internationalen Geophysikalischen Jahr Heft 2, Göttingen Vandenhoeck and Ruprecht, 1958.

Ludwig, G. H., and W. A. Whelpley, Corpuscular radiation experiment of Satellite 1959 Iota (Explorer VII), *J. Geophys. Research, 65,* 1119–1124, 1960.

Northrup, T. G., and E. Teller, Stability of the adiabatic motion of charged particles in the earth's field, *Phys. Rev., 117,* 215–225, 1960.

O'Brien, B. J., J. A. Van Allen, F. E. Roach, and C. W. Gartlein, Correlation of an auroral arc and a subvisible monochromatic 6300 A arc with outer-zone radiation on November 28, 1959, *J. Geophys. Research, 65,* 2759–2766, 1960.

Rothwell, P., and C. McIlwain, Magnetic storms and the Van Allen radiation belts: Observations with satellite 1958ε (Explorer IV), *J. Geophys. Research, 65,* 799–806, 1960.

Van Allen, J. A., and Wei Ching Lin, Outer radiation belt and solar proton observations with Explorer VII during March–April 1960, *J. Geophys. Research, 65,* 2998–3003, 1960.

Vestine, E. H., and W. L. Sibley, The geomagnetic field in space, ring currents and auroral isochasms, *J. Geophys. Research, 65,* 1967–1979, 1960.

(Manuscript received June 2, 1961.)

Journal of

GEOPHYSICAL RESEARCH

VOLUME 67   SEPTEMBER 1962   No. 10

The Morphology and Temporal Variations of the Van Allen Radiation Belt, October 1959 to December 1960

S. E. FORBUSH,[1] G. PIZZELLA,[2] AND D. VENKATESAN

Department of Physics and Astronomy
State University of Iowa, Iowa City

Abstract. The time variations of intensity in the 'horn' of the outer Van Allen radiation belt as measured by the omnidirectional GM counter 302 in Explorer 7 were investigated for the period October 1959 to December 1960. A consistent empirical relationship between intensity and scalar magnetic field B was derived for different values of the magnetic shell parameter L that made it possible to 'correct' for dependency of intensity on B and thus to examine the true time variations at any fixed L. The correlation between counting rates in the outer zone over North America and Australia at the same value of L improved decidedly after correction for B. The changes in the corrected intensity for several L values between 2.5 and 4.7 earth radii revealed some large variations (greater than a factor of 100) prominent only between $L = 2.5$ and $L = 3.2$, and some fluctuations by a factor of 10 only at the larger L values. The temporal variations of intensity between $L = 2.5$ and 3.5 over North America are similar to those between $L = 1.8$ and 2.5 over South Africa. The changes in intensity tend to be correlated negatively with U, the geomagnetic equatorial ring current field, for $L > 3.4$, and positively for $L < 3.4$. L_{max}, the value of L at which the maximum intensity occurs, tends to decrease with increasing U, the correlation coefficient between L_{max} and log U being -0.7. During the main phase of large magnetic storms the intensity for $L > 3.5$ is generally less, and for $L < 3.5$ it often tends to be greater, than before the storms. For $2.5 \leq L \leq 3.5$ nearly all the observations over Australia, and about half those over North America, were at B values equal to or greater than those at sea level over the region of South Africa for the same L values. Consequently, all the particles observed under these conditions over Australia and North America are lost as they drift over the region of South Africa. The observed intensities over Australia, in particular, measure the outflux of particles from the outer radiation belt in a time interval less than the longitudinal drift period. This result implies a very high rate of replenishment of energetic electrons in the outer belt.

INTRODUCTION

Period covered. A study of the first six weeks of Explorer 7 data [*Forbush, Venkatesan, and McIlwain*, 1961] suggested a possible correlation between temporal variations of intensity in the outer belt (as observed during the

[1] Visiting investigator from the Department of Terrestrial Magnetism, Carnegie Institution of Washington, Washington, D. C.
[2] Now at the Department of Physics University of California at San Diego, La Jolla, California.

excursions of Explorer 7 through the low-altitude horn of the outer belt over the North American continent) and U [*Kertz*, 1958], the equatorial geomagnetic field of the ring current, and the a_p measure for geomagnetic activity [*Solar-Geophysical Data, CRPL*, 1959–1961]. The contribution to intensity variations arising purely from changes in B (or in altitude) could not be evaluated at that time.

The present investigation over the period October 14, 1959, to December 31, 1960, has been made possible by the processing of the Explorer 7 data by Venkatesan and Pizzella. Data are available for each day from November 1, 1959, to August 31, 1960. Toward the end of the interval, especially during September and October, the lack of data is largely attributable to the low operating voltage of the batteries resulting from the unfavorable average aspect of the solar cells in relation to the sun.

Coordinate system adopted. On the basis of a theoretical analysis of spatial variations of the intensity of geomagnetically trapped radiation, the coordinates B and L were used. B is the scaler geomagnetic field, and L is the equatorial radius, in units of the earth's radius, of a magnetic shell on which most of the useful properties of the integral invariant are retained [McIlwain, 1961]. Moreover, L is practically constant throughout an 'invariant magnetic shell.' During periods of geomagnetic disturbances, the magnetic shells for L greater than 3 earth radii are probably distorted, but we can use B and L for magnetically quiet conditions as a reference [McIlwain, 1961].

Reduction of data. The primary counting rate data of the Geiger tube 302 at given times are available on IBM cards. About 250,000 such data points from the recording stations at Iowa City, Iowa; Blossom Point, Maryland; and Woomera, Australia, were analyzed. By means of the IBM 7070 computer at the State University of Iowa, the counting rates were calculated as functions of L, B, and time T.

By means of the smoothed orbital elements calculated at the Smithsonian Astrophysical Observatory (private communication, 1960) and the magnetic field program of *Jensen and Whitaker* [1960], duly adapted, the geographic coordinates and scalar magnetic field B were computed. For the evaluation of B a harmonic expansion consisting of the 48 coefficients of the earth's field was utilized. The computer program of McIlwain containing the definition of L in terms of the integral invariant I made possible the computation of the values of L corresponding to sequences of positions in space.

By using the primary IBM cards for an entire pass from a station, the orbital position, B, and L were computed at 1-min intervals during the pass, and, by linear interpolation, the time at which the satellite crossed the desired magnetic shell and the corresponding

counting rate at that time were obtained. The final output, about 75,000 IBM cards, contained the universal time, the geographic coordinates, L, B, and I, the ratio $B/B_{equator}$, counting rate R, log R, and a serial number designating each pass. For the numerous days on which the simultaneous availability of data from Iowa City and Blossom Point resulted in duplication of data, only the Iowa City data were used.

Counter characteristics and nature of measured radiation. The detection thresholds of the omnidirectional 302 counter [*Van Allen and Lin*, 1960; *Ludwig and Whelpley*, 1960] are 18 Mev for protons, 1.1 Mev for electrons (direct penetration), and 30 kev for X rays (5 per cent transmission). The calibration curve for correcting the apparent counting rate for dead time is reliable up to apparent counting rates of about 7000 counts per second. Few observed counting rates exceeded this.

It was initially presumed that in the outer zone the 302 counters were responding primarily to the bremsstrahlung generated by the impact of electrons in the energy range from a few tens to a few hundreds of kev [Van Allen, private communication, 1960]. Recent evidence from the Explorer 12 data [*O'Brien, Van Allen, Laughlin, and Frank*, 1962], however, has shown that the high counting rates of 302 tubes in the heart of the outer zone are due almost entirely to penetrating electrons of energy greater than 1.6 Mev. The Injun 1 data from an orbit similar to that of Explorer 7 [*O'Brien, Laughlin, Van Allen, and Frank*, 1962] support this finding as applying to the lower altitudes as well. Hence the present analysis probably refers to the intensity of penetrating electrons (those of energy greater than 1.1 Mev).

Definititon of symbols and units.

L, equatorial radius, in earth radii, of a magnetic shell.

B, scalar magnetic field in gauss.

$R(I)$, $R(W)$, and $R(J)$, observed counting rates in counts sec^{-1} from passes recorded respectively at Iowa City (or Blossom Point, Maryland); Woomera, Australia; and Johannesburg, South Africa.

$R(N)$ and $R(S)$, observed counting rates in counts sec^{-1} from northbound and southbound transits of the same pass at any one station.

$R_B(I)$, $R_B(W)$, and $R_B(J)$, above counting rates corrected to an arbitrary fixed value of B.

$R*_B(I)$, $R*_B(W)$, $R*_B(J)$, and $R*_B(JS)$, logarithmic average counting rate; e.g., daily average, $R*_B(I)$, is such that log $R*_B(I) =$ average of values of log $R_B(I)$ available for the day. (JS) refers to data from Johannesburg and some South American stations.

U, Equatorial geomagnetic field of ring current in gammas, positive southward.

INTENSITY CORRECTIONS TO CONSTANT SCALAR MAGNETIC FIELD B

Procedure. The orbit of Explorer 7 was such that the altitude and hence the scalar magnetic field B over Woomera and over Iowa City corresponding to any particular L exhibited a systematic variation of opposite phase with a period about 105 days. The range in B was about 0.1 gauss. During the time interval of a few weeks required for observing any considerable change in B at either Iowa City or

TABLE 1. Statistical Parameters for Each of Several Values of L

Parameters are derived by fitting the equation

$$200 \log [R(I)/R(W)] = A + b(B_W - B_I)$$

to 302 counting rates $R(I)$ and $R(W)$ (in counts sec^{-1}) for consecutive passes in the northern and southern hemispheres. B_I and B_W are the corresponding scalar magnetic fields (in gauss). All weight is given to $(B_I - B_W)$.

$10L$	λ_1	λ_2	n	$100r$	$b/10$	p.e. $(b/10)$	A	p.e. (A)	$s(y)$	$s(\Delta y)$	$s(\rho)$
25			99	51	114	13	4	4	77	66	2.1
29	221	259	88	59	120	12	55	8	116	94	3.0
29	260	290	88	59	146	10	54	9	121	98	3.1
29	290	325	87	71	159	12	14	9	151	106	3.4
29	Pooled		263	64	143	7			120	95	3.0
29	Combined		263	63	121	8	38		120	95	3.0
35	236	258	127	74	127	7	30	5	108	69	2.2
35	258	278	127	74	142	8	27	6	113	76	2.4
35	278	296	126	78	148	7	50	6	131	82	2.6
35	296	329	126	70	148	9	7	7	120	86	2.7
35	Pooled		506	74	141	4					
35	Combined		506	74	139		30		119	80	2.5
41	247	263	123	76	118	6	23	5	100	65	2.1
41	263	278	123	78	138	7	28	6	120	75	2.4
41	278	295	123	79	135	6	41	6	118	72	2.3
41	295	317	123	81	134	6	22	5	106	62	2.0
41	Pooled		492	79	132	3					
41	Combined		492	79	132		29		114	70	2.2
44	254	270	120	69	89	5	24	5	86	62	2.0
44	270	290	121	77	127	7	24	6	112	72	2.3
44	290	313	121	76	105	6	32	5	95	62	2.0
44	Pooled		362	74	108	5					
44	Combined		362	74	103		30		101	68	2.2
47	295	278	123	73	108	6	30	5	96	66	2.1
47	278	307	122	65	91	7	17	6	90	68	2.2
47	Pooled		245	60	100	5					
47	Combined		245	68	97		26		96	71	2.3
49	263	278	82	64	84	8	32	5	81	62	2.0
49	279	303	83	67	98	9	20	6	97	72	2.3
49	Pooled		165	66	92	6					
49	Combined		165	66	90		27		90	68	2.2
50	265	281	63	60	83	9	29	6	89	64	2.1
50	281	302	63	63	83	9	16	7	100	78	2.5
50	Pooled		126	62	86	7					
50	Combined		126	62	85		22		94	74	2.3

Woomera, large changes in counting rate not caused by B often occurred, and hence these could not be used to determine for Iowa City or Woomera separately the dependence of counting rate on B. However, when data for successive passes over Iowa City and Woomera are considered, the interval of time, 50 min, is short, in general, for any large time variation of intensity to occur. For any selected L, a graph of $\log [R(I)/R(W)]$ for successive passes over Iowa City (I) and Woomera (W) as a function of $(B_W - B_I)$ indicated that these were correlated. The counting rates at Iowa City for any given L corresponding to contiguous northbound and southbound passes (less than 15 min apart) were used to derive the correlation between $\log [R(N)/R(S)]$ and $(B_s - B_N)$. This difference in B, however, was much smaller in magnitude than $(B_W - B_I)$, and the correlation coefficients were also smaller.

Statistical results. In Table 1 are shown for each of several values of L the correlation coefficients r derived from data samples grouped according to the longitude interval, $\lambda_1 - \lambda_2$, within which all passes through the zone occurred in the Iowa City data. To establish a convenient scale of R and to avoid decimals, $200 \log R$ was used, and the tabulated constants apply to the equation

$$200 \log [R(I)/R(W)] = A + b(B_W - B_I) \quad (1)$$

If for convenience we put $y = 200 \log (R(I)/R(W))$ and $x = (B_W - B_I)$, then b for the least-squares fit is given by

$$b = r(s(y)/s(x)) \quad (2)$$

where $s(y)$ and $s(x)$ are the standard deviations of y and x about their sample means. b is also called the regression coefficient for the regression of y on x. A is determined from $A = \bar{y} - b\bar{x}$. For $(b/10)$ the estimated probable errors, p.e., are tabulated. Although these p.e. values are valid for comparing values of $(b/10)$ among different samples for the same L, they are underestimates for testing differences in $(b/10)$ from pooled or combined samples at different values of L, because the counting rates for different L values often come from the same pass and thus are not statistically independent. The same remarks apply also to the p.e. values of A given Table 1.

For each L the available samples in the table were pooled; that is, the sample values of A were accepted and the resulting values of the other parameters listed in Table 1 were determined from the pooled deviations from the sample means. Results are also listed for all samples combined into one sample for which one value of A is derived.

TABLE 2. Statistical Parameters for Each of Several Values of L
Parameters are derived by fitting the equation

$$200 \log [R_N/R_S] = A + b(B_S - B_N)$$

to 302 counting rates R_N and R_S (in counts sec^{-1}) for consecutive northbound and southbound passes in the northern hemisphere. B_N and B_S are the corresponding scalar magnetic fields (in gauss). All weight is given to $(B_S - B_N)$.

$10L$	λ_1	λ_2	n	$100r$	$b/10$	p.e. $(b/10)$	A	p.e. (A)	$s(y)$	$s(\Delta y)$	$s(\rho)$
29			68	66	148	14	4	4			
35	232	247	142	73	170	9	−8	2	56	38	1.5
35	247	298	130	60	168	13	36	3	66	53	1.8
35	Pooled		272	66	169	12					
41	246	252	127	61	152	12	−13	2	36	29	1.4
41	252	257	127	55	116	11	−12	2	38	32	1.4
41	258	271	126	42	105	14	−5	2	46	42	1.6
41	272	308	126	48	142	16	+15	2	44	39	1.6
41	Pooled		506	48	124	10					
47			101	23	90	26	−14	3	40	39	1.6
47			100	19	82	28	−10	4	53	52	1.8
47			100	32	157	30	−1	3	46	44	1.7
47	Pooled		301	23	102	17					

In general, the agreement in $(b/10)$ among the samples for the same L is reasonably good, indicating no important dependence on longitude, and differences in $(b/10)$ for samples pooled and combined are small.

The pooled values of b from Tables 1 and 2 are plotted in Figure 1. It is observed that b is substantially constant for $2.5 \leq L \leq 4.1$ and decreases continuously for L larger than 4.1. Values derived from data for consecutive northbound and southbound transits of the same pass at Iowa City are also given in Table 2. For later study the observed values of $200 \log R(I)$ and $200 \log R(W)$ were corrected to $B = 0.340$ gauss, using for each L the values of b from the lines marked 'adopted' in Figure 1.

The values of A in Table 1 are, for combined results, quite similar at all L values, and within samples the divergence of A values is not large. The average value for A for all combined samples is 26, indicating that, for the same L and B, $R_B(I)/R_B(W) = 1.35$. Thus the average counting rate corrected for B is about 35 per cent larger for the passes over the northern hemisphere between east longitudes of about 221° to 330° than for those over Woomera, Australia. This may be evidence for build-up of flux with increase of longitude east of the South African 'sink.'

Table 1 includes the sample standard deviations $s(y)$ of y with $y = 200 \log [R(I)/R(W)]$ and the standard deviation $s(\Delta y)$ of the residuals

after correction for $(B_W - B_I)$ is made. If we put $x = (B_W - B_I)$, and denote by y_c the values computed (or predicted from x), then

$$y_c = A + bx \qquad (3)$$

and

$$y_o = y_c + \Delta y \qquad (4)$$

in which y_o is the observed value of y. Thus it is convenient to regard y_o as the sum of two terms, the first of which exactly follows x (i.e., tracking component) and the second, Δy, is the residual (or nontracking component) not correlated with x.

Remarks on the meaning of correlation coefficients. If we put $s(y_c)$ for the standard deviation of y_c, then from (2) and (3)

$$s^2(y_c) = r^2 s^2(y) \qquad (5)$$

Also

$$s^2(\Delta y) = (1 - r^2) s^2(y) \qquad (6)$$

Then, of course, the sum of (5) and (6) gives the total variance, $s^2(y)$, of the observed values of y about their mean, and

$$\frac{s(\Delta y)}{s(y_c)} = \sqrt{\frac{1 - r^2}{r}} \qquad (7)$$

Putting $\sqrt{1 - r^2}/r = \delta$, it is seen that δ determines the ratio of the standard deviation of residuals to that of the correlated (or track-

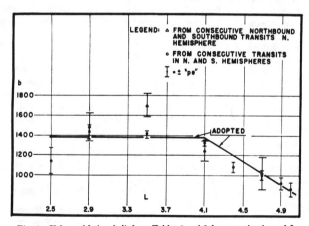

Fig. 1. Values of b (pooled) from Tables 1 and 2 for several values of L.

ing) component. A few values of δ, (r) are 0.1, (0.995); 0.2, (0.980); 0.5, (0.894); 1.0, (0.707); 1.5, (0.555); and 2.0, (0.447). In Table 1, for example, $r = 0.74$ for samples combined at $L = 3.5$, and so $\delta = 0.91$; thus the ratio of the standard deviation of residuals to that of the component of y_c which follows $(B_W - B_I)$ is about 0.9, indicating the importance of correcting for B.

Justification for use of equation 1 and independence of b on intensity level. The data of Table 1 do not indicate whether $200 \log [R(I)/R(W)]$ varies linearly with $(B_W - B_I)$ or whether the slope depends on the magnitude of the counting rates. To examine these questions, data for 504 consecutive passes at $L = 3.5$ (i.e., the same data used for $L = 3.5$ in Table 1 except for two pairs) were divided into three groups according to the magnitude of $[\log R(I) + \log R(W)]/2$, which, uncorrected for B, is a rough but sufficient measure of the intensity level averaged for Iowa City and Woomera, since $(B_I + B_W)/2$ is roughly constant for all consecutive passes. The means of $200 [(\log R(I) + \log R(W))/2]$ for the three groups were 312, 406, and 536, which correspond to mean values for $[R^*(I) + R^*(W)]/2$ of 36, 107, and 479 counts per sec, respectively, for the groups labeled low, medium, and high in Figure 2.

For each of these three groups of intensity levels the observed values of $200 \log [R(I)/R(W)]$ and $(B_W - B_I)$ were arranged in order of increasing $(B_W - B_I)$, and for each quartile of this array averages of $200 \log [R(I)/R(W)]$ and $(B_W - B_I)$ were derived and were plotted in Figure 2(A). Each point is thus the average for 42 pairs. The line (A) in Figure 2 is that for which the slope b and intercept A are given in Table 1 ($L = 3.5$ for the 506 pairs combined). This line is seen to fit about equally well the points for the three counting rate levels, and there is no important systematic departure from linearity. The points in Figure 2(B) were obtained by a similar procedure, with the important exception that the selection of the four subgroups was not according to magnitude of $(B_W - B_I)$ but according to the magnitude of $200 \log [R(I)/R(W)]$. The slope b_1 of the line in Figure 2(B) is about 2540 (from $b_1 = b/r^2 = 1390/0.74^2$; b and r from Table 1 at $L = 3.5$) and is the regression line with all weight on $200 \log [R(I)/R(W)]$, which is, of course, not

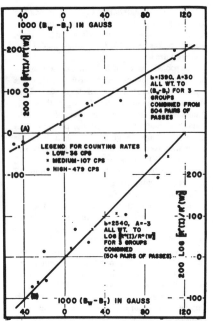

Fig. 2. Fit of regression line, $200 \log [R^*(I)/R^*(W)] = A + b(B_W - B_I)$, to group averages for three levels of counting rate at $L = 3.5$ from same data as for $L = 3.5$ in Table 1.

the slope to be used for correcting counting rates for B, if B is not subject to error. If from values of r and b in Tables 1 and 2 the slopes b_1 of the regression lines for all weight given to $200 \log [R(I)/R(W)]$ are calculated (i.e., $b_1 = b/r^2$), it will be found that the resulting values of b_1 vary by a factor of 10 or more if the values at $L = 4.7$ in Table 2 are used. Compared with this variation of b_1 the relatively small dispersion in the values of b in Tables 1 and 2 or in Figure 1 indicates that b is a good approximation to the true slope.

It should be noted that, in Tables 1 and 2, $s(\rho)$ in the last column is a rough measure for the standard deviation of $R(I)/R(W)$ obtained from the relation $\log s(\rho) = \log s(\Delta y)/200$.

COMPARISON OF COUNTING RATES IN NORTHERN AND SOUTHERN HEMISPHERES

In Figure 3, (A) shows the correlation $r = 0.54$ between counting rates $R^*(I)$ and $R^*(W)$ before correction for B, and (B) shows the correlation

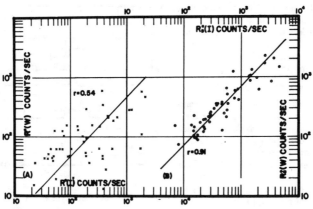

Fig. 3. Correlation between $R^*(I)$ and $R^*(W)$ averages at $L = 3.5$ for 10 sequential passes consecutive in the northern and southern hemispheres, (A) before and (B) after correction to $B = 0.340$.

$r = 0.91$ between counting rates $R^*(I)$ and $R^*_B(W)$ after correction to $B = 0.34$ gauss. The data for $L = 3.5$ have been used here as an example. The counting rate corrected to constant $B = 0.34$ is obtained from

$$\log_{B=0.34} R(I) = \log_B R(I)$$
$$- 1375/200 \ (0.34 - B)$$

where B is the scalar magnetic field at which the counting rate was observed. It may be noted that A in equation 1 was taken to be zero. Each point in Figure 3 is derived from the logarithmic average counting rate for ten pairs of passes, for each of which the passes over the two hemispheres occurred on successive transits (about 50 min apart). Thus the data for Figure 3 are from 500 of the 506 pairs from which the coefficient b shown in Table 1 for $L = 3.5$ was derived. Even if the data for the 500 single passes are used, the correlation coefficient r between log $R(I)$ and log $R(W)$ improves from 0.47 to 0.73 after correction to $B = 0.34$. Thus the improved agreement between the northern and southern hemispheres resulting from the correction for B can be seen. This is also indicated in Table 1 by comparing $s(\Delta y)$ with $s(y)$. $s(\Delta y)$ is approximately the same for all values of L in Table 1, which means that the agreement between $R_B(I)$ and $R_B(W)$ for passes consecutive between the northern (over Iowa City) and southern (over

Woomera) hemispheres is about the same for other values of L as for $L = 3.5$.

REPLENISHMENT OF TRAPPED PARTICLES LOST OVER SOUTH AFRICA

For $L \leq 3.5$ practically all values of B for the southern hemisphere passes (recorded over Woomera, Australia) were greater than 0.38 gauss. All trapped particles mirrored over Australia ($L \leq 3.5$; $B \geq 0.38$ gauss) have their mirror points on or below ground level in the neighborhood of 20°E geographic longitude and 58°S geographic latitude, south of Africa. Thus at $L \leq 3.5$ practically all the particles counted and responsible for $R(W)$ over Woomera would be completely absorbed when they drift eastward (if electrons) over the region south of Africa. However, only about half of the passes through the zone over the northern hemisphere (over Iowa City) occur at $B \geq 0.38$; consequently all trapped particles responsible for the counting rate for these values of B over Iowa City would also be completely absorbed when they drift south of Africa. This is true to the lower limit, $L = 2.5$, with which our study is concerned. Since, in addition, the agreement between $R_B(I)$ and $R_B(W)$, the intensities corrected for B at Iowa City and Woomera, is quite good ($r = 0.73$) for consecutive passes, it appears that particles absorbed over South Africa are replenished in a time shorter than the longitudinal drift period. This replenishment may arise from a lowering of the mirror points (i.e., to larger values of B). Some further evidence is given later in the comparison of intensities at $L = 2.5$ over South Africa and over the northern hemisphere (Iowa City).

GENERAL SURVEY OF THE OUTER ZONE

The top two curves of Figure 4 give an average picture of the intensity profile of the outer zone, as seen from the excursions of Explorer 7 through the horn of the outer zone, over the North American continent. These plots of $R^*_B(I)$ as a function of L, over the two periods November 1959 to December 1960 and November 1959 to August 1960, are quite similar and exhibit an I_{max}, or maximum of intensity, around $L = 3.5$. The general increase in intensity over all the L values due to the additional data over the period September to December 1960 can be seen.

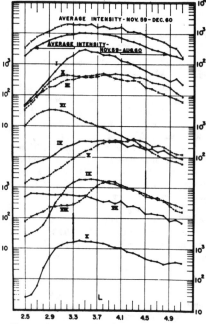

Fig. 4. Average profile of intensity $R^*_{B = 0.84}(I)$ during November 1959 to December 1960. Curves I to X are a comparison of 30-day average profiles of intensity $R^*_{B = 0.84}(I)$ during November 1959 to August 1960.

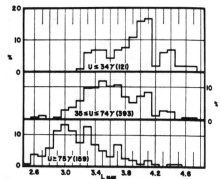

Fig. 5. Frequency histograms of L_{max} for three levels of U; number of passes is in parentheses.

The curves I to X in the figure exhibit the radical changes occurring in successive 30-day average intensity profiles during November 1, 1959, to August 26, 1960. Since the data after August 31, 1960, are meager, it was not possible to obtain such average profiles thereafter. In addition to the changes in the counting rates over the whole L range from month to month, it is observed that I_{max} occurs at various values of L in the range 2.9–4.3. Moreover, in curve VII, the intensity increases continuously over the L range 5.0–3.5 and is constant thereafter down to $L = 2.6$. Particularly striking is the change from the monthly profile in March 1960 (curve V) to one in April 1960 (curve VI) with an exceptional increase in the counting rates at the lower L values. The shift in the position of the intensity maximum from $L = 3.9$ in June 1960 to $L = 3.5$ in July 1960 (curves VIII and

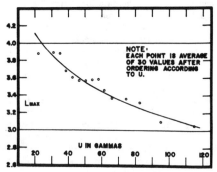

Fig. 6. Curve for regression of L_{max} on U.

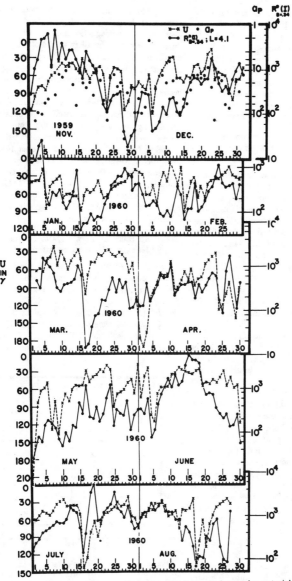

Fig. 7. Average counting rate $R^*_{B\,=\,0.34}(I)$ at $L = 4.1$ for each day, and equatorial ring current measure U, November 1, 1959, to August 27, 1960.

IX) with only slight change in the value of I_{max} is also of interest. The average daily sum of the a_p measure of geomagnetic activity for periods corresponding to curves V and VI is 230 and 626 γ, respectively ($\bar{U} = 43$ and 94 γ); and 298 and 346 γ ($\bar{U} = 47$ and 53 γ) for periods corresponding to curves VIII and IX, respectively. This suggestion of an association between geomagnetic activity and a shift in the L_{max} is considered in greater detail in the next section.

SHIFT OF THE MAXIMUM INTENSITY TOWARD SMALLER VALUES OF L DURING GEOMAGNETIC DISTURBANCES

Of all the data recorded at Iowa City and Blossom Point, only 674 passes extend over a range of L sufficient to locate unequivocally L_{max}, the L at which the maximum counting rate is observed. As was pointed out earlier, these data have been corrected to $B = 0.34$ gauss. The frequency distribution of L_{max} is shown by the histograms in Figure 5 for each of three groups chosen according to range of U (high, medium, and low), where U is the measure for the equatorial geomagnetic field of the ring current. The ordinates in Figure 5 indicate, for each range of U, the percentage of all L_{max} values for that group.

It is seen from Figure 5 that L_{max} tends to occur at lower values of L for groups with larger values of U (geomagnetically disturbed). Of the 674 passes considered, 224 had broad maxima; i.e., the counting rates were greater than 75 per cent of the maximum for at least 5 values of L (using intervals of 0.1 for L). For the 450 passes with sharp maxima the correlation coefficient r between L_{max} and log U is -0.7. Whereas the frequency distribution of L_{max} is approximately normal, that of U is not. But that of log U is approximately normal, and so, in accord with statistical practice, we use log U for correlation with I_{max}. Figure 6 shows the curve for the regression of L_{max} on U. This curve corresponds to that obtained (not shown) for the linear regression of I_{max} on log U, which is the same as the line fitted by least squares to L_{max} as a function of log U when all weight is given to log U. To obtain the points in Figure 6, 450 pairs of values of L_{max} and log U were ordered according to log U, were divided into 15 groups of 30, and were averaged.

The mechanism responsible for the shift of L_{max} to lower values during magnetic storms (large U) may be linked with that for the decrease in geomagnetic latitude of auroras during storms; the smallest values of L_{max} occur at about 2.6 (Figure 5), which corresponds to geomagnetic latitude about 51°, or roughly the latitude south of which auroras do not frequently occur.

We could have used the a_p measure of geomagnetic activity as the criterion for obtaining the three groups and a similar result would have been expected because of the correlation between U and a_p ($+0.82$ for the period October 1959 to July 1960). Nevertheless, U was used primarily to determine whether changes in the trapped radiation (as measured by the 302 Geiger counter) might be responsible for the ring current for the equatorial geomagnetic field of which U provides a measure.

COUNTING RATE AT $L = 4.1$ AND U FOR EACH 'DAY,' NOVEMBER 1, 1959, TO AUGUST 27, 1960

The data recorded at Iowa City were from passes occurring within successive 10-hour intervals interspersed by gaps of about 14 hours during which no data were received. The intensities $R^{*}{}_{B}(I)$ corresponding to L values 2.5, 2.9, 3.5, 4.1, and 4.7 were averaged for these 10-hour intervals for each 'day.' The average U and a_p for each day were derived from the three hourly

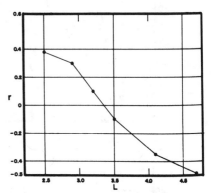

Fig. 8. Coefficient of correlation r at different L values between departures (numbering between 280 and 380) of daily means $R^{*}{}_{B}(I)$ and of log U (for corresponding passes) from averages for periods with data available on 20 days.

values for the intervals during which the passes occurred.

In Figure 7 the average intensity for each day at $L = 4.1$, corrected to $B = 0.34$ gauss, and average U for each day (plotted inverted) are shown for the period November 1, 1959, to August 27, 1960, during which data were more or less continuous. As an example, the average a_p for each day is plotted only for the period November 1 to December 31, 1959. It should be pointed out that each point may occasionally be a single value, or the average of as many as eight values (corresponding to the northbound and southbound transits of four passes), depending on the number of passes available for

the day. Usually, each point is the average of four or five values.

The correlation between U and intensity at $L = 4.1$ is obviously not high for the whole period, although there are intervals when the two curves follow each other closely. There is, however, a tendency for increases in U to be associated with decreases in intensity. A decrease in intensity of eastward longitudinally drifting electrons would, if this were the source of the ring current, increase the northward geomagnetic equatorial field of the ring current, and thus decrease U (U is positive southward). This is contrary, at $L = 4.1$, to the observed tendency. This result thus confirms the findings of

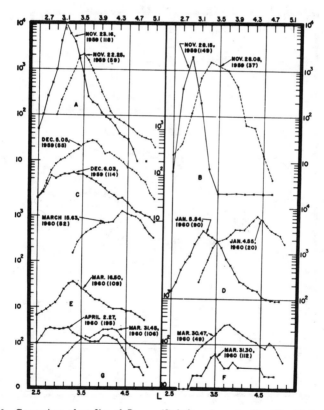

Fig. 9. Comparison of profiles of $R_{B = 0.34}(I)$ before storm (smaller U) with first profile available during storm. Ordinates: $R_{B = 0.34}(I)$ counts sec⁻¹. UT for pass given to nearest 0.01 day; U values in parentheses.

an earlier study over a limited period [*Forbush, Venkatesan, and McIlwain*, 1961] even after correcting for B. As will be indicated in the next section, the tendency at $L < 3.4$ is for increases in intensity to be associated with increases in U; nevertheless, there is no indication that any major contribution to the ring current arises from the longitudinal drifts of the electrons measured by the 302 GM counter.

CORRELATION BETWEEN DAILY AVERAGES OF INTENSITY AND U, AND MAGNETIC STORM EFFECTS ON INTENSITY PROFILES

Figure 8 indicates the coefficients of correla-

tion r for several values of L, between $R^*_B(I)$ and log U, for which the number of pairs of values available ranged between 280 and 380 for the various L values. The values of r apply to departures (280 to 380) of daily means of R^*_B and of log U from averages of 20 sequential daily means (since for some days no data were available, the actual interval was sometimes more than 20 days). Figure 8 indicates that r varies systematically from about -0.5 at $L = 4.7$ to $+0.4$ at $L = 2.5$, and that $r = 0$ near $L = 3.4$, which is about the median value of L_{max}. From Figure 8 it does not necessarily follow that $R^*_B(I)$ at $L = 2.5$, for example, is

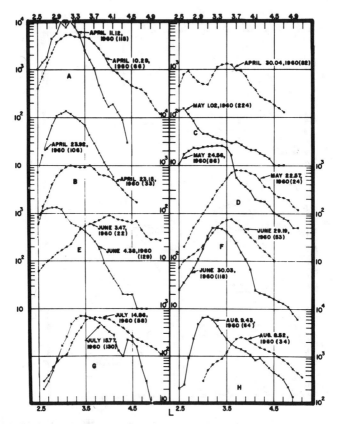

Fig. 10. Comparison of profiles of $R_{B = 0.34}(I)$ before storm (smaller U) with first profile available during storm. Ordinates: $R_{B = 0.34}(I)$ counts sec^{-1}. UT for pass given to nearest 0.01 day; U values in parentheses.

negatively correlated with $R^*_B(I)$ at $L = 4.7$; that is, a decrease in $R^*_B(I)$ at, say, $L = 4.7$ (following an effective increase in U) may or may not be accompanied, on the same pass, by an increase in $R^*_B(I)$ at $L = 2.5$.

For $L \geq 3.4$ the negative correlation between $R^*_B(I)$ and log U, shown in Figure 8, is in accord with indications based on intensity values uncorrected for B [Forbush, Venkatesan, and McIlwain, 1961].

Comparisons of intensity profiles, that is, $R^*_B(I)$ as a function of L, during the main phase of the storms (larger U) with those before the storm (smaller U) are shown for some 15 individual storms in Figures 9 and 10. During the main phase, the intensity compared with the prestorm value is less at higher values of L (in all but 1 of these 15 storms), and greater at the lower values of L in 10. Only in 7 does the I_{max} (the maximum intensity in the pass) appear

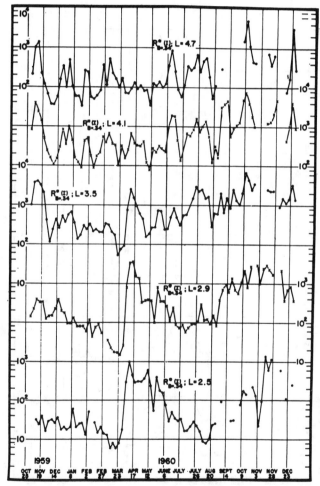

Fig. 11. Five-day averages of intensity $R^*_{B = 0.34}(I)$ at several values of L.

315

larger compared with the I_{max} before the storm. This together with the decrease of L_{max} with increasing U (Figures 5 and 6) indicates that the magnetic storm field may effect a redistribution of particles among the L shells and account for the decrease of L_{max} with magnetic disturbance. The reduction in observed intensity at higher L values of particles could also arise from dumping or from a change in the energy of the particles.

The very large increase observed at lower L values could result from additional particles not observed before.

LONG-TERM TIME VARIATIONS OF INTENSITY AT VARIOUS L VALUES DURING OCTOBER 1959 TO DECEMBER 1960

The 5-day averages of intensity plotted in Figures 11 and 12 give a composite picture of

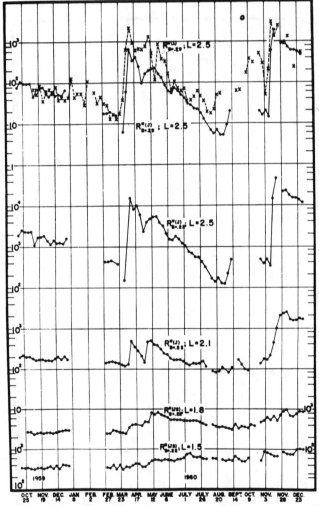

Fig. 12. Five-day averages of intensity $R^*_{B=0.29}(I)$, $R^*_{B=0.29}(J)$, $R^*_{B=0.22}(J)$, and $R^*_{B=0.22}(JS)$.

the long-term variations in intensity from $L = 4.7$ to $L = 1.5$ during the whole period of study. In Figure 12, $R^*_B(J)$ refers only to data from Johannesburg, South Africa, whereas $R^*_B(JS)$ refers to Johannesburg and also to Quito, Lima, Santiago, and Antofagasta in South America, the last four contributing about 40 per cent of data. Data from these stations are from smaller B values than those from Iowa City and Woomera and so were corrected by *Pizzella, McIlwain, and Van Allen* [1962] to $B = 0.22$ gauss by means of their empirical dependence of $R(J)$ or $R(JS)$ on B. For $L = 2.5$ this dependence was used to extrapolate $R_B(J)$ from $B = 0.22$ to $B = 0.29$; similarly $R_B(I)$ was extrapolated from $B = 0.34$ to $B = 0.29$ using the empirical relation obtained from the data at Iowa City and Woomera. The two curves uppermost in Figure 12 show generally comparable variations of intensities for 5-day averages, thus corrected to $B = 0.29$ gauss, at $L = 2.5$. For this L all passes over Australia, and about half the passes over North America, were at $B \geq 0.38$ gauss; consequently, all particles recorded there for such passes are absorbed in the region south of Africa where at sea level $B = 0.38$ gauss. The general similarity of these two curves is further evidence that such lost particles are replenished within intervals less than the longitudinal drift period, and that a continuous lowering of mirror points is probably involved.

Figure 11 indicates that the time variations of intensity at $L = 4.7$ and 4.1 are quite similar, as are those at $L = 2.9$ and 2.5. The curve for $L = 3.5$ shows a combination of the features of both groups. This difference between the higher and lower L groups is also indicated by the tendency, shown in Figure 8, for the intensity at $L > 3.4$ to decrease with increasing U and for the intensity at $L < 3.4$ to increase with increasing U, whereas at $L = 3.4$ the correlation with U is negligible. Also, in Figures 9 and 10, near $L = 3.5$, the intensity change from the prestorm value is not predominantly of either sign.

The changes of intensity by about a factor of 10 which occur, on an average, every 15 to 20 days at the higher L values are generally absent at $L \leq 2.9$. Minimum intensities tend to be about 100 counts sec^{-1} at $L = 4.1$ and about 60 counts sec^{-1} at $L = 4.7$ until late in August

1960, after which they are a different higher level. The more frequent fluctuations of intensity on the outer L shells might reasonably be expected as a result of impinging solar plasma. In contrast, the lower L values reveal monotonic decreases lasting months and sudden increases. At $L = 2.9$ and 2.5, an outstanding feature is the 100-fold increase of intensity in early April (when L_{Max} decreased from 4.2 to 3.0) and its slow decline over the following three months. The increase in April was at the end of an earlier monotonic decrease lasting for months. A second large increase in early November occurred at $L = 2.5$ and 2.1 and is not evident at $L = 2.9$, for which a 10-fold increase occurred in early September. Some of these events are discernible at $L = 1.8$, where the extremes of intensity for the whole period differ by a factor of 4, and at $L = 1.5$, by a factor of 3. The large increases at $L \leq 2.9$ may be due to trapping of fresh particles from the magnetized solar plasma and subsequent 'local' acceleration of trapped electrons within the geomagnetic field or to downward perturbation of mirror points of particles along lines of force in the outer zone, such particles not having been observed before at the altitude of measurement.

LONG-TERM VARIATIONS IN I_{max}, L_{max}, AND U, FOR 674 COMPLETE PASSES, AND $R^*_B(I)$

The long-term changes in the peak counting rate in the zone (I_{max}), the L shell at which this peak occurs (L_{max}), geomagnetic equatorial ring current field, U, and $R^*_B(I)$, the intensity at $L = 2.9$ corrected for B, are shown by the 5-day averages plotted in Figure 13. The numbers on the L_{max} curve correspond to the number of values that have been averaged for that point and for corresponding points on all curves except $R^*_B(I)$, for which all available data were used. All the 674 passes where I_{max} could be unequivocally determined have been used. To facilitate comparison, L_{max} is repeated at the bottom along with $R^*_B(I)$.

The bottom two curves indicate a strong tendency for the intensity at $L = 2.9$ to increase when L_{max} decreases, and vice versa.

The correlation between U and a_p is high enough to make it immaterial which is used for correlation with the other variables. The U and L_{max} curves indicate a tendency for an increase in U to be associated with a reduction of L_{max},

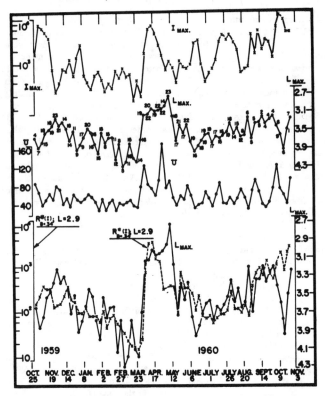

Fig. 13. Five-day averages of I_{max}, L_{max}, U, and $R^*_{B = 0.M}(I)$ from November 1, 1959, to October 29, 1960.

and vice versa (see also Figure 5). Between I_{max} and L_{max} the correlation is evidently negative.

OCCURRENCE OF TWO PEAKS ON INTENSITY PROFILES

The intensity profiles, $R(I)$ as a function of L, were examined for a subsidiary maximum, or second peak. Two peaks were observed in about 12 per cent of the profiles. One peak appeared between $L = 2.7$ and 4.4, about 2/3 of the cases between $L = 2.9$ and 3.5. The second peak occurred between $L = 3.6$ and 4.9, about 2/3 of the cases between $L = 3.9$ and 4.2. In about half these profiles the second-peak intensity was at least 75 per cent of the first-peak intensity. Three distinct peaks were rarely seen. Sometimes the double peaks persisted for 2 or 3 days.

SUMMARY AND CONCLUSIONS

1. For the period October 1959 to December 1960, the temporal variations and morphology of the outer Van Allen radiation belt were investigated for electrons of energy greater than about 1.1 Mev as measured by the SUI 302 GM counter in the altitude range 500 to 1100 km on Explorer 7.

2. Measured intensities were referred to McIlwain's B and L coordinates. For successive passes through the belt, at the same value of L, in the northern and southern hemispheres with respective counting rates $R(I)$ and $R(W)$, log $R(I)/R(W)$ was linearly correlated with $(B_w - B_I)$. On the basis of this dependence, the observed intensities $R_B(I)$, for example, were corrected to a constant B. The validity of and necessity for

the correction were demonstrated by the decided improvement in the correlation between $R(I)$ and $R(W)$, for consecutive passes, after such correction.

3. For passes through the zone at $L = 2.5$ over North America, the smallest values of B exceeded the sea-level values of B over South Africa. Nevertheless, the intensities in these two regions were comparable after correction to a value of B between the above extremes, despite the fact that the dependence of intensity on B in the two regions was quite different. The long-term variations in these intensities at $L = 2.5$ were also similar.

4. Over Australia, $2.5 \leq L \leq 3.5$, practically all particles were registered at $B \geq 0.38$ gauss, and hence these are completely absorbed when they drift eastward (if electrons) to the region near 20°E and 58°S geographic, where, near sea level, $B = 0.38$ gauss. Similarly, the particles registered on about half the passes for $2.5 \leq L \leq 3.5$ over North America are lost in the same region. The continuous maintenance of intensity over Australia for $2.5 \leq L \leq 3.5$ thus requires an efficient mechanism for replenishing the particles, lost south of Africa, in an interval less than the longitudinal drift period ($\sim$37 minutes for 1.1 Mev electrons at $L = 3.5$). The average counting rate corrected for B is about 35 per cent larger for passes over Iowa City than for passes over Woomera, Australia, for the same L value 3.5. This may be evidence for build-up of flux with increase of longitude east of the South African sink.

5. For 673 individual passes over North America with data complete from $L = 2.5$ to 5.0, and corrected for B, the values L_{max} for which the intensity was maximum showed a marked tendency to decrease with increasing U (southward equatorial geomagnetic field of the ring current). On the average, L_{max} decreased from $L = 4.0$ to $L = 3.0$ for an increase of 100 γ in U. A comparison of profiles (intensity versus L) before and during 15 large magnetic storms showed a decrease in L_{max} during the main phase of the storms. For these storms a reduction in intensity almost always occurred for L greater than about 3.5. For values of L less than this, the intensity often tended to be greater during the storms. Thus, associated with magnetic storms there is apparently a redistribution of particles among the L shells.

6. Variations of 5-day averages of intensity are similar at $L = 4.7$ and $L = 4.1$ but are quite distinct from those at $L = 2.9$ and $L = 2.5$, which are also similar. For the larger values of L the intensity varies by a factor of 10 or so within intervals of 2 or 3 weeks; these variations are not evident at $L = 2.9$ and $L = 2.5$. At these values of L the most prominent features are monotonic decreases lasting months and sudden increases. A 100-fold increase of intensity is seen early in April, followed by a gradual decrease over the next 2 or 3 months; at $L = 2.9$ this in turn is followed in September by a second rise to high values, which prevailed at least through November. The first of these features becomes very much smaller at $L = 2.1$ and $L = 1.8$ and is barely perceptible at $L = 1.5$. A 10-fold increase in early November was evident only at $L = 2.5$ and $L = 2.1$. The large increases at lower L values may be due to trapping of fresh particles from solar plasma, and subsequent local acceleration within the geomagnetic field or downward perturbation of mirror points of particles along lines of force, such particles not having been observed before at the altitude of measurement.

7. The correlation coefficient between departures of daily averages of intensity from means for 20 days and corresponding departure of log U changes systematically from -0.5 at $L = 4.7$ to $+0.4$ at $L = 2.5$, and is zero near $L = 3.4$. There is thus no clear evidence that any major contribution to the ring current arises from longitudinal drift of electrons of energy greater than 1.1 Mev measured by the SUI 302 GM counter.

8. In only about 12 per cent of the passes for which data were complete were two peaks observed; these sometimes persisted for 2 or 3 days.

Acknowledgments. We wish to thank Professor J. A. Van Allen for his continued interest and valuable comments. Our special thanks are due Dr. Carl McIlwain for his useful discussions and all valuable help connected with the computer and reduction of data. We wish to express our appreciation to all the people in the data reduction section who have helped to make such a study possible. Grateful acknowledgment is made to the Directors of the magnetic observatories at Apia, Samoa; Gnangara, Australia; Huancayo, Peru; Ibadan, Nigeria; and Binza, République du Congo, for generously providing magnetic data for the U measure, and to Mrs. L. Beach of the Department of Terrestrial Magnetism for calculating U values.

The research was assisted by a joint program of the Office of Naval Research and the Atomic Energy Commission under contract N9onr-93803. Support was also given by the National Aeronautics and Space Administration.

REFERENCES

Forbush, S. E., D. Venkatesan, and C. E. McIlwain, Intensity variations in outer Van Allen radiation belt, *J. Geophys. Research, 66*, 2275–2287, 1961.

Jensen, D. C., and W. A. Whitaker, A spherical harmonic analysis of the geomagnetic field (abstract), *J. Geophys. Research, 65*, 2500, 1960.

Kertz, Walter, Ein neues Mass für die Feldstärke des erdmagnetischen äquatorialen Ringstroms, *Abhandl. Akad. Wiss. Göttingen Math.-Physik. Kl., Beit. intern. Geophysik. Jahr,* Heft 2, Vandenhoeck und Ruprecht, Göttingen, 1958.

Ludwig, G. H., and W. A. Whelpley, Corpuscular radiation experiment of satellite 1959 (Explorer 7), *J. Geophys. Research, 65*, 1119–1124, 1960.

McIlwain, C. E., Coordinates for mapping the distribution of magnetically trapped particles, *J. Geophys. Research, 66*, 3681–3691, 1961.

O'Brien, B. J., C. D. Laughlin, J. A. Van Allen, and L. A. Frank, Measurements of the intensity and spectrum of electrons at 1000-kilometer altitudes and high latitudes, *J. Geophys. Research, 67*, 209–225, 1962.

O'Brien, B. J., J. A. Van Allen, C. D. Laughlin, and L. A. Frank, Absolute electron intensities in the heart of the earth's outer radiation zone, *J. Geophys. Research, 67*, 397–403, 1962.

Pizzella, G., C. E. McIlwain, and J. A. Van Allen, Time variations of intensity in the earth's inner radiation zone, October 1959 through December 1960, *J. Geophys. Research, 67*, 1235–1253, 1962.

Solar-Geophysical Data, Natl. Bur. of Standards, Central Radio Propagation Laboratory, Boulder, Colo., Dec. 1959–Feb. 1961.

Van Allen, J. A., and W. C. Lin, Outer radiation belt and solar proton observations with Explorer VII during March–April 1960, *J. Geophys. Research, 65*, 2998–3003, 1960.

(Manuscript received June 21, 1962.)

(GA34) SCOTT E. FORBUSH (Carnegie Institution of Washington, Dept. of Terrestrial Magnetism, Washington, D. C.), *Geomagnetic Field of the Equatorial Ring Current and Its Variation during Three Solar Cycles for Annual Means of All Days, Disturbed Days, and Quiet Days, and for the Single Quiet Day with the Smallest Ring Current in Each of Several Years.* The reliable determinations of the geomagnetic field of the equatorial ring current and its solar cycle variations are required for investigations of conductivity at great depths within the Earth and for comparison with ring current intensities calculated from the motion of geomagnetically trapped particles. Careful examination of the secular variation of the horizontal magnetic field, H, at Tucson and Huancayo (Peru) during the period 1924–1960 has resulted in a determination of the variation of annual means of H, from the equatorial ring current on magnetically quiet days for at least one solar cycle at each observatory. From these results a close correlation was found between annual means values of H for quiet days, and for disturbed minus quiet days, which permitted reliable determination of annual means, during three solar cycles, of H on quiet days, disturbed days, and all days. Even for the single quiet day, in each of several years, on which the ring current was minimum, the solar cycle variation is evident. For annual means of quiet days and of all days the ring current field is respectively about 40 and 60 per cent of that for disturbed days.

Time-Variations of Cosmic Rays.

By

Scott E. Forbush.

With 86 Figures.

First Part:

Results up to the International Geophysical Year.

A. Methods of observation.

1. Preliminary survey. Many investigators have undertaken systematic studies of time variations of cosmic-rays with the hope of finding some clue to their origin. The only definite evidence of this sort thus far obtained is that which indicates that the sun, on rare occasions emits charged particles in the cosmic-ray energy spectrum (below 10 BeV = 10^{10} electron volts if protons).

Nevertheless, researches on time variations have established the existence of several world-wide variations in cosmic-ray intensity (in addition to variations, more local in character, of meteorological origin). These include: variations during some magnetic storms, 27-day quasiperiodic variations, variations with sunspot cycle, a diurnal variation, and a variation in the amplitude of the diurnal variation with time.

In addition, observations in balloons with ionization chambers and in jet planes with neutron detectors have established variations in the primary spectrum during part of the solar cycle; and rocket and satellite-experiments have provided additional evidence (see the Second Part). For all of the established variations evidence indicates the sun is directly or indirectly responsible.

One of the early attempts to systematically search for time variations was made by A. Corlin [1] in Northern Sweden, at various intervals between October 1929 and July 1933. In spite of careful observations and analyses the results failed to indicate significant variations. In the light of present knowledge the reasons for this failure now appear twofold; first the larger world-wide variations which are associated with some magnetic storms seldom occur during years near the minimum of sunspot activity (such as those years studied by Corlin); and second the smaller variations, especially in the polar regions, are likely to be completely obscured, in ionization-chambers, by the large variability of atmospheric effects (connected with weather-like changes in the atmospheric structure) which arise because of variations in the height at which μ-mesons are produced. The greater this height the greater is the number of μ-mesons which decay (into rapidly absorbed electrons and neutrinos) before they reach the ground, and consequently the smaller is the measured intensity at the ground. These (local) effects for example, at Godhavn, Greenland amount to several percent but at Huancayo, Peru they are much smaller, while the world wide changes at Huancayo are about the same in magnitude as at Godhavn.

Thus with ionization chambers, and with Geiger counters, the world-wide variations are best observed at stations at low latitudes. On the other hand the increases observed during some solar flares were never observed near the equator until February 1956; and the magnitude of these increases varies markedly among stations at different longitudes. Thus, until recently, fortune has played an important part in the location of stations for continuous observations.

2. Instruments: ionization chambers (COMPTON-BENNETT). α) Stations. Due to their simplicity, reliability, and facility of maintenance, ionization-chambers have proved especially valuable for continuous operation over long periods of time. The longest existing series of continuous observations with the same type of instrument, has been obtained with Carnegie Institution of Washington (CIW) model C Compton-Bennett ionization chambers [2]. Table 1 indicates the location and elevation of stations where these have been in operation since the dates indicated therein.

Table 1. *Location of CIW Compton-Bennett cosmic-ray meters, 12 cm Pb shield.*

Station	Latitude deg.	Longitude deg.	Geomagnetic latitude deg.	Elevation m	Barometric coefficient c/o (mm Hg)$^{-1}$	Operation began
Godhavn, Greenland . .	69.2 N	53.5 W	79.9 N	9	− 0.18	October 1938
Cheltenham, Maryland .	38.7 N	76.8 W	50.1 N	72	− 0.18	March 1935
Climax, Colorado . . .	39.4 N	106.2 W	48.1 N	3350		September 1953[1]
Teoloyucan, Mexico . .	19.2 N	99.2 W	29.7 N	2285	− 0.345	February 1937[2]
Ciudad Universitaria, Mexico						September 1954
Huancayo, Peru	12.0 S	75.3 W	0.6 S	3350	− 0.30	June 1936
Christchurch, New Zealand	43.5 S	172.6 E	48.6 S	8	− 0.18	June 1936

[1] Previously operated intermittently at 3500 m.
[2] Not operated after 1945.

β) Description. The Compton-Bennett meter consists essentially of a spherical steel ionization chamber, volume 19.3 liters, filled with highly purified argon to 50 atmospheres pressure. Pure argon at high pressure gives larger ionization currents than any other gas; at 50 atmospheres the ionization is about 67 times that in air at normal pressure and temperature. The standard deviation of statistical errors arising from random arrival-times of individual rays is inversely proportional to the area of cross-section of the chamber. At Washington, from analysis of simultaneous hourly records from two Model-C meters, the standard deviation of statistical fluctuations in one hour's record from either of the instruments was found to be 0.7% of the total ionization due to cosmic-rays. Ions produced mainly by cosmic-rays in the argon of the main chamber are driven to the collector at the center of the sphere by a potential applied to the inner wall of the sphere. This is the main ionization current. A balancing current is supplied by ionization produced in a small auxiliary chamber, inside the sphere, by β-rays from metallic uranium (see Fig. 1). Turning the micrometer rod to which the uranium is attached permits varying the rate of entry of β-rays into the balance chamber as a consequence of changes in the amount of shielding. The balance current can thus be made about equal, and opposite in sign through choice of sweep field, to the average ionization current produced in the main chamber by cosmic rays. This permits ample sensitivity for recording changes in ionization due to cosmic rays on a photographic trace 60 mm wide. This method of balancing also compensates for changes in ionization which might result from changes in pressure or temperature of the gas. This compensation arises from the fact that the ionization current due to β-rays in the balance chamber is affected by changes in pressure and temperature in the same way as that due to cosmic rays in the main chamber.

γ) *Performance.* Direct tests involving two meters showed that the meters are not affected by temperature. Tests also show that at the higher pressures,

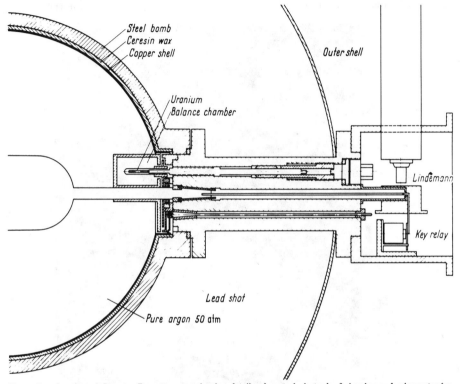

Fig. 1. Cross section of Compton-Bennett meter showing details of central electrode, β chamber and micrometer for adjusting position of uranium.

the balance is almost completely unaffected by changes in the pressure to which the chamber is filled. For this to be so, there must be negligible ionization in the main chamber due to α- and β-radiation from the walls, since this ionization will not increase with pressure as does that due to cosmic rays. Also, the dimension of the balance chamber must be so small that there is negligible absorption in the argon of the β-rays from the uranium. Shielding of the uranium with a thimble of aluminium 0.2 mm thick causes absorption of the α-particles and soft β-rays before they reach the balance chamber. There remain β-rays with a range of about 6 cm in argon at 50 atmospheres and some γ-rays. The β-rays traverse the balance chamber but are stopped within the 3 mm brass walls of the balance chamber. The γ-rays, which are not completely absorbed by the walls of the balance chamber, produce a small residual ionization in the main chamber which is proportional to the pressure and appears as a part of the zero correction.

It is known that the ionization in a closed chamber, filled with air at high pressure, increases with temperature because of less initial recombination of ions at the higher temperatures. In pure argon the initial recombination is small, and whatever small effect there may be is compensated by the balance arrangement. Changes in ionization current due to changes in the applied potential are also compensated by the balance chamber, since both chambers show the same degree of saturation at the voltages (250 V) used.

Batteries supply the sweeping potentials for the main chamber and the balance chamber. The batteries are connected across a 2.6 MΩ potentiometer, the variable center of which is grounded. In this way induced charges due to battery fluctu-

Fig. 2. Compton-Bennett meter at Cheltenham Magnetic Observatory.

ations are eliminated, provided the insulation of the batteries, potentiometer, and ionization-chamber walls is high compared to 2.6 MΩ. Elimination of induced charges on the collecting electrode from changes in the battery voltage also requires that the capacity coefficients C_1 and C_2 of the two chambers and the resistances r_1 and r_2 of the two arms of the potentiometer satisfy the relation: $r_1 C_1 = r_2 C_2$. By adjustment of the position of the balance chamber, C_1 and C_2 can be made very nearly equal. Balance is so effected that suddenly putting 500 V across the potentiometer induces a negligible deflection of the electrometer.

δ) *Recording.* The difference between the ionization currents in the two chambers is measured with a standard Lindemann electrometer housed in an airtight, dried chamber. At hourly intervals, the central electrode is grounded for about three minutes. The collecting electrode thus remains insulated or floating for about 57 min of each hour. Once every four hours, for about one minute, a relay connects the central electrode to a known potential. Thus the electrometer sensitivity is automatically recorded. Electrometer sensitivities of the order of one quarter of the maximum sensitivity attainable insure stability.

The shadow of the electrometer needle is projected through a compound microscope onto a strip of bromide paper 60 mm wide, moving 2 cm per hour for normal operation. A roll of paper 18 m in length is adequate for a month of continuous record. Visual observations can be made at any time from an image of the electrometer needle which is projected onto a ground glass scale.

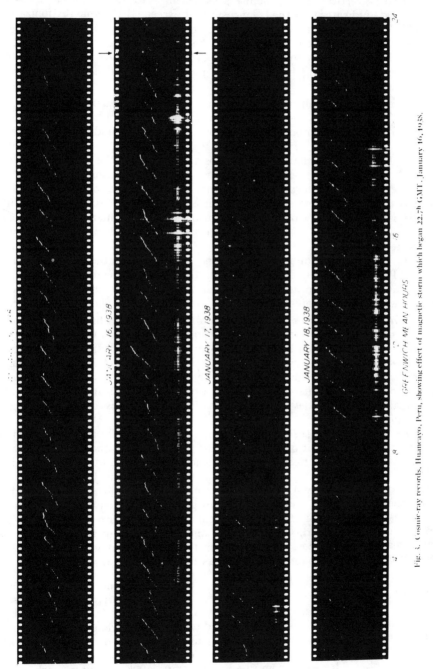

Fig. 3. Cosmic-ray records, Huancayo, Peru, showing effect of magnetic storm which began 22.7ʰ GMT, January 16, 1938.

In addition to the cosmic-ray ionization, the barometric pressure from an aneroid barometer is usually recorded. Notes may be written on the back of the bromide paper in dim light without affecting the records. Fig. 2 shows the meter at Cheltenham and Fig. 3 shows cosmic-ray records from Huancayo for four days. The dark vertical lines are hour-marks. The upward slope of the electrometer trace in Fig. 3 indicates that the balance current exceeds the cosmic ray ionization current. The meter is shielded by lead shot of uniform size which is equivalent to a layer of about 10.7 cm of solid lead. The total shielding, including the steel walls of the chamber, is equivalent to about 12 cm of lead, which practically effects complete shielding from any local radiations unless radioactive sources are brought near.

ε) *Evaluation.* The absolute ionization, i_a, is determined from the relation $i_a = (i_1 + i_2)/2$ in which i_1 is the ionization current when the same potential is applied to the main chamber and to the balance chamber, and i_2 is the current when these potentials are of opposite sign. If i_a and i_b are the ionization currents, respectively, in the main chamber (due to cosmic rays) and in the balance chamber, then $i_1 = i_a + i_b$ and $i_2 = i_a - i_b$. However, any residual ionization in the chamber contributes to i_a. To determine the residual ionization, the meters were taken into a coal mine 110 meters below the surface where the cosmic-ray intensity was estimated to be 0.2% of that at the surface. With the meter fully shielded, $i_a = (i_1 + i_2)/2$ and $i_b = (i_1 - i_2)/2$ were measured for different settings of the uranium micrometer.

In Fig. 3 an example of a burst at Huancayo is shown in the interval 2200 to 2300 GMT January 18, 1938. These records also show the decrease in ionization during the magnetic storm which started 2207 GMT January 16, 1938, and attained an intensity expressed as 8+ by the geomagnetic planetary index K_p.

ζ) *Other types.* Several other types of ionization chambers have been in operation for shorter periods of time. One of the more important of these is a 500 liter pressure chamber, shielded by 10 cm Fe, operated by the Physikalische Institut, under the direction of A. Sittkus [3] of the Universität of Freiburg. Results from this meter have been published quarterly since January, 1950 in the Sonnen-Zirkular of the Fraunhofer-Institut. The larger size of this instrument results in a smaller standard deviation of statistical fluctuations of hourly values, arising from sampling.

However, as will be seen later, the uncertainties in daily means of cosmic-ray intensity from ionization chambers arise not only from the sample size but also from atmospheric effects which influence the number of μ-mesons which decay before reaching the meter. In general, even for the Compton-Bennett meter (except at Huancayo), the latter are much greater than the former. For this reason the reliability of daily means is, in general, not likely to be improved by increasing the size of the ionization chamber unless corrections for the μ-meson effects are applied. On the other hand, for studies of the variability of the diurnal variation the statistical uncertainty of diurnal variations derived from larger ionization chambers will be less than from the smaller ones.

3. Geiger counter—telescopes. One particular experimental arrangement of Geiger counters of especial interest in measuring time variations is that used by Dolbear [4] and Elliott and earlier by Alfvén [5] and Malmfors in Stockholm to measure variations in different directions. Recordings were obtained with the axes of the telescopes in a N-S plane, and inclined 45° on both sides of the vertical. The axes of the telescopes could be similarly arranged in an E-W plane. Each telescope comprised three trays of ten counters each, with

25 cm spacing between trays. Counters 4 cm in diameter with an effective length of 40 cm were used. Triple coincidences between counters in the three trays were registered on electromechanical counters which were photographed every 15 minutes together with a clock and an aneroid barometer to provide time and barometric pressure. Two arrays were used, one with 35 cm of Pb in the train and the other with no absorber. With the axes of the arrays inclined 45° to the vertical the counting rates were about 7000 and 10000 coincidences per hour respectively for the arrangement with and without absorber. Such an arrangement has the advantage that atmospheric effects should be the same for both telescopes so that differences between the two are due to real effects dependent on direction.

4. Neutron monitors. It has been shown by Simpson [6], [7], [8] and others that the low energy nucleonic component produced within the atmosphere by the low energy portion of the primary particle spectrum, exhibits the largest

Table 2. *Location and elevation of continuously recording neutron monitors.*

Station	Geomagnetic latitude deg.	Elevation Feet
Chicago, Illinois	52 N	600
Climax, Colorado	48 N	11 000
Sacramento Peak, New Mexico . . .	42 N	9 800
Mexico City, Mexico	29 N	7 600
Huancayo, Peru	0.5 S	11 000

of the known geomagnetic latitude variations. At atmospheric depth 312 g cm^{-2} the variation in neutron intensity between the equator and 50° N was found to be about three times greater than that obtained in an ionization chamber. At sea level the ratio is even larger. A detailed analysis of neutron variations with latitude leads [7] to the conclusion that the ratio of the cross section for processes leading to neutron production to that for meson production increases rapidly with decreasing energy of primary particles. This means that neutron detectors are much more sensitive to changes in the low energy part of the primary spectrum than are ionization chambers which are mainly sensitive to the μ-meson component. For most of the time-variations the percentage changes from neutron counters are between 2.5 and 5 times greater than those from ionization chambers. The percentage increases in intensity associated with solar flares are of the order of twenty times greater than in ionization chambers [9]. Moreover, the neutron intensity is not affected by changes in the height of, say, the 100 mb pressure level in the atmosphere which can alter the meson intensity several per cent at the ground.

Several neutron piles have been constructed and put into operation by Simpson. These use proportional counters filled with enriched boron-10 trifluoride. The counters are surrounded by a lead-paraffin pile. Most of the neutrons are locally produced in the lead and slowed down to thermal energies by the paraffin, then captured by the boron-10. Table 2 indicates the location and elevation where neutron monitors are in continuous operation under the direction of John A. Simpson, University of Chicago. Further description of the apparatus is given in reference [6]. Also, complete details of the construction, maintenance, and operation of the neutron monitors are given in reference [6a]. These standardized neutron monitors have been adopted for observations during the International Geophysical Year (July 1957 to December 1958).

B. Atmospheric effects.
I. Barometer effects.

5. Ionization chambers. The barometric effect on cosmic-ray intensity as measured in ionization chambers has been known since 1928 when it was discovered by Myssowsky and Tuwin. The barometric coefficient is generally determined through the correlation between daily means of ionization and barometric pressure, an example of which, for 106 days of data from Cheltenham is shown in Fig. 4. The correlation coefficient for this sample is -0.84. The slope of the regression line which minimizes the sum of the squares of the differences between the observed values of ionization and those calculated from a linear relation between pressure and ionization is simply the product of the correlation coefficient and the ratio of the standard deviation of ionization values to the standard deviation of the pressure values.

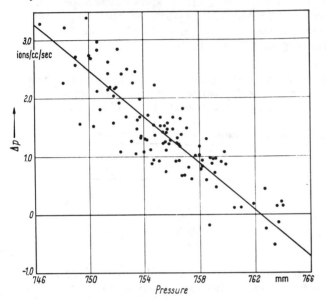

Fig. 4. Correlation between daily means (at Cheltenham, Maryland) of barometric pressure and departures from balance (Δ_p) 106 days, June 1— September 30, 1936.

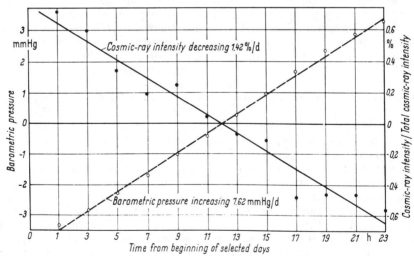

Fig. 5. Mean departures from average of bihourly values of cosmic-ray intensity and barometric pressure, derived from 15 selected 24-hour intervals with increasing pressure. October 7 to November 18, 1938, Godhavn, Greenland.

This procedure suffers the disadvantage that the daily means of ionization are often altered by world-wide changes, or by the passage of meteorological fronts over stations in temperate or polar latitudes. These effects can introduce appreciable errors into the barometric coefficient when it is calculated as described above. A better procedure, which mitigates the consequences of changes in ionization which are not due to pressure, is to select several intervals of 24 (or 48) hours length during each of which the barometric pressure is increasing (or decreasing), and to average, over all the selected intervals, the barometric pressure and ionization for each of the twelve bihourly divisions in each of the selected 24-hour intervals. A sample of results obtained at Godhavn (Greenland) is shown in Figs. 5 and 6 for 15 selected intervals with increasing pressure and in Figs. 7 and 8 for 15 selected intervals with decreasing pressure. Figs. 5 and 7 indicate that on the average the bihourly means of pressure and of ionization were changing linearly during the selected intervals. Figs. 6 and

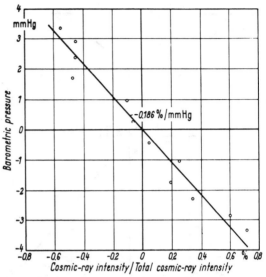

Fig. 6. Correlation between mean departures, from average, of bihourly values of cosmic-ray intensity and barometric pressure, derived from 15 selected 24-hour intervals with increasing pressure. October 7 to November 18, 1938, Godhavn, Greenland. Correlation coefficient $r = +0.982$.

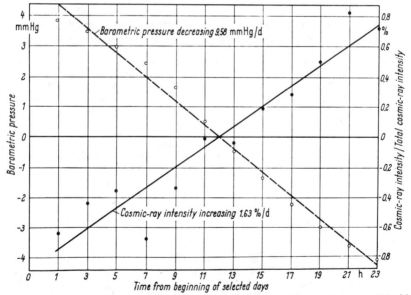

Fig. 7. Mean departure, from average, of bihourly values of cosmic-ray intensity and barometric pressure, derived from 15 selected 24-hour intervals with decreasing pressure. October 7 to November 18, 1938, Godhavn, Greenland.

8 indicate the correlation between corresponding averages of bihourly means of barometric pressure and ionization for the two sets of selected intervals. With this procedure the agreement between results from different

samples is generally quite close—and no systematic differences were found between results for increasing and decreasing pressure.

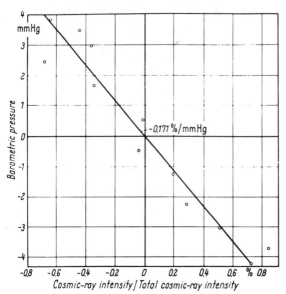

Fig. 8. Correlation between mean departures from average, of bihourly values of cosmic-ray intensity and barometric pressure from 15 selected 24-hour intervals with decreasing pressure. October 7 to November 18, 1938. Godhavn, Greenland. Correlation coefficient $r = +0.966$.

6. Special case: Huancayo. At Huancayo the daily mean barometric pressure changes only a very few mm of Hg during a month while the cosmic-ray ionization may ordinarily change a few percent. On the other hand the 12-hourly and 24-hourly waves in barometric pressure each have amplitudes of the order of one mm Hg at Huancayo. To mitigate the effect of the real changes in cosmic-ray intensity on the reliability of the determination of barometric coefficient, daily means were used only for selected intervals. The majority of intervals selected comprised four days; no interval was shorter than four days and few were as long as eight or nine days. The intervals selected were characterized by monotonic increases (or decreases) of the daily means of barometric pressure. For each interval, the departures, from the average of the interval, of each daily

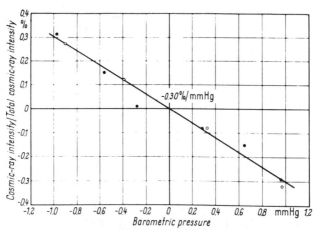

o = Average of 61 departures from mean of 61 selected intervals from March 1938 to March 1940.
● = Average of 30 departures from mean of 30 selected intervals from June 1936 to September 1937.

Fig. 9. Barometric coefficient for cosmic-ray intensity at Huancayo Observatory from correlation between average departures in cosmic ray- intensity and in barometric pressure from means for selected days. Days selected from weekly intervals with greatest changes in pressure. Regression line indicated, $r = 0.99$.

mean pressure and ionization were obtained. The departures in pressure were ranked (by magnitude and sign) and averages of all departures in each rank-interval were derived for pressure and for ionization. These are plotted in Fig. 9.

Each circled point is based on the average of 61 departures and each of the remaining points on the average of about 30 departures. The indicated slope of the line (− 0.30%/mm Hg) was determined by least squares giving all weight to the pressure departures. A similar determination, in which the ranking was based on the ionization departures according to size, resulted in a barometric coefficient of − 0.68%/mm Hg when all weight was given to the ionization departures. The large difference in the two barometric coefficients is due to the rather low ($r = − 0.70$) correlation between pressure and ionization departures for single days. The slope of the regression line which gives all weight to barometric pressure is that shown in Fig. 9. It is the coefficient actually used for Huancayo.

At Huancayo the phase of the 12-hour sine-wave in ionization is practically 180° from that in barometric pressure, which has an amplitude of about 1 mm Hg. The ratio of the amplitudes of these two waves gave an apparent barometric coefficient in good agreement with that shown in Fig. 9. The barometric coefficients indicated in Table 1 are those used in correcting the long series of observations published by the Carnegie Institution of Washington [10] for the period 1936 to 1959.

7. Counter telescopes. During 1951 and 1952 DAWTON [11] and ELLIOTT carried out experiments at Manchester to investigate time variations of the hard and soft component of cosmic radiation. Three counter trays each 40×40 cm were placed vertically above each other with a separation of 25 cm between trays. Between the two lower trays they placed a slab of lead 10 cm thick. Two-fold coincidences (C_{12}) between the upper two trays and between the bottom two trays (C_{23}) were recorded. The coincidence rates, C_{12} and C_{23}, thus respectively measured the intensities of the soft and hard components. Following DUPERIER'S procedure [12] they computed the partial correlation coefficients between the intensities and atmsopheric pressure, between the intensities and the height of the 100 mb atmospheric pressure level, and between the intensities and the temperature at the 100 mb level. From the correlation between the intensity, C_{23}, of the hard component and barometric pressure a pressure coefficient of − 0.17°/mm Hg was obtained. This is in good agreement with the values in Table 1 for Compton-Bennett meters at Godhavn, Cheltenham, and Christchurch with 12 cm Pb shielding.

8. Neutron monitors. As already indicated in Sect. 5, the intensities from neutron monitors are not affected by μ-meson decay effects. The only atmospheric effect of consequence is the barometric effect which is large, for which SIMPSON [6] and coworkers find the value 0.94%/mm Hg. They also show that their neutron barometric coefficient is essentially independent of latitude and of altitude for atmospheric depths greater than 600 g cm^{-2}. In order to obtain useful information from their very high counting rates, they take special precautions to obtain accurate values of barometric pressure to insure that the statistical reliability of the data is not limited by uncertainties in the barometric corrections.

II. μ-Meson decay effects and seasonal variations.

9. μ-Meson decay effects. Several investigators [13], [14] using ionization chambers have indicated negative correlations between the cosmic-ray intensity and temperature at the ground level. BLACKETT [15] was the first to correctly explain these effects as due to the instability of μ-mesons. He pointed out that the pressure level in the atmosphere where most of the μ-mesons were formed, now known to be in the region of the 100 mb pressure level, would be higher as the atmospheric temperature increased, and consequently from the greater height more μ-mesons would decay before reaching the ground thus resulting in a decrease of ionization. The correlations between ground temperature and cosmic-ray ionization were due entirely to the fact that both exhibit a seasonal variation. This was shown [16] from the fact that the variability of daily means of ground temperature within periods of a month or so exhibited no statistically significant correlation with the cosmic-ray ionization. Using the arrangement described in Sect. 7 Dawton and Elliott found from the partial correlations between: intensity and barometric pressure, intensity and height of the 100 mb level, and intensity and temperature at the 100 mb level, that an increase of 1 km in the height of the 100 mb level resulted in a decrease of about 4.0% $\pm$0.4% in the intensity of the hard component under 10 cm Pb. This effect of changing height of the 100 mb level is thus large enough to introduce serious errors of purely meteorological origin into the results at some stations. The consequences of this effect will be pointed out later in discussing the comparison of world-wide changes at several stations.

10. Seasonal variations. With the arrangement described above, in Sect. 7, Dawton and Elliott found (1950—1951) a seasonal wave of amplitude about 1.5%, with maximum near February. Their results also show a seasonal variation, of amplitude about 300 m, in the height of the 100 mb level at Manchester, with maximum about February. With the decay coefficient of — 4.0% per km, discussed in Sect. 9, an amplitude of about 1.2% in the hard component would be inferred from the height changes; this differs little from the observed amplitude of 1.5%, and the phases are consistent. It should be noted that Dawton and Elliott derived the decay coefficient, not through the correlation between monthly means of height of the 100 mb layer and intensity, but from the correlation between the departures of the daily means of these from their monthly averages. Thus it seems quite certain that most of the seasonal wave is due to the seasonal variation in height of the 100 mb layer.

In an analysis [17] of cosmic-ray data (ionization chambers) from several stations no seasonal wave was found for Huancayo. However, it was shown that if the seasonal waves at other stations were deducted from the data the residual variations were quite similar at all stations. In the harmonic dials of Fig. 10 the seasonal wave for each of several years is shown for Huancayo, and for Godhavn, Cheltenham, and Christchurch after deducting the monthly means for Huancayo from the monthly means at these stations. This procedure removes the world-wide variations [17]. For Cheltenham the seasonal wave is quite consistent for the several years indicated, and the amplitude[1], about 1.7%, of the average wave and its phase are in fair agreement with that obtained by Dawton and Elliott at Manchester. At Godhavn there is considerable variability among the twelve month waves for different years.

It is doubtful if the average seasonal wave for Huancayo is significant in view of the large variability which is probably due to the erratic nature of world-wide changes. If the monthly mean heights, averaged for ten years, of the 100 mb pressure level at Washington are plotted against the corresponding monthly mean values of ionization averaged for the same ten years, for Cheltenham minus Huancayo, the points do not fall on a straight line. The points deviate from a line in a manner suggesting that the differences, Cheltenham minus Huancayo, contain a seasonal wave not in phase with the seasonal wave in the height of the 100 mb layer at Cheltenham. This Lissajou characteristic disappears

---

[1] Amplitude denotes here, and throughout the chapter, the factor c in the sine-wave $c \sin(n\,t + \varepsilon)$; the difference of maximum and minimum is $2c$.

if a wave with amplitude about 0.3 % and maximum in October (as in Fig.10) is first deducted from Huancayo. Thus it seems likely that there is a small seasonal wave at Huancayo.

No data are yet available from neutron meters for long periods to determine if significant seasonal variations are present in neutron intensities. In any case

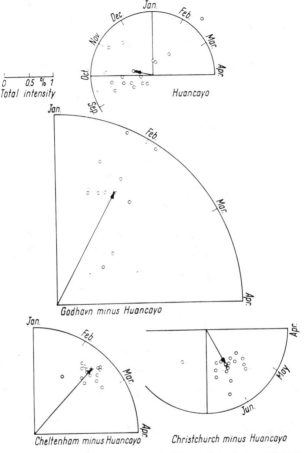

Fig. 10. Harmonic dials for 12-month waves at Huancayo and for Godhavn, Cheltenham, and Christchurch after deducting 12-month waves at Huancayo.

the neutron data are free of the μ-meson decay effects which contribute much to the variability of ionization-chamber data especially in temperate and polar latitudes. However, evidence will later be given indicating that variations in μ-meson decay effects contribute little to the variability of the daily means of cosmic-ray intensity measured at Huancayo with an ionization-chamber.

C. Diurnal variations.

I. Solar diurnal variations.

11. Methodology. α) *Point clouds in harmonic dials.* Many early investigations of time variations of cosmic-ray intensity have published results indicating diurnal variations. Most of these results have been in the form of averages with inadequate estimates for the reliability or the statistical reality of those averages. Those fields of geophysics which involve time variations and their variability demand the application of modern statistical procedures. Indeed in many cases these methods provide the only rational basis for deciding what is significant. No one has done more to foster the use of such techniques and to make clear the fruitfulness and power of such procedures than J. Bartels [18], [19]. It is not the purpose here to attempt any complete descriptions of his contributions but only to illustrate by means of examples their applicability to studies involving the diurnal variation of cosmic-ray intensity.

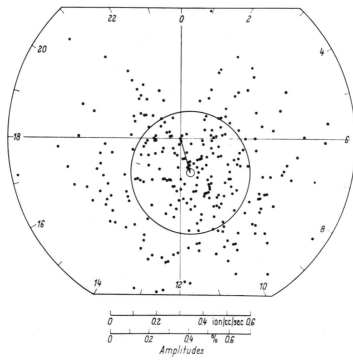

Fig. 11. 24-hour harmonic dial, apparent cosmic-ray intensity, 273 single days during April 20, 1935 to October 27, 1936, Cheltenham, Maryland (times of maximum in 75° West Meridian Mean Hours).

The harmonic dials in Figs. 11 and 12 show respectively the 24-hour and 12-hour waves for each day in a sample of 273 days of cosmic-ray ionization at Cheltenham corrected for barometric pressure. The vectors indicate the average 24-hour and 12-hour waves. The question to be answered is whether the average wave is statistically significant. The coordinates for the points in Figs. 11 and 12 are respectively a_1, b_1 and a_2, b_2 which derive from the harmonic analysis (corrected for non-cyclic change) of the 12 bihourly means of cosmic-ray ionization for each day. Each point in the dials is the end point of a vector, the length of which is the amplitude of the wave. The vector points to the time of the wave-maximum marked on the peripheral dial. The two-dimensional Gaussian frequency-distribution which best fits the cloud is in general elliptical [18]. The axes P_1, and P_2 of the probable ellipse, that is the ellipse which contains half the points inside it are given by:

$$P_1, P_2 = 0.833 \left\{ (\sigma_a^2 + \sigma_b^2) \mp [(\sigma_a^2 - \sigma_b^2)^2 + 4r^2 \sigma_a^2 \sigma_b^2]^{\frac{1}{2}} \right\}^{\frac{1}{2}}$$

in which σ_a^2 and σ_b^2 are the sample variances of a and b, and r is the coefficient of correlation between a and b. In case of circular symmetry (which applies to Figs. 11 and 12) $P_1 = P_2 = 0.833\,M$, with $M^2 = \sigma_a^2 + \sigma_b^2$. For the dial in Fig. 11:

$\sigma_{a_1} = 0.245$, $\sigma_{b_1} = 0.258$, $M_1 = 0.357$, and $C_1 = |\overline{a_1^2 + b_1^2}| = 0.169$, all in units of 1% of the total cosmic-ray ionization. The larger circle in Fig. 11 is the so-called probable-error circle for single days, its radius is $0.833\,M = 0.298\%$, and the number of points inside it is 138 or about half the total (273). If the deviations, from the average, of points for successive days are independent then the radius of the probable-error circle for the mean of 273 days should be $0.298/\sqrt{273} = 0.0180\%$. The mean amplitude $C_1 = 0.169$ is thus about 9.4 times the radius of its probable error circle. Thus for samples of 273 days from a population with $C_1 = 0$ the probability [19] that the average vector has a length $\geq C_1 = 0.169$ is $(\frac{1}{2})^{9.4^2} = 10^{-26}$, and the hypothesis $C_1 = 0$ is rejected. For the dial in Fig. 12 the corresponding probability is about 3×10^{-4} so that the hypothesis that the population value $C_2 = 0$ would also be rejected.

$\beta)$ *Random walk.* A slightly different approach is through the application of random walk theory discussed in a famous paper by BARTELS [19] which leads to the same conclusion, namely that both

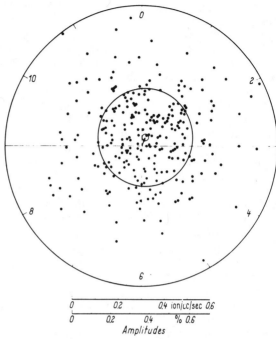

Fig. 12. 12-hour harmonic dial, apparent cosmic-ray intensity, 273 days during April 20, 1935 to October 27, 1936, Cheltenham, Maryland (times of maximum in 75° West Meridian Mean Hours).

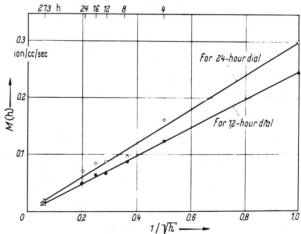

Fig. 13. Test for independence of points for successive days in 24-hour and 12-hour harmonic dials, April 20, 1935 to October 27, 1936, Cheltenham, Md.

waves are statistically significant. In either case the probabilities for reality depend upon the assumption of statistical independence of points in the dial derived from successive days. Fig. 13 tests this independence for both dials by indicating the dependence [19] on h, of the two dimensional standard deviation, $M(h)$, derived from means of h successive days, when values of $h = 1$, 4, 8, 12, 16, and 24 were used.

The lines in Fig. 13 indicate how $M(h)$ would decrease for independence. Thus in this sample there is no indication of lack of independence. In Fig. 13 $M(h)$ is given in units of ions cm^{-3} sec^{-1} in which units the total intensity was 84. It may be noted in Fig. 13 that $M_1 = 0.300$, and $M_2 = 0.246$ for the 24-hour and 12-hour waves respectively. From an analysis of the differences between hourly means from two Compton-Bennett meters operating simultaneously at Cheltenham, a value of 0.7% was found for the standard deviation of statistical fluctuations in hourly mean values from either instrument. It has been shown [19] that the two-dimensional standard deviation M, in harmonic dials derived from harmonic analysis of sets of r random ordinates with standard deviation ξ is given by: $M = 2\xi / \sqrt{r}$. For Cheltenham putting $\xi = 0.7\%$ and $r = 24$, $M = 0.286\%$ or $M = 0.240$ ions cm^{-3} sec^{-1}. This means that the values of M derived from the data in the harmonic dials of Figs. 11 and 12 can not be less than the value $M = 0.240$ ions cm^{-3} sec^{-1}. In these units the values $M_1 = 0.300$ and $M_2 = 0.246$ were obtained for the 24-hour and 12-hour clouds respectively. Thus practically all the scatter in the 12-hour dial can be ascribed to the statistical fluctuations of hourly means.

The scatter in the 24-hour dial can be conceived as due to two independent causes, one of which is that arising from statistical fluctuations in hourly values and the other from variations, from other causes, in the 24-hour wave from day to day. The standard deviation for the first cause was 0.286% and for the second the value 0.214%. If for example the standard deviation of statistical fluctuations in hourly values were only 0.15%, as it might well be in a sufficiently large ionization chamber, then from this cause alone the two-dimensional standard deviation for points in the harmonic dial would be: $2 \times 0.15 / \sqrt{24} = 0.061\%$. Suppose the variability of the wave itself from day to day, — a physical phenomenon which would be present even if the standard deviation of hourly values were negligible (i.e. an infinitely large chamber or counting rate), — were characterized by a two-dimensional standard deviation of say 0.214%, as above. The total standard deviation, from these two independent causes, for the points in the harmonic dial would be $\{0.214^2 + 0.061^2\}^{\frac{1}{2}}$ or about 0.222%. In this case it is quite evident that the mere specification of the statistical uncertainty of hourly values provides a practically useless lower limit for estimating the reliability of individual or average waves in harmonic dials. Nevertheless, this unfortunate practice continues.

12. 24-hour variation from counter telescopes. It was long realized that the small 24-hour wave in cosmic-ray intensity measured by ionization chambers and counters might be due entirely to atmospheric effects. Partly for this reason Alfvén [20] and Malmfors in Stockholm, Kolhörster [21] in Berlin and more recently Dolbear [22] and Elliott in Manchester using counter telescopes of the type described in Sect. 7 have measured the diurnal variation in different directions: With the axes of the two telescopes in the North-South plane and inclined 45° on opposite sides of the zenith. Elliott and Dolbear [23] made measurements over a period of a year, and confirmed the earlier results of Alfvén and Malmfors indicating the diurnal variation was significantly different for the North and South pointing telescopes. Since the particles recorded by the two sets of telescopes (N and S) pass through the same amount of atmosphere, it was concluded that at least part of the variation was due to causes outside the atmosphere. The difference between the variations in the two directions was interpreted as evidence for lack of isotropy of the primary rays.

This conclusion was later confirmed from the difference in diurnal variation obtained by ELLIOTT and DOLBEAR [24] with telescopes inclined in an East-West plane, both without absorber and with 35 cm of lead. They suggested that the anisotropy of the primary radiation appeared to be due entirely to solar influences, which as suggested by ALFVÉN [25] might arise from polarization effects in the so ar streams responsible for magnetic disturbances. They dichotomized data for 360 days according to the sum of the $8K$ indices (per day) of magnetic activity. For the "quiet" group this sum was ≤ 17, and for the other "disturbed" group it was ≥ 18; this division was selected to obtain the same number of days in each group. It was found that the amplitude of the average diurnal variation was about twice as large (about 0.2%), in the South minus North diurnal variation, on the geomagnetically "disturbed" days as on the "quiet" days.

13. World-wide variation in annual mean 24-hour wave from ionization chambers and counter telescopes. THAMBYAHPILLAI and ELLIOTT [26] were the first to point out that the local time of maximum for the yearly averages of the 24-hour wave, exhibited a large systematic variation over the period of 20 years or so for which data from various stations were available. Fig. 14 extends the results to include data for 1953 and 1954 for Huancayo and Cheltenham and Christchurch. It is evident that the variations indicated in Fig. 14 are undoubtedly real considering the general agreement between the results from different stations for the same year. The results from all the stations [except those for 1932, 1933, and 1934, from Hafelekar] are based on data which were corrected only for variations of barometric pressure. Those from Hafelekar were, in addition, unfortunately corrected for temperature. THAMBYAHPALLAI and ELLIOTT noted, from the results through 1952, the possibility of a 22 year wave in the phase. SARABHAI and KANE [27] claim, from observational data published by the Carnegie Institution that there are world-wide variations not only in the phase but also in the amplitude of the 24-hour diurnal variation.

14. Diurnal variation from neutron meters, its variability and comparison with ionization chamber results. FONGER [28] and SIMPSON [29], FIROR, and TREIMAN have compared the average diurnal variation obtained with neutron monitors at Climax with that from the large ionization chamber at Freiburg. The results are shown in Fig. 15 in which the observed values at Freiburg have been multiplied by 5. The authors estimated the standard deviations of random statistical fluctuations for each point, in Fig. 15, on the curve for Climax to be about 0.08% and that for points on the Freiburg curve to be about 0.023%. However, it seems quite improbable that the random statistical fluctuations in hourly values are the only cause for variability in the differences of corresponding hourly values on individual days at the two stations. For example it is quite possible that uncertainties in the corrections for μ-meson decay effects may introduce a variability into the diurnal variation at Freiburg, which should not affect the neutron data. In addition there were world-wide variations of appreciable magnitude during the period when the data for Fig. 15 were obtained. These world-wide variations will be simultaneous at all stations — thus introducing differences between the *local* time variations at the two stations. Thus (see also Sect. 11) the standard deviations for points in Fig. 15 may be grossly underestimated. In view of this, the differences between the two curves in Fig. 15 may not be significant. The similarity of the curves is direct experimental indication that the diurnal variation from pressure corrected data from ionization chambers is not seriously affected by other atmospheric effects.

Fig. 16a indicates the diurnal variation (d. v.) in neutron intensity [29] at Climax and at Huancayo averaged for 10 selected days in 1952 with *small* d. v.

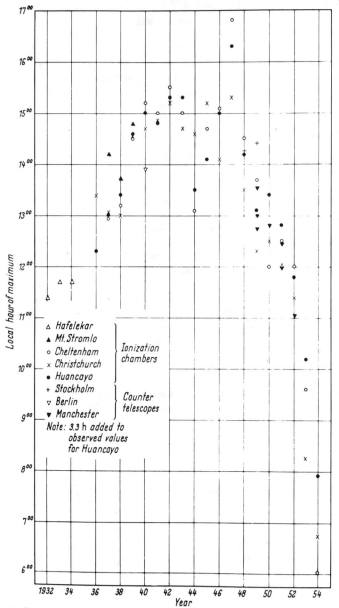

Fig. 14. Variation of local time of maximum of sketch 24-hour wave in cosmic-ray intensity.

at Climax while Fig. 16b indicates the averages for 10 selected days in 1952 with *large* d. v. at Climax. The results indicate *world-wide* variations in the amplitude of the diurnal variation on local solar time. Sittkus [30] has shown, from data registered by the large ionization chamber at Freiburg, that days with unusually large diurnal variation, 1 % or so, frequently occur on several successive days and that these occurrences exhibit a 27-day recurrence tendency. A comparison of these results and the data of EHMERT [*31*] in Weissenau with those of SEKIDO [*32*] in Tokyo indicated the world-wide occurrence of days with large diurnal variations *on local time.*

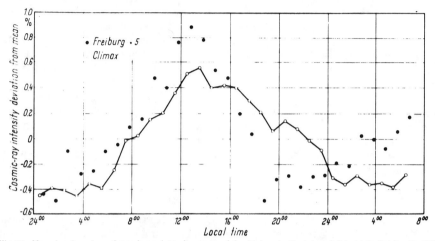

Fig. 15. Mean hourly values of cosmic-ray intensity averaged for 74 days in the interval July 14 to October 17, 1951 from neutron data at Climax and from ionization data at Freiburg (FONGER).

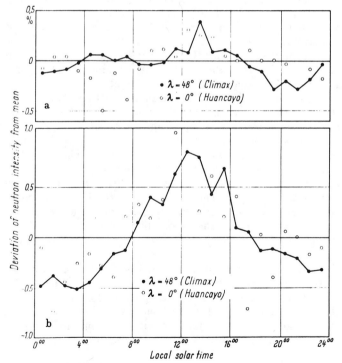

Fig. 16a and b. Diurnal variation (d. v.) in neutron intensity at Climax and Huancayo. a Average of 10 days with small d. v. at Climax. b For average of 10 days with large d. v. at Climax (FIROR, FONGER and SIMPSON).

15. Magnetic activity and the diurnal variation. In Sect. 12 it was indicated that the bihourly differences of cosmic-ray intensity from Northward and Southward pointing telescopes exhibited a diurnal variation which was greater on geomagnetically disturbed days than on quieter days. SEKIDO and YOSHIDA [32] concluded, from data with vertical counter telescopes, that the amplitude of the diurnal variation increased during magnetic storms and that the local time of maximum of the 24-hour wave came earlier. Also ELLIOTT and DOLBEAR [33] concluded from two magnetic storms (with associated decreases in cosmic-ray intensity) that the amplitude of the diurnal variation was much greater than normal. Finally YOSHIDA and KONDO [34] claim a 27-day recurrence tendency in the amplitude of the diurnal variation at Huancayo based on results for the period 1936—1940.

During these magnetic storms which are accompanied by large decreases in cosmic-ray intensity there are simultaneous world-wide variations (on universal time) within each storm. These variations may well have serious effects on the results for the diurnal variation on local time. Thus it would appear that it is difficult to establish the effects on the local time variation of cosmic-ray intensity during magnetic storms.

II. Sidereal diurnal variation.

16. Basis for sidereal variation. COMPTON and GETTING [35] first pointed out that if cosmic-rays originated outside our galaxy then a sidereal diurnal variation in intensity should arise as a consequence of the linear velocity of the earth due to the rotation of our galaxy and to the fact that the earth is far from the center of the galaxy. On the other hand no sidereal variation would be expected if the cosmic-rays originated in our own galaxy. The theory of the effect of galactic rotation was developed by COMPTON and GETTING without taking account of the deflection of charged particles by the earth's magnetic field. From rough estimates they concluded that the effect of deflections in the earth's magnetic field would decrease the amplitude of the sidereal diurnal variation to about one fifth of that computed in the absence of the earth's field.

VALLARTA, GRAEF, and KUSAKA [36] extended the theory of the galactic rotation effect to take account of the geomagnetic deflection of charged particles. Their investigation was confined to particles moving in the plane of the geomagnetic equator for which situation the equations of motion are integrable. For trajectories not in the geomagnetic equator and for non-equatorial latitudes they pointed out that the calculations are more difficult since the equations of motion are no longer integrable. They showed that the time of maximum for the 24-hour wave for all positive primaries depended rather critically, for particles in the geomagnetic equatorial plane, upon the energy distribution of primaries. With the number of primaries, all positive, varying inversely as the cube of their energy the maximum of the 24-hour wave was calculated to occur at about 13 hours sidereal time and with amplitude 0.17%. With the number of primaries, all positive, decreasing exponentially with energy, the calculated maximum occurred at 18 hours sidereal time with amplitude 0.24%. This comparison of observed and theoretical sidereal variations was only valid for observations made at the geomagnetic equator and then with rigor only for counter telescopes to insure that the particles counted arrive in or near the geomagnetic equatorial plane.

17. Search for a sidereal variation in the observations. Several investigators have reported 24-hour sidereal waves in cosmic-ray intensity. Most of these

results have presented only the average 24-hour sidereal wave, with no determination of its variability, without which a reliable test for the statistical reality
of the average wave is impossible. Using data for 595 days from a Compton-
Bennett meter at Cheltenham FORBUSH [37] found a 24-hour sidereal wave
with amplitude 0.03% and with maximum near 22 hours sidereal time.

To test whether this average amplitude was too large to be ascribed to chance
he used the two dimensional standard deviation derived from the scatter of points
for individual days in the harmonic dial for the 24-hour solar diurnal variation
in a sample of 273 days [38]. On this basis the average amplitude, 0.03%, of
the sidereal wave at Cheltenham was small enough to be ascribed to chance
and was, therefore, not regarded as real.

ELLIOTT and DOLBEAR [39] derived the 24-hour sidereal diurnal wave from
the south minus north variation from counter telescopes at Manchester as described in Sect.3. They found a 24-hour sidereal variation with amplitude about
0.03% and maximum near 8 hours sidereal time. This amplitude is about three
times its two-dimensional standard deviation as they derived it, which would
give a probability of about 10^{-4} of obtaining, by chance variations, an amplitude
$\geqq 0.03\%$. However, their standard deviation was estimated on the basis of
counting rate and as indicated in the last paragraph of Sect.11 this is certainly
an underestimate of the actual variability since it takes no account of the variability of the solar diurnal variation itself. ELLIOTT and DOLBEAR also derived
from nine years of data the 24-hour sidereal waves for Huancayo, and for the
average of Christchurch and Cheltenham. They obtained in both cases an amplitude of about 0.02%. The sidereal times of maxima for Huancayo and for the
average of Cheltenham and Christchurch were respectively 4 hours and 6 hours.

Fig.17 is a 24-hour sidereal harmonic dial based on data from Huancayo
for the interval 1937—1954. Each point in the dial was obtained by deducting
the yearly mean solar 24-hour wave from the average solar wave for one of the
indicated bimonthly intervals, and then transforming the coordinates of this
difference vector to a 24-hour sidereal dial by appropriate rotation of axes. The
standard deviations of the ordinates and abscissae of the points, from their means
shown by the end point of the average vector, were found to be respectively
0.047 and 0.044%. Since these two standard deviations are not significantly
different the cloud of points has circular symmetry. The standard deviation of
the distances between individual points and the end point of the average vector
is thus the geometric mean of these two standard deviations or 0.064%. This
corresponds to BARTELS' expectancy for single vector deviations from the mean
[19]. The expectancy for the averages of 108 statistically independent vector
deviations (corresponding to the 108 points in Fig.17) is then $0.064/\sqrt{108}$ or
about 0.0062%. The length of the average vector is 0.0155% so that $\varkappa =$
$0.0155/0.0062 \approx 2.5$ (see Sect.11 and Ref. [18]) and $e^{-\varkappa^2} = e^{-6.25} \approx 0.002$. Thus the
probability is about one in five hundred of obtaining by chance a vector as large or
larger than the average actually obtained in the sample of 108 in Fig.17. The
standard deviations derived for averages of three chronologically successive
vectors was found no larger than would be expected if these were independent,
so the above probability is not underestimated from this cause. The large circle
in Fig.17 is the so-called probable error circle; it contains approximately half
the points. The small circle is the probable error circle for the mean.

It is thus apparent that at Huancayo the 24-hour sidereal wave is small in
amplitude, about 0.015%; furthermore its phase is about opposite to that calculated by VALLARTA and GRAEF on the basis of positive primaries confined
to orbits in the plane of the geomagnetic equator. Finally, it must be emphasized

12*

that small systematic variations during the year in the phase and amplitude of the solar 24-hour wave can give rise to an apparent sidereal variation (see for example Fig. 14). *Thus from the 24-hour sidereal variation the evidence is certainly not convincing that any detectable fraction of cosmic-rays arrive at the earth from outside our galaxy.*

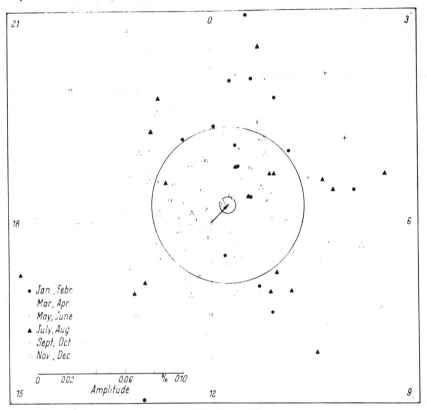

Fig. 17. Sidereal harmonic dial from data at Huancayo 1937—1954.

D. World-wide variations.

I. Variations associated with magnetic storms.

18. Some varieties of individual magnetic storm-effects. Fig. 3 (p. 163) is a reproduction of cosmic-ray records from Huancayo for the period January 15 to 18, 1938, and shows the decrease in cosmic-ray intensity during the magnetic storm which began at 22.7 hours G.M.T., January 16, 1938. Fig. 18 shows the bihourly means of cosmic-ray intensity averaged for three stations (all near the same longitude) and the daily mean horizontal magnetic intensity at Huancayo, Peru, for the period January 15 to 29, 1938. Fig. 19 indicates the correlation between the daily means of horizontal intensity at Huancayo and the daily means of cosmic-ray intensity for each of the three stations for which the average bihourly values were shown in Fig. 18. It is obvious in Fig. 19 that the ratio of the change in cosmic-ray intensity to that of horizontal magnetic intensity is more than twice as large for the storm beginning January 16 as for the storm following the sudden commencement on January 22. Fig. 20 compares the daily means of cosmic-ray intensity at three stations with those in horizontal intensity at Huancayo for the period January 11 to 31, 1938. Fig. 21 shows a similar comparison for the period April 16 to May 10, 1937.

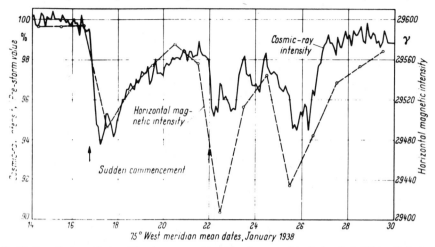

Fig. 18. Magnetic storm effects on bihourly mean cosmic-ray intensity averaged for Boston, United States, Cheltenham United States and Huancayo, Peru, and on daily mean magnetic horizontal intensity at Huancayo, Peru.

That not all magnetic storms are accompanied by decreases in cosmic-ray intensity is evident from Fig. 22 which shows no decrease in cosmic-ray intensity for a storm in which the daily mean horizontal magnetic intensity was depressed, on August 22, 1937, about 120 γ below normal (the geomagnetic planetary index Kp reached 8o).

Finally, Fig. 23 a and b shows the daily means in cosmic-ray intensity at Cheltenham and Huancayo together with the daily mean horizontal magnetic intensity for the period February 1 to 13, 1946. In this case the major decrease in cosmic-ray intensity preceded the major decrease in horizontal magnetic intensity by some three days. It may be of interest to indicate that a magnetic sudden commencement occurred at Huancayo and Watheroo toward the end of

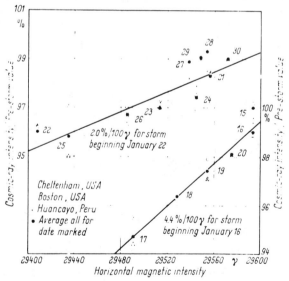

Fig. 19. Correlation between daily means of horizontal magnetic intensity at Huancayo, Peru and of cosmic-ray intensity at Boston, United States, Cheltenham, United States, and Huancayo, Peru, January 15–30, 1938.

February 2 (75° WMT) but it is conjectural whether this was in any way connected with the mechanism responsible for the decrease in cosmic-ray intensity (see also Sect. 22)

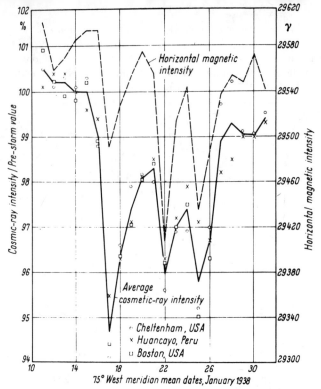

Fig. 20. Magnetic storm effects on daily mean cosmic-ray intensity at Boston, United States, Cheltenham, United States, and Huancayo, Peru, and Christchurch, New Zealand, and on magnetic horizontal intensity at Huancayo, Peru.

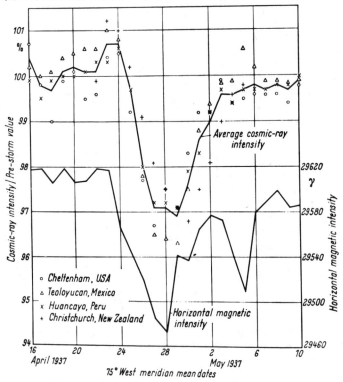

Fig. 21. Magnetic storm effects on daily mean cosmic-ray intensity at Cheltenham, United States, Teoloyucan, Mexico, Huancayo, Peru and Christchurch, New Zealand, and on magnetic horizontal intensity at Huancayo, Peru.

346

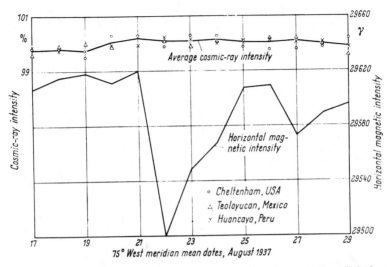

Fig. 22. Daily means horizontal magnetic intensity at Huancayo, Peru, and cosmic-ray intensity at Cheltenham, United States, Teoloyucan, Mexico, and Huancayo, Peru, showing no change in cosmic-ray intensity during magnetic storm beginning August 21, 1937.

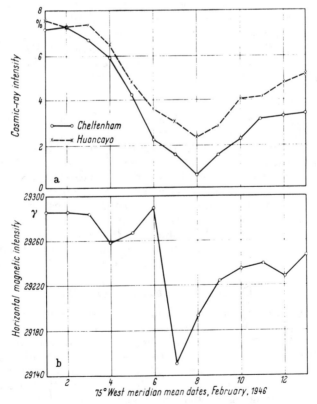

Fig. 23a and b. Daily means cosmic-ray intensity at Cheltenham and at Huancayo (a), and daily means magnetic horizontal intensity at Huancayo (b), February 1–13, 1946.

Additional examples of storms with and without associated decreases in cosmic-ray intensity may be seen in Fig. 28 in which are plotted the daily means of cosmic-ray intensity at Huancayo for the period 1937—1953 together with all the available daily means of horizontal magnetic intensity at Huancayo. Daily means for selected years for other stations are also included in the figure.

19. Cosmic-ray effects and the ring current. The decreases in daily mean horizontal intensity at Huancayo during the magnetic storms of January 16, 1938 and of August 21, 1937, shown in Fig. 20 and 22 respectively, were of about equal magnitude although the associated cosmic-ray effects were radically different. If the mechanism responsible for magnetic storms were in any way connected with that responsible for cosmic-ray decreases, this would require that in some respect the storm of January 16, 1938, should differ from the storm of August 21, 1937. An attempt [40] was therefore made to estimate for these two storms, the radius of the hypothetical ring current, concentric with the earth, and in the plane of the geomagnetic equator which would give rise to the main storm field changes. The magnetic potential of the storm field was assumed to be represented by a series of zonal harmonics, of odd degree since the field is known to be symmetrical with respect to the magnetic equator. Only two terms were used. These two terms in the expression for the external potential suffice to determine the ratio, R/a, of the radius, R, of the assumed ring current source, to the radius, a, of the earth. Data for both storms were obtained from several magnetic observatories and were corrected for the effect of ionospheric auroral zone currents. The results indicated that if the source were such a ring current then for neither of the two storms could R/a have been much less than two. However, for either storm R/a could have been indefinitely larger than two. This uncertainty in R/a arises from the fact that it is not possible to eliminate with certainty the magnetic effects of the S_D currents which flow in the ionosphere.

In any case when cosmic-ray decreases occur during magnetic storms, they are world-wide, having been observed not only near the equator, but also near the magnetic poles, at stations like Godhavn, geomagnetic latitude 80° N, and Thule, geomagnetic latitude 88° N. If cosmic-rays were excluded from reaching the earth by the magnetic field of a ring current it would certainly not be expected that they would be excluded from regions so near the geomagnetic pole as Godhavn and Thule. Theoretical calculations of the effects of the magnetic field of a ring current on the trajectories of cosmic-ray particles have given results which do not appear to explain the observed cosmic-ray effects [41]. The results of an investigation by Treiman [42] indicate that an increase in cosmic-ray intensity should arise from a decrease in the magnetic field at the equator if this decrease is due to a spherical current sheet concentric with the earth.

II. Geomagnetic activity effects.

20. Cosmic-ray intensity for magnetically quiet and disturbed days. Fig. 24a and b indicate, respectively, for Huancayo and Cheltenham, and for Huancayo and Godhavn, the correlation between the average difference of cosmic-ray intensity for the five magnetically disturbed days of each month less that for the five quiet days. It is evident from Fig. 24a and b that the frequency of positive values of the differences for Huancayo is only about one-fifth that for negative values, which indicates definitely that the cosmic-ray intensity tends to be less for the five magnetically disturbed days than for the five quiet days. It is also evident from these figures that the correlation between the differences for Huancayo and Godhavn is less than for Huancayo and Cheltenham. This, as will be shown later, is probably due to greater variations in vertical air-mass distribution at Godhavn as compared with those at Cheltenham. Fig. 25 indicates the relation

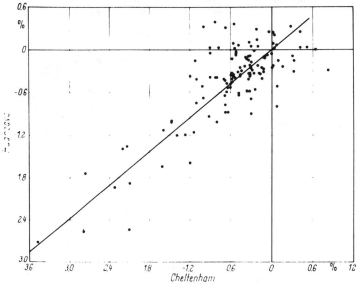

Fig. 24a. Average difference cosmic-ray intensity for five disturbed days less that for five quiet days in each month, April 1937 to December 1947, at Cheltenham and Huancayo.

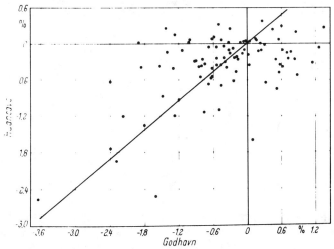

Fig. 24b. Average difference cosmic-ray intensity for five disturbed days less that for five quiet days in each month, January 1939 to December 1946, at Godhavn and Huancayo.

between the differences, disturbed minus quiet days, for cosmic-ray intensity and magnetic horizontal intensity at Huancayo. The differences are always negative for the horizontal intensity and preponderantly negative for cosmic-ray intensity at Huancayo.

The correlation coefficient for data of Fig. 25 would obviously be low. This is expected from the fact that the ratio between changes in cosmic-ray intensity to those in horizontal intensity is known to vary from one storm to another. Fig. 26 indicates the variation in the annual means, for disturbed minus quiet days, in cosmic-ray intensity at three stations and in horizontal intensity, H, at Huancayo (values for H were unavailable after 1947). It will be noted that the annual mean difference, disturbed minus quiet days, is always negative for cosmic-ray intensity at all three stations.

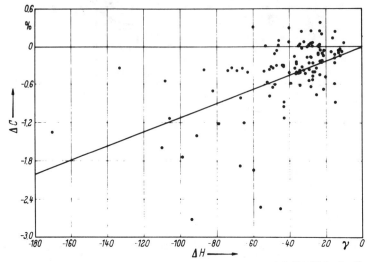

Fig. 25. Average difference for cosmic-ray intensity (ΔC) and for horizontal mangetic field (ΔH) for five disturbed days less that for five quiet days in each month, April 1937 to December 1946 at Huancayo.

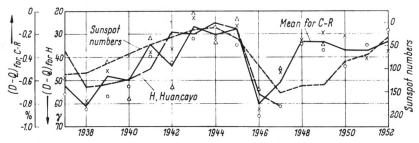

Fig. 26. Annual means for differences (D-Q) for magnetically disturbed less quiet days (5 each per month) for cosmic-ray intensity (C-R) and for horizontal magnetic intensity, H. Legend for (C-R): $\circ$ = Chettenham, $\times$ = Huancayo, $\triangle$ = Godhavn.

These facts together suggest that the mechanism responsible for the decrease in cosmic-ray intensity is connected with that responsible for mangetic storms. Alfvén [43] has proposed that the cosmic-ray decreases arise from deflection of cosmic-ray particles by magnetic fields carried away from the sun in the conducting streams which also give rise to magnetic storms.

21. Variability of daily means at Huancayo. Fig. 27 indicates the standard deviation of daily means from monthly means at Huancayo derived from pooling the variance of daily means from monthly means for each year 1937—1952. The curves show that standard deviations of departures from the monthly means are roughly four times smaller near sunspot minimum than near sunspot maximum. They are only slightly less when the five magnetically disturbed days of each month are excluded. This is probably due to the fact that in most months the variation arises principally from 27-day quasiperiodic variations. For 1944, the standard deviation of daily means from monthly means is about 0.21% (excluding the five magnetically disturbed days). Since this figure included the variability of the world-wide component, it is an upper limit for the combined effects of statistical fluctuation in the records and those from variations of μ-meson decay due to changes in vertical distribution of air-mass. It is thus evident that the latter effects are quite small at Huancayo, and that the daily means (relative to the mean of the month) at Huancayo are reliable to within at most 0.2% (that is, their standard deviation $s \leq 0.2\%$).

From a previous investigation [44] it was found that the world-wide changes at Teoloyucan, Mexico, were about 1/0.63 times those at Huancayo, Peru. For

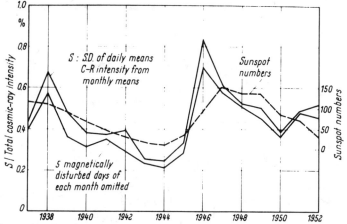

Fig. 27. Annual means: sunspot numbers and variability cosmic-ray intensity at Huancayo.

1937, daily means, with seasonal wave removed, were available from Teoloyucan for all months except January and November. These daily means for Teoloyucan were multiplied by 0.63 to reduce them to Huancayo. The difference between the daily mean at Huancayo and the reduced daily mean at Teoloyucan was found for each day of the ten months. The standard deviation of single differences about their average for the month was found to be 0.24% from pooling the ten samples of one month each. Assuming equal variance for statistical fluctuations at both stations, the standard deviation for the statistical fluctuations in single daily means is only 0.17%, which is slightly less than the figure of 0.2 derived from the fluctuations of daily means from the monthly means for Huancayo in 1944. The standard deviation for hourly values at Huancayo is about 0.6% (from differences between the same pair of hours on a number of quiet days) which gives for 24 independent hours a standard deviation of 0.12%. The standard deviation of the ten monthly mean differences about their average is about three times greater than would be expected from the fluctuations in the differences in daily means. This may indicate some small systematic change which would arise if the seasonal variations at Teoloyucan deviated from a pure 12-month wave or if there were a small seasonal wave at Huancayo. It will be shown later that variations arising from non-world-wide changes are much greater at the other stations than at Huancayo and Teoloyucan and that the data from Huancayo and Teoloyucan provide more reliable measures of the world-wide component than do those from the other stations. The absence of any large seasonal variation at Huancayo is further indication that the vertical distribution of air-mass there must vary little with season.

The curves for the standard deviation of daily means, in Fig. 27, are similar to that derived by MEYER [45] and SIMPSON from the variation of the amplitude of the 27-day variation, over the period 1936—1954. These were based on published ionization chamber data for Cheltenham, Christchurch, and Huancayo,

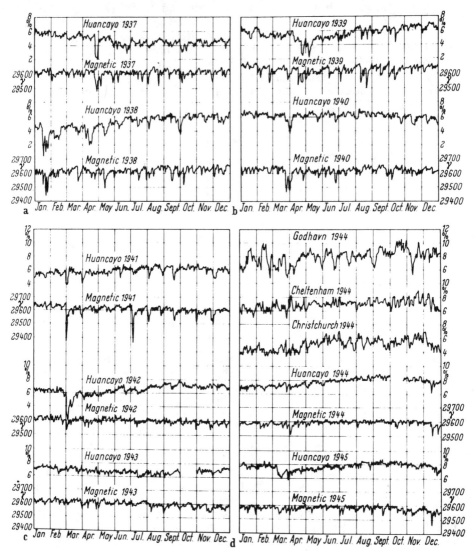

Fig. 28 a—f. Daily means cosmic-ray intensity for: Huancayo (75° WMT), Cheltenham (75° WMT), Godhavn (45° WMT), and Christchurch (172.5° EMT) and daily mean horizontal magnetic component at Huancayo (75° WMT).

for the period 1936—1946, and on neutron pile data at Climax and Chicago for the period 1951—1953. From the data at two stations, A and B, they derived the standard deviation of what was termed the tracking component of the variation. For a measure of the amplitude of the 27-day variation at station (A) they used the standard deviation, σ_A, of δ_A with $\delta_A = [I_A(t) - I_A(t-T)]/\bar{I}_A$ in which $I_A(t)$, and $I_A(t+T)$ are respectively the daily mean intensities on day t and $(t+T)$, and $\bar{I}_A$ is the average intensity. They used $T = 14$ days since as a consequence of the 27 day recurrence tendency σ_A was found [46] to have a maximum with $T = 14$ days. Similarly σ_B was determined for station B. For corresponding values of t the correlation coefficient r_{AB} between δ_A, and δ_B, was determined using data for six month periods. The standard deviation of the tracking component was taken as $|r_{AB}| \sigma_A$ which, except for the absolute value of r_{AB}, derives from the well-known relation for the standard deviation of the linearly dependent variations in A and B. This procedure has the advantage that it shows the 27 day

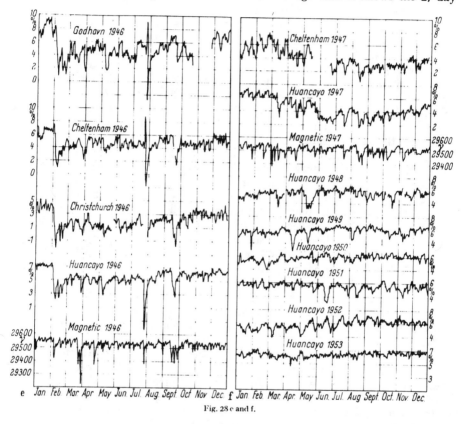

Fig. 28 e and f.

variation is correlated between the different stations and is thus a world-wide phenomenon. The world-wide nature of the 27-day variation had also been demonstrated in a different manner which will be discussed in the section on 27-day variations.

22. Variations in daily mean cosmic-ray intensity at Huancayo, 1937—1953 compared with variations in the earth's magnetic field and with cosmic-ray variations at other stations for selected years.

Fig. 28 is a graph of daily means of cosmic-ray intensity at Huancayo 1937—1953, together with daily means for Godhavn, Cheltenham, and Christchurch for selected years. These daily means have been corrected for bursts, barometric pressure and seasonal variation, also for a linear drift the magnitude and nature of which will be discussed in connection with the sunspot-cycle variation in cosmic-ray intensity. Inspection of the graphs of cosmic-ray intensity for Huancayo in Fig. 28 indicates a marked difference in the variability means in different years, which is particularly evident if the curves for 1944 are compared with those for 1946 and 1947; this variability was shown quantitatively in Fig. 27.

During 1946 and 1947, there were large variations at Huancayo, which in general follow those at Cheltenham, Godhavn (1946 only), and Christchurch (1946 only). There is, of course, the large increase at Godhavn, and Cheltenham on July 25, 1946, which occurred during a large solar flare on that date, and which is absent at Huancayo, and at Christchurch where the meter was out of operation for eight days starting July 23.

On the other hand, a comparison of the graphs for the four stations for the sunspot minimum year 1944 shows that the variability of the daily means is decidedly less at Huancayo than at the other three stations, and that the variability is greatest at Godhavn. Moreover, the major variations at Godhavn, Cheltenham, and Christchurch during 1944 were seen (by overlaying the original curves) to be essentially uncorrelated. At Cheltenham, and to some extent at Godhavn, the larger variations in 1944 (which were absent at Huancayo) occurred more often in winter than in summer. At Godhavn and at Cheltenham, it was found that the large variations in 1944 generally occurred during periods when the barometer was changing rapidly. These large variations are thus probably due to changes of the vertical air-mass distribution accompanying the movement of a front over the station and the consequent effects arising from meson decay. Although smaller variations occurring at Huancayo are often obscured at the other stations by this meteorological effect, it will be shown that the *averages* of a sample of such variations are very nearly the same at all four stations.

Fig. 28 also shows the daily mean values of the horizontal magnetic component (H) at Huancayo, 1937—1947, from which it can be seen whether decreases in H, which occur during magnetic storms, are accompanied by decreases in cosmic-ray intensity. From these graphs, a tabulation showed 48 cases (1937 to 1947) when from one day to the next a decrease, in H, of 75 γ or more occurred. In 36 of these cases, the change in cosmic-ray intensity at Huancayo was negative, although in only 22 cases was the decrease in cosmic-ray intensity greater than 0.4%.

The graphs were also used to tabulate the dates between which the daily means of cosmic-ray intensity at Huancayo decreased continuously (successive days with no change were included) for a total decrease of 1.0% or more. There were 92 such intervals from 1937 to 1947. The change, ΔH, in daily mean horizontal magnetic intensity at Huancayo from the first to the last day of each of the above-selected intervals was also tabulated; in 71 (out of 92) of the intervals, ΔH was negative. Examination of magnetograms for Huancayo (Peru) and Watheroo (Australia) indicated magnetic disturbance in most of the 21 cases for which ΔH was either zero or positive. It thus seems evident that during most of the periods when the cosmic-ray intensity at Huancayo is decreasing there is evidence for magnetic disturbance, which suggests that the cause of the cosmic-ray decreases is quite probably connected with the mechanism giving rise to magnetic disturbance.

Finally, in this connection, attention should be called to the graphs of daily means for cosmic-ray intensity and magnetic horizontal intensity, H, for February 1946 in Fig. 28. Between February 3 and 6, 1946, the five per cent decrease in cosmic-ray intensity at Huancayo was accompanied by only a small decrease in H, while the large decrease in H after February 6 was accompanied by only a small further decrease in cosmic-ray intensity. This can be better seen in Fig. 23 which was discussed in Sect. 18. There was at Huancayo and Watheroo a marked magnetic sudden commencement at $08^h 42^m$ (75° West Meridian Time) a few hours before the start of the decrease in cosmic-ray intensity. Attention should also be called to the fact that after Februray 6, 1946, both the cosmic-ray intensity and H at Huancayo remained low during the rest of the year. While it seems clear that most of the decreases in cosmic-ray intensity occur during periods of magnetic disturbance, no measurable characteristic of magnetic disturbance has yet been found which is quantitatively well correlated with changes in cosmic-ray intensity.

Attention has been called in Fig. 28 to the fact that the variability of daily means at Huancayo, obvious in 1944, is in general less than that for the other stations (see Fig. 28 d). It was also indicated that these variations at Godhavn, Cheltenham, and Christchurch were generally uncorrelated and could well obscure small variations which can be reliably seen at Huancayo on account of the lack of appreciable μ-meson decay effects arising from variations in the height of the 100 mb pressure level.

To determine whether small variations at Huancayo can be traced in the averages of several effects at the other stations, Fig. 28 was used. From these plots of daily means for Huancayo ten intervals of 20 days length were selected with each interval exhibiting a variation similar to that shown for Huancayo

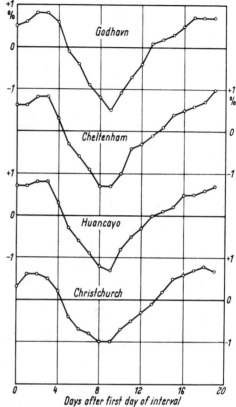

First Day of 10 Intervals: Apr 27, 1939; Mar 25, 1940; May 1, 1946; Sep 14., 1946; Mar 7, 1947; Jul 10, 1947; Aug 10, 1947; Aug 27, 1947; Apr 5, 1949; Jul 30, 1949.

Fig. 29. Cosmic-ray intensity variations averaged for ten intervals of 20 days, each with comparable variation at Huancayo.

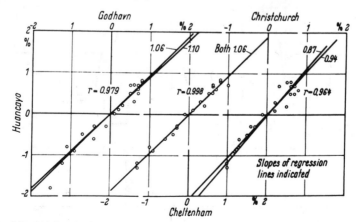

Fig. 30. Correlation (r) between variations at Huancayo and those at Godhavn, Cheltenham, and Christchurch from data in Fig. 29.

in Fig. 29. Fig. 29 indicates the variation averaged for the same ten intervals for each of the three stations. Fig. 30 indicates the correlation between the averaged variation for Huancayo and that for each of the other three stations. The correlation coefficients and slopes of the regression lines are also indicated. The smaller of the two slopes results from the assumption of no statistical error in the means for Huancayo. Except for Christchurch, the factors are in fair agreement with those derived in an earlier study of world-wide changes.

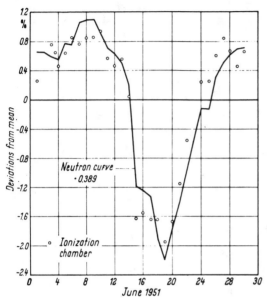

Fig. 31. Comparison neutron daily means (New Mexico) multiplied by 0.389, with those from ionization chamber at Huancayo. Standard deviation of differences is 0.25%.

23. Sample comparisons of variations of daily means from ionization chambers and neutron monitors.
Fig. 31 is a comparison of the variation of daily means for June 1951 from the Compton-Bennett meter at Huancayo and those published by Simpson [47] from neutron counters at Sacramento Peak, New Mexico. The standard deviation (s. d.) of the differences between daily means from the Compton-Bennett meter at Huancayo and those from the neutron counters (multiplied by 0.389) at Sacramento Peak is about 0.25%. The series is

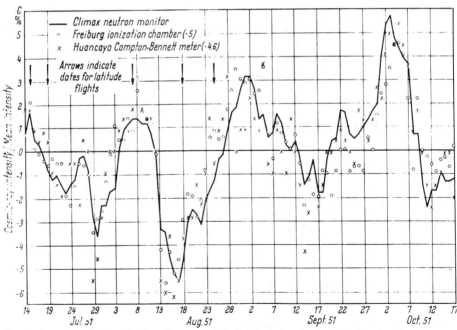

Fig. 32. Comparison of daily means from Huancayo and Freiburg ionization chambers with those from Climax neutron monitor, July–October, 1951.

too short to determine whether there are systematic changes in background in either instrument involved. The occurrence of any such changes would result in increasing the s.d. If the value of 0.17% is accepted (see Sect. 21) for the s.d. of single daily means from the monthly average at Huancayo, then from the value of 0.25% for the s.d. of differences between the Huancayo daily means and those for neutrons in June 1951, the s.d. of daily means for the latter is found to be about 0.19%, in the reduced neutron units. Or, the s.d. of the neutron daily means would be $0.19/0.389 = 0.49\%$. Since this figure is many times greater than would be expected in view of SIMPSON's high neutron counting rate, it is evident that during the period of this comparison one of the two instruments was subject to variations, either real or instrumental, which did not affect the other. Since the Instituto Nacional de la Investigacion Cientifica and the University of Mexico, Mexico, D.F., are now collaborating (since September 1, 1954) with the Department of Terrestrial Magnetism in the operation of a Compton-Bennett meter at the University of Mexico, Mexico, D.F., it will be possible in future to compare results from it with those from the neutron monitor which is in operation there by SIMPSON's group.

Fig. 32 shows a comparison published by FONGER [48] of daily means from the neutron monitor at Climax and from the large ionization chamber at Freiburg to which has been added the daily means for Huancayo. While the changes at the three stations are in generally good agreement, it may be noted that the ratio of changes at Climax to those at Huancayo is decidedly greater than that indicated for Fig. 31. This indicates that further comparisons are needed before it is certain whether or not the ratio is constant.

III. Variation with sunspot cycle.

24. Ionization chamber results. α) *Evidence in long series of observation.* When all available annual means, from Godhavn, Cheltenham, Huancayo, and Christchurch, of cosmic-ray ionization, corrected for bursts, and barometric pressure, were examined, a secular decrease was obvious in the results for Christchurch. Since there was no evidence in the results for Cheltenham of any significant secular change, other than the sunspot variation as shown in Fig. 33, no correction for drift was applied to the data for Cheltenham. By comparing results for the other stations with those for Cheltenham, the following linear changes were found: Christchurch, -1.40% yr^{-1}; Godhavn, -0.25% yr^{-1}; and Huancayo, $+0.40\%$ yr^{-1}. The annual means in Figs. 33 and 34 have been corrected for the above linear changes, which are assumed to be instrumental and probably arise from decay of radioactive contamination in the main chamber or in the balance chamber of the meters.

The agreement between the annual means of cosmic-ray intensity for the four stations, or their average, and that for annual mean sunspot numbers is evidence that the mechanism responsible for these changes in cosmic-ray intensity involves some phenomenon associated with solar activity. It is known that some magnetic storms are accompanied by large decreases in cosmic-ray intensity, and it was shown that most of the major decreases which occur during intervals of a few days are associated with magnetic storms or periods of magnetic disturbance. Thus, there arises the question of whether these decreases are mainly responsible for the variation of cosmic-ray intensity with sunspot numbers shown in Fig. 33. To answer this question, the variation of annual means of cosmic-ray intensity at Huancayo for all days (as used in Fig. 33) is compared in Fig. 34 with that for international magnetic quiet days and with that for international magnetic disturbed days. It is evident from Fig. 34 that the variation of annual means for all days, which in Fig. 33 follows the curve of sunspot numbers, is very little different from that for quiet days and not greatly different from that for disturbed days. Thus, *the main features of the variation of cosmic-ray intensity*

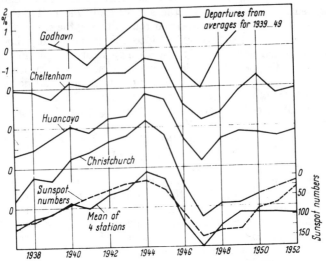

Fig. 33. Annual means cosmic-ray intensity at four stations.

with sunspot numbers persist for long periods (six months or more) and are not ascribable to transient decreases accompanying some magnetic storms.

Further evidence of effects that persist for long periods of time is indicated in Fig. 28d (p. 188), in which the curve for daily means of cosmic-ray intensity at Huancayo shows a gradual increase of about 1.5% from January 1944 to September 1944 during a period in which there was no very large transient decrease in cosmic-ray intensity and no great magnetic storm. It thus appears that the transient decreases in cosmic-ray intensity which occur during some magnetic storms and magnetically disturbed periods are superimposed upon a variation with sunspot cycle.

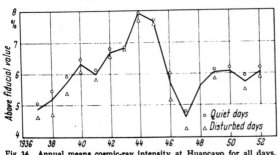

Fig. 34. Annual means cosmic-ray intensity at Huancayo for all days, international quiet days and international disturbed days.

β) *Comparison with geomagnetic data.* Fig. 35 shows the variation of the monthly means of cosmic-ray intensity for four stations after removing the seasonal wave and linear trend. Also shown is the monthly mean horizontal magnetic component at Huancayo corrected for a linear estimate of secular change. It is evident that the horizontal intensity and cosmic-ray intensity were both markedly lower throughout 1946 and 1947 than in 1944. This fact suggests the possibility that the same mechanism may be responsible for both effects. In this connection, it should be mentioned that Vestine [49], from a long series of magnetic data from many observatories, found evidence for an 11-year variation in the horizontal component of the earth's field. In deriving the latitude distribution for the three geomagnetic components of the storm-time field, D_{st} (disturbed minus quiet days) he found that the eastward geomagnetic component of D_{st} was zero, on the average. However, he points out that the yearly average of the east component for all days is not only not zero but varies during the sunspot cycle, indicating that the cause of this variation (and probably also that for horizontal intensity on all days) may be distinct from that for D_{st}. Thus, this unexplained variation in the earth's field is possibly connected with the sunspot variation in cosmic-ray intensity.

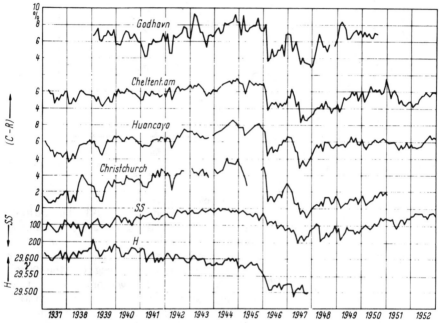

Fig. 35. Monthly means: cosmic-ray intensity (C-R) at four stations; sunspot numbers (SS) and magnetic horizontal intensity (H) at Huancayo.

The variation of monthly means is further compared in Fig. 36. To effect this comparison, the monthly means for Huancayo were categorized in six intervals of 1% (3% to 9%). Monthly means for each of the other stations were averaged for each of these six categories. These group means are reasonably well fitted by the straight lines shown in Fig. 36. These lines thus approximate the regression lines obtained by assuming the monthly means at Huancayo are free of statistical errors, which are presumed present only in the means for the other stations. The factor of 1.23 for the ratio of changes in Christchurch to those at Huancayo is roughly 20% greater than that found earlier from a shorter series of data. It is also greater than that obtained in the last paragraph of Sect. 22. For Cheltenham and Godhavn, the factors shown on Fig. 36 are more nearly consistent with those derived earlier and with those derived in Sect. 22.

γ) *Comparison with the 100 mb level.* The question arises whether variations with solar cycle in the height of the 100 mb pressure level might explain, through μ-meson decay effects, the sunspot variation in cosmic-ray ionization. The average seasonal wave in cosmic-ray intensity at Cheltenham has an amplitude of 1.45% and is opposite in phase to the average seasonal wave, amplitude 260 m, in the height of the 100 mb pressure level derived from U. S. Weather

13*

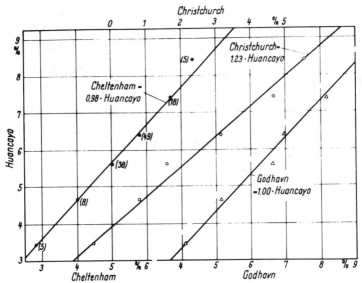

Fig. 36. Correlation between averages of groups of months for Huancayo and each of three other stations. Numbers in parenthesis indicate number of months in each group.

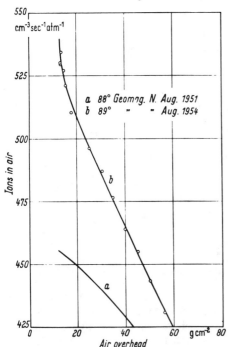

Fig. 37. On an expanded scale the ionization vs. air mass overhead is shown for the two years, 1951 and 1954, near the geomagnetic pole. The increased slope in 1954 at the lowest pressures is evidence for low energy particles (150 MeV if protons) present in 1954 that were absent in 1951 (Neher and Stern).

Bureau radiosonde data at Washington for the period 1944—1953. The ratio of these amplitudes gives a 5.6% decrease in cosmic-ray intensity per kilometer increase in height of the 100 mb level. The range in annual mean heights of the 100 mb level (1944—1953) was only 75 m and the annual means exhibited no variation similar to the solar cycle. Thus the sunspot variation in cosmic-ray intensity does not arise from this cause.

25. The "knee" of the cosmic-ray latitude curve. At sea level the "knee" of the cosmic-ray latitude curve has been placed between geomagnetic latitudes 40 and 45°. The "knee" at sea level is doubtless due to atmospheric absorptions. However, at high altitudes divergent results have been obtained by different investigators. Neher [50] points out that these discordant results may arise from the fact that the data were taken at different times. Fig. 37 shows Neher's results for the ionization as a function of air mass overhead for high altitude (95 000 ft) flights made in the summers of 1951 and 1954 near the geomagnetic pole. The increase of ionization for a given decrease in air overhead is markedly greater for the 1954 curve than for the 1951 curve. This effect was found on each of the five 1954 flights which reached sufficient altitude.

360

Fig. 38 shows, for the indicated atmospheric depths, the variation of ionization with latitude for 1951 [51] and 1954. It is evident that large changes in the radiation took place in the intervening three years at the northern latitudes at these high altitudes. While the change in 1951 from 58° to 68° was less than 1% at 20 g cm^{-2}, in 1954 a change of about 6% was found at the same depth, covering the same range of latitude. This difference is indication that low energy particles which were present in 1954 were absent in 1951. From geomagnetic theory it is found that 150 MeV protons can arrive vertically at geomagnetic latitudes north of 65°. The evidence from Fig. 38 is that particles (if protons) were present in 1954 in the primary radiation down to 150 MeV. The increase in the ionization from 68° to 90° shown in the curves of Fig. 38, amounting to 12 ions cm^{-2} sec^{-1} atm^{-1} of air at 20 g cm^{-2}, was the same in 1954 as in 1951. Since no new particles, admitted by the opening of the STÖRMER cones, could reach the instrument north of 68°, this increase is ascribed to the opening of the shadow cones [51] for both occasions. From Fig. 37 the continued increase in the slope of the 1954 curve at 15 g cm^{-2} probably indicates that particles were present which had ranges

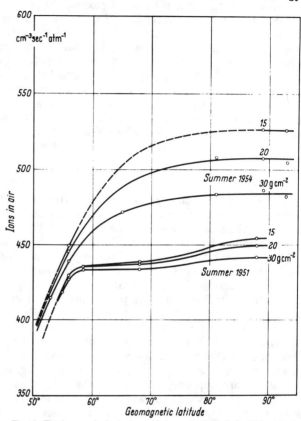

Fig. 38. The increase in ionization with increasing latitude in 1954 up to at least 68° N shows a lack of cut-off of primary particles down to at least 150 MeV, if protons (NEHER and STERN).

equal to and less than this value. These results indicate that there was no cutoff of the primary particles down to at least 150 MeV (assuming protons) in the summer of 1954 at northern latitudes. In contrast, the apparent cutoff for protons in 1951 was estimated at 800 MeV. These experiments provide additional evidence of an inverse relationship between solar activity and cosmic-ray intensity. During the summer of 1954 the sun was at its lowest ebb in 22 years and particles of low energy were arriving at the earth which at other times were excluded.

26. Changes in the low energy particle cut-off and primary spectrum. In a communication kindly sent in advance of publication the authors, PETER MEYER and J. A. SIMPSON, at the University of Chicago show from measurements with a nucleonic component detector, in B 47 jet aircraft at 310 g cm^{-2} atmospheric depth, that the low rigidity cut-off for particles in the primary cosmic-ray spectrum has decreased within the period 1948 through 1951. They determine that this decrease corresponds to a northward change, between 1948 and 1951, of 3° in the "knee" for the nucleonic component. This change is accompanied by a change in the primary spectrum for particle rigidities less than about 4 Bv and by an increase in the total primary intensity. They find that if the differential primary intensity at low rigidities is given by $j = C/(p/z)^2$ for 1948 then the spectrum for 1951 through 1954 is approximately $j = C'/(p/z)^{2.7}$, where p/z is proportional to the magnetic rigidity of particles with charge ze. The total change of intensity due to changes in spectrum and low rigidity cut-off is more than 13%. Thus the decrease in low rigidity cut-off discovered by Neher (see preceding Sect. 25) to occur between 1951 and 1954 appears to be an extension of the decrease in low rigidity cut-off found by Meyer and Simpson to occur between 1948 and 1951. These authors conclude that neither known terrestrial magnetic fields, assumed geoelectric fields, nor a solar magnetic dipole (even if changing with time) could produce the observed effects. They suggest that solar system fields may be found which will prevent low energy particles present within the galaxy from entering the solar system near the earth's orbit.

IV. 27-day variations.

27. Waves of 27-days period in cosmic-ray intensity, magnetic activity and horizontal intensity. Figs. 39a and b are harmonic dials [52] for the departures of individual 27-day waves from the average 27-day wave for the daily American magnetic character figure (from 0.0 for very quiet to 2.0 for intense storm), and for cosmic-ray intensity at Huancayo, respectively. Figs. 39c and d were obtained, respectively, by rotating the vectors in a to vertical and by rotating vectors for corresponding intervals in b (of 27 days) through the same angle. Statistical tests show that the probability, P, of obtaining an average vector as large or larger than that in Fig. 39d in a sample of 34 vectors from a population in which the components of the vectors are independent and random, with standard deviations estimated from the 34 vectors in d is only about 2×10^{-6}. This indicates that the vectors in a as well as in b definitely tend to have similar phases. Furthermore, the phases in c and d of Fig. 39 indicate that the phases in a and b are opposite, that means that the maxima of the 27-day waves in cosmic-ray intensity tend to occur near the minima of the 27-day waves in magnetic activity, as measured by the American character-figure.

Fig. 40 indicates the results of a similar comparison between magnetic horizontal intensity and cosmic-ray intensity at Huancayo. For c of Fig. 40, the probability, P, is 7×10^{-5}, indicating correlation between the phases of the vectors in a and b of Fig. 40. Here it will be noted that the maxima of the waves in cosmic-ray intensity and those in horizontal intensity tend to be in phase, which is consistent with the results in Fig. 39, since low horizontal intensity occur at times of high magnetic activity.

28. World-wide and quasi persistent nature of 27-day waves. Using a procedure similar to that in Figs. 39 and 40 the 27-day waves were shown [52] to be

Fig. 39a—d. Harmonic dials for departures from average of 27-day waves in American character-figure (a) and in cosmic-ray intensity at Huancayo, Peru (b), computed from 34 rotations (intervals of 27 days) beginning June 13, 1936; harmonic dials (c) and (d) obtained respectively by rotating vectors in (a) to vertical and by rotating vectors for corresponding intervals in (b) through the same angle.

Fig. 40a—c. Harmonic dials for departures from average of 27-day waves in magnetic horizontal intensity at Huancayo, Peru (a), computed from 34 rotations (intervals of 27 days) beginning June 13, 1936; harmonic dials (b) and (c) obtained respectively by rotating vectors in (a) to vertical and by rotating vectors for corresponding interval in (b) of Fig. 39 through

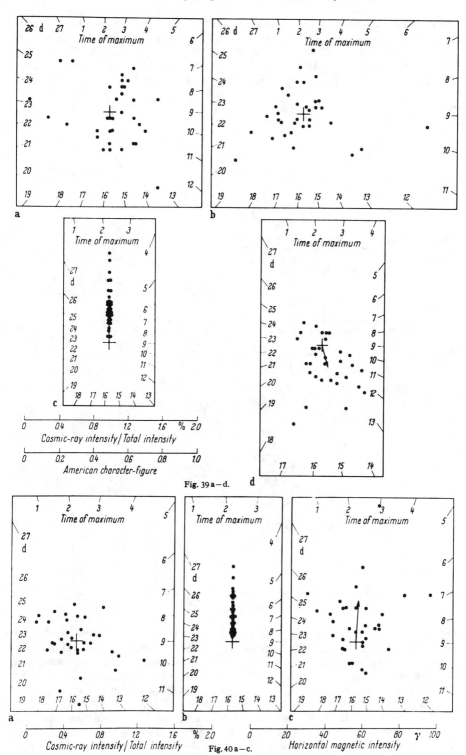

Fig. 39 a—d.

Fig. 40 a—c.

363

correlated at Cheltenham, Huancayo, and Christchurch in a sample of 34 intervals of 27-days. Figs. 41 and 42 indicate the 27-day waves at Cheltenham, for 158 rotations (intervals of 27 days) and at Huancayo for 169 rotations. These figures exhibit the range in amplitude (and phases) obtained from individual rotations. The large circles are the so-called probable error circles, inside which fall about

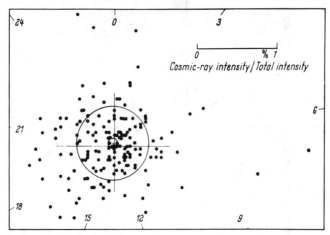

Fig. 41. 27-day harmonic dial for cosmic-ray intensity at Cheltenham, 158 rotations.

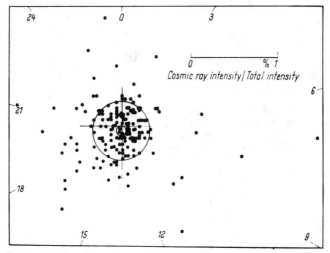

Fig. 42. 27-day harmonic dial for cosmic-ray intensity at Huancayo, 169 rotations.

half the points. Their radii for Figs. 41 and 42 are respectively about 0.45% and 0.35%. The larger radius for Cheltenham is doubtless due to the fact that at Cheltenham the variability in the 27-day waves due to μ-meson decay effects is greater than at Huancayo. The small circles in the figures are the so called probable error circles of the *average* 27-day vectors. In both cases the averages are so small that no statistically real persistent 27-day wave is indicated.

The correlation between the 27-day waves summed for 5 successive rotations, at Huancayo and Cheltenham is shown in Fig. 43 as derived from a total of 155 rotations. Incidentally a test made [52] for a *persistent* 27-day wave in the difference of 27-day waves, Cheltenham minus Huancayo, indicated nothing large enough to be regarded as statistically significant. Such a wave would only be expected from a sufficiently large solar magnetic dipole inclined to the sun's axis of rotation.

The quasi persistent nature of the 27-day waves is evident in Fig. 44 which shows the BARTELS' [53] characteristic diagram as derived from the data of Figs. 41 and 42. In a harmonic dial let the two-dimensional standard deviation of points about their average be M, then if the deviations for a successive chronological sequence of points are statistically independent, the standard deviation $M(h)$ for means of h successive points has the expected value $M(1)/\sqrt{h}$, or with $c(h) = M(h)\sqrt{h}$, $c(h)$ has the expectation $M(1)$. For pure persistence, or complete lack of independence, $c(h)$ increases linearly with $\sqrt{h}$. For quasi persistence $c(h)$ at first increases with $\sqrt{h}$ but for larger values of $\sqrt{h}$, approaches asymptotically a constant value, indicating that for sufficiently large h the means of h successive deviations are essentially independent. In Fig. 44 it is evident that the ordinates, which are proportional to $c(h)$ increase little for h greater than about 3 or 4. Thus the 27-day waves are quasi-persistent.

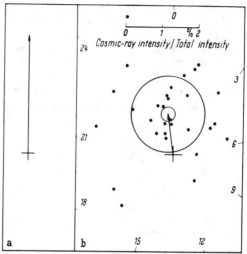

Fig. 43a and b. Correlation between 27-day waves in cosmic-ray intensity at Huancayo and Cheltenham; (a) average vector (sum for 5 successive rotations) for Huancayo after rotating each to vertical, (b) harmonic dial for Cheltenh; m after turning each vector (sum for 5 rotations) for corresponding interval through same angle as for (a).

29. 27-day changes in the nucleonic component and its latitude variation. Fig. 32 (p. 192) showed the variation in daily mean values from the neutron

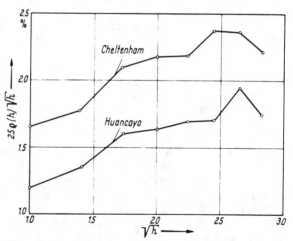

Fig. 44. Radius $\varrho(h)$ of probable error circle for means of h successive 27-day rotations as function of $\sqrt{h}$ indicating quasipersistence in 27-day waves.

monitor at Climax. FONGER [48] determined the autocorrelation function, $r(T)$ for the Climax neutron curve of Fig. 32. From unity with $T=0$ (T is the lag in days), $r(T)$ decreases to a minimum of about -0.4 for T near 14 days. From $T>14$ days $r(T)$ rises to a maximum of $+0.4$ for $T=28$ days after which it decreases to a second minimum of about -0.4 near $T=42$ days. While the series of data is rather short for such purposes $r(T)$ clearly indicates the 27-day recurrence tendency.

A most important contribution to the better understanding of the 27-day variation has been made by Simpson [54] who made an extensive series of measurements on the latitude variation of the nucleonic component at atmosphere depth of 312 g cm^{-2}. The measurements were made in a type RF-80 jet aircraft with which data over the geomagnetic latitude, λ, interval 40° to 53° (where the change of intensity with latitude is large) could be obtained in a few hours. However, complete latitude curves were obtained over a fixed route between $\lambda = 40°$ and 65° N. Referring to Fig. 32 the vertical arrows show the dates on which one of the series of latitude flights was made. It will be seen that the latitude flights were planned to take place near maxima and minima on the curves of daily means from neutron ground station monitors. In this way the flights were made

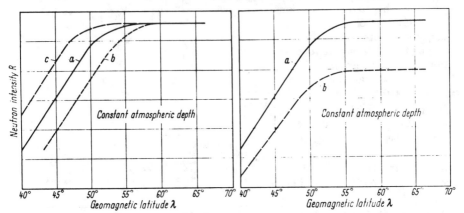

Fig. 45. The predicted behavior of neutron intensity as a function of latitude based upon the assumption that the primary cosmic-radiation intensity variation is produced by a *geomagnetic field variation* (Simpson).

Fig. 46. The predicted behavior of neutron intensity as a function of latitude based upon the assumption that the cosmic radiation intensity variation is produced by a *change in the primary flux with time* (Simpson).

at times between which there were large intensity variations. The purpose of the experiments was to determine whether the recurring 27-day variations are the result of changes in primary particle flux or of variations of the geomagnetic field.

Fig. 45 shows the latitude variation to be expected if the 27-day variation is due to variations of the geomagnetic field. Simpson [55], Fonger and Treiman have obtained the relationship between the counting rate R of a neutron detector located at atmospheric depth x, geomagnetic latitude λ and the vertical, differential primary flux $j_Z(p/Z, t)$ of particles of momentum p and charge Z. To relate these functions they defined the specific yield of neutrons as a function $S_Z(p/Z, x)$ which is experimentally determined from the time averaged parameters j_Z and R to yield the neutron counting rate at depth x arising from a unit flux of vertically incident primary particles of charge Z and rigidity p/Z. Thus, they found [55]:

$$R_v(\lambda, x, t) = \sum_Z \int\limits_{[p/Z]_\lambda}^{\infty} S_Z(p/Z, x)\, j_Z(p/Z, t)\, d(p/Z)$$

where $[p/Z]_\lambda$ is the cut off for vertical arrival at λ and $R_v(\lambda, x, t)$ is the counting rate due only to those primaries which arrive from the vertical direction per unit solid angle at time t. It was shown that R_v and the observed rate R are related in good approximation by the Gross transformation. A variation of R_v may thus be produced either by a variation of the lower limit of the integral which is determined by parameters of the geomagnetic field or by a variation of j_Z

366

in the integrand, namely, a variation of j_z with time. Thus in Fig. 45, a is a typical latitude curve, curve b shows approximately how the latitude curve will appear

for a variation $+ [\delta p/Z]_\lambda$ and curve c that for a variation $- [\delta p/Z]_\lambda$. The integral counting rate, R_v, is unchanged by variations of $[p/Z]_\lambda$ above the knee of the curve.

On the other hand, if the fast neutron latitude curve at time t is represented as curve a, Fig. 46, then if at time t_1 a variation occurs to produce a fractional change of intensity $- \delta R/R$, which variation for simplicity was made independent of latitude, then at t_1 the neutron latitude curve will appear as curve b, Fig. 46. For the special case where the function $S_z(p/Z, x)$ vanishes for finite value of $j(p/Z, t)$ the observed cutoff of the latitude curve will be determined by S rather than j, and again, if j undergoes a variation of intensity with time, the counting rate of the detector will change with time above and below the cutoff determined by S_z. From the equation above for R_v the maximum value for the integral at time t is obtained by setting the lower limit equal to the minimum particle rigidity $[p/Z]_{min}$ observed in the primary spectrum. Hence, for all values of $[p/Z]_\lambda < [p/Z]_{min}$ the integral is a constant. Now if $j_z(p/Z, t)$ undergoes a variation the observed counting rate will change for observations at latitudes corresponding to $[p/Z]_\lambda < [p/Z]_{min}$ as well as for latitudes corresponding to $[p/Z]_\lambda \geqq [p/Z]_{min}$. Thus, an observer within the atmosphere would measure a change in secondary particle intensity above and below the cutoff of the latitude curve.

Fig. 47 shows a typical set of latitude observations. As already indicated the data in the latitude interval

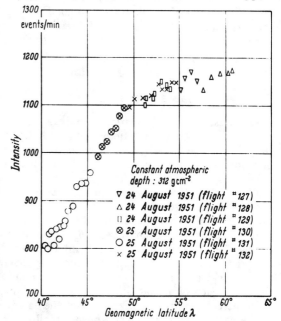

Fig. 47. The experimental neutron intensity data which are used to establish the latitude curve for 25 August 1951 (curve c in Fig. 48). The magnitudes of the standard deviations are given by the size of the flight indication symbols (SIMPSON).

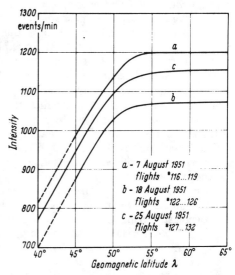

Fig. 48. The latitude curve c derived from Fig. 47 and similar curves (a) and (b). The extrapolated portions of the curves are dashed (SIMPSON).

$40°$ to $53°$, where $dR/d\lambda$ is large, were obtained within a few hours. The additional data required to construct the latitude curve, such as that between $\lambda = 55°$ to $65°$

are corrected for the intensity at the time of the flight between 40° and 55°. From other sets of data for August 7 and 18 similar to that in Fig. 47, the curves in Fig. 48 are derived. It is evident when Fig. 48 is compared with Figs. 45 and 46 that *the changes due to the 27-day variation are produced by changes of primary particle flux rather than by variations of geomagnetic field intensity.*

E. Solar flare effects.

I. Large increases of cosmic-ray intensity associated with solar flares.

30. Results from ionization-chambers. During some 17 years of continuous registration of cosmic-ray ionization with Compton-Bennett meters at several stations four unusual increases have been observed [56]. The sudden increase in cosmic-ray intensity which began at 10^{h}45^m GMT, November 19, 1949, was the largest yet recorded. The other three increases occurred on February 28, 1942, March 7, 1942 (observed also by Ehmert [56a] in Germany), and July 25, 1946. All were registered with Compton-Bennett ionization chambers completely shielded with 12-cm Pb. Three of the increases began during intense chromospheric eruptions or solar flares [56]. While no solar flare was actually observed during the increase of cosmic-ray intensity on March 7, 1942, a radio fadeout occurred very near the time the increase in cosmic-ray intensity began. The fadeout, which occurred only on the day-light side of the earth definitely indicates the occurrence of a solar flare. The terrestrial magnetic effect of such solar flares is an increase, on the daylight side of the earth, in the normal diurnal variation in the earth's field [57]. The known small diurnal variation [58] in cosmic-ray intensity excludes the possibility that the increases were due to changes in the earth's external magnetic field resulting from an augmentation of the magnetic diurnal variation. The evidence thus indicates [56] that the four increases in cosmic-ray intensity were probably due to charged particles accelerated by some mechanism [59] on the sun. Unless the particles responsible for the increases were charged, it would be difficult to explain either the simultaneous occurrence of the increases on both the daylight and dark hemispheres or the absence of the increases at the equator.

The sudden increases in cosmic-ray intensity on February 28 and March 7, 1942, are shown in Fig. 49, in which the curves are drawn through the bihourly means, after correcting these to constant barometric pressure. It is evident that neither increase occurred at Huancayo and that the increase on February 28 did not occur at Teoloyucan. The decrease in cosmic-ray intensity during the magnetic storm following the sudden commencement on March 1 is evident at all the stations.

The observations of the sudden increase of July 25, 1946 at Godhavn and Cheltenham is shown in Fig. 50. Again, no increase occurred at Huancayo, although the decrease during the subsequent magnetic storm is evident there. In Fig. 51 the increase observed [60] at geomagnetic latitude 88° N on July 25, 1946, with a Millikan-Neher electroscope is compared with that observed at Godhavn with the Compton-Bennett meter.

The sudden increase in cosmic-ray intensity at the time of the solar flare on November 19, 1949, is shown in Fig. 52. This is the first instance when an increase in cosmic-ray intensity accompanying a solar flare has been recorded at a mountain station and at sea-level stations. The increase, in percent of the total cosmic-ray ionization, is obviously very much greater at Climax than at Cheltenham. In fact, if the ordinates on the curve showing the increase at Cheltenham are multiplied by 4.2, the resulting points, shown in Fig. 53, lie on the curve for Climax. It may also be noted in Fig. 52 that the increase on November 19 was not followed

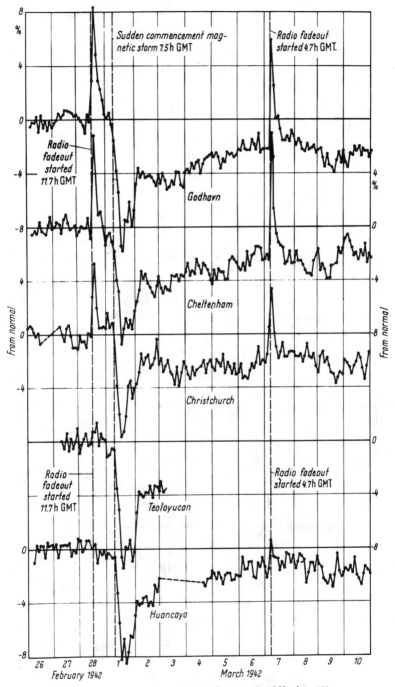

Fig. 49. Increases of cosmic-ray intensity, February 28 and March 7, 1942.

by a decrease in cosmic-ray intensity during the magnetic storm which began about 18ʰ GMT on November 19 (see also the discussion of events on that day in [60a]).

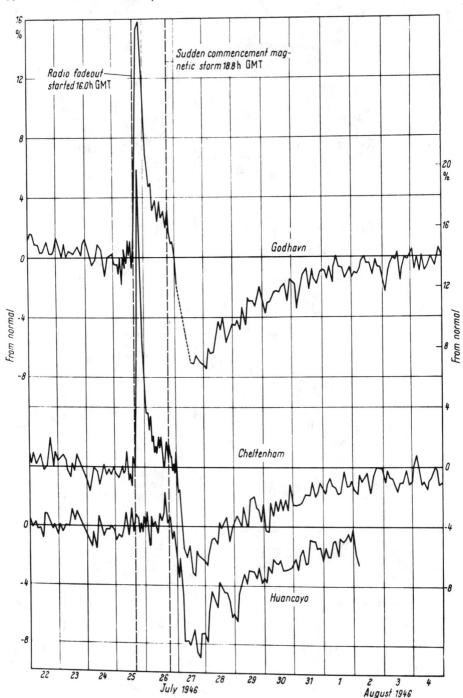

Fig. 50. Increase of cosmic-ray intensity, July 25, 1946.

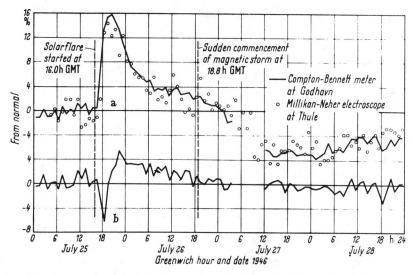

Fig. 51 a and b. Cosmic-ray ionization 1946, July 25—28. (a) Hourly means at Thule and at Godhavn, Greenland. (b) Hourly means at Godhavn less hourly means at Cheltenham.

In addition to complete shielding by 12-cm Pb, the meter at Climax, during the flare of November 19, 1949, was under a rectangular iron shield 4 ft long, 1 ft wide, and 16.5 cm thick. The absorption mean free path for nucleons of medium energy in iron is approximately 240 g cm^{-2}. Taking the dimensions of the shield into account and figuring the zenith angle distribution for a radiation exponentially absorbed with an absorption coefficient of about 145 g cm^{-2}, it is estimated that the increase at Climax on November 19, 1949, would have been 15% greater without the iron shield. Thus it is estimated [61] that the maximum of the increase on November 19 at Climax would have been about 207% without the iron shield, instead of the uncorrected 180% as shown in Figs. 52 and 53. This correction also results in a factor of 4.8 for the ratio of the percentage increase at Climax on

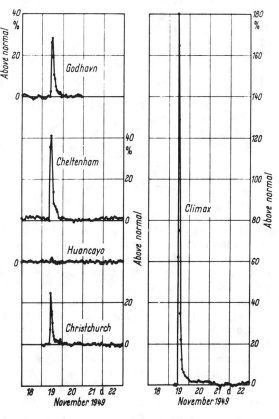

Fig. 52. Increase of cosmic-ray intensity, November 19, 1949.

371

November 19 relative to that at Cheltenham instead of 4.2 as is indicated in Fig. 53.

31. Flare increases due to nucleonic-component. Since the total ionization at Climax (under 12-cm Pb) is about 2.5 times that at Cheltenham, and since the percentage increase on November 19, 1949, was about 4.8 times that at Cheltenham, the actual magnitude of the increase on that date at Climax was about 12 times greater than at Cheltenham [61]. Since the difference in the atmospheric layer is equivalent to 340 g cm⁻², the radiation responsible for the increase

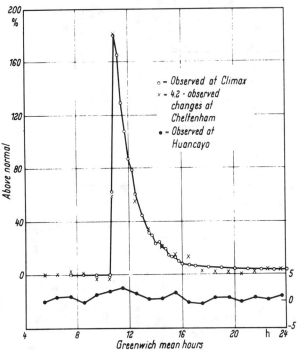

Fig. 53. Cosmic-ray intensity, November 19, 1949.

during the flare has an absorption coefficient of about 137 g cm⁻². This is just about the rate at which the nucleonic component responsible for star production in photographic emulsions, increases with altitude. The increase in total ionization under 12-cm Pb by a factor of 2.5 from Cheltenham to Climax is mainly due to mesons. It is thus evident that the magnitude of the flare effect increases too rapidly with altitude to be ascribed to ordinary mesons. The latitude effect in chambers under 12-cm Pb, due principally to mesons, is small, whereas the flare effect exhibits a strong dependence on latitude (being zero at the equator). This also indicates that ordinary mesons contribute negligibly to the flare effect.

The results on the latitude variation of the proton and neutron intensity suggested that the cross section for nucleon production relative to that for meson production decreases rapidly with increasing energy of primary particles. This is in accord with the conclusion that the increase in intensity during the solar flare of November 19, 1949, was due principally to the nucleonic component generated in the atmosphere by relatively low energy primaries, and not to ordinary mesons. At Climax, under 12-cm Pb, probably not more than about 10% of the total ionization is normally due to local radiation originating from the nucleonic component. If this radiation is produced entirely by particles in the same band of energy as those responsible for the increase of 207% in ionization on November 19, 1949, then the number of primary particles, reaching there per unit time, in that band of energy, must have increased to at least 20 times the normal value. Three sets of triple coincidence counters and one set of four-fold coincidence counters, located above the meter at Climax and arranged to record air showers, were in continuous operation during the period of the increase in ionization on November 19, 1949. There was no evidence of any significant increase in the rate of air showers during this period.

The conclusion in the last paragraph that the increase of cosmic-ray intensity on November 19, 1949 was due to the nucleonic component generated in the atmosphere by relatively low energy primaries and that the increase in this component must have been at least twenty fold is confirmed by the results in Fig. 54. These observations were obtained by ADAMS [62] with a neutron counter in a large "pile" of high purity graphite. The hourly mean counting rate increased to a maximum 550% above normal. This was about 12 times the corresponding maximum hourly increase in ionization recorded at Cheltenham, and about 40 times the maximum increase recorded on three fold-coincidence meson telescopes pointing north and south, 45° to the horizontal, at Manchester [62]. The increase in neutron intensity at Manchester was nearly 70 times that observed by CLAY [63] in an ionization chamber at Amsterdam. Data for the solar

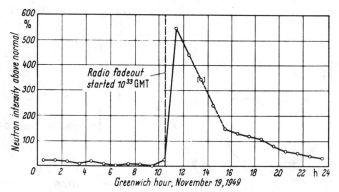

Fig. 54. Increase in neutron intensity at Manchester, November 19, 1949 (ADAMS).

flare increases which have been observed at many stations have been carefully summarized by SEKIDO and YOSHIDA [64] who from its analysis determined that the maximum of the flare-effect occurs between 3 and 8 hours local time. In two review articles ELLIOTT [65] and BIERMANN [66] have also summarized some of the observations of the four large flare effects.

II. Expected geographical distribution of solar-flare increases if these come from the sun.

32. Theoretical impact zones. When all the available observations of the four large solar-flare increases in cosmic-ray intensity are summarized, as it has for example been done by SEKIDO, and YOSHIDA [64], it is quite evident that the magnitude of the observed increase depends not only on latitude but also on the local time at which the increase occurred. EHMERT [67] by utilizing some of the trajectories of cosmic-ray particles integrated by STÖRMER showed that the distribution of the increase on February 28, 1942, observed in America but not in Germany, indicated positive particles arriving from the direction of the sun. SCHLÜTER [68] has integrated twenty trajectories of cosmic-ray particles using the method originated by STÖRMER, in which particles are assumed to be initially moving along a line parallel to the sun-earth line. He concludes from an examination of the impact points on the earth of these orbits that for positive particles from the sun there should be a sharp maximum near 0900 local time, and he regards the observation of one of the large increases in Germany in the afternoon as in disagreement with the theory. Using STÖRMER's Nullbahnen, i.e., orbits which would pass through the dipole if extended, as representative of all orbits striking the earth, Schlüter discusses the seasonal change in the impact points on the earth for particles from the sun, as well as the existence of forbidden zones near the poles of the earth for solar particles. The observation of the increases at Godhavn, Greenland ($\lambda = 80°$) he also regards as in disagreement with the theory, since this station lies in such a forbidden region (see also Fig. 51).

Firor [69] has estimated the geographical distribution of the solar flare increase in cosmic-ray intensity. He showed the existence of a background or latitude zone within which the magnitude of the increase would have no strong dependence on local time. He also made rough estimates of the relative intensities in the different zones, within which the magnitude of the increase depends strongly upon local time. In addition Firor considered the effect of a finite source larger than the visible sun. To determine the latitude and longitude of the impact zones for particles arriving from a point source Firor utilized the available data for about 80 orbits which arrive vertically thus insuring that the results will be of greatest importance for detectors deep in the atmosphere for which vertically incident particles are most effective. The trajectories

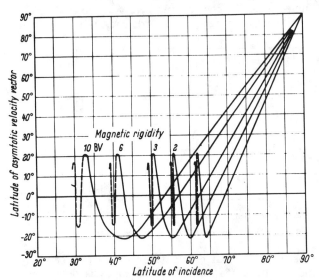

Fig. 55. Geomagnetic latitude of impact on the earth of particles arriving vertically with magnetic rigidities of 1 — 10 BV as a function of the geomagnetic latitude of the source (Firor).

used involved those calculated by Störmer [70], Dwight [71], and Schlüter [68], together with those obtained from the model experiments of Malmfors [72] and Brunberg [73]. To complete the picture required interpolation between the known orbits and in some cases extrapolation beyond the known orbits. In this way curves in Fig. 55 were obtained for the geomagnetic latitude of impact points on the earth for particles arriving vertically with each of several different magnetic rigidities between 1 and 10 BV, as a function of the geomagnetic latitude (latitude of asymptotic velocity vector) of the source. The lower limit of the latitude at which flare increases have been observed indicates that few of the particles involved had magnetic rigidities of more than 10 BV, corresponding to cutoff at geomagnetic latitude 30°.

Curves obtained for the geomagnetic longitude of impact points on the earth, for magnetic rigidities between 1 and 10 BV, as a function of the source latitude are shown in Fig. 56. Here it will be noted that for rigidities between 1 and 10 BV the curves do not depend critically on rigidity and they form a narrow band about one hour wide. For particles from the sun these zones will be several hours wide. Fig. 56 which does not involve rigidity as a parameter greatly facilitates the construction of impact zones for various situations. From Fig. 55 the latitudes on the earth of possible impact points are readily obtained for a particular rigidity and for a particular source latitude. For the same latitude of the source the long-itudes on the earth of the impact points are readily obtained from Fig. 56.

In Fig. 57 are plotted as smooth curves the impact points in the nothern hemisphere for particles leaving a point source far from the earth and in the plane

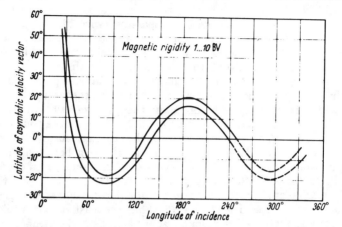

Fig. 56. Geomagnetic longitude of impact on the earth of particles with magnetic rigidities of 1—10 BV. Curves shown are the envelope of the individual curves for each rigidity (FIROR).

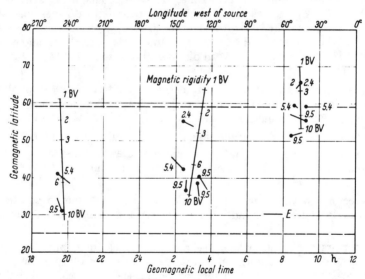

Fig. 57. Impact points on the earth for particles arrpoaching in the equatorial plane. The solid lines show the impact points for particles arriving vertically. The circles give the impact points for some orbits which arrive with large zenith angles. The line attached to each of the latter shows the azimuth and zenith angles of arrival. This line is the projection on the earth's surface of the velocity vector, length "E", at the time of arrival. The longitude scale in hours local time assumes the source of particles to be at noon local time. The dashed lines indicate the band of latitudes which is filled with impact points of 1 to 10 BV particles (FIROR).

of the geomagnetic equator, and arriving at the earth with vertical or near vertical incidence. Also shown are the impact points of some orbits integrated by SCHLÜTER which arrive with large zenith angles. The line attached to each SCHLÜTER orbit in Fig. 57 is a projection on the earth's surface of the velocity vector of the particle at the time of arrival. The length marked "E" in the figure

14*

gives the reference length of the velocity vector. Since the source is to be identified with the sun the lower scale is in hours, with the source at local noon.

In Fig. 57 the impact points fall into three groups. The group near 0900 local time, results from orbits that remain north of the equatorial plane. The group at 0400 results from orbits that pass once through the equatorial plane, and the group at 2000 comes from orbits passing twice through the equatorial plane, and so on. There are in fact for a particular source latitude, within limits, an infinite number of intersections with the impact curve of Fig. 56, since as STÖRMER [74], [75] pointed out this curve oscillates through the ordinate 0° for increasing longitudes. This results in groups of impact point, not shown in Fig. 57, which have latitude distributions similar to the group at 2000 hours. Fig. 57 was constructed for the source in the geomagnetic equatorial plane. Figs. 55 and 56 may be used to estimate impact points for the source not in the equatorial plane. Moving the source south of the equator causes the two morning impact groups to move closer together, in the nothern hemisphere, while the evening group occurs earlier. This is reversed if the source is moved north of the equator. In this way the effect of a finite particle source upon the impact points can be calculated. For a source 10° in diameter the lines of Fig. 56 become zones about an hour wide. The zone at 0900 has the least spread, while the zone at 2000, and presumably all the similar zones are the widest. Since each of the latter is at a greater longitude than the preceding one, this infinity of impact zones fills the whole range of local times between geomagnetic latitudes 25° to 60° for 1 to 10 BV particles. This results in three distinct zones one at 0900, and one at 0400, and the background zone. FIROR [69] finds that the expected magnitude of the increase for the above three zones, respectively, should be roughly in the ratio: 7:2:1 for a source 15° in dimᵃter and in the ratio 7:3:1 for a source 30° in diameter.

The theoretical *results* thus far obtained indicate that if the sun, while near the geomagnetic equator, emits a pulse of particles, with rigidities up to 10 BV, the following effects should have been recorded by identical detectors:

1. Detectors located at latitudes less than 25° would receive no new particles.

2. Detectors between 25° and 35° would record an increase in counting rate. This increase would be due to the background zone and would have no strong longitude (local time) dependence.

3. Detectors above 35° would see an increase as in 2, but in addition those detectors at local times around 0400 would see an increase about three times as large as in 2.

4. Detectors above 50° would see an increase as in 2, but detectors near 0400 or 0900 would see additional increases, up to seven times as large as in 2. The highest latitude at which detectors would see increases, either the background type as in 2 or the strongly local time dependent increase as in 3 and 4, would depend on the lowest rigidity particles to which the detector was sensitive.

33. Comparison with observations. The observed increases were compared with the predicted distribution by constructing an impact zone diagram for the position of the sun at the time of the increase, and marking on the diagram the location of stations reporting the increase.

Figs. 58, 59, and 60 are the resulting diagrams respectively for the flare increases which occurred November 19, 1949, February 28, 1942 and March 7, 1942. The stations reporting one or more of the flare events are summarized in the following table which includes references to reports of the events:

In Figs. 58, 59, and 60 the density of the crosshatching indicates roughly the relative intensities to be expected in the different zones. The lack of symmetry

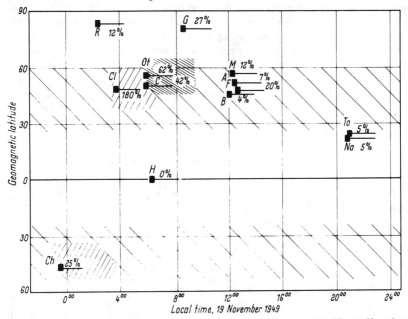

Fig. 58. Impact zones on the earth at the time of one of the large cosmic-ray increases — 1100 UT on 19 November 1949. The cross-hatching indicates the positions of the zones with the density of crosshatching giving roughly the relative intensities predicted for the different zones. The cosmic-ray stations are indicated by solid squares; the lines attached to each station show its motion during the increase. Near each station is an identifying letter (see Table 3), p. 215 and the percentage increase observed (Firor).

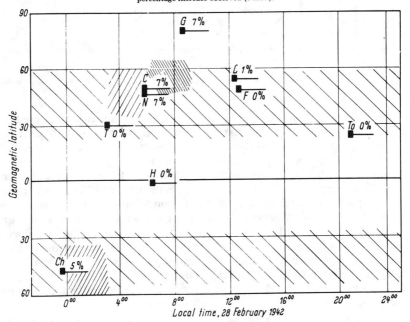

Fig. 59. Impact zones and observations at cosmic-ray stations at 1100 UT on 28 Febrauary 1942. Symbols same as in Fig 58 (Firor).

between the zones in the northern and southern hemispheres is an example of the seasonal effect mentioned earlier. The small squares are the positions of the cosmic-ray stations reporting the increase (or its absence), with the line attached to each indicating its motion during the increase. Near each station is marked the percentage increase observed using for the most part short time intervals (5 to 15 min). The zones were constructed assuming a source 15° in diameter.

In Fig. 58 there is evident a tendency for the stations in the morning impact zones to show the largest increase. In particular the station at Ottawa, in the 0900 zone, recorded an increase about five times as large as that at Manchester,

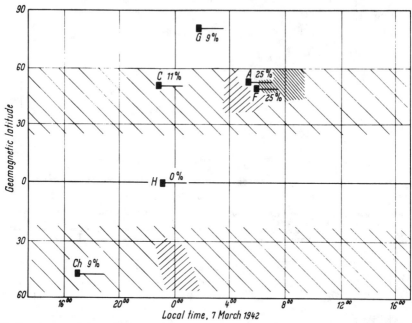

Fig. 60. Impact zones and observations at cosmic-ray stations at 0400 UT on 7 March, 1942. Symbols same as in Fig. 58 (Firor).

located in the background zone. This is consistent with the ratio seven predicted above. The 0900 and 0400 zones may be compared by using Cheltenham and Christchurch, giving a ratio of about two, consistent with the predicted 7:3. The 20% increase at Freiburg seems large compared with the other European observations and is probably due to the fact that this figure is derived from a very short time interval at the peak intensity. Similarly, the 7% at Amsterdam is a one hour average and should be made somewhat larger for comparison with other stations. Two German stations, not shown for lack of room, gave increases of 15% for each of two similar counter telescopes and 17% for a shielded ion chamber. The very large increase at Climax can in part be attributed to the high altitude of that station. A reasonable altitude correction to sea level lowers the percentage increase to about the same value as at Cheltenham [61]. The high latitude stations, Resolute and Godhavn, recorded increases although they were not in any impact zone.

Fig. 59 shows the impact zones and results observed on February 28, 1942. In this case the increases marked in the figure are for one hour averages centered on the half hour. It is seen (with Godhavn again the possible exception) that the distribution of the increase is again consistent with positive particles arriving at the earth from the direction of the sun. All of the increases for this event began roughly an hour after the flare began. The ratio of the increases in and out of the morning impact zones is, in this case, seven or greater.

Table 3. *Observations of large cosmic-ray increases*

Station	Symbol used in Figures	λ (degrees)	Altitude (a) (meters)	References			
				28 February 1942	7 March 1942	25 July 1946	19 November 1949
Amsterdam	A	54 N		b	b	c	d
Bagneres	B	46 N	550	—	—	—	e
Bargteheide		54 N		—	—	—	f
Cheltenham	C	50 N		g	g	g	g
Christchurch	Ch	48 S		g	g	—	g
Climax	Cl	48 N	3500	—	—	—	g
Darmstadt		50 N		—	—	—	f
Freiburg . . . . }	F	48 N		h	h	—	f
Friedrichshafen . . . . . . }							
Godhavn	G	80 N		g	g	g	g
Huancayo	H	1 S	3350	g	g	g	g
London	L	54 N		i	i	—	—
Manchester	M	57 N		—	—	j	k
Nagoya	Na	25 N		—	—	—	l
Norfolk	N	49 N		m	—	—	—
Ottawa	Ot	56 N		—	—	—	n
Resolute	R	83 N		—	—	—	n
Teloyucan	T	30 N	2285	g	—	—	—
Tokyo	To	25 N		o	o	—	p
Weissenau		49 N		—	—	—	f
Mt. Wilson		43 N	1800	—	—	q	—
Thule		88 N		—	—	r	—

a Altitude listed only for mountain stations.
b CLAY, JONGEN and DIJKER: Proc. Kon. Ned. Akad. Wet. **52**, 923 (1949).
c CLAY, J.: Proc. Kon. Ned. Akad. Wet. **52**, 899 (1949).
d CLAY, J., and H. F. JONGEN: Phys. Rev. **79**, 908 (1950).
e DAUVILLIER, A.: C. R. Acad. Sci., Paris **229**, 1096 (1949).
f MÜLLER, R. et al.: J. Atmosph. Terr. Phys. **1**, 37 (1950).
g FORBUSH, S. E., M. SCHEIN and T. B. STINCHCOMB: Phys. Rev. **79**, 501 (1950).
h EHMERT, A.: Z. Naturforsch. **3a**, 264 (1948).
i DUPRERIER, A.: Proc. Phys. Soc. Lond. **57**, 468 (1945).
j DOLBEAR, D. W. N., and H. ELLIOT: Nature, Lond. **159**, 58 (1947).
k See references [62] and [63].
l SEKIDO, KODAMA and YAGI: Rept. Ionos. Res. Japan **4**, 207 (1950).
m BERRY, E. B., and V. F. HESS: Terr. Magn. Atm. Electr. **47**, 251 (1942).
n ROSE, D. C.: Canad. J. Phys. **29**, 227 (1951).
o NISHIMURA, J.: J. Geomag. Geoelectr. **2**, 121 (1950). — SEKIDO, YOSHIDA and KAMIYA: Rept. Ionos. Res. Japan **6**, 195 (1952).
p MIYAZAKI, WADA and KONDO: Rept. Ionos. Res. Japan **4**, 176 (1950).
q NEHER, H. V., and W. C. ROESCH: Rev. Mod. Phys. **20**, 350 (1948).
r GRAHAM, J. W., and S. E. FORBUSH: Phys. Rev. **98**, 1348 (1955).

In Fig. 60 is shown the result, based on hourly averages, for the flare of March 7, 1942. Here also the largest increase occurred in the morning impact zone.

For the increase on July 25, 1946 no results were reported from a station in the morning impact zones at this time. The increase was seen by five stations (see Table above) with roughly the same amplitude of about 15%. No increase was seen at Huancayo. These observations again agree with the predicted distribution.

Thus from Figs. 58, 59, and 60, it is evident that the results for middle and low latitude stations are consistent with predictions based on the assumption that the flare increases resulted from particles with magnetic rigidities of less than 10 BV coming from a source about 15° in diameter, centered at the sun. However, the increases at high latitudes reported for Resolute, Godhavn, and Thule (see Fig. 51) are not consistent with the predictions. The explanation for the increases at these high latitudes has yet to be found.

34. Effects of small flares. Since the model described above gave reasonably consistent agreement with observations for large solar flare increases it was used by Firor [69] as a basis for investigating the question of whether small increases might also be statistically detected during small solar flares. The model was used to determine the range in local time during which the flare effects at Climax would be expected to have the greatest amplitude. Data from the Climax neutron monitor were used since these are the most sensitive to the low energy particles from flares. The reported flares, with importance 1+ or more, were dichotomized so that the first group contained flares the starting time of which occurred when Climax was in a morning impact zone, which was considered four hours wide. The second group contained the flares which started when Climax was at least an hour away from the edge of a morning impact zone. The cosmic-ray data were averaged on "flare time" for each group. Some indication was obtained for an increase of 0.8% lasting an hour or so during the time of occurrence of 12 flares in 1951 in the first group. An increase of about 0.5% was similarly indicated for four flares of 1952 which were in the first group. In neither case was an increase evident in the second group. In neither case for the first group was the increase greatly in excess of the indicated standard deviations and the exact, manner in which these were obtained is not described.

Firor [69] also compared the diurnal variation in neutron intensity at Climax on days with large flare indices with that for days with small flare indices. The results indicated, for days with the larger flare indices, a slightly greater average intensity during the period when Climax was in a morning impact zone. Firor also found a correlation of 0.6 between the flare index and the size of the cosmic-ray increase from a sample of 16. Thus, the probability is about $1/_{50}$ that the sample was from a population with zero correlation.

While these results suggest the possibility of small increases during small solar flares the effects are not so large relative to the variability as to be thoroughly established without more rigorous tests of their statistical reality.

F. Summary of results up to the Geophysical Year.

35. Observational procedures and meteorological effects. Experience has shown that ionization chambers have the merit of simplicity and reliability for registering changes in cosmic-ray intensity over long periods of time. This advantage is offset to a considerable extent by the influence of meteorological factors on the recorded intensity. In addition to barometric effects such measurements are affected by the height of the region of μ-meson production, where the barometric pressure is about 100 mb. Since this height varies seasonally, corresponding more or less regular seasonal variations are introduced into the measurements. Irregular variations in this height during the passage of meteorological fronts also affect the measurements. Unless meteorological data are regularly available to heights of 16 km or more, no reliable corrections for these latter effects are possible. Near the equatorial zone these effects are quite small. Geiger-counter apparatus is similarly affected. Neutron monitors, on the other hand, are affected only by barometric pressure and while the pressure coefficient is large, the advantage of freedom from other meteorological influence is great. In addition neutron monitors are more sensitive to changes in the lower energy portion of the primary cosmic-ray spectrum. While the time variations from neutron monitors are generally similar to those found from ionization chambers the magnitude of percentage changes is generally four of five times larger for the former.

36. Time variations. The solar 24-hour diurnal variation has been established Its amplitude undergoes world-wide changes. A large world-wide secular change in its phase is evident. There are indications that the amplitude of the diurnal variation changes with magnetic activity but the mechanism involved is not understood. The 24-hour sidereal variation is extremely small, and since an apparent sidereal variation could arise from other causes it is not regarded as significant.

Large decreases in cosmic-ray intensity occur during some magnetic storms but storms also occur without decreases in cosmic-ray intensity. Furthermore, there are some decreases in cosmic-ray intensity without marked magnetic activity. Nevertheless, in a statistically significant majority of cases, the cosmic-ray intensity is less on the five international magnetically disturbed days of each month than on the five quiet days. The variability of the daily means of cosmic-ray intensity is significantly less near sunspot minimum than near sunspot maximum. There is a world-wide sunspot cycle variation in cosmic-ray intensity with maximum near sunspot minimum. At high altitudes and near the geomagnetic pole, particles of lower energy are detected near sunspot minimum than at other times, and there is evidence that the "knee" of the latitude curve moves northward as the sunspot numbers approach the minimum.

A world-wide quasi-persistent (q.p.) 27-day variation in cosmic-ray intensity is established with minima near the times of maximum of the 27-day (q.p.) wave in magnetic character figure. These times of minimum are near those for the 27-day (q.p.) waves in the horizontal component of the earth's magnetic field at the equator. The character of the latitude variations, measured with neutron detectors in jet planes at times of maximum and minimum of the 27-day (q.p.) variations shows that these variations result from changes in primary particle flux.

Sudden increases in cosmic-ray intensity during a few chromospheric eruptions are caused by charged particles evidently accelerated by some solar mechanism. The large variation in observed intensity of these events with geographical position is generally consistent with calculation, if the particles (magnetic rigidity up to 10 BV) are assumed to come from the neighborhood of the sun. On the other hand, the increases observed near the geomagnetic poles during these events are not yet explained. Finally, there is some inconclusive evidence that increases in cosmic-ray intensity may occur during small solar-flares or chronospheric eruptions.

Thus most of the established time-variations in cosmic-rays are the direct or indirect result of solar activity, and a better comprehension of the mechanisms responsible will doubtless advance understanding of other phenomena related to solar activity. The results of observation and experiment in many related fields of geophysics executed in the International Geophysical Year (July 1957 to December 1958) have provided valuable material for interpreting the mechanisms which cause the time variations in cosmic-rays; this will be discussed in the Second Part.

Second Part:

Results obtained in the International Geophysical Year and afterwards.

37. Introduction. The first part reviewed in some detail much of what had been learned about temporal variations of cosmic-ray intensity until about the end of 1955 from the time of the discovery of cosmic radiation, which resulted from the historic balloon flight made by V. F. HESS [*76*] fifty years ago on August 17, 1912. Since 1955 research activities concerned with temporal variations and other geophysical aspects of cosmic radiation have increased enormously. This was due principally to the interest and opportunities generated by the *International Geophysical Year* (IGY), July 1957 to December 1958, and the *International Geophysical Cooperation Year* (1959). Many of these activities have continued and will continue during the forthcoming *International Quiet Sun Years* (IQSY), planned for the 24 months January 1964 to December 1965.

During the IGY cosmic ray measurements were carried out at 128 stations. The stations [*77*] included: 67 with neutron monitors; 31 ionization chambers; 67 with meson telescopes plus 18 with narrow angle telescopes; 13 using emulsions; 8 using balloons, aircraft, or rockets; and several stations using other instruments. The location of these stations is given in [*77*]. In addition, *artificial earth satellites* and *space probes* have since provided opportunities for measuring cosmic-ray intensity in new environments. Many of the IGY and IGC cosmic-ray data are available from the IGY world data centers where all observational results have been collected.

For the IGY the choice of a period near an expected maximum in the cycle of solar activity was especially fortunate, since during this interval sunspot numbers and magnetic activity reached the highest values yet recorded. During the IGY and since, the results obtained in each of several areas of investigation have provided results important to the understanding of phenomena in other related areas. It is now established that most, if not all, of the temporal changes of cosmic-ray intensity are the consequence of solar phenomena. *The solar cycle variation in cosmic-ray intensity and the so-called Forbush decreases, which generally are associated with magnetic storms, both arise from electromagnetic phenomena which originate on the sun and modulate or change the primary cosmic ray intensity.* Electromagnetic phenomena associated with *solar flares* also accelerate charged particles which occasionally increase the apparent cosmic ray intensity in ground level detectors. At high altitude and high latitude where effects from lower energy solar particles can be measured, such *solar flare events* are much more frequent. The number density and energy spectrum of these "solar cosmic-ray particles" has been measured during several of such events. The energy spectrum of primary cosmic-ray particles has been determined at different epochs of the solar cycle and also during Forbush decreases of intensity. These decreases have been observed in interplanetary space showing that they are not due to the geomagnetic storm field. Programs, for the study of cosmic-ray intensity variations and their causes, initiated during the IGY and subsequently, have led to the discovery of the *Van Allen geomagnetically trapped radiation belts*, and to the discovery of *strong bursts of X-rays coincident with visible zenith aurora*.

G. Solar cycle variations.

I. Results from ionization chambers and neutron monitors.

38. Comparison with sunspot numbers. The longest series of continuous registration of cosmic-ray intensity is that obtained from ionization chambers [78]. Fig.61 shows the monthly means of cosmic-ray ionization at Huancayo,

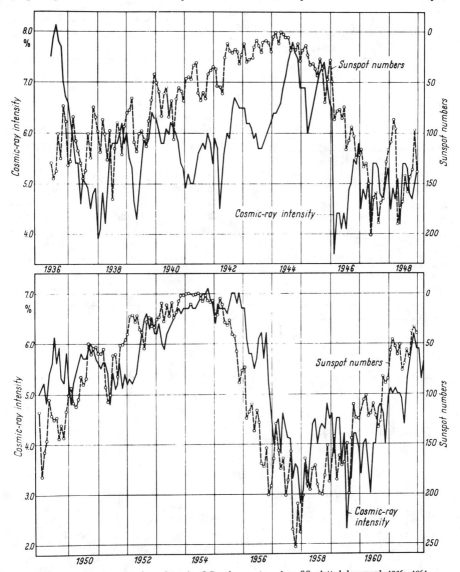

Fig. 61. Monthly means cosmic-ray intensity C-R and sunspot-numbers SS, plotted downward. 1936—1961.

and monthly mean sunspot numbers for the period from June 1936 to March 1962. For this interval the correlation coefficient between annual means of sunspot numbers, (SS), and cosmic-ray intensity, $(C-R)$ is -0.85. Fig.61 indicates, at least in 1955, that following sunspot minimum the increase in *SS* numbers continues for some months without an accompanying decrease in *C-R* at Huancayo.

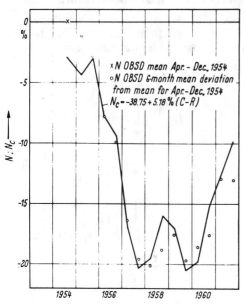

Fig. 62. Six-month means neutron intensity, *N*, observed at Ottawa (Rose) and computed, N_c, from cosmic-ray *C-R* ionization at Huancayo. April 1954 to December 1961.

Fig. 62 shows the variation in 6-month means of *C-R* ionization at Huancayo and of neutron intensity at Ottawa. After 1954 the neutron intensity at Ottawa appears to start decreasing some months earlier than the ionization at Huancayo.

39. Variations for quiet and disturbed days. Fig. 63 shows the variation, of cosmic-ray ionization at Huancayo from 1937 to 1961 for international magnetically quiet and for international magnetically disturbed days (five each per month). The annual mean cosmic-ray intensity, (*C-R*), at Huancayo is practically always less for disturbed days although the solar cycle variation is similar for disturbed and quiet days. Also shown in Fig. 63 are the yearly means of the southward geomagnetic field of the equatorial ring current field, ERC, for 1939 to 1945 from the values

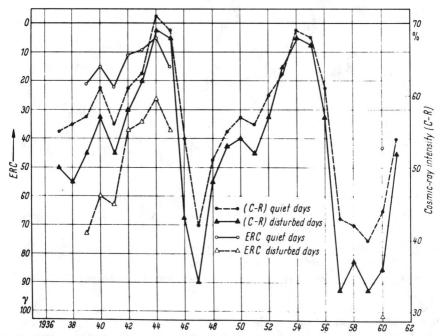

Fig. 63. Yearly means for magnetically quiet and disturbed days for *C-R*, 1937—1961 and for ERC 1939—1945 and 1960.

published by Kertz [79] and for 1960, similarly derived. Fig. 63 indicates that relative to the amplitude of the solar cycle variation for quiet days the difference, disturbed minus quiet days, is several times larger for ERC than for cosmic-ray intensity. The large decrease in *C-R* from January to February 1946, in Fig. 61, is conspicuous for the fact that the monthly mean in February

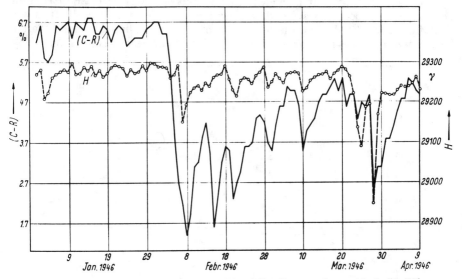

Fig. 64. Daily means (75° West Meridian Time) *C-R* and *H* at Huancayo. January 1 to April 9, 1946.

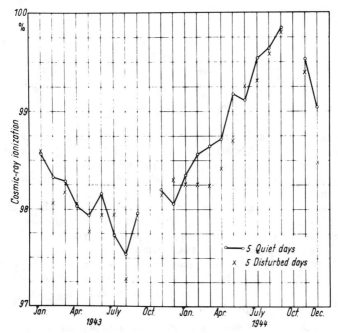

Fig. 65. Means cosmic-ray ionization at Huancayo for 5 quiet and for 5 diturbed days by months for 1943 and 1944.

385

was the lowest until 1957, and for the fact that a monthly mean as high as that for January 1946 did not occur thereafter until 1952.

The exceptional decrease in cosmic-ray intensity in Februray 1946 is shown in greater detail in Fig. 64 in which daily means are plotted together with those for horizontal magnetic field, H, at Huancayo Peru, for the period January 1 to April 9, 1946. In Fig. 64 it is evident that the conspicuous decrease in H from February 6 to 7 occurred a few days *after* the large decrease in C-R. It is not clear from Fig. 64, whether the low value of C-R prevailing for months after February 8, 1946 was due to a sequence of independent decreases (each occurring

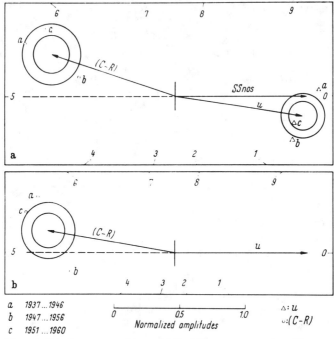

Fig. 66a and b. Harmonic dials 10-year waves. (a) u and C-R relative to sunspot numbers (SS). (b) C-R relative to u.

before the post perturbation recovery from preceding decreases) or whether the initial decrease in early February was the result of some large scale solar plasma emission which resulted in a depression of intensity, which even without the subsequent superimposed decreases having faster recoveries, would have continued for some months.

Fig. 65 shows the means of cosmic-ray intensity for the five quiet and for the five disturbed days in each month in 1943 and 1944. The steady rise of intensity from November 1943 to September 1944 appears definitely not be to the consequence of recovery from storm time decreases after August 1944 since the latter are quite small. The slow increase from November 1943 to September 1944, which continues in spite of whatever small stormtime decreases occurred during the interval, indicates a large scale mechanism the effect of which changes very slowly, with time constant of the order of a year.

40. Ten-year waves in cosmic-ray intensity, sunspot numbers, and magnetic activity. Fig. 66 shows harmonic dials for 10-year waves in cosmic-ray intensity (C-R), at Huancayo, SS numbers, and geomagnetic activity as measured by

BARTELS [80] *u*-measure. In Fig. 66a times of maxima for the 10-year waves in (*C-R*) and *u* are relative to that for the weighted mean for *SS* numbers. In Fig. 66b the times of maxima for *C-R* vectors are relative to that for *u*. For the interval 1937—1960 the period of the solar cycle was very near 10 years. The vectors in Fig. 66 are weighted means. Points lettered a and b (see legend) were given weight one and those numbered c half weight. The weighted mean amplitude for each of the indicated vectors is normalized to unity on the scale shown. In addition, in Fig. 66, the amplitude for each of the three rotated vectors (end points lettered: a, b, and c) was corrected for amplitude deviations from the weighted mean amplitude for the vector rotated to zero time of maximum, using the approximately linear relation between amplitudes. The so-called prob-

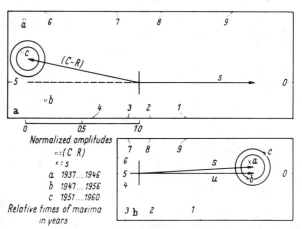

Fig. 67a and b. Harmonic dials 10-year waves. (a) *C-R* relative to *s*.
(b) *s* relative to *u*.

able error circles are indicated for the weighted averages and for single waves. The sizes of these error circles must be regarded as lower limits. For a "cloud" of only 2.5 points (1.5 degrees of freedom) the dispersion can not be reliably estimated and the statistically sound procedures developed by BARTELS [81] can not be used, in the usual manner, with so few data. Thus, to obtain some estimate for the error circles for (*C-R*), for example in Fig. 66b, the regression of yearly means of *C-R* on yearly means of *u* was derived ($r = -0.86$). From this regression line yearly values (*C-R*) were computed from *u* and the standard deviation, σ, of yearly differences between the observed and computed values of (*C-R*) were derived.

Regarding these differences as random the estimated radius, ϱ_1, of the error circle [81] for single vectors was obtained from: $\varrho_1 = (0.833 \times 2\sigma/\sqrt{10})$ in which 10 is the number of ordinates used in the harmonic analysis. Similarly the radii of the error circles in Fig. 66a were obtained. If the above differences are not due mainly to random effects then the size of the circles is underestimated. Tests made indicated the differences were essentially random although these tests with such a limited number of differences are not very reliable. At any rate this procedure gives some indication of the uncertainty in the 10-year waves indicated in Fig. 66.

The results indicate that the average 10-year wave in *u* has its maximum about 3 months after the maximum in the 10-year wave for *SS* numbers. Similarly, the minimum in *C-R* occurs about 6 months after the maximum for *SS* numbers, or about 3 months after the maximum for *u*.

In a similar manner Fig. 67 shows the phase of the 10-year wave in cosmic-ray intensity at Huancayo, *C-R*, relative to *s*, and of *s* relative to *u*, for which *s* is for cosmic-ray intensity the yearly pooled standard deviation of daily means from monthly means [78]. Fig. 67a indicates that the 10-year wave in cosmic-ray intensity has its minimum about 4 months after the maximum for *s*. Fig. 67b

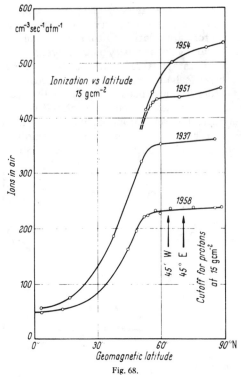

Fig. 68.

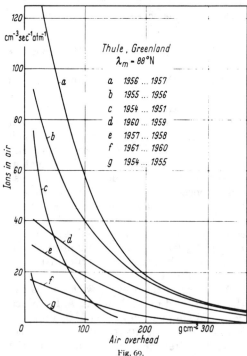

Fig. 69.

indicates no significant difference in phase between the waves for s and u. Together these results also indicate that the 10-year wave in cosmic-ray intensity may be due in part at least to a mechanism with a much longer time constant than that for the Forbush-type decreases, the effects of which are principally measured by s, for which the 10-year wave is essentially in phase with u.

II. Results from high altitude observations.

41. Ionization chambers. Using carefully calibrated and standardized ionization chambers NEHER [82] has measured cosmic-ray ionization at high altitude near the geomagnetic pole over Thule, Greenland in each of most years in the period 1951 to 1961. In addition he made latitude surveys in four different epochs of the solar cycle. Fig. 68 shows the results obtained, under 15 g cm⁻² of air, in four different years. Also shown in Fig. 68 are the latitudes at which protons coming in at 45° W and 45° E of the zenith, and having a range in air of 15 g cm⁻², are eliminated by the earth's magnetic field. Except for the earliest of these, flights were also made from a base station at Bismark, North Dakota, to monitor the radiation. Temporal changes observed from the base station flights were used to make appropriate corrections to the roving station.

The sharp "knee" which is a prominent feature of the curve for 1937, 1951, and 1958, does not appear in the curve for 1954. NEHER [83] concluded that the knee was also absent in 1955. He showed [84] that the latitude of the "knee"

Fig. 68. Results of four latitude surveys. The knee moves only over about 2° of latitude, whereas the ionization changes by a factor of 2. In 1954 the knee apparently disappeared (after NEHER).

Fig. 69. These curves represent the way in which the radiation that changes from one year to another is absorbed in the atmosphere (after NEHER).

in 1951 and in 1958 differed by only 2° although in 1951 the ionization, at 15 g cm^{-2}, above the knee was about twice that in 1958. To study the absorption in the atmosphere of the particles responsible for the changes from one year to another NEHER plotted the differences for successive years as in Fig. 69. These data are taken from curves for the averages of the flights for any one year. From Fig. 69 it is evident that the decrease in the intensity is quite different from one curve to another. For example, the radiation responsible for the increase from 1951 to 1954, curve c, was quite absorbable compared with some other years. With the approach of sunspot maximum the radiation that was removed by the modulation mechanism became increasingly energetic.

Comparing the ratio of percentage changes in neutron intensity at Uppsala to those in the ionization chamber at Huancayo for several rapid decreases (Forbush effects) SANDSTRÖM [85] et al. concluded that this ratio probably decreased from sunspot minimum to sunspot maximum. This also would indicate that *the radiation removed by the solar cycle modulation mechanism becomes increasingly energetic near sunspot maximum.*

Fig. 70 shows the ionization at different atmospheric depths obtained by NEHER [84] over Thule,

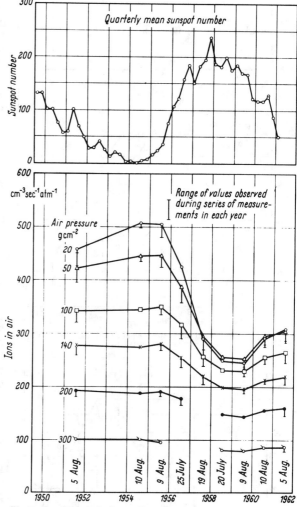

Fig. 70. Cosmic-ray ionization at Thule, Greenland at definite atmospheric depths for various years. Note the nearly anticorrelation with Zürich sunspot numbers (after NEHER).

Greenland. It is clear that the magnitude of the solarcycle variation (see sunspot numbers at the top of Fig. 70) increases very rapidly with diminishing air overhead.

During sunspot minimum in 1954 NEHER [86] concluded, from atmospheric absorption and rigidity requirements, that protons with energy $\leqq 150$ MeV were required to explain his results at high latitude and high altitude. The knee of the latitude variation, at least for protons down to 150 MeV, was completely absent in the summer of 1954.

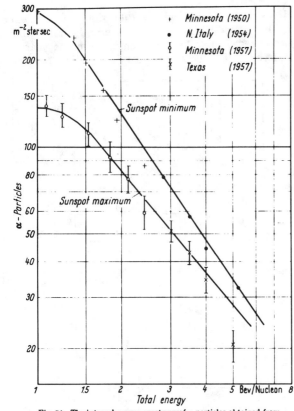

Fig. 71. The integral energy spectrum of α-particles obtained from scattering measurements (after Ney).

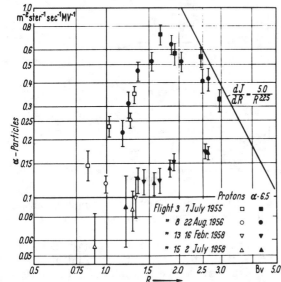

Fig. 72. Rigidity spectra, obtained from flights from Minneapolis, Minnesota, for protons and α-particles at two epochs of solar cycle after multiplying α-particle flux by 6.5 (after McDonald).

At 20 g cm⁻² the ionization in 1958 and 1959 was only about 50% of that in 1954 and 1955. The corresponding change in the neutron intensity in Ottawa was about 20% and in the ionization chamber at Huancayo only about 4% (see Fig. 62). Neher [84] et al. concluded that the largest negative correlation between sunspot numbers and ionization at 15 g cm⁻² obtained when ionization was correlated with sunspot numbers observed from 9 to 12 months earlier.

The cosmic-ray group at the University of Minnesota [87], [88] has made numerous balloon flights with ion chambers since 1956. They found that the solar cycle variation of ionization at 10 g cm⁻² was about three times the corresponding variation in neutron monitors near sea level.

42. Nuclear emulsions, other detectors and spectrum changes.

Neher used the results of his high altitude latitude surveys in 1954 to estimate the proton spectrum that would account for the increased ionization between latitudes 56° and 89°. Meyer and Simpson [89] used neutron monitors in high altitude latitude surveys with aircraft and from the results showed that the exponent in the power law spectrum, as well as the total intensity, were both greater near sunspot minimum in 1954 than in 1948 or 1956.

Changes in the low-energy part of the spectrum during the solar cycle are probably best derived from measurements on *primary α particles* in nuclear emulsion packages carried to high altitudes in balloons.

Although there are numerous experimental difficulties in separating the proton component from other singly charged components it is possible to discriminate between α particles and the background of secondaries. This procedure has been used by the group at the University of Minnesota [90] to determine the flux and energy distribution of primary α particles at different epochs of the solar cycle.

Fig. 71 shows their integral energy spectrum for α particles at two different epochs in the solar cycle in the energy range from $1-5$ MeV/nucleon. The total flux of α particles near sunspot minimum is about twice that near sunspot maximum. The slope of the integral energy spectrum near sunspot maximum is definitely less than near sunspot minimum, which agrees with the results from neutron monitors [89]. Investigations of the differential energy spectrum of α particles near sunspot minimum and near sunspot maximum have shown that the solar cycle modulation of cosmic-ray intensity results in a decrease in flux of particles in the energy spectrum at least up to 30 BeV/nucleon.

The spectrum of both protons and α particles has been measured by Mc-DONALD [91], [92] using a combination of ČERENKOV and scintillation counter detectors. Fig. 72 shows his results for the differential rigidity spectra for protons and for α particles at two epochs of the solar cycle. After multiplying the α particle flux by 6.5, which is the ratio of protons to α particles in the primary cosmic rays, the spectra for protons and α particles are alike near sunspot minimum and near sunspot maximum, although the flux of α particles and of protons, for rigidity $R < 3.0$ Bv, is only about one quarter as great in 1958 as in 1955, and 1956. Since particles of the same rigidity are modulated in the same way a magnetic modulation mechanism is indicated to account for the changes observed. The *solar cycle variations* in cosmic-ray intensity are most likely *produced by the magnetic fields in plasma clouds ejected from the sun*. When the earth is inside such a cloud the cosmic-ray intensity is reduced due to deflection of primary trajectories in these magnetic fields which may extend throughout a large part of the planetary system.

H. Forbush decreases.

I. Results from continuous monitors.

43. Ionization chambers. Sect. 22 described the variations of daily means of cosmic-ray ionization from several stations during the period $1937-1953$. Fig. 28a—f (p. 188—9) showed daily means of cosmic-ray intensity for Huancayo for the period $1937-1953$ together with daily means of horizontal magnetic field H at Huancayo. Daily means of ionization from Godhavn and Christchurch were plotted for selected years, mainly to show that at these two stations there are large fluctuations of ionization which do not appear at Huancayo. These undoubtedly are due principally to variations in the height of the 100 mb pressure level where μ-mesons are predominantly produced. In several months near sunspot minima in 1944 and 1954, the standard deviation of daily means from monthly means was as low as 0.16%.

During the period January to September 1954, 53 intervals of three consecutive days were selected such that within each interval the daily mean barometric pressure was essentially the same. The variance of D was computed with: $D = [(M_1 + M_3)/2 - M_2]$ in which, $M_{1,2,3}$ are daily means of ionization, corrected for bursts (not for barometric pressure), for the three consecutive days of each interval. This value of D eliminates any linear change over the three day interval;

15*

thus the values of D arise principally, if not entirely near sunspot minimum, from statistical fluctuations. Thus if s_1 designates the standard deviation of statistical fluctuations for means for single *days* then the standard deviation

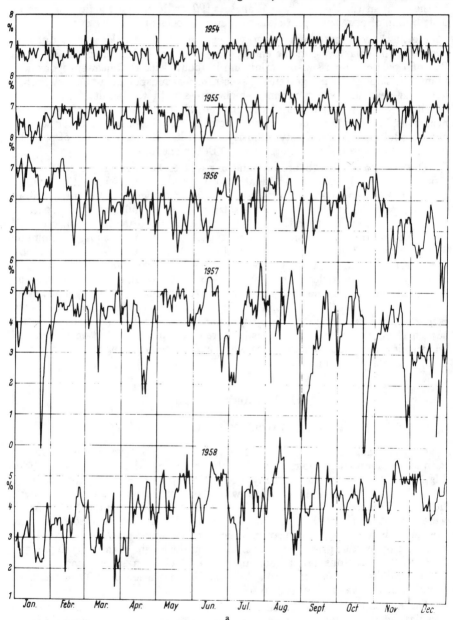

Fig. 73a and b. Daily means cosmic-ray intensity Huancayo (ordinates in percent from fiducial value). The daily means for 1954 refer to days between midnights of 75° West Meridian Time, those for 1955—1961 are for Greenwich days.

of D is $s_1 \sqrt{3/2}$, from which s_1 is obtained. From the above 53 intervals the value 0.14% was obtained for s_1. In a similar manner the standard deviation for *hourly* means (uncorrected for pressure) was found to be 0.59%.

392

At Cheltenham the standard deviation for hourly means was also obtained experimentally from differences between simultaneous hourly values in two identical ionization chambers. The value obtained was 0.70%. The total ionization at Cheltenham is about 0.63 times that at Hunacayo so that the standard

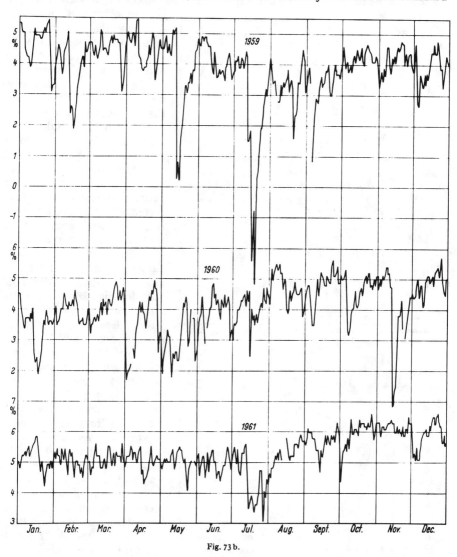

Fig. 73 b.

deviation of statistical fluctuations at Huancayo would be expected to be smaller by the factor, $0.63^{\frac{1}{2}} = 0.80$, or about 0.56% which is essentially the same as the value, 0.59% obtained at Huancayo for hourly values, or about 0.12% for daily means. For 1954 the yearly pooled standard deviation of pressure corrected daily means from monthly means is 0.21%. Since any real changes of cosmic-ray intensity within months increase this standard deviation it is evident that, within months, real variations must be quite small in 1954. Moreover, changes due to meteorological effects, other than the small changes in barometric pressure, at Huancayo must be quite small also.

The above procedure is described in detail since it may be used as well for data from neutron counters to compare with the standard deviation, $N^{-\frac{1}{2}}$ in per cent, derived from the counting rate. Appreciable discrepancies between these two values have been used to reveal instrumental malfunction.

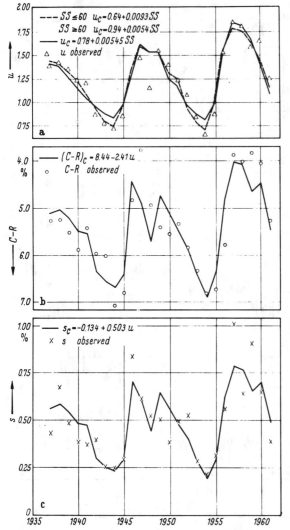

Fig. 74 a—c. 1937—1961 comparison of yearly means, observed and computed. (a) u with u_c computed from SS Nos. (b) C-R with C-R computed from u. (c) s with s_c computed from u.

The corrections for instrumental drift at Huancayo and the possibility of a seasonal wave with amplitude about 0.27% have been discussed by Forbush [78]. Figs. 73 a and b show the daily means of cosmic-ray intensity from the ionization chamber at Huancayo corrected for barometric pressure and instrumental drift for each day in the interval 1954—1961. In these figures the ordinate scale distance for 1% is twice that used in the First Part for Figs. 28 a—f. Fluctuations of daily means are obviously least in 1954 and are much greater (about 5 times) in 1957. In 1961 fluctuations are considerably less than in 1960.

Fig. 74 c compares the observed yearly pooled standard deviation, s (crosses), with that computed from the linear regression of s on yearly means of u, for which $r = +0.84$. The geomagnetic activity index u was computed from the interdiurnal variability of daily means of horizontal magnetic intensity as described by Bartels [80]. Similarly, Fig. 74 b compares observed annual means of cosmic-ray ionization, C-R, at Huancayo with values computed from the linear regression of annual means of $(C$-$R)$ on u for which $r = -0.86$. Fig. 74 a compares observed annual means of u with those computed from the regression of u on sunspot numbers, SS. The solid line indicates values of u computed from a single regression line ($r = +0.90$) for annual means. The dashed curve shows u computed from two regression lines as indicated.

For the intense magnetic storm of July 15, 1959, Fig. 75 shows the bihourly means of cosmic-ray ionization, C-R, at Huancayo, and three hour means for the southward (positive downward) geomagnetic field, ERC, of the equatorial ring current computed by the method of KERTZ [79]. For many of the magnetic storms in the interval 1955—1961 the correspondence between cosmic-ray intensity and ERC may likely be less marked than for the July 1959 storms. In fact, the intense storm of August 1937 (see Fig. 22, p. 183) was not accompanied by any detectable decrease in cosmic-ray ionization. The July 1959 storms [107], however, suggest a large scale solar emission of plasma clouds.

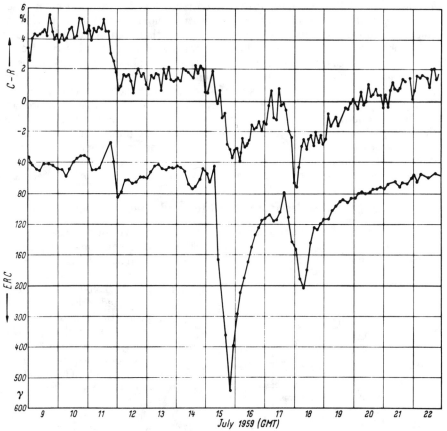

Fig. 75. Bihourly means cosmic-ray ionization, C-R, at Huancayo and three hour means of southward geomagnetic equatorial ring current field, ERC, July 9—22, 1959.

44. Neutron monitors. The numerous Forbush decreases during the period 1955—1961 derived from neutron monitors in general show changes somewhat similar, except for magnitude, to those shown in Fig. 73. For the IGY period July 1, 1957 to December 31, 1958, monthly means, daily means, and bihourly means, of pressure corrected neutron intensity at some 72 stations are shown separately graphed in publications of the IGY National Committee of Japan [93].

LOCKWOOD [94] has made an analysis of the large and rapid cosmic-ray Forbush decreases in cosmic radiation which occurred from 1954—1959. Data from the IGY network of neutron monitor stations were used together with other observations, to obtain the changes in the primary rigidity spectrum, onset times and the existence of anisotropies during the decreases. The decreases analyzed were preceded within 3 hours by a sudden commencement magnetic storm and most occurred from 6—36 hours after a polar cap absorption event [95] preceded by solar flares. Since the primary rigidity spectrum changes during the solar cycle, Lockwood determined the rigidity spectrum *before* each decrease in order to obtain the changes in rigidity spectrum produced *during* the decreases.

In general, the observed decreases were found to diminish with increasing values of the primary vertical cutoff rigidity in the range 1—15 BV, although for the event of December 19, 1957, the decrease was found to be independent of rigidity in the energy region to which neutron monitors respond. In addition, among the several events analyzed, there were other significant differences in the slope of the curves for percentage decreases as a function of vertical cutoff rigidity. The decrease in intensity during July 1959 as shown in Fig. 75 represents an extreme case of modulation near the recent solar activity maximum. According to Webber [96] the integral intensity of all particles at the top of the atmosphere during the period of minimum cosmic-ray intensity in July 1959 was probably only about one-fifth that at sunspot minimum. He estimates that at 15 BV the intensity was at least 30% less than at sunspot minimum. Webber [96] also discusses other Forbush decreases in detail and gives curves showing the rigidity dependence for these events. He gives a critical review of each of the several models to account for the solar cycle variation and the Forbush decreases.

II. Results from satellites and space probes.

45. Satellites. That the mechanism to account for Forbush decreases does *not* depend on the geomagnetic storm time field (i.e., the equatorial ring current field) was demonstrated by the results obtained by Simpson [97] and colleagues at the University of Chicago. In the Explorer VI satellite they used a triple coincidence, proportional counter system which measured protons with energy greater than 75 MeV or electrons with energy greater than 13 MeV. The triple coincidence detector is not affected by bremsstrahlung from electrons trapped in the geomagnetic field. Fig. 76 compares the triple coincidence rate in the satellite, at distances $\geq 35 \times 10^3$ km from the earth, with the counting rate of the neutron monitor at Climax, Colorado, during the period August 14 to August 22, 1959. For the period August 19 to 21 (excluding passages through the trapped

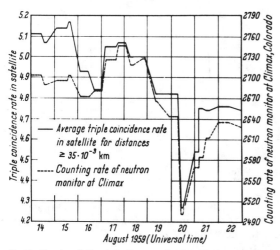

Fig. 76. Cosmic-radiation intensity as a function of time showing on August 20, 1959 a Forbush-type intensity decrease of ~15% within five hours. The nucleonic component monitor provides a measure of primary intensity changes at the earth. The light-dashed and solid lines are used to connect the periods for which satellite data are available (after Fan et al.).

radiation zone) they find a correlation coefficient of $+0.96$ between triple coincidence rates in the satellite and neutron intensity at Climax. The relative change in the satellite detector was about twice that in the neutron monitor. The relative intensity changes in the primary intensity outside the atmosphere were estimated to be about twice those in the neutron monitor. This estimate was based on a comparison of Forbush decreases in neutron monitors in high altitude balloons with those observed in the Climax neutron monitor for that period of the solar cycle. The authors thus conclude that the magnitude of the decrease observed out to distances of 7.5 earth radii were essentially the same as those estimated for the changes in primary radiation near the earth.

46. Space probe results. On the space probe Pioneer V SIMPSON [98] and colleagues at the University of Chicago had equipment identical with that used on Explorer VI which was described in the preceding paragraph. Fig. 77 shows their results obtained between March 27 and April 3, 1960, together with magnetometer results obtained by COLEMAN [99] et al. A decrease of 28% in the triple coincidence counter rate occurred on March 31, and April 1, 1960, during an intense magnetic storm. The authors estimate that the relative decrease of intensity in the Pioneer V detector, on April 1, at about 5×10^6 km from the earth was about 30% greater than the relative changes observed in the neutron monitor at Climax and extrapolated to the top of the atmosphere. They conclude that the larger decrease measured in Pioneer V was due to the removal of cosmic-ray primaries with magnetic rigidities below the cutoff (2.4 BV) for the Climax monitor but not below the threshold of the detector in Pioneer V. They found that this diminution of low energy primaries persisted for more than 30 days after April 1, although the higher energy particle flux had returned much earlier to the level observed before the March 31 storm (see also Fig. 73 f). From

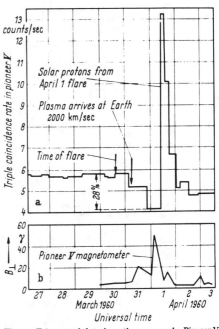

Fig. 77. Telemetered data from the space probe Pioneer V at distances $4-5.5 \times 10^6$ km from the earth. The time for the first arrival of solar protons April 1 was determined by the onset of enhanced ionization from protons at the polar cap. The magnetometer measurements in Pioneer V are published by COLEMAN et al. (after FAN et al.).

analysis of neutron monitor data extending to the geomagnetic equator where the cutoff reaches 15 BV they found independent evidence that the intensity decrease in this case was more strongly dependent upon particle magnetic rigidity than for the event of August 19—20, 1959 observed in Explorer VI.

From the interval between the solar flare on March 30, 1960 and the sudden commencement of the magnetic storm the velocity of the solar plasma was estimated as 2×10^8 km · sec^{-1}. FAN [98] et al. estimate that about 28% of the particle flux detected by Pioneer V disappeared in less than 20 hours, with half of this decrease occurring in less than six hours. From these data and the computed radial velocity of 2000 km/sec for the advancing "front" of solar plasma, a penetration of about 0.3 astronomical units behind the front was required to reduce the intensity by 14% — half the full intensity decrease. For other events of this kind where the rate of decrease may be as high as 5—6% per hour the corresponding depth of penetration may be smaller by a factor of 3. This requires the rapid appearance of enhanced interplanetary magnetic field intensities.

The magnetometer results in Fig. 77 show magnetic field intensities 10 to 20 times those for the quiescent field. These results directly indicate the existence of magnetic fields frozen in conducting plasma ejected from solar flares. The authors conclude that *the cosmic-ray decrease is caused by the convective removal of galactic primaries by particle collisions with advancing large scale magnetic field irregularities such as a shock front.* From Fig. 78 it may be seen that by 0600 April 1 the advancing region which produced the full decrease of intensity had passed outward beyond Pioneer V and the earth.

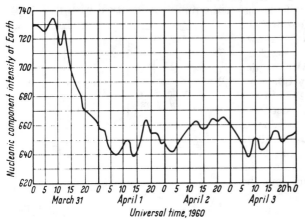

Fig. 78. The changes in galactic cosmic-ray intensity for particles above 2.4 BV magnetic rigidity. The neutron monitor is located at Climax, Colorado.

Fig. 77 shows that solar protons arrived at the Pioneer V detector about one hour after the April 1 flare. The solar particle increase is shown in Fig. 77 for protons >75 MeV (but less than 1 BeV since no increases were observed by neutron monitors at the earth). Maximum intensity was reached within 50 minutes. From this the authors conclude that the interplanetary magnetic field conditions behind the advancing front are either smooth and radial or weak, irregular fields ($B_{rms} < 5 \times 10^{-6}$ Gauss), otherwise the low energy solar protons would not penetrate to the orbit of the earth in less than one hour.

I. Solar flare effects.

I. Results from ground level monitors.

47. Ionization chambers. In Sects. 30 and 31, the four increases in cosmic-ray intensity which had been observed in the period 1937—1955 were discussed and results shown in Figs. 49—54. The fifth and *largest increase* recorded in ionization chambers to date (December 1962) occurred on *February 23, 1956*. In the Cheltenham ionization chamber the maximum intensity was about 85% above the pre-flare value [*100*]. The increase began about 18 min after the solar flare was first observed at 0330 GMT February 23, 1956. From a large shielded ionization chamber at Derwood, Maryland, the increase was determined for one minute intervals during the first hour of the increase and for six minute intervals thereafter. Fig. 79 shows the curve from the Derwood meter and points from the ionization chambers at Godhavn, Cheltenham, Mexico, Huancayo, and Christchurch.

This is the only occasion, from 1937—1962, that any particles accelerated in a solar flare acquired sufficient energy, greater than 15 BeV for protons, to be *detected at the equator*: The neutron monitor intensity at Huancayo, as reported by Simpson and his colleagues [*101*] registered a maximum of 20% above the pre-flare value.

48. Neutron monitors: the event of February 23, 1956. The first observational evidence that the flux of particles responsible for solar flare effects is much greater for particles in the lower energy band to which neutron detectors are sensitive than in the band to which ionization chambers respond, was obtained during the solar flare of November 19, 1949 by FORBUSH et al. [61]. The solar flare increase of November 19, 1949 observed in an ionization chamber at Climax,

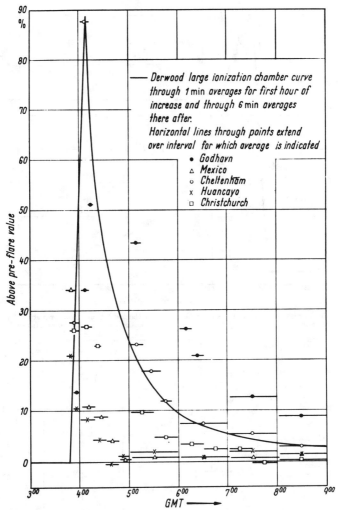

Fig. 79. Cosmic-ray increase during solar flare of February 23, 1956.

Colorado (altitude 3500 m) was nearly five times that observed in an ionization chamber at Cheltenham near sea level.

From this result, discussed in Sect. 31, it was predicted that for the flare of November 19, 1949, a neutron monitor should have shown an increase of the order of twenty times that in the ionization chamber at Cheltenham. This prediction was confirmed by the results obtained in a neutron pile by ADAMS at Manchester [62], see Sect. 31, Fig. 54, who observed an increase of 550% or 12 times that observed in the ionization chamber at Cheltenham.

The first solar flare increase to be observed in several neutron monitors was that on February 23, 1956. This increase was recorded by 72 instruments which included 17 ionization chambers (with 10—12 cm Pb shielding), 14 neutron monitors and 41 counter telescopes.

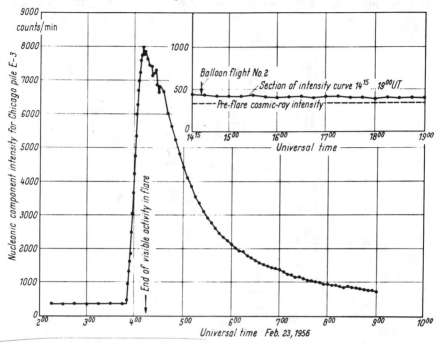

Fig. 80. The intensity increase of secondary neutrons generated in the atmosphere from primary solar-flare protons. These observations were obtained with a neutron monitor pile at Chicago, Ill. (after SIMPSON et al.).

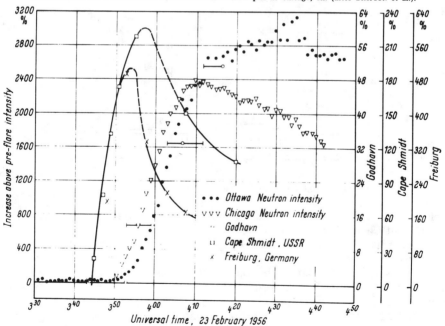

Fig. 81. Detectors located in different parts of the geomagnetic field respond to different energies of solar protons provided they come from a point source — the impact zone effect (after SIMPSON et al.).

Fig. 80 shows the increase observed in the neutron monitor at Chicago. From an analysis of data from this and several other stations the group at the University of Chicago discovered that one of the characteristics regarding the arrival of the first particles from this flare is that high-energy particles appear to arrive ahead of low-energy particles with a spread in arrival time of the order of 10—15 min for an energy-range of 10 BeV for protons. This dispersion effect is shown in Fig. 81 which shows examples of prompt and delayed onset times, taken from a world-wide distribution of cosmic-radiation intensity recorders. The primary flare particle spectrum derived by SIMPSON et al. [*101*] by taking account of the different cut off energies imposed by the earth's magnetic field at different locations is shown in Fig. 82. The spectra are rather similar at the three different times indicated although the intensities differed by an order of magnitude. The differential rigidity spectra of Fig. 82 follow approximately the power law N^{-7}. These investigators obtained independent evidence for the character of the spectrum of flare particles at the lower particle rigidities from neutron detectors flown in balloons from 9—12 hours after the start of the increase.

The development of the February 23, 1956, event is described by SIMPSON [*102*]

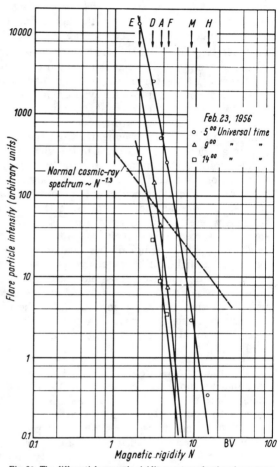

Fig. 82. The differential magnetic rigidity spectrum for the solar protons after particle storage or trapping in the solar system had taken over. Here the particle magnetic rigidity $N = p\,c/Z\,e$, where p is momentum, c is velocity of light, $Z\,e$ is the particle charge, and N is measured in volts. The scale is given in BV = billion volts = 10^9 V (after SIMPSON et al.).

as follows. The apparent source is a relatively large area which includes the sun. Initially, the particles arriving are of the highest energies in the flare spectrum. Subsequently, the whole sky becomes "illuminated" with arriving particles extending to lower energies, and for these low-energy particles and late-arriving high-energy particles, there develops a remarkable isotropy in space near the earth. Following the onset of isotropy, the particle intensity gradually diminishes with the flare particle spectrum essentially unchanged over many hours, as shown in Fig. 82, while the particles escape from the magnetic fields which store them in the solar system. The intensity at the earth is observed to diminish to its pre-flare level within a period of 20 hours. These phenomena suggest three intervals of time in the development of the cosmic-ray flare:

1. Beginning with the initial release of high-energy particles and ending at the time when the solar particles reach a maximum intensity at the earth, the particles come from a limited source direction. This interval of time is of the order 10—30 minutes, depending upon the solar event.

2. A brief period of transition sets in when particles begin to arrive from directions other than the source—suggesting that particles arrive later as a result of scattering or passage through indirect magnetic channels in the interplanetary magnetic fields connecting the earth to the solar region.

3. At late times when isotropy has been established, all evidence for the release of energy in the solar flare region has vanished. But the influx of cosmic-ray particles at the earth continues for many hours. This fact and the fact that no source direction persists for even the highest energy particles strongly support the view that the sun accelerates particles only during a short interval of time and that these particles are trapped and stored in the interplanetary magnetic fields only to be lost subsequently from the solar system or arrive at the earth. The decay mode of the particles in the vicinity of the earth opens the possibility for determining the characteristics of the storage or trapping in magnetic fields. This kind of evidence for the storage of charged particles is at present the strongest evidence for the existence of interplanetary magnetic fields. From the arrival of solar protons at the equator over Huancayo, Peru, and their detection in balloons where the cutoff for protons is as low as 1—2 BV, it is clear that the energy range certainly exceeds 1—24 BV.

Dorman [103] using all the available data for the event of February 23, 1956, has contributed an extensive analysis and discussion of the implications of the results, including possible acceleration mechanisms for the solar flare particles. Besides the diffusion model of Simpson et al. [101] to explain the persistence of the cosmic-ray increase after the disappearance of the flare, other models have been proposed. These are critically discussed by Webber [96]. A model involving the "magnetic tongue" from the sun is discussed by Cocconi et al. [104].

49. Other events. Carmichael [105] and Steljes showed that the small solar flare increase on July 17, 1959, was observed only in neutron monitors at latitudes where the cutoff rigidity was less than about 1.1 BV. Fig. 83 shows the increase in neutron intensity at several stations plotted against Quenby [106] and Webber cutoff rigidities. Only the point for Sulphur Mountain in Fig. 83 is quite off the curve and it is doubtless due to the high altitude of the Sulphur Mountain Station. This increase was only about 5% at Churchill and since it occurred during a large Forbush decrease (see Fig. 75), it was only detected from differences of intensities between neutron monitors at quite high latitudes and the neutron monitor at Uppsala.

The same data as in Fig. 83 when plotted against cutoff rigidities determined for the earth's *eccentric* dipole resulted [105] in points with very large deviations from any smooth curve that could be drawn through them. It is probable that so small an increase could not have been reliably ascribed to the solar flare without the availability of cutoff rigidities as determined by Quenby and Webber's procedure [106]. This method takes account of the actual field of the earth in deriving cutoff rigidities for any points on the earth. Cutoff rigidities similarly determined accounted very well [96] for the cosmic-ray equator determined by Simpson in high altitude air-craft using neutron monitors.

The cosmic-ray events of July 1959 have been discussed by several investigators. These and other geophysical events of closely related interest were the

subject of a symposium at Helsinki in 1960. The collected papers presented at
this symposium have been published [*107*].

Three other large increases of cosmic-ray intensity have been recorded in
neutron monitors. The first of these occurred on May 4, 1960, and the other two
on November 12 and November 15, 1960 [*108 a*]. The increase in neutron intensity
is shown at Ottawa in Fig. 84, together with the cosmic-ray ionization at Huancayo which shows no increase but only the large decrease associated with the
magnetic storm for which the geomagnetic equatorial ring current field, ERC,
is shown. By an ingenious procedure, McCRACKEN [*108*] has analyzed the neutron

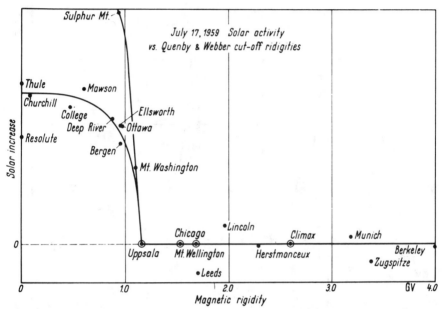

Fig. 83. The size of the July 17 solar cosmic-ray increase plotted against the cutoff rigidity of the observing station according to Quenby and Webber. The circled points are base stations where it was assumed that there was no solar increase. A vertical scale has not been indicated. It is such that the size of the solar increase at Churchill between 0200 and
1600 on July 17 is 5.4% of the average counting rate on the previous day (after Carmichael et al.).

data from several stations to derive the direction from which the increases came.
He determines the solid angle containing all the asymptotic directions of particles
which contribute to the counting rate of the detector. From a careful analysis
he identifies those neutron monitor stations which have small cones of acceptance
and determines a weighted mean asymptotic direction by giving a weight for
each rigidity which is proportional to the fraction of the total counting rate
which it contributes. He has also shown for certain stations in high latitude
that standard neutron monitors at these stations will record identical percentage
enhancements when an isotropic flux of solar particles is incident upon the earth.

Differences in the observed magnitude of events at these different stations
then indicate anisotropy in the radiation. Moreover, when the radiation is aniso-
tropic he determines whether there is some asymptotic direction about which
the radiation fluxes are symmetrical. For the solar flare increase of May 4, 1960,
he finds that the direction of maximum intensity was persistently inclined
about 50° to the west of the sun. For the solar flare event of November 12, 1960,
the direction of maximum intensity was also about 50° to the west of the sun
during the first 4 hours of the increase, from 1400—1900 GMT. Near 1900 GMT

an abrupt increase of intensity occurred soon after which the radiation was found to be isotropic. For the increase of November 15, 1960, the radiation became isotropic very rapidly after an initial phase of very marked anisotropy.

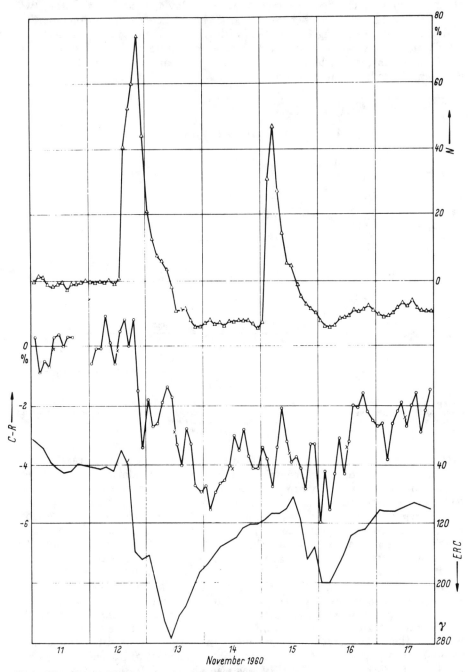

Fig. 84. Bihourly means (GMT) in per cent from prestorm values for neutron intensity, N, at Ottawa (after Rose), cosmic-ray ionization, C-R, at Huancayo, and three hour means of southward geomagnetic equatorial ring current field, ERC, November 11–17, 1960. Note different ordinate scales for N and C-R.

These time variations of the anisotropy and their important significance in determining the magnetic regime, and changes in it, through which the solar cosmic rays must pass to reach the earth are discussed by McCracken [108].

II. Results from other methods of detection.

50. Riometers. The magnitudes of solar flare events observed in neutron monitors have been enormously greater than in ionization chambers, due to their much greater sensitivity to low energy primaries, down to rigidities of about 1 BV (for ground level instruments). Nevertheless, solar flare events are not often detected by ground level neutron monitors. The discovery in 1957 of solar proton beams of low energy, however, has shown that the neutron monitor usually does not indicate the arrival at the top of the atmosphere of solar cosmic rays. That incoming solar streams of heavy ions would produce, and did produce, in the February 1956 case, great ionization of the ionospheric D-layer and consequent absorption of radiowaves was first realized by Bailey [109].

For the period of the solar flare of February 23, 1956, Bailey [109] investigated oblique-incidence signal intensities and simultaneous observations of the background cosmic noise at very high frequency for a number of high latitude communication links employing the ionosphere scatter mode of propagation. During the flare and for some time afterward, all paths were in the dark hemisphere. Simultaneously with the arrival of solar cosmic rays he observed a sharp enhancement of signal intensity which he attributed to solar protons. In summarizing his conclusions Bailey [109] made the following statement:

"... it seems necessary to recognize a new class of signal-intensity enhancement for waves propagated by ionosphere scattering and a new kind of high-altitude absorption phenomenon. It would seem inappropriate to identify the observed absorption effects as merely a special case of the well-known polar blackout absorption. It is clear that an event such as that reported is rare. To the extent that such events are associated generally with outbursts of solar cosmic rays, they may on the basis of very meager statistics be expected to occur about once in four years. Actually smaller events may occur more frequently, but are likely to be associated with important flares. The particle velocities thought necessary to account for the absorption effects are of the order of a tenth of cosmic-ray particle velocities, and the associated particle energies are correspondingly lower. The sun is, therefore, more likely to eject absorption-producing particles than particles having cosmic-ray energies."

These predictions have been verified by subsequent events registered in high latitudes from *riometers*, which register on a frequency of about 27 megacycles the intensity of galactic radio emission. Following the solar flare of July 29, 1958, Leinbach and Reid [110] observed a large attenuation of cosmic radio noise which they ascribed to absorption resulting from increased ionization in the ionospheric D-region. The effect differed from cosmic noise absorption produced in the auroral zone in that absorption was observed at stations well inside the auroral zone and in addition, the absorption exhibited a latitude cutoff. They concluded that this absorption resulted from enhanced ionization produced near 60 km altitude by ions from the solar flare, as had been previously suggested by Bailey [109] and by Little and Leinbach [111]. Penetration to 60 km altitude requires proton energies of 20—30 MeV.

From signal strength recordings obtained since 1952 from several VHF high latitude scatter paths Bailey and Harrington [112] have detected 44 such events. They list occasions when the polar cap absorption extends much further southward during the main phase of magnetic storms. During some of these occasions Ney et al. [113] have also observed protons with energy well below that for which protons are normally excluded by the earth's field at the latitude of their balloon flights.

51. Balloons, satellites. From May 1957 through July 1959, Reid and Lein- bach [*114*] observed 24 so-called Type III absorption or PCA (polar cap absorp- tion) events on riometer records. Detailed observations of solar cosmic rays from balloon observations, by the Minnesota group and others, were made possible by taking advantage of the riometer indications of the arrival of solar particles, and launching balloons as soon thereafter as possible. The nature of the solar particles arriving at the earth was established by Anderson [*115*], who measured an increase in charged particle intensity above Churchill, Canada, and Fairbanks,

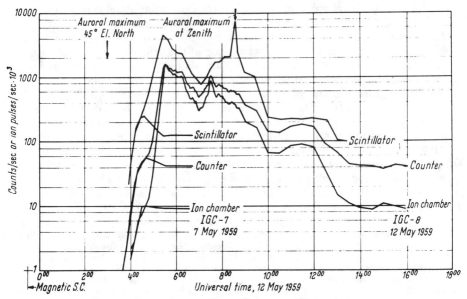

Fig. 85. Total counting rate of sodium iodide scintillator (50-kV threshold), single Geiger counter, and integrating ioni- zation chamber on balloon flight IGC-8. For comparison, note same instruments on flight IGC-7, five days earlier. Solar particles are first detectable at 0430 UT at a balloon depth of 100 g/cm². The balloon reached ceiling at 0530 UT. Note large selective response of scintillator to auroral x-rays at auroral maximum. The solar particles temporarily increased during the aurora at 0730 UT (after Ney et al.).

Alaska, on August 22 and 23, 1957. They measured with ion chambers the vari- ation in counting rate as a function of altitude following the solar flare and compared this counting rate with normal ones on days of no solar activity. The variation of counting rate with altitude allowed them to infer a proton energy spectrum and to set some limits on the flux of primary electrons. Anderson deduced a differential number energy spectrum $N(E)\, dE = K E^{-5} dE$ and found that the results were consistent with protons arriving in the energy range from 100—400 MeV. Rothwell and McIlwain [*116*], in experiments with Explorer IV, found large solar flare increases occurring at 0432 UT on August 16 and at 0005 UT on August 22. They measured the counting rates in shielded and un- shielded Geiger counters during the period of the increases and concluded that the incoming beam must consist largely of protons. The first observations of incoming solar cosmic rays in emulsions were made by Freier [*117*] et al. in the event which occurred on March 26, 1958.

Ionizing radiation recorded in balloon altitudes in high latitudes (in northern Sweden and in North America) have been discussed by Pfotzer et al. [*116a*].

One of the more spectacular events from which detailed information was obtained concerning solar cosmic rays was that of May 12, 1959, reported by

NEY et al. [*113*]. Fig. 85 shows the results of their observations derived from four balloon flights made on May 12, 1959. Preceding these observations a 3 + solar flare of 3 hours' duration occurred at 2000 GMT on May 10 and gave rise to a very strong solar noise storm. At 2340 GMT, May 11, there occurred a magnetic sudden commencement followed by a magnetic storm accompanied by a large decrease of cosmic-ray intensity. At Huancayo, as seen from Fig. 73 b (p. 229), the daily mean ionization on May 12 was about 5% less than on May 11. The riometer at College, Alaska, indicated absorption starting about 0100 GMT on May 11. From the latter part of May 11, the riometer indicated absorption exceeding 17 db for nearly one day. In Fig. 85 the ion chamber curve shows that the ionization increased to 160 times that normally due to cosmic rays. From emulsion tracks NEY [*113*] et al. estimated the vertical flux of solar cosmic-ray particles at 1000 times the normal cosmic-ray flux.

Fig. 86 shows the proton rigidity spectrum obtained from measurements in emulsions for the May 12 event and also that for another event on March 26, 1958. The figure also shows the proton rigidity spectra near solar maxima and minima.

52. Summary on solar proton events. From this and other similarly observed solar proton events NEY [*118*] summarizes the general features that have been established for solar proton events as follows:

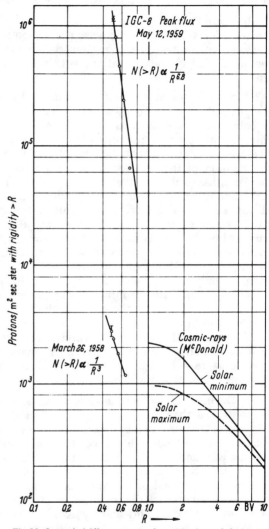

Fig. 86. Integral rigidity spectrum of protons observed during the solar flares of May 12, 1959 and March 26, 1958. The cosmic-ray proton spectrum during solar maximum and solar minimum is that measured by McDonald (F. B. McDonald and W. R. Webber to be published). F. B. McDonald (to be published). (After Ney et al.)

a) The atmospheric effects are produced by protons primarily in the energy range from 30 MeV to something in excess of 500 MeV.

b) α particles appear to be present with approximately the same rigidity spectrum as the incoming protons.

c) The arrival of the proton beam heralded by the onset of cosmic noise absorption may occur within hours of the onset of the solar flare. The longest

16*

delay observed was the case of the storm of March 28, 1958, in which the arrival of the proton beam was coincident with the Forbush decrease and the magnetic storm. This occurred 20 hours after the flare. In contrast to this long delay, half of the events have delay times less than 5 hours.

d) The measurements at high latitudes of incoming particle fluxes confirm the suspicion that the cosmic noise absorption shown by the riometer is a good measure of the intensity of arrival of the incoming beam and that the incoming protons die off quasiexponentially with a time constant of the order of several days.

e) At latitudes as far south as Minneapolis, the events are very much more complicated. Even through particles may be arriving at high latitude, as shown in balloon measurements as well as riometer indications, the particles do not enter the atmosphere at Minnesota until the onset of the geomagnetic storm. The reason for this is that the normal cosmic ray cutoffs at Minnesota exclude the majority of the protons of the energies contained in the solar beam. It is, therefore, impossible for these particles to enter at this latitude without a perturbation of the earth's magnetic field. Although the cutoff energy for protons at Minnesota is of the order of 300 MeV, during the magnetic storm the solar particles with energies down to 50 MeV are able to enter. The time decay of the particle intensity at this low latitude, however, seems to be entirely governed by the behavior of the magnetic field as affected by the incoming solar stream. In the event in May 1959, the particle intensities persisted at their very high value for only several hours although the riometer showed that they continued to persist at high latitude for a number of days. Direct observations of the high latitude cosmic rays and the corresponding protons at Minnesota were made in the July 1959 event during which it became quite clear that the particle fluxes could arrive at Minnesota in essentially full intensity as long as the magnetic field was perturbed, but when the earth's magnetic field returned to normal the particles were not allowed to enter in Minnesota but were still arriving at the latitude of Churchill, Canada.

f) The incoming solar beams appear to be almost entirely positively charged. Certainly less than 10% of the particles in these beams are electrons. Since α particles of the same rigidity appear to be present, the question arises as to whether electrons accelerated with the positive particles can lose their energy in some specific way and therefore not be allowed to reach the earth. One possible explanation for the absence of the electrons would be that these electrons are trapped in the magnetic field of the solar corona and lose their energy by synchrotron radiation, thereby emitting radio waves. It has been established now by THOMPSON and MAXWELL [119] and by KUNDU and HADDOCK [120] that the radio emission from the sun at the time of the flare acceleration of the cosmic rays is of a particular kind which could, in fact, be attributed to synchrotron radiation of electrons.

g) The incoming proton beams, several hours after the beginning of the event, seem to be isotropic at the top of the atmosphere. This conclusion must be taken with some reservation, however, until direct measurements are made simultaneously at satellite altitudes and at balloon altitudes.

WEBBER [96] has provided a detailed comprehensive analysis and survey of individual solar flare increases including associated phenomena, general characteristics, and a discussion of specific models for the propagation and control of solar flare effects.

References.

See also Vol. XLVI, Cosmic Rays, of this Encyclopedia.

[1] CORLIN, A.: Ann. Obs. Lund Nr. 4 (1934).
[2] COMPTON, A. H., E. O. WOLLAN and R. D. BENNETT: Rev. Sci. Instr. 5, 415 (1934).
[3] SITTKUS, A.: Z. Naturforsch. 1, 204 (1946).
[4] DOLBEAR, D. W. N., and H. ELLIOTT: J. Atmosph. Terr. Phys. 1, 215 (1951).
[5] ALFVÉN, H., and K. G. MALMFORS: Ark. Mat. Astronom. Fys. 29 A, No. 24 (1943).
[6] SIMPSON, J. A., W. FONGER and S. B. TRIEMAN: Phys. Rev. 90, 934 (1953).
[6a] SIMPSON, J. A.: Cosmic radiation neutron monitor. Institute of Nuclear Studies, University of Chicago.
[7] SIMPSON, J. A.: Phys. Rev. 81, 895 (1951).
[8] SIMPSON, J. A.: Phys. Rev. 83, 1175 (1951).
[9] ADAMS, N., and H. J. BRADDICK: Phil. Mag. 41, 501 (1950).
[10] Cosmic-ray results from Huancayo Magnetic Observatory, Peru, June 1936—December 1946. Including summaries from observatories at Cheltenham, Christchurch, and Godhavn through 1946. Researches of the Department of Terrestrial Magnetism, Vol. XIV, Carnegie Institution of Washington Publication 175, Washington, D.C. (1948). See Vols: XX (1956) and XXI (1961) for subsequent results through 1959.
[11] DAWTON, D. I., and H. ELLIOTT: J. Atmosph. Terr. Phys. 3, 295 (1953).
[12] DUPERIER, A.: Proc. Phys. Soc. Lond. A 62, 684 (1949).
[13] HESS, V. F., H. GRAZIADEI u. R. STEINMAUER: Sitzgsber. Akad. Wiss. Wien 144, 53 (1935).
[14] COMPTON, A. H., and R. N. TURNER: Phys. Rev. 52, 799 (1937).
[15] BLACKETT, P. M. S.: Phys. Rev. 54, 973 (1938).
[16] FORBUSH, S. E.: Terr. Magn. 42, 1 (1937).
[17] FORBUSH, S. E.: Phys. Rev. 54, 975 (1938).
[18] BARTELS, J.: Terr. Magn. 37, 291—302 (1932).
[19] BARTELS, J.: Terr. Magn. 40, 1—60 (1935). — The contents of these papers are partly reproduced in Chap. 16 and 19 of CHAPMAN, S., and J. BARTELS: Geomagnetism. p. 543—605. Oxford Univ. Press 1940 (reprinted 1951).
[20] ALFVÉN, H., and K. G. MALMFORS: Ark. Mat. Astronom. Fys. 29 A, No. 24 (1943).
[21] KOLHÖRSTER, W.: Phys. Z. 42, 55 (1941).
[22] DOLBEAR, D. W. N., and H. ELLIOTT: J. Atmosph. Terr. Phys. 1, 215 (1953).
[23] ELLIOTT, H., and D. W. N. DOLBEAR: Proc. Phys. Soc. Lond. 63 A, 137 (1950).
[24] ELLIOTT, H., and D. W. N. DOLBEAR: J. Atmosph. Terr. Phys. 1, 205 (1951).
[25] ALFVÉN, H.: Phys. Rev. 75, 1732 (1949).
[26] THAMBYAHPILLAI, T., and H. ELLIOTT: Nature, Lond. 171, 918 (1953).
[27] SARABHAI, V., and R. P. KANE: Phys. Rev. 90, 204 (1953).
[28] FONGER, W.: Phys. Rev. 91, 351 (1953).
[29] SIMPSON, J. A., J. FIROR, W. FONGER et S. B. TREIMAN: Recueil des travaux de l'observatoire du Pic-du-midi, Série: Rayons cosmiques No. 1, Congr. Internat. sur le rayonnement cosmique, 4 (1953).
[30] SITTKUS, A.: Recueil des travaux de l'observatoire du Pic-du-midi, Série: Rayons cosmiques No. 1, Congr. Internat. sur le rayonnement cosmique, 11 (1953).
[31] EHMERT, A., u. A. SITTKUS: Z. Naturforsch. 6a, 618 (1951).
[32] SEKIDO, Y., and S. YOSHIDA: Rep. Ionosph. Res. Japan 4, 37 (1950).
[33] ELLIOTT, H., and D. W. N. DOLBEAR: J. Atmosph. Terr. Phys. 1, 205—214 (1951).
[34] YOSHIDA, S., and I. KONDO: J. Geomag. Geoelectr. 6, 15 (1954).
[35] COMPTON, A. H., and I. A. GETTING: Phys. Rev. 47, 817 (1935).
[36] VALLARTA, M. S., C. GRAEF and S. KUSAKA: Phys. Rev. 55, 1—5 (1939).
[37] FORBUSH, S. E.: Phys. Rev. 52, 1254 (1937).
[38] FORBUSH, S. E.: Terr. Magn. 42, 1 (1937).
[39] ELLIOTT, H., and D. W. N. DOLBEAR: J. Atmosph. Terr. Phys. 1, 205 (1951).
[40] FORBUSH, S. E.: Terr. Magn. 43, 203 (1938).
[41] HAYAKAWA, S., J. NISHIMURA, T. NAGATA and M. SUGIURA: J. Sci. Res. Inst. Japan 44, 121 (1950).
[42] TREIMAN, S. B.: Phys. Rev. 89, 130 (1953).
[43] ALFVÉN, H.: Cosmical electrodynamics. Oxford: Clarendon Press 1950.
[44] FORBUSH, S. E.: Phys. Rev. 54, 978 (1938).
[45] MEYER, P., and J. A. SIMPSON: Phys. Rev. 96, 1085 (1954).
[46] FONGER, W. H.: Phys. Rev. 91, 351 (1953).
[47] SIMPSON, J. A.: Phys. Rev. 94, 426 (1954).
[48] FONGER, W. H.: Phys. Rev. 91, 351 (1953).

[49] VESTINE, E. H., L. LAPORTE, I. LANGE and W. E. SCOTT: The geomagnetic field, its description and analysis. Washington, D.C., Carnegie Inst. Pub. 580 (1947).
[50] NEHER, H. V., and E. A. STERN: Phys. Rev. 98, 845 (1955).
[51] NEHER, H. V., V. Z. PETERSON and F. A. STERN: Phys. Rev. 90, 655 (1953).
[52] FORBUSH, S. E.: Trans. Washington Meeting. Int. Union Geod. Geophys., Assoc. Terr. Magn. Electr., Bull. No. 11, 438 (1940).
[53] BARTELS, J.: Terr. Magn. 37, 1 (1935).
[54] SIMPSON, J. A.: Phys. Rev. 94, 426 (1954).
[55] SIMPSON, J. A., W. H. FONGER and S. B. TREIMAN: Phys. Rev. 90, 934 (1953).
[56] FORBUSH, S. E.: Phys. Rev. 70, 771 (1946).
[56a] EHMERT, A.: Z. Naturforsch. 3a, 264 (1948).
[57] McNISH, A. G.: Terr. Magn. 42, 109 (1937).
[58] FORBUSH, S. E.: Terr. Magn. 42, 1 (1937).
[59] FORBUSH, S. E., P. S. GILL and M. S. VALLARTA: Rev. Mod. Phys. 21, 44 (1949).
[60] GRAHAM, J. W., and S. E. FORBUSH: Phys. Rev. 98, 1348 (1955).
[60a] MÜLLER, R., et al.: J. Atmosph. Terr. Phys. 1, 37 (1950).
[61] FORBUSH, S. E., M. SCHEIN and T. B. STINCHCOMB: Phys. Rev. 79, 501 (1950).
[62] ADAMS, N.: Phil. Mag. 41, 503 (1950).
[63] CLAY, J., and H. F. JONGEN: Phys. Rev. 79, 908 (1950).
[64] SEKIDO, Y., and S. YOSHIDA: Rep. Ionosph. Res. Japan 7, 147 (1953).
[65] ELLIOTT, H.: Progress in cosmic ray physics, Chap. VIII. Amsterdam: North Holland Publishing Company 1952.
[66] BIERMANN, L.: In W. HEISENBERG (ed.), Kosmische Strahlung. Berlin-Göttingen-Heidelberg: Springer 1953.
[67] EHMERT, A.: Z. Naturforsch. 3a, 264 (1948).
[68] SCHLÜTER, A.: Z. Naturforsch. 6a, 613 (1951).
[69] FIROR, J.: Phys. Rev. 94, 1017 (1954).
[70] STÖRMER, C.: Astrophys. Norv. 1, 1 (1934).
[71] DWIGHT, K.: Phys. Rev. 78, 40 (1950).
[72] MALMFORS, K. G.: Ark. Mat. Astronom. Fys., Ser. A 32, No. 8 (1945).
[73] BRUNBERG, E.: J. Geophys. Res. 58, 272 (1953).
[74] STÖRMER, C.: Astrophys. Norv. 1, 115 (1934).
[75] STÖRMER, C.: Terr. Magn. and Electr. 22, 23 (1917).
[76] HESS, V. F.: Sitzgsber. Wien. Akad. II a, 2001-32 121 (1912).
[77] Annals of the International Geophysical Year, Vol. VIII, pp. 209—214. London: Pergamon Press 1959.
[78] FORBUSH, S. E.: J. Geophys. Res. 63, 651 (1958).
[79] KERTZ, WALTER: Ein neues Maß für die Feldstärke des erdmagnetischen äquatorialen Ringstroms. Abh. der Akad. der Wiss. in Göttingen, Math.-physik. Kl. Beiträge zum Internat. Geophysikalischen Jahr, H. 2. Göttingen: Vandenhoeck and Ruprecht 1958.
[80] BARTELS, J.: Terrestrial magnetic activity and its relation to solar phenomena. Terr. Magn. 37, 1 (1932).
[81] BARTELS, J.: Random fluctuations, persistence and quasipersistence in geophysical and cosmical periodicities. Terr. Magn. 40, 1 (1935).
[82] NEHER, H. V., and H. R. ANDERSON: J. Geophys. Res. 67, 1309 (1962).
[83] NEHER, H. V.: Phys. Rev. 107, 588 (1957).
[84] NEHER, H. V.: J. Geophys. Res. 66, 4007 (1961).
[85] SANDSTRÖM, A. E., and S. E. FORBUSH: J. Geophys. Res. 63, 876 (1958).
[86] NEHER, H. V.: Phys. Rev. 103, 228 (1956).
[87] NEY, E. P., J. R. WINCKLER and P. S. FREIER: Phys. Rev. Letters 3, 183 (1959).
[88] WINCKLER, J. R., L. PETERSON, R. HOFFMAN and R. ARNOLDY: J. Geophys. Res. 64, 597 (1959).
[89] MEYER, P., and J. A. SIMPSON: Phys. Rev. 106, 568 (1957).
[90] FREIER, P. S., E. P. NEY and C. J. WADDINGTON: Phys. Rev. 114, 365 (1959).
[91] McDONALD, F. B.: Phys. Rev. 116, 462 (1959).
[92] McDONALD, F. B., and W. R. WEBBER: Phys. Rev. 115, 194 (1959).
[93] Cosmic-ray intensity during the IGY, Nos. 1, 2, and 3, March 1959, December 1959, and April 1960, National Committee for the IGY, Science Council of Japan, Veno Park, Tokyo, Japan.
[94] LOCKWOOD, J. A.: J. Geophys. Res. 65, 3859 (1960).
[95] REID, G. C., and H. LEINBACH: J. Geophys. Res. 64, 1801 (1959).
[96] WEBBER, W. R.: Progress in cosmic ray physics, Vol. 6 (in press).
[97] FAN, C. Y., P. MEYER and J. A. SIMPSON: Phys. Rev. Letters 4, 421 (1960).
[98] FAN, C. Y., P. MEYER and J. A. SIMPSON: Phys. Rev. Letters 5, 269 (1960).

[99] COLEMAN jr., P. J., C. P. SONETT, D. L. JUDGE and E. J. SMITH: J. Geophys. Res. 65, 1856 (1960).
[100] FORBUSH, S. E.: J. Geophys. Res. Letter 61, 155 (1956).
[101] MEYER, P., E. N. PARKER and J. A. SIMPSON: Phys. Rev. 104, 768 (1956).
[102] SIMPSON, J. A.: Symposium on Astronomical Aspects of Cosmic Rays, Suppl. Series. Astrophys. J., Suppl. No.44, 44, 369—422 (June 1960).
[103] DORMAN, L. I.: Cosmic ray variations, State Publishing House for Technical Literature, Moscow, U.S.S.R. 727 pp. (1957).
[104] COCCONI, G., K. GREISEN, P. MORRISON, T. GOLD e S. HAYAKAWA: Nuovo Cim. 10, Suppl. 8, 161 (1958).
[105] CARMICHAEL, H., and J. F. STELJES: International Union of Geodesy and Geophysics, Symposium on the July 1959 Events and Associated Phenomena, Helsinki 1960, Monograph No.7, p.10 (November 1960).
[106] QUENBY, J. J., and W. R. WEBBER: Phil. Mag. 4, No.37, 90—113 (January 1939).
[107] International Union of Geodesy and Geophysics, Symposium on the July 1959 Events and Associated Phenomena, Helsinki, July 1960, Monograph No.7 (November 1960).
[108] McCRACKEN, K. G.: J. Phys. Soc. Japan 17, 310, Suppl. A-II, Internat. Conference on Cosmic Rays and the Earth Storm, Part II (1962).
[108a] EHMERT, A., u. G. PFOTZER: Mitt. Max-Planck-Inst. Aeronomie, Nr.8, 54 (1962).
[109] BAILEY, D. K.: J. Geophys. Res. 62, 431 (1957).
[110] LEINBACH, H., and G. C. REID: Phys. Rev. Letters 2, 61 (1959).
[111] LITTLE, C. G., and H. LEINBACH: Proc. I.R.E. (Inst. Radio Engrs.) 46, 334 (1959).
[112] BAILEY, D. K., and J. M. HARRINGTON: J. Phys. Soc. Japan 17, 334, Suppl. A-II, Internat. Conference on Cosmic Rays and the Earth Storm, Part II (1962).
[113] NEY, E. P., J. R. WINCKLER and P. S. FREIER: Phys. Rev. Letters 3, 183 (1959).
[114] REID, G. C., and H. LEINBACH: J. Geophys. Res. 64, 1801 (1959).
[115] ANDERSON, K. A.: Phys. Rev. Letters 1, 335 (1958).
[116] ROTHWELL, P., and C. E. McILWAIN: Nature, Lond. 184, 138 (1959).
[116a] PFOTZER, G., A. EHMERT and E. KEPPLER: Mitt. Max-Planck-Inst. Aeronomie, Nr.9A and B (1962).
[117] FREIER, P. S., E. P. NEY and J. R. WINCKLER: J. Geophys. Res. 64, 685 (1959).
[118] NEY, E. P.: Ann. Rev. Nuclear Sci. 10, 461 (1960).
[119] THOMPSON, A. R., and A. MAXWELL: Nature, Lond. 185, 89 (1960).
[120] KUNDU, M. R., and F. T. HADDOCK: Nature, Lond. 186, 610 (1960).

JOURNAL OF GEOPHYSICAL RESEARCH   VOL. 72, No. 19   OCTOBER 1, 1967

Letters

A Variation, with a Period of Two Solar Cycles, in the Cosmic-Ray Diurnal Anisotropy

SCOTT E. FORBUSH

Department of Terrestrial Magnetism
Carnegie Institution of Washington
Washington, D. C. 20015

Annual means, 1937–1965, of the cosmic-ray diurnal anisotropy component, in the asymptotic direction 128°E of the sun, are well-fitted by a wave with a period of 20 years, which is twice the solar cycle period of 10 years for the interval 1937–1965. This variation is independent of magnetic activity, and when removed the residual variations combine with those in the asymptotic component 38°E of the sun (in which there is no 20-year wave) to give resultant variations, which are principally in the asymptotic component 90°E of the sun. Yearly means of this resultant component are well correlated ($r = +0.75$) with magnetic activity, U_0, and, on the average, vanish for $U_0 = 0$. U_0 is the absolute value of the southward geomagnetic component of the so-called equatorial ring current [*Forbush*, 1966]. The amplitude of the 20-year wave is 60% of the amplitude for the 1937–1965 average diurnal anisotropy 90°E of the sun. The wave passes through a zero in the middle of 1958 near the ·time shown by *Babcock* [1960] for the reversal of the sun's general magnetic field.

In the Archimedean spiral streams from the sun *Ness and Wilcox* [1964] found a decided tendency for the magnetic field, in the plane of the ecliptic, to be oriented parallel or antiparallel to the theoretical direction (135°) proposed by *Parker* [1958]. This is approximately the direction of the cosmic-ray diurnal anisotropy component with the 20-year variation. *Wilcox and Ness* [1965] found that, within sectors, the field directions were consistent with the outward or inward solar fields over the solar region from which the sectors originated. These facts may bear on the mechan-

ism for the 20-year variation in the cosmic-ray diurnal anistropy.

These results are based on the pressure-corrected ion-chamber data at Cheltenham-Fredericksburg (USA) and at Huancayo (Peru) for the period 1937–1965, and at Christchurch (New Zealand) for the period 1937–1961. To eliminate the so-called local temperature effect that remains after correcting to constant barometric pressure, the deviation of each yearly mean vector from a 25-year mean, in the 24-hour LMT harmonic dial, is obtained for each station. These yearly deviation vectors for Huancayo and for Christchurch are normalized to those for Cheltenham by a clockwise rotation of 37° and 20°, respectively, and by multiplying respective amplitudes by 0.91 and 1.00. Components of these yearly deviation vectors at

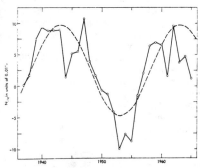

Fig. 1. Twenty-year wave fitted to deviations, N_{-18}, of yearly means from the 1937–1961 average of the diurnal anisotropy component in the asymptotic direction 128° E of the sun.

413

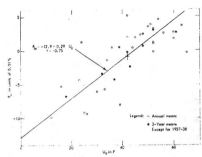

Fig. 2. Correlation between magnetic activity, U_0, and the deviations, P_{36}, of yearly means from the 1937-1961 average, of the diurnal anisotropy component in the asymptotic direction 90° E of the sun.

Cheltenham and of the normalized deviation vectors for Huancayo and for Christchurch are taken on a right-handed set of axes, P_0 and N_0, with the P_0 axis making an angle θ, in degrees, positive clockwise, with the observed 25-year average vector for Cheltenham, which has its maximum at 13.8 hours local solar time in the 24-hour harmonic dial.

Statistical analysis shows that the values of P_0 and of N_0 among the three stations are in remarkably good agreement and that the so-called temperature effect is quite effectively eliminated. Thus, yearly mean values of P_0 and of N_0 averaged for all stations are used hereafter.

Figure 1 shows the 20-year wave fitted to N_{-16}, the yearly deviations from the 25-year mean, of the diurnal anisotropy 128°E of the sun. The amplitude of the 20-year wave is zero in P_{-16}, the yearly deviations from the 25-year mean of the diurnal anisotropy 38° E of the sun. The deviations $(N_{-16}-W)$, of points from the wave in Figure 1 are correlated ($r = +0.75$) with U_0 and with yearly means of P_{-16} ($r = +0.65$), which are also correlated with U_0 ($r = 0.59$). The components parallel, P_{36}, and perpendicular, N_{36}, to the asymptotic direction 90°E of the sun are obtained from P_{-16} and $(N_{-16}-W)$. The mean value of N_{36} is zero, and the correlation between N_{36} and U_0 is negligible.

The correlation ($r = +0.75$) between P_{36} and U_0 is shown by the open circles in Figure 2. Making use of these correlations the values

of N_{-16} in Figure 1 are corrected to $U_0 = 0$ (no magnetic activity). These corrected values of N_{-16} are well fitted in Figure 3 by the 20-year wave used in Figure 1. This indicates that the 20-year variation is independent of magnetic activity. Using the above correlations, and assuming that the total diurnal anisotropy 90°E of the sun vanishes for $U_0 = 0$ (i.e., no magnetic activity), the resulting amplitude of the diurnal anisotropy 90°E of the sun, averaged for 1937-1965, is 0.12%.

As indicated above, the 1937-1965 mean of N_{36} is zero, and the correlation between yearly means of N_{36} and of U_0 is not significant. Also the variance of annual means of N_{36} is only 23% of that for P_{36}. This is the basis for assuming heretofore that the variability of P_{36} arises from variations, from the 25-year average, of the component in the diurnal anisotropy 90° E of the sun. Similarly the time of maximum of the 1937-1965 average vector, say C_{18}, (amplitude 0.12%) for the total diurnal anisotropy is taken as 18.0 hours local asymptotic time. For each station, the local solar time of maximum of the vector C_{18} determines the additive correction required for geomagnetic deflection if, in fact, the local asymptotic time of maximum for C_{18} is 18.0 hours.

For Godhavn (not otherwise used herein), Cheltenham, Christchurch, and Huancayo these corrections, in hours, are, respectively, 1.2, 1.7, 2.7, and 4.1. These values are reasonably compatible with those obtained from calculations of geomagnetic deflections by *Venkatesan and Dattner* [1959] and by *McCracken et al.* [1965]. This indicates that the local asymptotic time of maximum for C_{18} is near 18.0 hours.

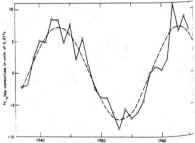

Fig. 3. Twenty-year wave from Figure 1 fitted to N_{-16}, corrected to $U_0 = 0$.

For Huancayo the difference, in the 24-hour harmonic dial on local solar time, between the observed 1937–1965 average diurnal variation vector and the vector ($C_{18}/0.91$) gives the average vector for the so-called temperature effect. This is in fair agreement with preliminary values from a few yearly mean differences between the diurnal variation in the ion-chamber and that from Simpson's IGY neutron monitor.

These results comprise a self-consistent, statistically satisfactory, account of the variability of yearly means of the diurnal anisotropy and of its average for the period 1937–1965. Details will appear in a later publication.

REFERENCES

Babcock, Horace W., The magnetism of the sun, *Sci. Am., 202*, 53–62, 1960.

Forbush, S. E., and L. Beach, The absolute geomagnetic field of the equatorial ring current, *Carnegie Inst. Wash. Year Book, 65*, 28–36, 1966.

McCracken, K. G., V. R. Rao, B. C. Fowler, M. A. Shea, and D. F. Smart, *IQSY Instruction Manual No. 10*, 1965.

Ness, N. F., and J. M. Wilcox, Solar origin of the interplanetary field, *Phys. Rev. Letters, 13*, 461–464, 1964.

Parker, E. N., Dynamics of the interplanetary gas and magnetic fields, *Astrophys. J., 128*, 664–676, 1958.

Venkatesan, D., and A. Dattner, Long term changes in the daily variation of cosmic-ray intensity, *Tellus, 11*, 116–129, 1959.

Wilcox, J. M., and N. F. Ness, Quasi-stationary corotating structure in the interplanetary medium, *J. Geophys. Res., 70*, 5793–5805, 1965.

(Received May 22, 1967.)

Monte Carlo experiment to determine the statistical uncertainty for the average 24-hour wave derived from filtered and unfiltered data[1]

Scott E. Forbush

Department of Terrestrial Magnetism, Carnegie Institution of Washington, Washington, D.C.

AND

S. P. Duggal and Martin A. Pomerantz

Bartol Research Foundation of the Franklin Institute, Swarthmore, Pa., U.S.A.

Received June 21, 1967

To test whether the diurnal variation is more reliably determined from filtered data, a daily harmonic analysis is made before and after filtering an adequate sequence of synthetic bihourly values containing only random noise and a 24-hour wave of constant phase and amplitude. For each of three filters it is shown, empirically, that the statistical uncertainty of the 24-hour wave from N days of such filtered data does not differ significantly from that from N days of unfiltered data. The filters were of different bandwidths and each was designed to pass the 24-hour wave.

Ables *et al.* (1966) emphasized that numerical filtering of cosmic-ray data would result in a more reliable determination of the diurnal and semidiurnal variation than is given by harmonic analysis. For data containing persistent waves that are not harmonics of the diurnal variation, filters would effect, as they indicate, a significant improvement. Bartels (1935, pp. 44–45) showed the effect of such persistent nonharmonic waves on the amplitudes obtained from harmonic analysis, and how such persistent waves might be detected. In the power spectrum shown in Fig. 1 by Ables *et al.* (1935) for Deep River neutron data, the diurnal and semidiurnal waves are persistent, confined to a narrow frequency band, and superposed on the noise spectrum. There is no evidence for persistent variations with frequencies that are not harmonics of the diurnal wave. To approximate such data artificially we take a pure diurnal wave of constant amplitude and phase and superpose random normal noise. We then inquire whether the statistical uncertainty for the average diurnal variation (24-hour wave) is less when obtained from such data filtered than from harmonic analysis of the unfiltered data.

Figure 1(A) shows a sample of bihourly values, y_{tj}, for the bihours t ($t = 0, 1, 2, \ldots, 11$) for each day, j, for "days" 10 to 34 for which

$$(1) \quad y_{tj} = [50 \cos(15° + t_j \times 30°) + \xi_{tj}],$$

where ξ_{tj} is a randomly sampled normal deviate from a population with $\sigma = 100$ and $\mu = 0$. Figures 1 (B) and (C) show respectively a sample output for "days" 10 to 34, from "filtering" y_{tj} with filters I and II. Their amplitude response, symmetrical about $f = 1.00$, is shown in Fig. 2, together with the response for daily harmonic analysis. Filters I, II, and III produce no phase shift. The actual filtering is done digitally by convolving the input signal with the transform of the filter response. The procedures and parameters for the filters are referred to later.

The harmonic coefficients a_1 and b_1 were computed for the 24-hour wave for each day from the unfiltered and from the filtered data.

Bartels (1935) showed that, for single days, the expectancy, $M(1)$, for vectors C_ν, of frequency ν/day, derived by harmonic analysis of r daily random normal ordinates is independent of ν (provided that for r even the coefficient $a_{r/2}$ is excluded) and is given by:

$$(2) \quad M(1) = 2\xi/\sqrt{r} = \{s(a_1)^2 + s(b_1)^2\}^{1/2}$$

in which ξ is the standard deviation of the

[1] Presented at the Tenth International Conference on Cosmic Rays, held in Calgary, June 19–30, 1967, MOD-72A.

Canadian Journal of Physics. Volume 46, S985 (1968)

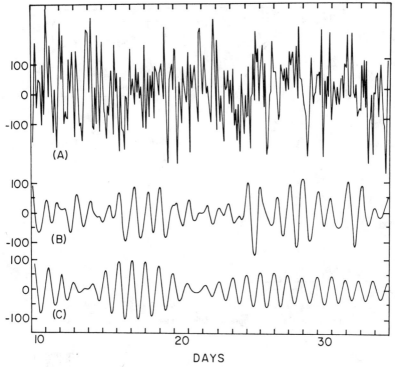

Fig. 1. (A) Sample for "days" 10–34 of bihourly values for pure 24-hour wave plus random noise. (B) Output from (A) using filter I. (C) Output from (A) using filter II.

population from which the r ordinates are randomly drawn and $|C_1| = \{a_1{}^2 + b_1{}^2\}^{1/2}$. The standard deviations of a_1 and b_1 are $s(a_1)$ and $s(b_1)$. For the problem under consideration, $\sigma(a_1) = \sigma(b_1)$ and $r = 12$ (bihourly values).

If, for successive days, the deviations from the average wave are statistically independent, Bartels (1935) showed that

(3) $$M(h) = M(1)/\sqrt{h},$$

in which $M(h)$ is the expectancy for means of C_v for h successive days. Equation (3) may be written:

(4) $$M(h)\sqrt{h} = M(1).$$

If, in a graph of $M(h)\sqrt{h}$ vs. $\sqrt{h}$, $M(h)\sqrt{h}$ is (within statistical uncertainties) constant, or independent of $\sqrt{h}$, then the vector deviations for successive days are statistically independent. When these deviations are not

statistically independent (but quasi-persistent) $M(h)\sqrt{h}$ at first increases with $\sqrt{h}$ and for larger $\sqrt{h}$ asymptotically approaches $M_e(1)$, the effective expectancy for single days; that is, when h becomes large enough so that means of C_1 for h successive (nonoverlapping) days are essentially independent.

Figure 3 shows the computed values of $M(h)\sqrt{h}$ as a function of $\sqrt{h}$, for the unfiltered data and for the filtered data using filters I and II. The horizontal lines in the graphs indicate the value $M(1)$ computed from (1).

In the bottom graph of Fig. 3 it should be emphasized that from $\sqrt{h} = 1$ to $\sqrt{h} = 2$ there is no significant increase in $M(h)\sqrt{h}$. The number of samples for computing $M(h)$ is N/h with N the number of days available. On account of sampling fluctuations, h should not be much greater than about $N/50$ if the

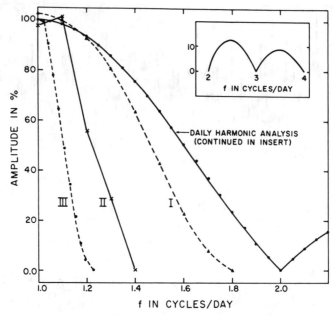

FIG. 2. Amplitude response for daily harmonic analysis and for filters I, II, and III when >1%.

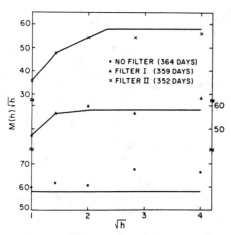

FIG. 3. $M(h)\sqrt{h}$ vs. $\sqrt{h}$ for 24-hour waves from daily harmonic analysis before and after filtering bihourly values comprised of a pure 24-hour wave plus random noise.

expectancy is to be determined within about 10%. For Fig. 3 at $\sqrt{h} = 2.83$ ($h = 8$) N/h is about 45. Thus, for the unfiltered data the vector deviations (from the average in a 24-hour harmonic dial) may safely be considered statistically independent.

For filters I and II, $M(h)\sqrt{h}$ clearly increases as $\sqrt{h}$ increases from 1 to 2 or 3, and in both cases $M(h)\sqrt{h}$ approaches the value 57.7, computed from (1), which is indicated by the horizontal line in the graphs. The increase in $M(h)\sqrt{h}$ with $\sqrt{h}$ for filter III is shown in Fig. 5(A) using the 113 "days" of available data. For the filtered data this clearly shows that, in the 24-hour harmonic dial, the vector deviations (from the average) for successive days are definitely not statistically independent, but quasi-persistent.

Consequently, if the expectancy $M(h)$, for the means of C_1 from h days, is computed for these filtered data from $M(1)$ using (3), then $M(h)$ is seriously underestimated and the reliability for the average vector from N days of filtered data is thus overestimated.

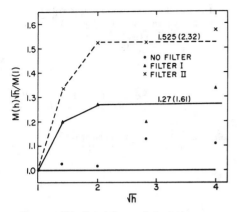

FIG. 4. $M(h)\sqrt{h}/M(1)$ vs. $\sqrt{h}$ for 24-hour waves from daily harmonic analysis before and after filtering bihourly values comprised of a pure 24-hour wave plus random noise.

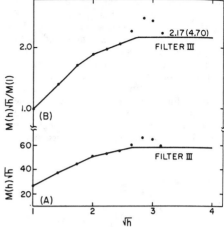

FIG. 5. (A) $M(h)\sqrt{h}$ vs. $\sqrt{h}$ and (B) $M(h)\sqrt{h}/M(1)$, for 24-hour waves from daily harmonic analysis after filtering bihourly values comprised of a pure 24-hour wave plus random noise.

In Figs. 3 and 5(A) the asymptotic value for $M(h)\sqrt{h}$ is 57.7 for all three filters and for the unfiltered data. Thus, 57.7 is the *effective expectancy*, $M_e(1)$, for single days and the expectancy for the average wave from N days is $57.7\sqrt{N}$, which is the same whether derived from N days of filtered or unfiltered data. Thus for the data used here the statistical reliability of the average wave from N days of data is not improved by filtering.

It should be emphasized that for the filtered data the expectancy for the average wave from N days *must not* be computed from $M(1)/\sqrt{N}$, with $M(1)$ derived from single days since this erroneously assumes that deviations for successive days are statistically independent. Figures 4 and 5(B) show $M(h)\sqrt{h}/M(1)$ as a function of $\sqrt{h}$ for filters I, II, and III, and for the unfiltered data. If h_e denotes the square (shown in parentheses) of the asymptotic value approached by $M(h)\sqrt{h}/M(1)$, then $N/h_e = N_e$ is the effective number of independent days among the N days available. The expectancy for the average wave for N days could be properly obtained for the filtered data from $M(1)/\sqrt{N_e} = M(1)/\sqrt{(N/h_e)}$, which is the same as $M_e(1)/\sqrt{N}$.

Thus, if the data contain only random noise

and a pure persistent 24-hour (or 12-hour) wave (or if the frequency spectrum is narrow for the diurnal or semidiurnal wave) and no persistent waves with nonharmonic frequencies, then harmonic analysis from filtered data results in an average wave which has the same statistical uncertainty as that derived from harmonic analysis of the unfiltered data. In cosmic-ray monitors finite counting rates are subject to random errors, and to other uncertainties which are effectively random (i.e. meteorological effects from rapid pressure changes, etc.).

TABLE I

Filter parameters after Martin (1959)

Filter No.	N	h	r_{c1}	r_{c2}
I	18	0.030	0.075 00	0.091 66
II	60	1/110	0.075 00	0.091 66
III*	60	1/110	0.083 33	0.083 33

*Cosine-type filter.

The parameters for the filters are given in Table I in accord with the notation used by Martin (1959). This reference describes the numerical calculation of the gain function (or response as plotted in Fig. 1) and of the weighting function, or cosine transform of the filter function, which when convolved with the input (a time-consuming process even for digital computers) gives the filtered output time function. From Table I ($2N + 1$) gives the number of bihourly values for which the weighting function (symmetrical about the center of the interval) was computed and used. r_{c1} and r_{c2} indicate the limits between which the amplitude response is essentially

flat at 100%. The corresponding frequencies in cycles/day are given by $12r_{c1}$ and $12r_{c2}$. Outside the limits ($r_{c1} - 2h$) and ($r_{c2} + 2h$) the response is negligible.

REFERENCES

ABLES, J. G., McCRACKEN, K. G., and RAO, U. R. 1966. Proc. Intern. Conf. Cosmic Rays, London, 2, 208.
BARTELS, J. 1935. Terrest. Mag. 40, 1.
MARTIN, M. A. 1959. IRE Trans. on Space Electronics and Telemetry, 33.

Journal of

GEOPHYSICAL RESEARCH

Space Physics

VOLUME 74   JULY 1, 1969   No. 14

Variation with a Period of Two Solar Cycles in the Cosmic-Ray Diurnal Anisotropy and the Superposed Variations Correlated with Magnetic Activity

SCOTT E. FORBUSH

Department of Terrestrial Magnetism
Carnegie Institution of Washington
Washington, D. C. 20015

Annual means of the diurnal anisotropy from 1937 to 1967 are shown to result from the addition of two distinct diurnal components. One component, with maximum in the asymptotic direction 128° east of the sun, contains a well-determined wave, W, with a period of two solar cycles. W passes through zero in 1958 when the sun's poloidal field reversed. The remaining component with W eliminated, has its maximum in the asymptotic direction 90° east of the sun. Annual means of this component, with maximum at 18.0 hours local asymptotic time, are well correlated ($r = +0.75$) with magnetic activity and determine a solar cycle variation with minimum near sunspot minimum and amplitude about two-thirds that of W. These results derive from a statistical investigation of the variability of annual means of the diurnal variation from ion-chamber data at Cheltenham-Fredericksburg, Huancayo, and Christchurch. The absolute, or total, diurnal anisotropy and the atmospheric diurnal temperature effect are in reasonable agreement with those derived independently through a comparison between the diurnal anisotropy from ion-chamber data and from Simpson's 1953-1966, IGY neutron monitor data at Huancayo.

INTRODUCTION

An impressive agreement between the observed diurnal variation derived from neutron monitors and that predicted by the *Axford* [1965]-*Parker* [1964] model was shown by *Mc-Cracken and Rao* [1965]. They concluded that the diurnal anisotropy was invariant during the period 1957–1965. However, statistical tests by *Duggal et al.* [1967] showed that for the available neutron monitor data the amplitude of the diurnal anisotropy (and of the asymptotic component 90° east of the sun) was significantly less in 1965 than in 1958. These were, respectively, years of maximum and minimum cosmic-ray intensity. They also showed, from ionization chamber data for the Carnegie Institution of Washington stations, that minima of the amplitude of the annual means of the diurnal anisotropy occurred in 1944 and 1954 when maxima of cosmic-ray intensity were observed. In ionization chamber data from Cheltenham-Fredericksburg, Christchurch, and Huancayo for the period 1937 to 1959, *Forbush and Venkatesan* [1960] found relatively large variations in the annual means of the diurnal anisotropy that were similar at these three stations. They observed that these variations suggested the possibility of a 22-year variation. *Thambyahpil-*

lai and Elliot [1953] noted the possibility of a 22-year variation in the time of maximum for the yearly mean 24-hour wave in ionization-chamber data for the interval 1932–1952.

To investigate critically the nature and reality of these apparent variations requires a more detailed appropriate statistical analysis of the diurnal anisotropy from the available ionization chamber data from 1937 to 1966. The use of digital filtering was considered since Abels et al. [1965] indicated this would improve the statistical reliability of the diurnal anisotropy. However, it was shown by Forbush et al. [1968] that filtering did not reduce the statistical uncertainty of harmonic coefficients for the diurnal variation derived from synthetic data comprised of random noise superposed on a constant diurnal wave. Appropriate analyses resulted in procedures that reduced the uncertainties in the diurnal anisotropy, at Huancayo, due to quasi-systematic uncertainties in barometric pressure as a result of friction in the recording barographs.

BAROMETRIC CORRECTIONS AND DATA USED

For the available annual means of the 24-hour wave in pressure-corrected ionization (less bursts) the Fourier coefficients a_1 and b_1 in Table 1 are referred to 75° WMT for Huancayo (HU) and Cheltenham-Fredericksburg (CH), to 172.5° EMT for Christchurch (CC), and to 45° WMT for Godhavn (GO). These coefficients are corrected for the use of bihourly means according to Bartels [1935]. The effect of the so-called noncyclic change in these coefficients is negligible. The barometric coefficients used were −0.33%/mm Hg for Huancayo and −0.18%/mm Hg for the other stations.

The approximate amplitudes, in mm Hg, for the annual mean 24-hour wave in barometric pressure are 1.0, 0.5, 0.35, and 0.1, respectively, for HU, CH, CC, and GO. Thus, the amplitudes, in per cent, for the annual mean 24-hour wave in ionization resulting from that in pressure are about 0.33, 0.09, 0.06, and 0.02, respectively, for HU, CH, CC, and GO. For a given error in phase due to any friction in the barographs, the resulting error in pressure-corrected ionization is greater for HU and CH than for the other stations. For Huancayo this effect was discussed by Forbush and Venkatesan [1960] who showed that the effect at Chelten-

ham was negligible. In March 1957 the Instituto Geofísico del Peru replaced the ordinary barograph previously used for correcting ionization chamber results at Huancayo with a microbarograph. Recently Dr. John A. Simpson kindly made available the 1953–1966 pressure data from the barograph used with his IGY neutron-monitor at Huancayo. These data were all analyze by Forbush [1968] to investigate the behavior of the different barographs and to re-determine the barometric coefficient at Huancayo. This analysis indicated the essential validity of the procedure adopted before Simpson's pressure data were obtained, to derive the pressure-corrected Fourier coefficients in Table 1 for Huancayo.

In this procedure constant corrections +8 and +34, in units of 0.01%, were added, respectively, to the Fourier coefficients $a_1(I_p)$ and $b_1(I_p)$ for each annual mean 24-hour wave, 1937–1968, in ionization (less bursts) uncorrected for pressure.

The effect of friction was found to be greatest and variable from year to year in the microbarograph. In data from Simpson's barograph there was no evidence for friction, and the constant corrections of +8 and +32 (in units of 0.01%) based on these data may be compared, respectively, with the previously adopted values of +8 and +34 given above. The analysis by Forbush [1968] shows that at Huancayo the 12-hour wave in ionization uncorrected for pressure results entirely from that in pressure. This analysis also revealed an increase, after about 1955, of about 10% in the sensitivity used to reduce the ionization chamber results for Huancayo, possibly due to a change in the zero of the meter used to read the electrometer calibration voltage.

NORMALIZATION AND COMPARISON OF THE
VARIABILITY OF ANNUAL MEANS OF THE
DIURNAL VARIATION AT THREE STATIONS

The values a_1, b_1 in Table 1 for year y and station s specify the pressure-corrected diurnal variation by a vector $\mathbf{C}_{ys}$ in a 24-hour harmonic dial. The deviation, $\Delta\mathbf{C}_{ys}$, for year y at station s from the vector mean for n years is then

$$\Delta\mathbf{C}_{ys} = \mathbf{C}_{ys} - \left(\sum \mathbf{C}_{ys}/n\right) \qquad (1)$$

If the annual mean 24-hour wave due to the temperature effect at a given station is constant

TABLE 1. Fourier Coefficients for 24-Hour Wave in Pressure-Corrected Cosmic-Ray Ionization, Origin Standard Mean time, in Units of 0.01%

	Huancayo		Cheltenham[†]		Christchurch		Godhavn	
	a_1	b_1	a_1	b_1	a_1	b_1	a_1	b_1
1936[*]	−12	+15	−17	− 1	−12	− 5		
1937	−11	+12	−12	− 3	−11	− 3		
1938	−13	+10	−12	− 4	−15	− 4		
1939	−17	+ 7	−14	−11	−14	−11	−5	− 6
1940	−17	+ 5	−10	−11	−16	−14	−5	− 6
1941	−16	+ 4	−10	−10	−14	−13	−4	− 6
1942	−15	+ 2	− 8	−11	− 9	−11	−5	− 6
1943	−17	+ 3	−12	−12	−13	−11	−6	− 9
1944	− 7	+ 7	− 7	− 2	− 6	− 5	−4	+ 1
1945	−11	+ 6	− 8	− 7	− 8	− 9	−5	− 5
1946	−16	+10	−10	−10	−13	− 8	−6	−14
1947	−17	+ 4	− 5	−16	−10	−12	−7	− 7
1948	−12	+10	−14	−11	−12	− 5	−8	−16
1949	−14	+13	−14	− 9	−15	− 1	−6	−17
1950	−16	+13	−12	0	−14	− 2	−7	−11
1951	−19	+16	−15	− 2	−16	0	−9	− 5
1952	−12	+18	−15	0	−13	+ 2	−5	0
1953	− 7	+21	−11	+ 8	− 4	+ 7	−7	0
1954	+ 4	+15	0	+ 6	− 1	+ 5	0	+ 4
1955	− 4	+20	− 8	+ 4	− 7	+ 8	−5	+ 2
1956	−13	+13	−17	− 3	−13	+ 1	−9	− 4
1957	−14	+11	−13	− 8	−15	− 1	−9	− 4
1958	−14	+ 2	−10	− 8	−10	− 7	−4	− 6
1959	−16	+ 3	− 9	−10	−10	− 7		
1960	−14	+ 2	−10	−10	−10	− 6		
1961	− 7	+ 9	− 7	− 3	− 6	− 6		
1962	−16	+ 2	−12	−13				
1963	− 9	+ 7	−11	− 7				
1964	− 8	+ 6	−10	− 9				
1965	− 6	+ 7	− 4	− 2				
1966	−14	+ 2	− 9	− 7				
1967	−11	+ 5	− 5	− 7				
1968	−15	+ 2	− 8	− 9				

[*] HU: June 1936 to May 1937; CH: January to December 1936; CC: April 1936 to March 1937.
[†] Fredericksburg after October 1956.

from year to year, then it is eliminated from ΔC_{y_s}. The unknown average of the n annual means of the 24-hour wave, with the diurnal temperature effect removed, is also eliminated. However, it is shown later that this average wave can be determined reasonably well. To compare and normalize the vectors, ΔC_{y_s}, for n years ($y = 1, 2, \dots n$) at two stations, each vector ΔC_{y_1} for station 1 is rotated through a particular angle θ_{y_i}, so that its time of maximum is at zero hour SMT and each vector ΔC_{y_2} for station 2 is rotated through the same particular angle θ_{y_i} for the corresponding year. The 24-

hour harmonic dial of Figure 1 shows the sequentially cumulated sum (to the year indicated by the two digits) of the rotated vectors for Huancayo, Christchurch, and Godhavn when the vectors ΔC_{y_s} for Cheltenham are rotated to zero hour SMT. Or, equivalently, if in (1) we put $s = 1$ for Cheltenham and $s = 2$ for the second station and let

$$\sum_{y=1}^{n} R_{y2}$$

indicate the summation of rotated vectors for the second station, then

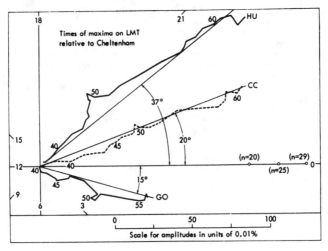

Fig. 1. Twenty-four-hour summation harmonic dial for deviations from mean of n years rotated to 0 hour for Cheltenham and through the corresponding angle for Godhavn ($n = 20$), Christchurch ($n = 25$), and Huancayo ($n = 29$).

$$\sum_{\nu=1}^{n} \mathbf{R}_{\nu 2}$$

$$= \sum_{\nu=1}^{n} \frac{[\Delta\mathbf{C}_{\nu 1} \cdot \Delta\mathbf{C}_{\nu 2}\mathbf{i} + \Delta\mathbf{C}_{\nu 1} \times \Delta\mathbf{C}_{\nu 2}\mathbf{j}]}{|\Delta\mathbf{C}_{\nu 1}|} \quad (2)$$

in which $\mathbf{i}$ and $\mathbf{j}$ are unit vectors, respectively, in the direction of the axes through 0 and 6 hours in Figure 1. For Cheltenham, $s = 1$, the summation in Figure 1 is simply

$$\sum_{\nu=1}^{n} |\Delta\mathbf{C}_{\nu 1}|$$

The summation vector for Cheltenham is shown by the points marked $n = 20$, 25, and 29 since the number of years (n in equation 1) for which data were available (in Table 1) was not the same for all stations. Incidentally, the data for 1936, 1966, 1967, and 1968 in Table 1 were not available when Figure 1 was made. On Figure 1 are indicated the values in degrees, adopted for the difference in phase between the corresponding yearly deviation vectors, $\Delta\mathbf{C}_{\nu s}$, at Cheltenham and at the other stations. Thus, to normalize the phase of yearly deviation vectors to the phase of those for Cheltenham, each yearly deviation vector was rotated by a constant angle; 37° clockwise for Huancayo, 20° clockwise for Christchurch, and 15° counter-

clockwise for Godhavn. An harmonic dial was made analogous to that in Figure 1 except that yearly deviation vectors, $\Delta\mathbf{C}_{\nu s}$, for Christchurch were rotated to zero hour time of maximum and those for the other stations through the corresponding angle. From this figure (not shown) the magnitude of the summation vector (25 years) for Cheltenham was 0.91 times that for Huancayo, and 1.00 times that for Christchurch in Figure 1. Similarly the magnitude of the summation vector for Cheltenham (20 years) was 1.78 times that for Godhavn. This procedure avoids using the amplitude of the summation vector from yearly deviation vectors rotated to 0 hour (i.e., for Cheltenham in Figure 1). Since these amplitudes are in fact the sum of scalar amplitudes of individual vectors, this sum is too large owing to statistical errors in the individual vectors. Thus, to normalize the amplitudes to those for Cheltenham, the amplitude of each yearly deviation vector, $\Delta\mathbf{C}_{\nu s}$, in (1), was multiplied by 0.91, 1.00, and 1.78, respectively, for Huancayo, Christchurch, and Godhavn.

From the yearly deviation vectors at Huancayo and at Christchurch, with phases and amplitudes normalized to those for Cheltenham, and from the yearly deviation vectors for Cheltenham, components P_s and N_s are taken on

424

the axes P_θ and N_θ in Figure 2. In Figure 2, θ is the angle between the P_θ axis and the observed average vector for the 25 years, 1937–1961, at Cheltenham. Yearly values of P_0 and N_0 (for $\theta = 0$) are plotted in Figure 3. The solid lines indicate, for each year, the average P_0, or N_0, from the three or two stations for which data were available.

Figure 3 shows good agreement between the values of P_0 and N_0 from the different stations. The statistical significance of the variations in Figure 3 among years and between stations is obtained from a simple analysis of variance that also provides the residual variance for a single station-year that determines the confidence limits shown in Figure 3. Table 2a lists the variances: $s^2(\Delta P_0)$, $s^2(\Delta N_0)$ and $M^2 = s^2(\Delta P_0) + s^2(\Delta N_0)$ for yearly differences of P_0 and of N_0 for the indicated pairs of stations. Since M^2 differs little for the three pairs, the variance at the three stations is essentially homogeneous, and valid results may be obtained from the analysis of variance.

Table 2b summarizes the results of the anal-

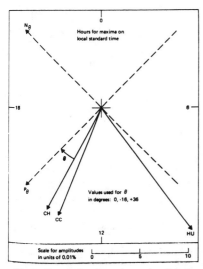

Fig. 2. Twenty-four-hour harmonic dial for average, 1937–1961, vector for Cheltenham, Christchurch, and Huancayo, and axes for components P_θ, N_θ of normalized deviation of yearly means from 25-year mean.

ysis of variance for values of P_0, and N_0 for the 25 years, 1937–1961, at three stations. The variance among stations (row 2) is zero, which results from normalizing the yearly deviation vectors at Huancayo and Christchurch (in equation 1) to those at Cheltenham. Each of the variances: $s^2(P_0)$, $s^2(N_0)$, and $M^2 = s^2(P_0) + s^2(N_0)$ is greater than 20 times that for the corresponding residual variance in the last row of Table 2b. This indicates that it is practically certain that the variations, among years, in P_0 and in N_0 are due to real variations and not to statistical uncertainties in the station values. For a single station-year mean of the diurnal variation the residual variance, $M^2 = 6.0$ ($M = 0.024\%$), is 2.6 times the underestimated value $M^2 = 2.3$ derived from the 'counting rate,' and only 1.5 times the more realistic value 4.0 ($M = 0.020\%$) based on the variability on 273 days for Cheltenham that was obtained earlier by Forbush [1937]. This shows that the yearly values of P_0 (and of N_0) in Figure 3, for the three stations agree remarkably well. This in turn means that the yearly mean diurnal temperature effect must be nearly constant from year to year at each station.

THE 20-YEAR WAVE IN THE DIURNAL ANISOTROPY AND STATISTICAL PROCEDURES FOR DETERMINING THE SUPERPOSED VARIATIONS ASSOCIATED WITH MAGNETIC ACTIVITY

Since the period of the sunspot cycle from 1937 to 1966 was quite close to 10 years, the fundamental solar period during this interval is taken as 20 years. A wave with this period was fitted to each of the solid graphs in Figure 3. These waves in N_0 and P_0 show that the amplitude of the 20-year wave in the component P_{-16} (for $\theta = -16°$ in Figure 2) should be zero as is actually found. Figure 4 shows the 20-year wave fitted to N_{-16}. It is shown later that N_{-16} is the component of the diurnal anisotropy in the asymptotic direction 128°E of the sun, and it is the amplitude of this component that exhibits the 20-year variation.

In Figure 4 let the vertical distance between the yearly means of N_{-16} (open circles) and the wave be designated $(N_{-16} - W)$. In Figure 5a, these yearly values of $(N_{-16} - W)$ are plotted as a function of the corresponding yearly means of

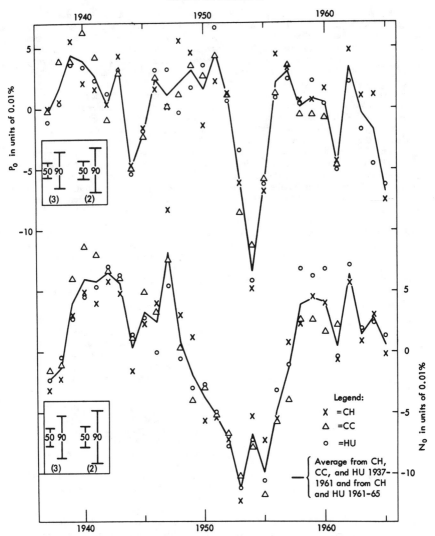

Fig. 3. Yearly means of P_0 and of N_0 for Cheltenham-Fredericksburg (CH), Christchurch (CC), and Huancayo (HU) and average for year from stations as indicated. Inserts indicate 50% and 90% confidence limits for means of (3) and (2) stations.

U_0. U_0 is the horizontal geomagnetic component (positive southward) at the geomagnetic equator due to the so-called equatorial ring current (ERC) here used as a measure of geomagnetic activity. For annual means, the zero level for

U_0 has been determined [*Forbush and Beach,* 1967] with an uncertainty of about 2 or 3 gammas ($1 \gamma = 10^{-5}$ oe). It is assumed that annual means of U_0 provide an indirect measure of irregularities of the magnetic field that are

TABLE 2a. Variances, $s^2(\Delta P_0)$ and $s^2(\Delta N_0)$ for Difference between Pairs of Annual Means of P_0 and of N_0 (in units of 0.01%) from n years at Cheltenham, Christchurch, and Huancayo

Pair	n	$s^2(\Delta P_0)$	$s^2(\Delta N_0)$	M^2
CH-CC	25	5.2	8.0	13.2
CH-HU	29	6.5	6.3	12.8
CC-HU	25	4.1	5.9	10.0

From (a): Variance homogeneous.

TABLE 2b. Analysis of Variance Results from Yearly Means, 1937-1961, at CH, CC, and HU, for P_0 and N_0 (in units of 0.01%)

	d. f.	$s^2(P_0)$	$s^2(N_0)$	M^2
Among years	24	53.0	87.5	140.5
Among stations	2	0.0	0.0	0.0
Residual	48	2.5	3.5	6.0

From (b): Residual variance $M^2 = s^2(P_0) + s^2(N_0) = 6.0$ ($M = 0.024\%$).

For single station-year mean, variance among years $> 20 \times$ residual variance.

Note; Std. Dev. (0.66%) for hourly 'counting rate' results in underestimated $M^2 = 2.3$ ($M = 0.015\%$) for one station-year.

From variability of 24-hour waves on 273 single days at Cheltenham $M^2 = 4.0$ ($M = 0.020\%$) for one station-year

basic to the corotation model for the diurnal anisotropy.

In Figure 5b yearly means of P_{-16}, in which there is no 20-year wave, are shown as a function of U_0. Figure 5c shows yearly means of P_{-16} as a function of those for $(N_{-16} - W)$. The dashed lines, in Figure 5, through the center one of the three crosses are the two regression lines the slopes of which are given in Table 3 in the rows with $w_x = \infty$ or $w_x = 0$.

In Figures 5a, b, and c the upper cross indicates the average, for each of the coordinates, for 15 years with sunspot numbers greater than 55. Similarly the lower cross is the average for 14 years with sunspot numbers less than 55.

This dichotomy using sunspot numbers results in averages that are not biased by the 'errors' in either coordinate of Figure 5. Considering the number of years involved, the 'errors' in these averages are assumed negligible, which is the basis for adopting the solid lines in Figure 5 for the most probable relation between the coordinates. These parameters for the solid lines are given in Table 3 together with a value, w_x, for the weight of x relative to that ($w_y = 1$)

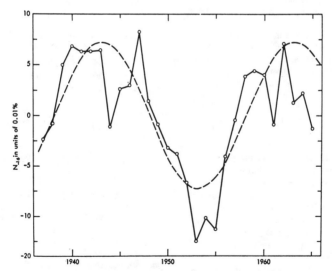

Fig. 4. Twenty-year wave fitted to yearly means of N_{-16}, the component in the asymptotic direction 128°E of the sun. In P_{-16}, the asymptotic component 38°E of the sun, there is no 20-year wave.

TABLE 3. Correlation Coefficients, r, between Yearly Means of x and y and Weighted Least-Squares Parameters in $y = a + bx$ for Weight w_x of x Relative to w_y

Figure	Y	x	r	w_x	a^*	$b\dagger$
5a	$(N_{-16} - W)$	U_0	0.75	∞	$- 8.7$	0.186
				0	-14.7	0.330
				0.110	-10.3	0.226
5b	P_{-16}	U_0	0.59	∞	$- 6.4$	0.146
				0	-17.6	0.415
				0.182	$- 7.8$	0.180
5c	P_{-16}	$(N_{-16} - W)$	0.65	∞	$+ 0.3$	0.646**
				0	$+ 1.1$	1.530**
				2.55	$+ 0.4$	0.794**
7	P_{36}	U_0	0.75	∞	-10.7	0.236
				0	-18.4	0.421
				0.174	-12.9	0.288

* All values of a are in units of 0.01%.
† All values of b are in units of 0.01% per gamma except double-starred values for Figure 5c, which are dimensionless.

for y. Using these values of w_x in fitting a line to the individual points in Figure 5, by minimizing the sum of the weighted squares of the residuals, results in the parameters listed in Table 3 that were determined from the solid lines in Figure 5 as described above. These weights, w_x, are required only to determine the most probable values corresponding to the observed points and the 'errors' or differences between these.

For convenience a summary is given of the procedure for determining the weighted least-squares parameters and the most probable values corresponding to those observed.

Thus, suppose in the absence of observational errors that y is a strictly linear function of x given by

$$y = a + bx \qquad (3)$$

When x and y are both subject to random errors, Uhler [1923] gives procedures for determining the constants that minimize the sum of weighted squares of the residuals for the case when the weights for x and y (inversely proportional to the variance of the errors) are constants. It is, however, more convenient to first determine the slope, b, for each of the two regression lines and then obtain the slope, m, for the actual weights.

Thus, let w_x equal the weight of the x coordinates relative to that ($w_y = 1$) for the y coordinates. If in (3) we put

$b = \alpha$ for $w_x = \infty$ (all error in y)
and

$b = 1/\beta$ for $w_x = 0$ (all error in x)
then

$$\alpha = rs_y/s_x = \frac{\sum (x - \bar{x})(y - \bar{y})}{\sum (x - \bar{x})^2} \qquad (4)$$

and

$$1/\beta = s_y/rs_x = \frac{\sum (y - \bar{y})^2}{\sum (x - \bar{x})(y - \bar{y})} \qquad (5)$$

In (4) and (5) r is the correlation coefficient, s_y and s_x are the sample values of the standard deviations of y and of x, and $\bar{x}$ and $\bar{y}$ are the sample means.

For a particular value w_x let $b = m$, then from the results given by Uhler [1923], it can be shown that

$$m = \tfrac{1}{2}[- A \pm (A^2 + 4w_x)^{1/2}] \qquad (6)$$

with

$$A = \left[(w_x/\alpha) - \left(\frac{1}{\beta}\right) \right] \qquad (7)$$

Using this value of m, a in (3) is obtained from

$$a = \bar{y} - m\bar{x} \qquad (8)$$

If m is known, then w_x is determined from

$$w_x = [(m/\beta) - m^2] \Big/ \left[\frac{m}{\alpha} - 1\right] \qquad (9)$$

The values of w_s in the third row of each section in Table 3 were determined from (9) using for α the value under b with $w_s = \infty$ and for $1/\beta$ the value under b with $w_s = 0$. It remains to determine the most probable values of the coordinates consistent with w_s.

In a graph of y as a function of x let λ equal the ratio of the number of units of x per unit distance on the x scale to that for y. In such a graph a line with slope $-(\lambda w_x/m)$, drawn through an observed point intersects the weighted least-squares line, $y = (\bar{y} - m\bar{x}) + mx$, at a point the coordinates of which are the most probable values of x and y, [Uhler, 1923]. Thus, for Figure

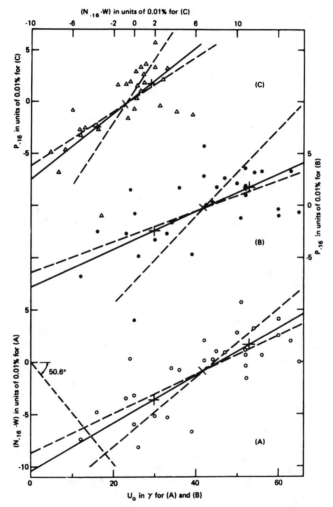

Fig. 5. Correlations: Magnetic activity (a) U_0 and N_{-16} (with 20-year wave, W, removed); (b) U_0 and P_{-16}; and (c) N_{-16} (W removed) and P_{-16}. Regression lines are dashed, and weighted least-squares lines are solid.

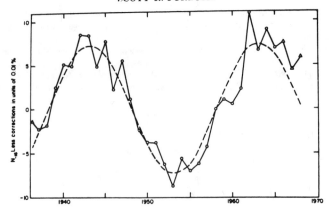

Fig. 6. Twenty-year wave from Figure 4 and yearly means of N_{-16} less corrections for U_0 derived from Figure 5a.

$5a$ $\lambda = 2.5$, and lines parallel to the dashed line, with slope $-(2.5 \times 0.110/0.226) = -1.216$ (which makes an angle of $50.6°$ with the x axis), are drawn through each observed point. Each of these lines intersects the solid line at a point the coordinates of which are the most probable values corresponding to the observed coordinates.

Thus determined, the most probable value $(N_{-16} - W)_i^*$, for year i, indicates the effect of U_0 on N_{-16}. Consequently, the yearly values of N_{-16} may be corrected for variations in U_0 by subtracting values of $(N_{-16} - W)^*$ for the corresponding year. These corrected values of N_{-16} plotted in Figure 6, are quite well fitted by the same 20-year wave shown in Figure 4.

The standard deviation of the departures of the points from the wave in Figure 6 is 1.8 in units of 0.01%. In Table 2b the two-dimensional variance, M^2, of residuals (M in units of 0.01%) for one-station year is 6.0, from which it follows that the one-dimensional standard deviation of statistical uncertainties in the mean of three stations is 1.0 (in units of 0.01%). Comparison of this with the value 1.8 above indicates that the wave in Figure 6 fits the points reasonably well.

The first and the last three points (triangles) in Figure 6 were added recently and were not available when the 20-year wave was fitted to the remaining points. The wave passes through zero at 1958.7 close to the time found by Bab-cock [1959] for the reversal of the sun's polar magnetic field. An informal communication from Mt. Wilson Observatory stated that no indication of reversal of the sun's polar field had appeared by the end of 1968. This possibility accounts for the observed point for 1968 being considerably above the wave. In any case the period of the wave may vary several years as does the interval between sunspot minima. A further test of the influence of the sun's polar magnetic fields on the diurnal anisotropy will doubtless be provided in the next few years by noting in Figure 6 whether the ordinates of the points observed after 1968 change sign when the sun's polar field is next observed to reverse.

It may be recalled that the amplitude of the 20-year wave is greatest in the component of the diurnal anisotropy in the asymptotic direction $128°$ east of the sun. Thus, for positive ordinates of the 20-year wave in Figure 6 (e.g. 1944 and 1964) the time of maximum of this component occurs when the 'direction of viewing' is approximately along the Archimedean spiral stream, predicted from the model of Parker [1958], and away from the sun. Wilcox and Ness [1965] found a decided tendency for the magnetic field, in the plane of the ecliptic, to be oriented parallel or antiparallel to the direction of the solar streams. These facts may bear on the mechanism responsible for the 20-year variation.

THE DIURNAL ANISOTROPY WITH THE 20-YEAR WAVE REMOVED AND ITS DEPENDENCE ON MAGNETIC ACTIVITY

From Table 3 the equations for the solid lines in Figures 5a, b, and c, with U_0 in γ and $(N_{-16} - W)$ and P_{-16} in units of 0.01% are, respectively

$$(N_{-16} - W) = -10.3 + 0.226 U_0 \qquad (10)$$

$$P_{-16} = -7.8 + 0.180 U_0 \qquad (11)$$

$$P_{-16} = 0.4 + 0.794(N_{-16} - W) \qquad (12)$$

Referring to Figure 2, equation 12 shows that if the components $(N_{-16} - W)$ and P_{-16} from (10) and (11) are referred to axes P_{36} and N_{36} (since $\theta = [16° + \cot^{-1} 0.794] \doteq 36°$), the component N_{36} should be negligible and independent of U_0. From (10) and (11) these components, with U_0 in γ and P_{36} and N_{36} in units of 0.01%, are

$$P_{36} = -12.9 + 0.288 U_0 \qquad (13)$$

and

$$N_{36} = -0.2 - 0.003 U_0 \qquad (14)$$

With W removed (14) shows that the component in the direction of N_{36} (Figure 2) is essentially independent of U_0. Also the correlation coefficient between annual means of N_{36} (W removed) and U_0 is zero.

In Figure 7 annual means of P_{36} are plotted as a function of U_0. The upper and lower crosses are averages, respectively, for the 15 years with sunspot numbers greater than 55 and for the 14 years with sunspot numbers less than 55. The central cross is the mean for all years. The equation given in Figure 7 for the line through the crosses corresponds with the parameters in the bottom row of Table 3 and is the same as equation 13. In the second and third rows from the bottom of Table 3 are given the parameters for the two regression lines that are not shown in Figure 7. The basis for adopting the solid line in Figure 7 for approximating the most probable linear relation between P_{36} and U_0 is the same as that described above in connection with Figure 5a. The determination of w_s from (9) is analogous to that discussed in connection with Figure 5.

In the corotation theory no diurnal anisotropy would be expected in the absence of large-scale magnetic field irregularities (in the solar streams) with sources on the sun. We assume that the absence of such irregularities over a period of a year would be associated with the absence of magnetic activity, and that the equatorial ring current would also vanish. Thus (in

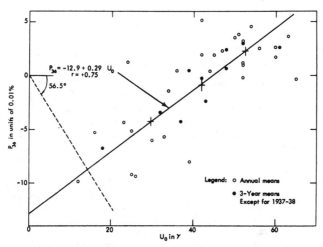

Fig. 7. Correlations between annual means of U_0 and P_{36}, the asymptotic component 90°E of the sun.

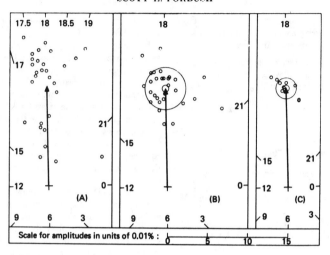

Fig. 8. Twenty-four-hour harmonic dials on local asymptotic time: (a) yearly means averaged for 3 stations (normalized to CH and W removed); (b) from (a) corrected for deviations of U_0 from $\bar{U}_0$; (c) 3-year averages from (b).

Figure 7) we assume that the total component, say T_{36}, in the direction of P_{36}, vanishes for $U_0 = 0$. The validity of this assumption is tested later by comparing the diurnal anisotropy from the ion chamber with that from the neutron monitor at Huancayo. From the equation in Figure 7

$$T_{36} = [P_{36} + 12.9] = 0.29U_0 \qquad (16)$$

with U_0 in γ and the other quantities in units of 0.01%.

Based on the corotation theory we further assume that the 29-year average diurnal anisotropy, after removing the 20-year wave, W, has its maximum at 18.0 hours local asymptotic time as shown for the vector average in Figure 8a. This assumption is equivalent to adopting, for the deflection of trajectories in the earth's magnetic field, the following additive corrections, in hours, to the observed local mean time of maximum of the diurnal variation (if the 24-hour wave due to the temperature effect is removed): Cheltenham-Fredericksburg, 1.9; Christchurch, 3.1; Huancayo, 4.3; and Godhavn, 1.4. The corresponding correction for the IGY neutron monitor at Huancayo is 5.8 hours.

In Figure 8a the points indicate the amplitude and local asymptotic time of maximum for

yearly means of the diurnal anisotropy (W removed) for the 29 years 1937–1965. Only 9 of the 29 vectors in Figure 8a have times of maximum outside the limits 18.0 ± 0.5 hours. Much of the variability in the amplitudes of the yearly vectors is associated with variations in the yearly means of U_0 as shown in Figure 7.

THE DIURNAL ANISOTROPY WITH W REMOVED:
(A) CORRECTED FOR MAGNETIC ACTIVITY AND
(B) THE 0-HOUR AND 18-HOUR ASYMPTOTIC
COMPONENTS AND THE AVERAGE SOLAR
CYCLE VARIATION IN THESE

The slope of the dashed line in Figure 7 is based on the parameters in the section of Table 3 that refers to Figure 7. Thus, the slope is: $-\lambda \ w_x/m = -(2.5 \times 0.174/0.288) = -1.51 = -\tan 56.5°$. The procedure for obtaining λ and the slope is described shortly after equation 9.

In Figure 7 a line through each point and parallel to the dashed line intersects the solid line at a point the ordinate of which is the most probable value (taking account of the relative weights of the two coordinates) of P_{36} corresponding to the value of U_0 for the year involved. The difference between this value of P_{36} and that for the 29-year average, shown by the cross, is used to 'correct' each yearly value of T_{as} (see equation 16) for deviations of yearly values of U_0 from the 29-year mean (41.7 γ). These 'corrected' values of T_{as} and values of N_{as} (W deducted) for the corresponding year are plotted in the harmonic dial of Figure 8b. The standard deviations of the yearly values for the 18-hour and 0-hour components are, respectively, 2.1 and 2.2 in units of 0.01%. Bartels' [1935] expectancy M (or the two-dimensional standard deviation) is given by $M = [2.1^2 + 2.2^2]^{1/2} = 3.06$ in units of 0.01%. The radius, ρ_1, for the so-called probable error circle for single years is: $\rho_1 = 0.833\,M = 2.55$ in units of 0.01% and for the mean of the 29 years $\rho_{29} = 2.5/(29)^{1/2} = 0.47$ in units of 0.01%, as shown in Figure 8b. In Figure 8c are plotted points for 3-year averages from 8b except for the two-year average for 1937 and 1938. For single 3-year averages the standard deviations of the 18-hour and 0-hour components are, respectively, 1.02 and 1.17, giving $M = 1.56$ in units of 0.01%.

This is essentially the same as M for single years divided by $(3)^{1/2}$ or 1.77 indicating that successive yearly means are practically independent statistically, and consequently that the 'probable error' estimate for the means of all years is valid. The larger circle in Figure 8c is the so-called 'probable error' circle for single 3-year means and has the radius $1.56 \times 0.833 = 1.30$ in units of 0.01%. The inner probable error circle for averages of ten 3-year means has (in units of 0.01%) the radius $1.29/(10)^{1/2} = 0.41$, which is practically the same as that (0.47) for the inner circle in Figure 8b computed from M for single years assuming deviations for successive years are statistically independent. It may be noted that the radius of the inner circle of Figure 8c is only about 0.006 times the standard deviation of hourly values at one station.

In Figure 9a are the plotted yearly means (W removed) of the 0.0-hour asymptotic component, N_{as}. In Figure 9b the connected open circles are yearly means of the 18.0-hour asymp-

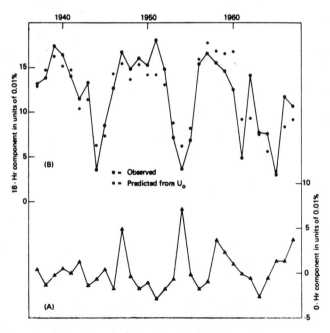

Fig. 9. Yearly mean diurnal anisotropy components: (a) with maximum at 0.0 hr and (b) with maximum at 18.0 hr local asymptotic time (W removed).

totic component (P_{36} + 12.9) from Figure 7, and the solid circles are the most probable values for this component as predicted from U_o (i.e. the ordinate of the point of intersection of a line through the data point and parallel to the dashed line of Figure 7). As indicated above in connection with Figure 8b the standard deviation for a single yearly mean (for the average of 3 stations) of the 0.0-hour asymptotic component is 0.022%. Thus, in Figure 9a only a few of the points deviate significantly from zero.

Figure 10 shows, as a function of years from sunspot minima (1944, 1954, 1964), the average solar cycle variation (period 10 years during the 29-year interval 1937–1965) in T_{36}, N_{36} (W removed), sunspot numbers, 0.29 U_o, and (T_{36} − 0.29 U_o). T_{36} and (N_{36} − W) are, respectively, the 18.0-hour and 0.0-hour asymp-totic components. The twenty-year wave (W) in the component 128°E of the sun has been removed before computing T_{36} and N_{-36}. Figure 10 indicates no evidence for a solar cycle variation in N_{36}. The solar cycle variation in T_{36} follows more closely that in U_o than either U_o or T_{36} follows the variation in sunspot numbers.

Figure 11a is the harmonic dial for the 20-year wave in the amplitude of the component, N_{-36} (W removed), of the diurnal anisotropy in the asymptotic direction 128° east of the sun (maximum at 20.5 hours local asymptotic time). Figure 11b is the harmonic dial for the 10-year wave in U_o and in T_{36}, the asymptotic component 90° east of the sun. There is no significant 20-year wave in U_o. The amplitude of the 20-year wave in the component N_{-36} (128°E of the sun) is about 1.4 times that of the 10-year wave in T_{36} (90°E of the sun).

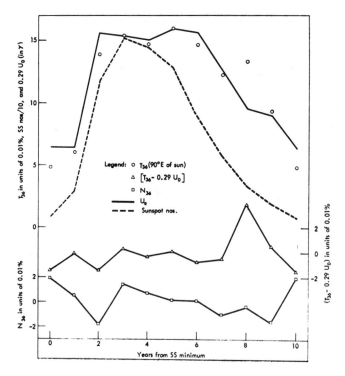

Fig. 10. Average over 2.9 solar cycles, 1937–1965, for T_{36}, U_o, sunspot numbers (T_{36} − 0.29 U_o), and N_{36}, W removed.

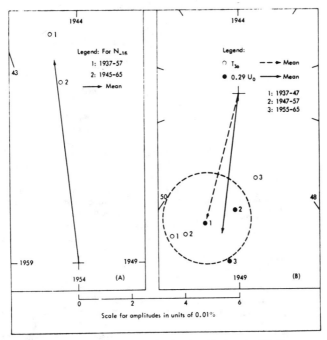

Fig. 11. Harmonic dials: (*a*) 20-year wave in N_{-16} (128°E of sun); (*b*) 10-year wave in T_{36} (90°E of sun) and 0.29 U_0. Time for maxima relative to 1944.

COMPARISON OF THE DIURNAL ANISOTROPY
FROM THE ION CHAMBER WITH THAT FROM
SIMPSON'S IGY NEUTRON MONITOR AT
HUANCAYO AND THE DIURNAL ATMOS-
PHERIC TEMPERATURE EFFECT

Annual means of the pressure-corrected diurnal variation were available from Simpson's IGY neutron monitor at Huancayo for the thirteen years, 1953–1966, excepting 1964. Deviations of yearly means of the observed diurnal variation from this average for these 13 years were derived from neutron monitor data and from ion chamber data. Figure 12 shows the 24-hour harmonic dial for these thirteen neutron monitor departures with each yearly departure vector referred to a set of axes parallel (through 0.0 hour) and normal (through 6.0 hours) to the ion chamber departure vector for the corresponding year.

From the summation vector in Figure 12 the time of maximum is found to occur about 1.6

hours (24°) earlier, on the average, for the neutron diurnal variation departure vectors than for those from the ion chamber.

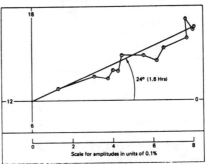

Fig. 12. Twenty-four-hour summation dial for neutron departures of yearly mean diurnal variation from 13-year average, relative to ionization departures rotated to 0.0 hr time of maximum at Huancayo (*W* not removed).

SCOTT E. FORBUSH

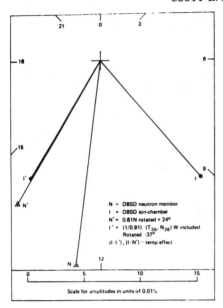

Fig. 13. Twenty-four-hour harmonic dial on 75° WMT for mean of diurnal variation for 13 years (1953-1966 except 1964) from neutron monitor and ion-chamber at Huancayo.

In the 24-hour harmonic dial of Figure 13 the vectors with endpoints N and I are, respectively, the observed averages for the indicated 13 years, for the neutron monitor, and for the ion chamber. N' is the endpoint of the vector $0.81 N$ after this has been rotated 24° (1.6 hours) clockwise. This rotation assumes that the difference between the time of maximum for the *total* observed neutron vector and that from the ion chamber, in the absence of any temperature effect, is the same as the difference between the times of maximum derived from *deviation* vectors as described above. In these deviation vectors the atmospheric temperature effect in the ion chamber data is eliminated.

To obtain the vector with endpoint I' in Figure 13, means of P_{36} and N_{36} (see Figure 2) were computed from the 13 years 1953-1966, excepting 1964, *without deducting* the 20-year wave in N_{-16}. Adding 12.9 to this value of P_{36} gives the corresponding mean for T_{36} in units of 0.01% (see equation 16). These 13-year mean components, $T_{36} = 10.7$ and $N_{36} = +0.7$, in

units of 0.01% are those referred to axes with $\theta = 36°$ in Figure 2. Taking into account the angle (28.5°) between the vector **CH** and the vertical axis in Figure 2 and of the fact that the vector T_{36}, N_{36} is to be rotated 37° *counterclockwise* (see Figure 1) and its amplitude multiplied by 1/0.91, we find the vector I' of Figure 13. This is the best estimate of the 13-year average vector that would have been observed at Huancayo without the atmospheric temperature effect. In other words, if we rotate the vector I' in Figure 13, 37° clockwise, multiply its amplitude by 0.91, and take components on axes with $\theta = 36°$ in Figure 2, we obtain T_{36} and N_{36} (W not deducted). These last two operations are just those described earlier for computing P_{36} and N_{36} from deviations of yearly mean vectors (at Huancayo) from a 25-year mean. The vector I' thus derives from deviations of observed annual means at Huancayo, Cheltenham, and Christchurch (through 1961).

Figure 13 shows that the time of maximum for I' is practically identical with that for N'. Although the amplitude of N' is somewhat greater than that of I', the agreement between I' and N' shows that at Huancayo the diurnal anisotropy derived herewith from ion chamber data only is reasonably consistent with that from the neutron monitor.

It remains only to describe how the factor 0.81 indicated in Figure 13 was obtained. This is shown in Figure 14 in which yearly means of $T_{36}(I)_{CH}$ from the ion chambers (normalized

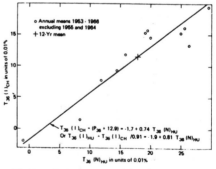

Fig. 14. Annual means of T_{36}, the component of the diurnal anisotropy in the asymptotic direction 90°E of the sun, as derived from ion chamber data, as a function of that from the IGY neutron monitor at Huancayo.

436

to Cheltenham) are plotted as a function of $T_{36}(N)_{HU}$ from the neutron monitor at Huancayo. The correlation coefficient, r, between these is $r = +0.94$, and the line in Figure 14 is drawn through the mean point (cross) with a slope 0.74, which is the mean of those (0.69 and 0.78) for the two regression lines, since it is not certain how the two coordinates should be weighted. Since $T_{36}(I)_{HU}$ as shown on Figure 14 is $T_{36}(I_{CH})/0.091$ the variations in $T_{36}(I)_{HU}$ are 0.81 times the variations in $T_{36}(N)_{HU}$. The line in Figure 14 indicates for $T_{36}(N)_{HU} = 0$ that $T_{36}(I)_{CH}$ is -1.7 in units of 0.01%. Thus, the absolute values $T_{36}(I)_{CH}$ or $T_{36}(I)_{HU}$ are in reasonable accord with those from the IGY neutron monitor at Huancayo. It should be noted that the relative magnitude of these has been derived from comparing the variability of the two and was not obtained in a way to make the agreement perfect. In addition, the value 12.9 in Figure 14, or in equation 16, is somewhat uncertain as is also the slope of the line in Figure 14. In the light of these considerations the agreement is quite satisfactory between the diurnal anisotropy from the ion chamber data and from the neutron monitor data at Huancayo.

In Figure 13 the vector $(I-I')$ is the atmospheric temperature vector obtained from ion chamber data only before neutron diurnal variation results at Huancayo were available. $(I-N')$ is the temperature effect obtained from the difference between the observed ion chamber average for the indicated 13 years and that for the neutron

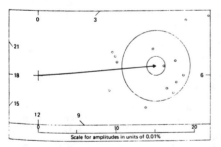

Fig. 15. Twenty-four-hour harmonic dial on 75° WMT for yearly means of diurnal temperature effect at Huancayo from differences between vectors from ion chamber and normalized vectors from neutron monitor.

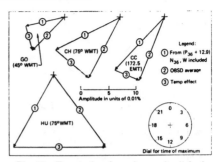

Fig. 16. Twenty-four-hour harmonic dials for diurnal temperature effect at Huancayo, Cheltenham, Christchurch, and Godhavn.

monitor at Huancayo normalized as indicated in Figure 13.

Figure 15 is a 24-hour harmonic dial on 75° WMT for the temperature effect at Huancayo derived from yearly differences between observed yearly mean vectors from the ion chamber and from normalized (as in Figure 14) neutron vectors. With differences in units of 0.01%, the two-dimensional variance, M^2, for single differences in Figure 15 is $M^2 = 27.6$. From the analysis of variance summarized in Table 2 the value $M^2 = 6.0$ was found for a single station year. For Huancayo $M^2 = 6.0/(0.91)^2 = 7.2$, corrected for the normalizing factor, 0.91. Thus, the two-dimensional variance for a single year from the normalized neutron data (using the multiplying factor 0.81) is obtained from the difference, $27.6 - 7.2 = 20.4$. This is about three times that for the ion chamber. For the period, 1953–1966 excepting 1964, the average number of complete-days data used for the diurnal variation was 298 for the ion chamber as compared with 188 for the neutron monitor. This may account for the larger variance obtained as above for statistical fluctuations in the diurnal variation from the neutron monitor.

Figure 16 is a 24-hour harmonic dial on the indicated standard meridian times, showing the results for the atmospheric temperature effect at four stations. The vectors marked (1) are obtained from the ion chamber vectors with components T_{36} and N_{36} in which the 20-year wave W has not been removed. These vectors are then referred to the indicated standard meridian time and their amplitudes corrected

437

for the normalizing factors used for each station. The procedure follows that described for Huancayo in connection with Figure 13. In Figure 16 for all stations except Godhavn the results are based on averages for the 25 years 1937–1961. For Godhavn the results are based on the 20 years 1939–1958. The amplitudes in units of 0.01% and the local mean solar time of maximum in hours for the atmospheric temperature vectors (3) in Figure 16 are, respectively, as follows: Godhavn: 4.5, 10.4; Cheltenham 7.7, 8.5; Christchurch: 4.8, 7.4; and Huancayo: 16.6, 6.0.

SUMMARY

In annual means of the diurnal anisotropy the component with maximum at 18.0 hours, local asymptotic time, is subject to relatively large variations that are well correlated with magnetic activity. This component exhibits a solar cycle variation with maximum near sunspot maximum and a range, on the average, of about 0.10% from minimum to maximum.

Superposed on the above variations are those in the component with maximum at 20.5 hours (or 8.5 hours) local asymptotic time. This component varies approximately sinusoidally, about its zero average value, with a period twice that of the solar cycle and with amplitude about 0.07% (range 0.14%); it passes through zero in 1958.7 close to the time when the sun's poloidal magnetic field reversed. When this component is positive, as, for example, near its maximum in 1944 and 1964, the 'direction of viewing' is approximately along the Archimedean spiral and away from the sun.

Acknowledgments. Successful operation of cosmic-ray ionization chambers over a long period at several stations has been effected only by the wholehearted, unselfish cooperation of several organizations and individuals. For this, grateful appreciation is expressed to the U. S. Coast and Geodetic Survey and the staff of its magnetic observatory at Fredericksburg, Virginia; to the Government of Peru and the Director of the Instituto Geofísico del Peru and staff at its magnetic observatory in Huancayo, Peru; to the Department of Scientific and Industrial Research and staff of its magnetic observatory at Christchurch, New Zealand; and to the Danish Meteorological Institute and staff of its magnetic observatory at Godhavn, Greenland.

Grateful appreciation is acknowledged to Mrs. L. Beach of the Department of Terrestrial Magnetism, for efficient, reliable reduction of records, processing of data, and other assistance.

REFERENCES

Abels, J. G., K. G. McCracken, and U. R. Rao, The semidiurnal anisotropy of the cosmic radiation. *Proc. Intern. Conf. Cosmic Rays, London,* 1, 208, 1965.

Axford, W. I., The modulation of galactic cosmic rays in the interplanetary medium, *Planetary Space Sci., 13,* 115, 1965.

Babcock, Harold D., The sun's polar magnetic field, *Astrophys. J., 130,* 364, 1959.

Bartels, J., Random fluctuations, persistence, and quasi-persistence in geophysical and cosmical periodicities, *Terrest. Magnetism Atomospheric Elec., 40,* 1, 1935.

Duggal, S. P., M. A. Pomerantz, and Scott E. Forbush, Long-term variation in the magnitude of the diurnal anisotropy of cosmic rays, *Nature, 214,* 154, 1967.

Forbush, S. E., On diurnal variation in cosmic-ray intensity, *Terrest. Magnetism Atmospheric Elec., 42,* 1, 1937.

Forbush, S. E., Barometric pressure corrections to the cosmic-ray ionization at Huancayo and the change in apparent ion-chamber sensitivity after about 1955, to be included in *Cosmic-Ray Results, Carnegie Inst. Wash. Publ. 175,* vol. 22 (in publication), 1968.

Forbush, Scott E., and Liselotte Beach, The absolute geomagnetic field of the equatorial ring current, *Carnegie Inst. Wash. Yr. Book, 65,* 28–36, 1967.

Forbush, Scott E., S. P. Duggal, and Martin A. Pomerantz, Monte Carlo experiment to determine the statistical uncertainty for the average 24-hour wave derived from filtered and unfiltered data, *Can. J. Phys., 46,* S935, 1968.

Forbush, Scott E., and D. Vénkatesan, Diurnal variation in cosmic-ray intensity, 1937–1959, at Cheltenham (Fredericksburg), Huancayo, and Christchurch, *J. Geophys. Res., 65,* 2213, 1960.

McCracken, K. G., and U. R. Rao, A survey of the diurnal anisotropy, *Proc. Intern. Conf. Cosmic Rays, London, 1,* 213, 1965.

Parker, E. N., Dynamics of the interplanetary gas and magnetic fields, *Astrophys. J., 128,* 664, 1958.

Parker, E. N., Theory of streaming of cosmic rays and the diurnal variations, *Planetary Space Sci., 12,* 735, 1964.

Thambyahpillai, T., and H. Elliot, World wide changes in the phase of the cosmic-ray solar daily variation, *Nature, 171,* 918, 1953.

Uhler, H. S., Method of least squares and curve fitting, *J. Opt. Soc. Am., 7,* 1043, 1923.

Wilcox, J. M., and N. F. Ness, Quasi-stationary corotating structure in the interplanetary medium, *J. Geophys. Res., 70,* 5793, 1965.

(Received February 24, 1969.)

JOURNAL OF GEOPHYSICAL RESEARCH, SPACE PHYSICS   VOL. 75, No. 7, MARCH 1, 1970

The Variation with a Period of Two Solar Cycles in the Cosmic Ray Diurnal Anisotropy for the Nucleonic Component

S. P. DUGGAL, S. E. FORBUSH, AND M. A. POMERANTZ

Bartol Research Foundation of the Franklin Institute
Swarthmore, Pennsylvania 19081

It has been shown previously that annual means of the diurnal anisotropy determined from 30 years of ionization chamber data result from the superposition of two distinct independent components. One component, W, has its maximum (or minimum) in the asymptotic direction 128° east of the sun-earth line, with amplitude that varies sinusoidally with a period of two solar cycles. The other, V, has its maximum in the asymptotic direction 90° east of the sun-earth line; it is well correlated with magnetic activity and varies with the sunspot cycle. For the observed component of the total diurnal anisotropy 90° east of the sun (which contains all of V and 79% of W), a correlation coefficient of 0.94 was found between annual means from the Huancayo neutron monitor and from ionization chambers. The resulting implication that V and W were each about 1.35 times greater in the neutron monitor data is explicitly established. Analysis of the diurnal anisotropy, from a geographical distribution of neutron monitors normalized to Churchill, shows that there V and W are each about twice the value for ionization chambers. These results reveal that the variational spectrum is the same for V and for W within the experimental uncertainties. Thus, it is valid to derive variational spectra from total vectors.

INTRODUCTION

In testing theoretical models of the diurnal anisotropy, the predictions of calculations are compared with determinations of the free-space amplitude and phase based on neutron monitor observations [*Parker*, 1964, 1967; *Axford*, 1965; *Gleeson*, 1969]. Thus far, the contribution to this problem of the variation, W, with a period of two solar cycles in the cosmic ray diurnal anisotropy, has not been evaluated. For this purpose, knowledge of the variational spectrum of this wave is required.

To determine W from available neutron monitor data, which encompass an interval less than two solar cycles, we must invoke the results of the statistical analysis of the diurnal anisotropy made by *Forbush* (1967 and 1969, hereafter designated 1 and 2) with ionization chamber data recorded over a period of thirty years. In this analysis, annual means of the diurnal anisotropy from Simpson's neutron monitor data at Huancayo were found to be 1.35 times those from ionization chamber data that were normalized to Cheltenham, based on 12 available pairs of annual means (1953–1966), between which the correlation coefficient was 0.94. This implied a similar ratio for the amplitude of W

and of V. The objectives of the present work are to consider this implication in detail and to examine its consequences, in the light of results based on the analysis of data from a geographical distribution of neutron monitors.

DEFINITION OF SYMBOLS

The discussion of an explicit determination of the variation with a period of two solar cycles from neutron monitor data is facilitated by the definition of a number of symbols, based on the earlier results in 1 and 2 relating to the elimination of the diurnal temperature effect, the normalization of phases and amplitudes to those at Cheltenham, the determination of the total diurnal anisotropy with the diurnal temperature effect eliminated, and the statistical procedure for determining empirically the corrections for geomagnetic deflection.

In this paper, we use the following symbols to designate annual mean diurnal anisotropy vectors and their components on the 24-hour harmonic dial in local asymptotic time

M and N: diurnal anisotropy from ionization chamber (mesons) and from neutron monitor (nucleons) data, respectively.

θ: asymptotic direction, degrees east of the sun-earth line.

t: midpoint of year over which diurnal anisotropy is averaged (e.g. if calendar year is 1958, $t = 1958.5$).

$\mathbf{T}_0(M, t)$: total observed vector (with atmospheric temperature effect thoroughly eliminated) averaged for three stations (two after 1961) with phase and amplitude normalized to Cheltenham.

$\mathbf{T}_i(N, t)$: total vector at station i for nucleonic component during year t.

i: Hu for Huancayo. Ch average for stations in Table 2 normalized to Churchill.

$T_0(M, \theta, t)$: component of $\mathbf{T}_0(M, t)$ in the asymptotic direction θ.

$T_i(N, \theta, t)$: component of $\mathbf{T}_i(N, t)$ at station i in the asymptotic direction θ.

W: that constituent of the diurnal anisotropy the amplitude of which varies about zero mean with a period of two solar cycles, or 20 years, in the interval 1937–1966.

$W(M, \theta)S(t)$: amplitude of W in the asymptotic direction θ as function of t, where $W(M, \theta)$ is maximum for $\theta = 128°$, and $S(t) = \sin(2\pi/20)(t - 1958.7)$, $V(M, \theta, t) = T_0(M, \theta, t) - W(M, \theta)S(t)$, $T_c(M, \theta, t) = W(M, \theta)S(t) + V_c(M, \theta, t)$, where $V_c(M, \theta, t)$ is the most probable value of $V(M, \theta, t)$ computed from weighted least squares line based on the correlation ($r = +0.75$) between $V(M, \theta, t)$ and $U_0(t)$.

$U_0(t)$: annual mean equatorial geomagnetic field (positive southward) of the ring current.

$W_i(N, \theta, t)$: yearly value of component of W anisotropy for neutrons.

(N/M): ratio of anisotropy observed with neutrons to that observed with mesons, determined from mean of slopes of the two regression lines including origin as an observed point.

THE DIURNAL ANISOTROPY

In 1 and 2 it was shown that the annual means of the cosmic ray diurnal anisotropy, (with the atmospheric temperature effect thoroughly eliminated) from ionization chamber data, resulted principally from the superposition of two distinct independent diurnal components, with asymptotic directions of maximum 128° and 90° east of the sun-earth line. Thus, for year t

$$\mathbf{T}_0(M, t)$$

$$= W(M, 128)S(t) + V_c(M, 90, t) \qquad (1)$$

The amplitude $W(M, \theta)$ varied sinusoidally about zero mean with a period of two solar cycles, or 20 years, during the interval 1937–1967. Maxima of $W(M, \theta)S(t)$ occurred at 1943.7 and 1963.7. When $W(M, 128)S(t)$ is positive, as, for example, between 1938.7 and 1948.7, the asymptotic direction of viewing is approximately along the Archimedean solar spiral stream and away from the sun. $W(M, 128)S(t)$ passed through a zero in 1958.7, when the sun's poloidal field reversed.

The component $V(M, 90, t)$ is obtained from

$$V(M, 90, t)$$

$$= T_0(M, 90, t) - W(M, 90)S(t) \qquad (2)$$

The correlation coefficient was $r[V(M, 90, t), U_0(t)] = 0.75$, where $U_0(t)$ is the absolute value of the geomagnetic equatorial ring current field used as an indirect measure of large scale magnetic field irregularities in the solar streams.

The 20-year waves $W(M, 90)S(t)$ and $W(M, 180)S(t)$, respectively, are shown as the dashed curves in Figures 1 and 2. The statistical reliability of the 20-year wave is discussed in 2, with other cogent evidence from which its phase and amplitude are determined within small limits. In Figure 1 the axis of the sine wave

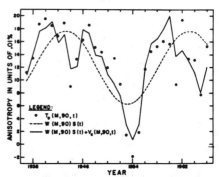

Fig. 1. W, the component of the diurnal anisotropy whose amplitude varies with a period of two solar cycles (dashed curve), and comparison of annual means of the total observed vector (points) with the sum of the two components of the diurnal anisotropy, W and V, (solid curve) for mesons, in the asymptotic direction 90° east of the sun-earth line.

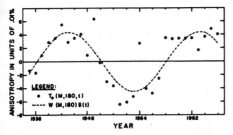

Fig. 2. Comparison of the W anisotropy (dashed curve) with annual means of the total observed vector (points), for mesons, in the asymptotic direction 180° east of the sun-earth line.

is at 12.0 on the ordinate scale, because this is the average value of $V(M, 90, t)$ for the period 1937–1966. Figure 2 shows that during this period the variation of $T_0(M, 180, t)$ is approximated by $W(M, 180)S(t)$.

In Figure 1 the points indicate annual means of $T_0(M, 90, t)$. These represent averages for Cheltenham-Fredericksburg, Huancayo, and Christchurch until 1962, after appropriately normalizing the phase and amplitude of the annual means (diurnal temperature effect eliminated) at the latter two stations to those at Cheltenham-Fredericksburg. The solid curve is drawn through annual means of $T_c(M, 90, t)$, computed as indicated above. The fact that the solid curve and the points are in good accord indicates a satisfactory statistical agreement (see 1 and 2) between $T_0(M, 90, t)$ and $[W(M, 90)S(t) + V_c(M, 90, t)]$. Incidentally, Figures 1 and 2 show that the small amplitudes of the diurnal anisotropy for the 1953–1955 solar minimum epoch, as compared to those for 1944 or 1964, were largely due to W.

From 12 available pairs of annual means in the interval 1953–1966, the correlation coefficient $r\ [T_0(M, 90, t), T_{Hu}(N, 90, t)] = 0.94$ was found in 2. Rewriting equation 2

$$T_0(M, 90, t)$$

$$= W(M, 90)S(t) + V(M, 90, t) \quad (3)$$

and because $T_0(M, 90, t)$ contains 79% of the maximum possible contribution from W (see equation 6 and Table 1), the aforementioned high correlation implies a relation similar to equation 3 for the diurnal anisotropy as ob-

served with the neutron monitor at Huancayo. Thus

$$T_{Hu}(N, 90, t)$$

$$= W_{Hu}(N, 90)S(t) + V_{Hu}(N, 90, t) \quad (4)$$

The mean of the slopes of the two regression lines in the graph of $T_{Hu}(N, 90, t)$, as a function of $T_0(M, 90, t)$, was found in 2 to be 1.35. This implied that the ratio between the corresponding terms on the right side of (3) and (4) probably does not differ appreciably from 1.35. We shall now determine these ratios explicitly from the comparison of neutron monitor and ionization chamber data. Furthermore, data from additional nucleonic intensity detectors, with their maximum response at lower primary rigidities, will be analyzed to investigate the energy dependence of the 20-year wave.

ANALYSIS

The Fourier coefficients for the annual mean diurnal anisotropy at each station were computed from unfiltered data, because *Forbush et al.* [1968] showed that using filtered data does not decrease the statistical uncertainty in the harmonic coefficients.

The rotation of the earth around the sun produces an anisotropy in cosmic rays from a direction 90° west of the sun-earth line. The amplitude of this anisotropy [*Compton and Getting*, 1935; *Gleeson and Axford*, 1968] is given by

$$R_1 = (2 + \gamma)v/c \quad (5)$$

where v is the orbital velocity of the earth, c is the velocity of light, and γ is the exponent in the differential energy spectrum of cosmic rays. The theoretical correction to the observed amplitude introduced by this effect is only about 0.03%. Furthermore, as in the case of the corotational anisotropy, its actual magnitude is probably smaller than the predicted amplitude. Therefore, although the procedure for determining $T_0(M, t)$ automatically removed any Doppler effect, this effect has not been eliminated from the neutron monitor data.

Huancayo neutron monitor. Using the above results, we can now separate $W_{Hu}(N)$ and $V_{Hu}(N)$ and determine the magnitude of each.

For the component $T_0(M, \theta, t)$ in any di-

TABLE 1. Ratios (N/M) for different combinations of the parameters in equation 6 from the correlation $r[T_{Hw}(N, \theta, t), T_0(M, \theta, t)] = r\{[kW(N, 128, t) + lV(N, 90, t)], [kW(M, 128)S(t) + lV(M, 90, t)]\}$ for indicated values of θ from data for the period 1953–1966.

Figure No.	$\theta°$	k^1	l^2	r	$(N/M)^4$
3	180	0.62	0.00	0.71	1.53
4	90	0.79	1.00	0.91	1.35
5	90³	0.00	1.00	0.90	1.40
6	38	0.00	0.62	0.75	1.36

[1] $k = \cos(\theta - 128)$.
[2] $l = \cos(\theta - 90)$.
[3] from folded data.
[4] (N/M) is the mean of the slopes of the two regression lines.

rection θ, it follows from equation 1 that

$$T_0(M, \theta, t)$$

$$= kW(M, 128)S(t) + lV(M, 90, t) \qquad (6)$$

where $k = \cos(\theta - 128)$ and $l = \cos(\theta - 90)$, and similarly for N, assuming that $W(N)$ has the same phase as $W(M)$.

Table 1 summarizes separate determinations of (N/M), involving the indicated values of the parameters k and l in equation 6. Note that, for $k = 0$, the ratio refers exclusively to the V component, and for $l = 0$, the ratio is that for the amplitudes of W. In the second row of the table (N/M) is the value for the indicated combination of both components.

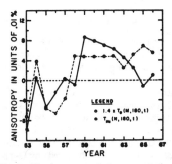

Fig. 3. Yearly means of the asymptotic component 180° east of the sun-earth line during 1953–1966, as observed by the Huancayo neutron monitor (solid points) and ionization chambers (open points).

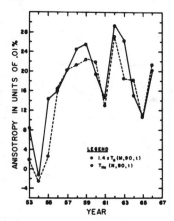

Fig. 4. Yearly means of the asymptotic component 90° east of the sun-earth line during 1953–1966, as observed by the Huancayo neutron monitor (solid points) and ionization chambers (open points).

Figure 3 shows the yearly means of the asymptotic component 180° east of the sun during the period 1953–1966, as deduced from the Huancayo neutron monitor (N) and ionization chamber (M) data. For the latter, the points correspond to those in Figure 2, which shows the agreement with $W(M, 180)S(t)$. The correlation coefficient between pairs of points in Figure 3 is $r = 0.71$, indicating that $W_{Hw}(N, 180, t)$ is approximated by $1.4W(M, 180)S(t)$. The correlation with W removed from the meson data is negligible.

As Figure 4 shows, the components perpendicular to the aforementioned direction are also well correlated ($r = 0.91$). In this case, (N/M) derives from the ratio of the linear combination of the components W and V indicated in Table 1.

We would now like to determine the relative amplitude of V for the nucleonic component with respect to the meson component. The ratio $V_{Hw}(N, \theta, t)/V(M, \theta, t)$ can be obtained by a folding process that makes use of the fact that the amplitude of W was zero in 1958, if we assume, based on Figure 3, that $W(M)$ and $W(N)$ are in phase. Figure 5 shows the correlation between *means* for N and for M for the indicated pairs of years, chosen relative to the zero of W so that its contribution is neg-

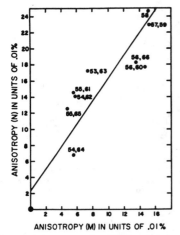

ANISOTROPY (M) IN UNITS OF .01%

Fig. 5. Plot, for neutrons versus mesons, of means of the amplitudes of the anisotropy in the direction 90° east of the sun-earth line for pairs of years (t_1, t_2) for which the net contribution of W to the mean is negligible. The origin (circled dot) is regarded as a point in the correlation analysis. Ordinate: $[T_{Hu}(N, 90, t_1) + T_{Hu}(N, 90, t_2)]/2$ Abscissa: $[T_0(M, 90, t_1) + T_0(M, 90, t_2)]/2$.

ligible. From this figure $(N/M) = 1.40$ is the ratio for the V component, which has its maximum in the asymptotic direction 90° east of the sun (cf. Table 1). The value of $(N/M) = 1.36$, determined from the correlation between $T_{Hu}(N, 38, t)$ and $T_0(M, 38, t)$ in Figure 6, also applies to the V component, because W vanishes for $\theta = 38°$.

In Table 1, all values of (N/M) agree within the estimated uncertainties of each. This shows that the ratio (N/M) for W and for V are the same within their uncertainties. The fact that the values of (N/M) are evidently independent of k and l indicates that the asymptotic phase of $W(N, \theta, t)$ is essentially the same as that of $W(M, \theta, t)$.

Geographical distribution of neutron monitors. Although the available data cover only about half a cycle of W, we may now attempt to evaluate the 20-year wave for a high-latitude neutron monitor by invoking certain assumptions based upon the preceding analysis. For this purpose, six stations distributed around the earth were selected (Table 2).

Factors for amplitude and phase were deter-

mined for each station by normalizing its mean vector of anisotropy for the period 1958–1966 to that at Churchill. These factors were used to normalize the yearly vector for each station. The normalized vectors were then averaged to provide the composite neutron monitor vector for Churchill, $\mathbf{T}_{ch}(N, t)$.

Inasmuch as the data were restricted to only nine years, the procedures described above are not applicable. Instead, $W_{ch}(N, 90, t)$ can be estimated from the following relationship:

$$W_{ch}(N, 90, t)$$
$$= T_{ch}(N, 90, t) - \alpha V(M, 90, t) \quad (7)$$

where $\alpha = V_{ch}(N, 90, t)/V(M, 90, t)$.

It can be shown from Figures 1 and 2 that the total anisotropy $\mathbf{T}_0(M, t)$ from 1958–1966 was considerably greater (about 3 times) than the contribution from W. Hence

$$\alpha \approx |\mathbf{T}_{ch}(N, t)|/|\mathbf{T}_0(M, t)| \quad (8)$$

From the data recorded during the entire period 1958–1966, we find the value $\alpha = 2.0$.

Figure 7 is a plot of $W_{ch}(N, 90, t)$, for each of the indicated years determined from equation 7 with $\alpha = 2$, as a function of $W(M, 90)S(t)$. The slope of the regression line (abscissa free from error) is $\beta = 2.0 \pm 0.6$,

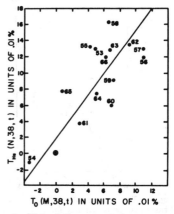

T_0 (M,38,t) IN UNITS OF .01%

Fig. 6. Plot of the asymptotic component in the direction 38° east of the sun-earth line, for which the contribution of W vanishes, from neutron data as a function of that from ionization chambers. The origin (circled dot) is regarded as a point in the correlation analysis.

TABLE 2. Neutron monitor data included in the analysis.

Station	Geog. Lat.	Geog. Long., East	Altitude, m	Cut-off, GV	Years
Thule	76.6°	291.6°	260	0	58–66
Churchill	58.8°	265.9°	11	0.2	58–66
Mt. Washington	44.3°	288.7°	1917	1.2	58–66
Rome	41.9°	12.5°	60	6.3	58–65
Huancayo	−12.0°	283.1°	3400	13.5	53–66
Mawson	−67.6°	62.9°	0	0.2	58–66

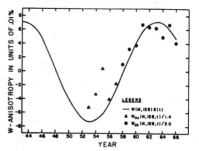

Fig. 8. Comparison of yearly values of the W anisotropy 128° east of the sun-earth line from neutron monitor data (points) with the wave from ionization chamber data (line).

computed from the correlation coefficient $r = 0.80$.

Now

$$W_{Ch}(N, 180, t) = T_{Ch}(N, 180, t) \qquad (9)$$

since ideally $V(N, 180, t) \doteq 0$ (cf. Figure 9 (A) of 2). From the above 90° and 180° components, $W_{Ch}(N, 128, t)$ is readily obtained. Yearly values of $W_{Ch}(N, 128, t)/2$ from 1958 to 1966 are plotted with a solid circle in Figure 8. $W_{Hu}(N, 128, t)/1.4$ from 1953 to 1957 are also shown by a solid triangle. The points in Figure 8 are seen to be well fitted by the solid curve for $W(M, 128)S(t)$. This agreement persists even if the Huancayo data are not included in the determination of the composite neutron monitor vector for the years 1958 to 1966.

CONCLUSIONS

Figure 8 indicates that the 20-year wave in the cosmic ray solar diurnal variation occurs over the energy range covered by ground-based neutron monitors and ionization chambers. Furthermore, at a station viewing the equatorial plane, and characterized by a low geomagnetic threshold rigidity, the amplitude of $W(N)$ is larger than that of $W(M)$. Thus, for example, the amplitude for neutron monitor data normalized to Churchill is twice that for ionization chamber data normalized to Cheltenham. Finally, it has been shown that the ratio (N/M) for V is the same as that for W.

Considered together, these results indicate that the variational spectra of W and V are the same within the experimental uncertainties. The precision can be expected to improve significantly with the addition of data at least through the next solar minimum.

In the light of these results, it is valid to deduce the variational spectrum of the corotational anisotropy from the total vectors observed during the period 1953–1966 [*Jacklyn et al.*, 1969].

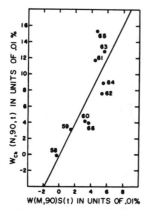

Fig. 7. The component of the W anisotropy 90° east of the sun-earth line for neutron monitors normalized to Churchill as a function of the amplitude of the W wave in the same direction for ionization chambers normalized to Cheltenham.

Acknowledgments. This work was supported by the National Science Foundation and Air Force Cambridge Research Laboratory.

REFERENCES

Axford, W. I., The modulation of galactic cosmic rays in the interplanetary medium, *Planet. Space Sci.*, *13*, 115, 1965.

Compton, A. H., and I. A. Getting, An apparent effect of galactic rotation on the intensity of cosmic rays, *Phys. Rev.*, *47*, 817, 1935.

Forbush, S. E., A variation, with a period of two solar cycles, in the cosmic-ray diurnal anisotropy, *J. Geophys. Res., 72,* 4937, 1967.

Forbush, S. E., Variation with a period of two solar cycles in the cosmic ray diurnal anisotropy and the superposed variations correlated with magnetic activity, *J. Geophys. Res., 74,* 3451, 1969.

Forbush, S. E., S. P. Duggal, and M. A. Pomerantz, Monte Carlo experiment to determine the statistical uncertainty for the average 24-hour wave derived from filtered and unfiltered data, *Can. J. Phys., 46,* S985, 1968.

Gleeson, L. J., and W. I. Axford, The Compton-Getting effect, *Astrophys. Space Sci., 2,* 431, 1968.

Gleeson, L. J., The equations describing the cosmic ray gas in the interplanetary region, *Planet. Space Sci., 17,* 31, 1969.

Jacklyn, R. M., S. P. Duggal, and M. A. Pomerantz, The spectrum of the cosmic ray solar diurnal modulations, *Proc. 11th International Cosmic Ray Conference,* Budapest, 1969.

Parker, E. N., Theory of streaming of cosmic rays and the diurnal variation, *Planet. Space Sci., 12,* 735, 1964.

Parker, E. N., Cosmic ray diffusion, energy loss, and the diurnal variation, *Planet. Space Sci., 15,* 1723, 1967.

(Received October 15, 1969;
revised December 1, 1969.)

Cosmic Ray Diurnal Anisotropy 1937–1972

SCOTT E. FORBUSH

Department of Terrestrial Magnetism, Carnegie Institution of Washington
Washington, D.C. 20015

A previous investigation by Forbush (1969) showed that annual means of the cosmic ray diurnal anisotropy from 1937 to 1967 resulted from the addition of two distinct diurnal components. One, w, has its maximum (or minimum) in the asymptotic direction 128° east of the sun and is well approximated by a wave W with a period of 2 solar cycles. Wave W passes through zero in 1958, when the sun's poloidal field reversed. The remaining component with W eliminated has its maximum in the asymptotic direction 90° east of the sun. Annual means of this component, with its maximum at 18.0 hours local asymptotic time, are well correlated ($r = +0.75$) with magnetic activity and determine a solar cycle variation, the minimum being near sunspot minimum and the amplitude about two-thirds that of W. During the interval 1937–1967 or so, the 'period' of W was 20 years (that of the sunspot cycle was 10 years). If W were strictly periodic, its next change of sign after 1958 would have occurred in 1968. The present analysis shows this reversal of sign was delayed until 1971, near the time found for the latest reversal of the sun's polar magnetic field by Dr. Robert Howard of the Hale Observatories. These results derive from a statistical investigation of the variability of annual means of the diurnal variation from ion chamber data at Cheltenham-Fredericksburg, Huancayo, and Christchurch. The absolute, or total, diurnal anisotropy and the atmospheric diurnal temperature effect obtained are in reasonable agreement with those derived independently through a comparison of the diurnal anisotropy from ion chamber data and that from Simpson's 1953–1966 IGY neutron monitor data at Huancayo.

It was shown by *Forbush* [1969] that annual means from 1937 to 1967 of the diurnal anisotropy of cosmic ray intensity, which undergo relatively large variations in phase and amplitude, are well approximated by the sum of two distinct diurnal components: T_{se} and w. The maximum (or minimum) of w is in the asymptotic direction 128° east of the sun, having an amplitude well fitted by a wave W, about zero mean, a period of 20 years (2 solar cycles), and an amplitude 0.072% of the total cosmic ray intensity. For the component w the 'direction of viewing' is along the solar stream Archimedean spiral away from the sun when w is positive, as it was, for example, in 1944 and 1964.

Component T_{se} is the component with its maximum in the asymptotic direction 90° east of the sun. Annual means of T_{se} are correlated ($r = +0.75$) with magnetic activity U_0. The 10-year wave or solar cycle variation in T_{se} is in phase with that in U_0, the minimum being near sunspot minimum and the amplitude being

about two-thirds that of W. There is no 20-year wave in U_0. Thus, although the corotation theory of *Axford* [1965]–*Parker* [1964] accounts for T_{se}, a separate mechanism is required for W.

Wave W passed through zero in 1958, when *Babcock* [1959] found that the sun's north polar field reversed. If W were strictly periodic with a period of 20 years, its next passage through zero would have occurred in 1968. However, as was pointed out by *Forbush* [1969], the data through 1968 showed no indication of such a change of sign. Since the 'period' of the sunspot cycle varies by a few years, variations of several years may be expected in the interval between reversals of the sun's polar magnetic field.

The present investigation, based on data through 1972, shows that w passed through zero in 1971, near the time found for the most recent reversal of the sun's north polar field by Dr. Robert Howard of the Carnegie Institution of Washington, Hale Observatories [*Howard*, 1972].

SUMMARY OF METHODS OF ANALYSIS

The procedures used in the analysis by *Forbush* [1969] were described in detail and will only be summarized here together with pertinent aspects of the results from 1937 to 1965, which are included for continuity with those obtained after 1965 by a necessary modification of procedure.

The basic data from 1937 to 1972 are the annual mean 24-hour harmonic coefficients a_1 and b_1 in Table 1 for the pressure-corrected 24-hour waves from Carnegie Institution of Washington ionization chambers. The 24-hour pressure-corrected wave in data from such meson detectors contains the much-discussed, notorious, and rather uncertain so-called temperature effect. This temperature effect is later shown to be quite satisfactorily eliminated at each station in the departures of annual mean

TABLE 1. Fourier Coefficients for 24-Hour Wave in Pressure-Corrected Cosmic Ray Ionization

	Huancayo		Cheltenham*		Christchurch		Godhavn	
	a_1	b_1	a_1	b_1	a_1	b_1	a_1	b_1
1936†	−12	+15	−17	−1	−12	−5		
1937	−11	+12	−12	−3	−11	−3		
1938	−13	+10	−12	−4	−15	−4		
1939	−17	+7	−14	−11	−14	−11	−5	−6
1940	−17	+5	−10	−11	−16	−14	−5	−6
1941	−16	+4	−10	−10	−14	−13	−4	−6
1942	−15	+2	−8	−11	−9	−11	−5	−6
1943	−17	+3	−12	−12	−13	−11	−6	−9
1944	−7	+7	−7	−2	−6	−5	−4	+1
1945	−11	+6	−8	−7	−8	−9	−5	−5
1946	−16	+10	−10	−10	−13	−8	−6	−14
1947	−17	+4	−5	−16	−10	−12	−7	−7
1948	−12	+10	−14	−11	−12	−5	−8	−16
1949	−14	+13	−14	−9	−15	−1	−6	−17
1950	−16	+13	−12	0	−14	−2	−7	−11
1951	−19	+16	−15	−2	−16	0	−9	−5
1952	−12	+18	−15	0	−13	+2	−5	0
1953	−7	+21	−11	+8	−4	+7	−7	0
1954	+4	+15	0	+6	−1	+5	0	+4
1955	−4	+20	−8	+4	−7	+8	−5	+2
1956	−13	+13	−17	−3	−13	+1	−9	−4
1957	−14	+11	−13	−8	−15	−1	−9	−4
1958	−14	+2	−10	−8	−10	−7	−4	−6
1959	−16	+3	−9	−10	−10	−7		
1960	−14	+2	−10	−10	−10	−6		
1961	−7	+9	−7	−3	−6	−6		
1962	−16	+2	−12	−13				
1963	−9	+7	−11	−7				
1964	−8	+6	−10	−9				
1965	−6	+7	−4	−2				
1966	−14	+2	−9	−7				
1967	−11	+5	−5	−7				
1968	−15	+2	−8	−9				
1969	−15	+2	−12	−11				
1970	−16	+3	−13	−8	−11	−6		
1971	−19	+12	−18	−1	−12	+2		
1972	−14	+11	−16	+2	−10	+1		

Origin is standard mean time; units are in 0.01%.
* Fredericksburg after October 1956.
† Huancayo, June 1936 to May 1937; Cheltenham, January to December 1936; Christchurch, April 1936 to March 1937.

vectors (in a 24-hour harmonic dial) from the average vector for 25 years. These annual mean departure vectors in the 24-hour harmonic dial for Huancayo, Peru, and Christchurch, New Zealand, were separately normalized to those at Cheltenham-Fredericksburg by clockwise rotations of 37° (and multiplication by 0.91) and 20° (and multiplication by 1.00), respectively. Differences in magnetic deflection in the earth's field most likely account for the difference in rotations.

From the yearly deviation vectors at Huancayo and Christchurch, phases and amplitudes being normalized to those for Cheltenham, and from the yearly deviation vectors for Cheltenham, components P_0 and N_0 are arbitrarily taken on the axes P_0 and N_0 in Figure 1. In Figure 1, θ is the angle between the P_0 axis and the observed average vector for the 25 years, 1937–1961, at Cheltenham. Yearly values of P_0 and N_0 (for $\theta = 0$) are plotted in Figure 2. The solid lines indicate for each year the average P_0 or N_0 from the two or three stations for which data were available.

Figure 2 shows good agreement between the values of P_0 and N_0 from the different stations. The statistical significance of the variations in Figure 2 among years and between stations is obtained from a simple analysis of variance that also provides the residual variance for a single station-year that determines the confidence limits shown in Figure 2. Table 2a lists the variances $s^2(\Delta P_0)$, $s^2(\Delta N_0)$, and $M^2 = s^2(\Delta P_0) + s^2(\Delta N_0)$ for yearly differences of P_0 and N_0 for the indicated pairs of stations. Since M^2 differs little for the three pairs, the variance at the three stations is essentially homogeneous, and valid results may be obtained from the analysis of variance.

Table 2b summarizes the results of the analysis of variance for values of P_0 and N_0 for the 25 years, 1937–1961, at three stations. The variance among stations (Table 2) is zero, which results from normalizing the yearly deviation vectors at Huancayo and Christchurch to those at Cheltenham. Each of the variances $s^2(P_0)$, $s^2(N_0)$, and $M^2 = s^2(P_0) + s^2(N_0)$ is more than 20 times that for the corresponding residual variance in the bottom row of Table 2b. Thus variations among years in P_0 and N_0 are almost certainly due to real variations and not to statistical uncertainties in the station

values. For a single station-year mean of the diurnal variation, the residual variance $M^2 = 6.0$ ($M = 0.024\%$) is 2.6 times the underestimated value $M^2 = 2.3$, derived from the 'counting rate,' and only 1.5 times the more realistic value 4.0 ($M = 0.020\%$), based on the variability on 273 days for Cheltenham that was obtained earlier by *Forbush* [1937]. Thus the yearly values of P_0 (and of N_0) in Figure 2 for the three stations agree remarkably well. This agreement in turn means that the yearly mean diurnal temperature effect must be nearly constant from year to year at each station.

DETERMINATION OF THE TWO SEPARATE AND DISTINCT COMPONENTS w AND T_{20} OF THE DIURNAL ANISOTROPY

Since the period of the sunspot cycle from 1937 to 1965 was quite close to 10 years, the fundamental solar period during this interval is taken as 20 years. A wave with this period was fitted to each of the solid graphs in Figure

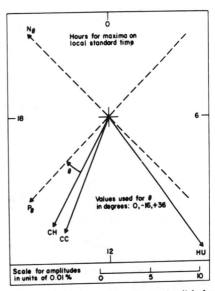

Fig. 1. Twenty-four-hour harmonic dial for average, 1937–1961, vector for Cheltenham, Christchurch, and Huancayo and axes for components P_0, N_0 of normalized deviation of yearly means from 25-year mean.

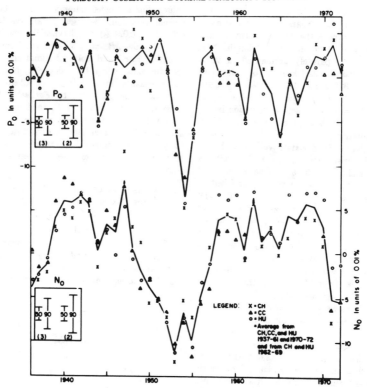

Fig. 2. Yearly means of P_0 and N_0 for Cheltenham-Fredericksburg (CH), Christchurch (CC), and Huancayo (HU) and average for year indicated. Inserts indicate 50 and 90% confidence limits for means of 3 and 2 stations.

2. When data through 1965 are used, these waves in N_0 and P_0 show that the amplitude of the 20-year wave is a maximum in the component N_{-16} and is zero in the component P_{-16} (for $\theta = -16°$ in Figure 1), as is actually found. Part B of Figure 3 shows the 20-year wave W fitted to N_{-16}. It is shown later that N_{-16} is the component of the diurnal anisotropy in the asymptotic direction 128° east of the sun, and it is the amplitude of this component that exhibits the 20-year variation.

In part B of Figure 3 let the differences between the yearly means of N_{-16} and the wave be designated as $N_{-16} - W$. Between these yearly values of $N_{-16} - W$ and corresponding yearly means of U_0, the correlation coefficient r is 0.75 (Table 3); U_0 is the horizontal geo-

magnetic component (positive southward) at the geomagnetic equator due to the so-called equatorial ring current, here used as a measure of geomagnetic activity. For annual means, the zero level for U_0 has been determined [Forbush and Beach, 1967] with an uncertainty of about 2 or 3 γ (1 γ = 10^{-5} oe). It is assumed that annual means of U_0 provide an indirect measure of irregularities of the magnetic field that are basic to the corotation model for the diurnal anisotropy of Axford [1965]-Parker [1964]. Table 3 also shows the correlation coefficient between P_{-16}, in which there is no 20-year wave, and U_0 and that between P_{-16} and $N_{-16} - W$. The last three columns in Table 3 list the relative weights and the parameters for the most probable linear relation between the indicated variables, both of which are subject to 'error.' On a graph of the observed points and the preceding 'most probable line,' the most probable values for the two coordinates for a given year are those of the intersection of this line with a second line through the observed point. The slope of the second line is determined by the relative weights of the two coordinates, as is discussed in detail by *Forbush* [1969].

From parameters a and b in the first two rows of Table 3 it follows that, if the components $N_{-16} - W$ and P_{-16} are referred to axes P_{36} and N_{36} ($\theta = 36$ in Figure 1), then both parameters for the most probable linear relation between N_{36} and U_0 are essentially zero, indicating that N_{36} vanishes and is independent of U_0. From the parameters in the first two rows of Table 3 those in the last and third rows are readily found.

It is assumed that P_{36} is the component explained by the corotation theory of *Axford* [1965]–*Parker* [1964] and that in accordance with this theory it is in the local asymptotic direction 90° east of the sun, i.e., with its maximum at 18.0 hours local asymptotic time (LAT). In the corotation theory, no diurnal anisotropy is expected in the absence of large-scale interplanetary field irregularities (in the solar streams) with sources on the sun. We assume that the absence of such irregularities over the period of a year would be associated with the absence of magnetic activity and that U_0 would vanish. The value of a in the last row of Table 3 shows that, for $U_0 = 0$, $P_{36} = -12.9$ in units of 0.01%. If we write

$$T_{36} = 12.9 + P_{36} = 0.288 U_0 \qquad (1)$$

then the total component T_{36} (W deducted) of the diurnal anisotropy with its maximum at 18.0 hours vanishes for $U_0 = 0$ according to the assumption above. Thus was determined the additive constant in (1), which was eliminated (along with the temperature effect) in taking departures of annual means from a 25-year mean. Annual means of the total diurnal anisotropy (W not deducted) thus derived solely from ion chamber data agreed most satisfactorily, as is shown by *Forbush* [1969] with properly normalized annual means (1953–1966) from Simpson's IGY neutron monitor at Huancayo, indicating the validity of the procedure used and the assumptions made.

TABLE 2a. Variances $s^2(\Delta P_0)$ and $s^2(\Delta N_0)$ for Difference between Pairs of Annual Means of P_0 and N_0 from n Years at Cheltenham, Christchurch, and Huancayo

Pair	n	$s^2(\Delta P_0)$	$s^2(\Delta N_0)$	M^2
Cheltenham-Christchurch	25	5.2	8.0	13.2
Cheltenham-Huancayo	29	6.5	6.3	12.8
Christchurch-Huancayo	25	4.1	5.9	10.0

P_0 and N_0 are in units of 0.01%.
From this table the variance is homogeneous.

In part C of Figure 3 the open circles (1937–1965) indicate the values of N_{-16} in part B of Figure 3 corrected to the 25-year mean of U_0. The yearly correction, subtracted from N_{-16} for each year, is the most probable value of $N_{-16} - W$ obtained from the weighted least squares line for the relation between $N_{-16} - W$ and U_0, when the relative statistical weights of these two variables (both subject to error) are taken into account, as is described in detail by *Forbush* [1969]. These corrected values of N_{-16} are seen in part C of Figure 3 to be well fitted by W (1937–1965).

DETERMINATION OF W_0 IN LIEU OF W AFTER 1965

This procedure just described is valid only if the period (and amplitude) of W remain fixed. If the period of W is assumed to remain 20

TABLE 2b. Analysis of Variance Results from Yearly Means, 1937–1961, at Cheltenham, Christchurch, and Huancayo for P_0 and N_0

	d.f.	$s^2(P_0)$	$s^2(N_0)$	M^2
Among years	24	53.0	87.5	140.5
Among stations	2	0.0	0.0	0.0
Residual	48	2.5	3.5	6.0

P_0 and N_0 are in units of 0.01%.
From this table the residual variance $M^2 = s^2(P_0) + s^2(N_0) = 6.0$ ($M = 0.024\%$). For a single station-year mean, the variance among years is >20 times the residual variance. Note that the standard deviation (0.66%) for hourly counting rate results in underestimated $M^2 = 2.3$ ($M = 0.015\%$) for 1 station-year. From variability of 24-hour waves on 273 single days at Cheltenham, $M^2 = 4.0$ ($M = 0.020\%$) for 1 station-year.

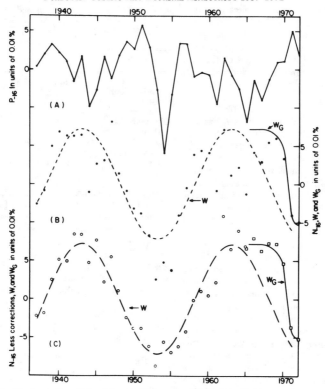

Fig. 3. Yearly means derived from yearly averages of P_0 and N_0 in Figure 2: A, for P_{-16}; B, for N_{-16}; and C, for N_{-16} less corrections for U_0; W is the 20-year wave fitted to the points in the interval 1937–1965, and W_G is the 'guesstimated' wave after 1965.

years from 1966 to 1971, the corrected points (not shown in part C of Figure 3) fall far above W, as might be expected if the period, or intervals between the zeros of W, had lengthened, as appears to have occurred between the last two maximums of the sunspot period (Figure 4, part B). Thus, to determine the new W, say, W_G, beginning with 1966, a modified analogous procedure is required that does not involve assumptions about its period.

To facilitate description of this procedure, attention is called to the solid circles in part B of Figure 3 for the period 1937–1965. These circles represent the most probable values of the 18.0-hour (LAT) component (W removed), or $T_{36}{}^*$ from (1), by means of the most probable values for P_{36}, $P_{36}{}^*$, obtained by using

relative weights in the relation between P_{36} and U_0 according to the general procedures outlined above. From 1966 to 1972, values for $P_{36}{}^*$ and $T_{36}{}^*$ are similarly obtained except that P_{36} (W removed) is replaced by V, from which the unknown wave component w is eliminated with

$$V = P_{36}{}' - N_{36}{}' \tan 52° \qquad (2)$$

in which the observed components $P_{36}{}'$ and $N_{36}{}'$ include the yearly contributions w of the unknown 'wave,' which is assumed to be in the same direction as W (i.e., 128° east of the sun). It was indicated earlier by means of the parameters in Table 3 that, when W is deducted (1937–1965), N_{36} (i.e., its most probable value) vanishes and is independent of U_0. Part

TABLE 3. Correlation Coefficients r between Yearly Means of x and y and Weighted Least Squares Parameters in $y = a + bx$ for Weight w_x of x Relative to w_y

y	x	r	w_x	a, 0.01%	b, 0.01%/γ
$N_{-16} - W$	U_0	0.75	0.110	−10.3	0.226
P_{-16}	U_0	0.59	0.182	−7.8	0.180
P_{-16}	$N_{-16} - W$	0.65	2.55	+0.4	0.794*
P_{36}	U_0	0.75	0.174	−12.9	0.288

* Dimensionless.

A of Figure 4 also shows that yearly values of the 0.0-hour component (W deducted) (i.e., N_{36}) are negligible within statistical uncertainties. Thus the observed component N_{36}' is used in (2) to correct the observed component P_{36}' for the contribution from the unknown w. For the period 1927–1965 the correlation between yearly values of V from (2) and P_{36} (W removed) shows that these procedures give essentially equivalent results and also that V may be used in place of P_{36} in the bottom row of Table 3. The most probable values of V and P_{36} are practically identical from 1927 to 1965. Thus from 1966 to 1972 the yearly values of V are substituted for P_{36} in (1) to

give the open squares in part B of Figure 4 from 1966 to 1972. Similarly, the most probable values of V (from its linear relation with U_0) in place of those for P_{36} in (1) result in the solid circles in part B of Figure 4 from 1966 to 1972, representing the most probable or 'predicted' values of T_{36} for the total 18.0-hour component.

Yearly values of w, which are to be fitted by W_G from 1966 to 1972, are then determined from yearly values of the component N_{-16} by using the relation

$$w = (N_{-16} - V^* \cos 38°) \qquad (3)$$

in which V^* is the most probable value of V for the year and 38° is the angle between the

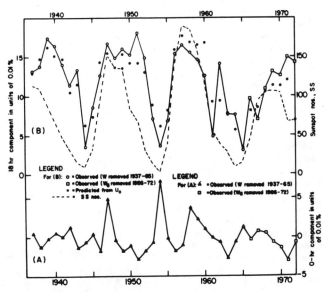

Fig. 4. Yearly mean total diurnal anisotropy components, W being removed from 1937–1965 and W_G from 1966–1971: A, with maximum at 0.0 hour; B, with maximum at 18.0 hours LAT.

axes P_{36} and N_{-18} (Figure 1). These values of w indicated by the open squares (1966–1972) in part C of Figure 3 are fitted by the curve W_o for the 'guesstimated' variation from 1966 to 1972 of the component that prior to 1966 was well fitted by W; W_o is also plotted in part B of Figure 3.

The open squares and W_o in part C of Figure 3 indicate that W_o became quite definitely negative in 1971 and 1972. Dr. Robert Howard of the Hale Observatories of the Carnegie Institution of Washington finds that about the middle of 1971 the sun's north polar magnetic field changed from negative to positive [*Howard*, 1972]. In 1958, when W changed from negative to positive, Dr. Harold Babcock of the Hale Observatories found that the sun's north polar field changed from positive to negative [*Babcock*, 1959].

RESIDUALS AFTER REMOVING W (OR W_o) AND THE MOST PROBABLE VALUES OF THE 18.0-HOUR COMPONENT T_{36}

Figure 5 shows a 24-hour harmonic dial on LAT for annual mean residuals after deducting W (1937–1965) or W_o (1966–1971) and the most probable values for T_{36} obtained from its relation with U_o in (1). The 0.0-hour component is that shown in part A of Figure 4, and the 18.0-hour component is that observed less that predicted in part B of Figure 4. The radius of the 'probable error' circle in Figure 5 is 2.4 (in units of 0.01%). Tests indicate that residuals for successive years are statistically independent and that a residual as large as or larger than that for 1954 is expected in about 1 in 24 samples of 35 years each, and thus the deviation for 1954 may be regarded as being statistical with a probability of 0.04.

The two-dimensional variance in Figure 5 is $M^2 = 8.6$ when the two coordinates are in units of 0.01%. As was mentioned earlier, the residual two-dimensional variance for a single station-year is 6.0 in the same units, or 2.0 for the mean of three stations. The 2.0 value results from instrument counting rate uncertainties, variation in annual mean temperature effect, and inhomogeneities in the data available.

If this variance is deducted from that for Figure 5, the remaining variance is 6.6, and the radius of the resulting probable error circle becomes about 2.1 (in units of 0.01%) for single

years, and thus it represents the uncertainties in the yearly values for W (or W_o) and for T_{36} that were deducted. The variance among years (Table 2b) is 140.5, or about 21 times the remaining variance of 6.6. Together, these results show that the observations are well approximated by the vector sum of the wave component 128° east of the sun and the component 90° east of the sun estimated from U_o.

SUMMARY

An earlier investigation [*Forbush*, 1969] showed that annual means of the cosmic ray diurnal anisotropy were well approximated by the sum of two distinct components w and T_{36}, owing to different mechanisms. Component T_{36} is the component with its maximum at 18.0 hours LAT, its amplitude being dependent on magnetic activity. Component w is the component with its maximum at 20.5 hours LAT (128° east of the sun). From 1937 to 1965 the amplitude of w varied sinusoidally about zero mean and was well fitted by a wave W with a

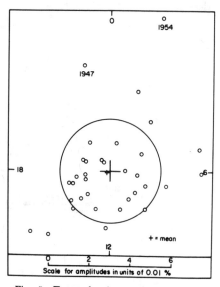

Fig. 5. Twenty-four-hour harmonic dial on LAT for annual mean residuals after deducting W (1937–1965), W_o (1966–1971), and corrections for U_o. The 0-hour component is from part A of Figure 4, and the 18-hour component is the observed minus predicted from part B of Figure 4.

period of 20 years (or two sunspot periods during that interval). Wave W changed from negative to positive in 1958, when the sun's north polar magnetic field changed from positive to negative. The present investigation, based on cosmic ray data through 1972, shows that after 1965 the period of W changed (not surprisingly, since the sunspot cycle also lengthened) and that the most recent reversal (from positive to negative) in the w component occurred in 1971, again consistent with the observation by *Howard* [1972] that in 1971 the north polar field of the sun changed from negative to positive.

Schatten and Wilcox [1969] interpret the wave W in the diurnal anisotropy component 128° east of the sun in terms of enhanced magnetic reconnection between the nearby galactic field and field lines in the polar regions of the heliosphere during half of the 20-year (approximately) cycle. This interpretation leads to the conclusion that the component parallel to the solar rotation axis of the nearby galactic field is directed northward.

Acknowledgments. Successful operation of cosmic ray ionization chambers since 1937 is due to wholehearted gratuitous cooperation of organizations and individuals. For this, grateful appreciation is expressed to the U.S. Coast and Geodetic Survey and the staff of its geomagnetic center at Fredericksburg, Virginia; to the government of Peru and the Director of the Instituto Geofisico del Peru and the staff of its magnetic observatory in Huancayo, Peru; and to the Department of Scientific and Industrial Research and the staff of its geophysics division at Christchurch, New Zealand.

Special appreciation is wholeheartedly expressed to Mrs. L. Beach of the Department of Terrestrial Magnetism for indispensible, efficient, and reliable reduction of records, processing of data, and other assistance.

* * *

The Editor thanks H. Carmichael and J. M. Wilcox for their assistance in evaluating this paper.

REFERENCES

Axford, W. I., The modulation of galactic cosmic rays in the interplanetary medium, *Planet. Space Sci., 13*, 115, 1965.

Babcock, H. D., The sun's polar magnetic field, *Astrophys. J., 130*, 364, 1959.

Forbush, S. E., On diurnal variation in cosmic ray intensity, *Terr. Magn. Atmos. Elec., 42*, 1, 1937.

Forbush, S. E., Variation with a period of two solar cycles in the cosmic ray diurnal anisotropy and the superposed variations correlated with magnetic activity, *J. Geophys. Res., 74*, 3451, 1969.

Forbush, S. E., and L. Beach, The absolute geomagnetic field of the equatorial ring current, *Carnegie Inst. Wash. Yearb., 65*, 28, 1967.

Howard, R., Polar magnetic fields of the sun 1960-1971, *Solar Phys., 25*, 5, 1972.

Parker, E. N., Theory of streaming of cosmic rays and the diurnal variations, *Planet. Space Sci., 12*, 735, 1964.

Schatten, K. H., and J. M. Wilcox, Direction of the nearby galactic magnetic field inferred from a cosmic-ray diurnal anisotropy, *J. Geophys. Res., 74*, 4157, 1969.

(Received July 13, 1973; accepted September 11, 1973.)

REVIEWS OF GEOPHYSICS AND SPACE PHYSICS, VOL. 20, NO. 4, PAGES 971-976, NOVEMBER 1982

Random Fluctuations, Persistence, and Quasi-Persistence in Geophysical and Cosmical Periodicities: A Sequel

S. E. FORBUSH, S. P. DUGGAL, M. A. POMERANTZ, AND C. H. TSAO

Bartol Research Foundation of the Franklin Institute, University of Delaware, Newark, Delaware 19711

The method of Chree analysis (superposed epochs) has been used in many disciplines, including geophysics, astrophysics, and solar-terrestrial relationships. However, procedures to test the statistical significance of the results obtained thereby have not been available heretofore. Claims for statistical reality of average variations (from Chree analyses or otherwise) are unacceptable without testing the assumption that deviations from the average variation are random and sequentially independent. In many phenomena this assumption is not valid. One objective of this paper is to expand established methods for the analysis of variance in order to provide a statistical procedure for testing the Chree analysis result from data with nonrandom deviations from average. In addition, the statistical method developed by Bartels (1935) to determine the quasi-persistency of deviations from average signals in the form of sine waves is also applied to the Chree analysis problem. The two alternative procedures for evaluating the significance of Chree analysis results (or variations otherwise obtained) are then compared to determine the circumstances under which one of them may be preferable.

INTRODUCTION

The method of Chree analysis or superposed epochs [*Chree*, 1912, 1913; *Chapman and Bartels*, 1940] was originally conceived for studying geophysical time variations. For example, in the first publications of this procedure, *Chree* [1913] reported a 27-day recurrence tendency in geomagnetic data. In this study he selected 5 days (during each month) which were characterized by the largest daily range in the magnetic horizontal force at Kew Observatory in England. The average variation on these days, covering the period 1890–1900, as well as on 5 days before and 35 days later demonstrated that the disturbances on selected days were followed by similar variations roughly 27 days later, as shown in Figure 1. For a later period, 1906–1911, he used international 'character' figures to obtain the second similar plot in this figure.

In recent years this technique has been used in several disciplines either for testing the relationship between two diverse phenomena or to search for periodicities in the data. Unfortunately, thus far, proper statistical tests have not been utilized for assessing the statistical reality of 'signals' revealed by Chree analysis. Hence the literature is full of examples in which the authors either have subjectively accepted the signal revealed by Chree analysis as real or have performed simple, statistical tests for the signal by assuming that the signal is imbedded in random data.

However, a serious problem in this and other techniques utilized for investigating time series arises from the fact that the data (deviations from average) often are not random or sequentially independent. In this case, if quasi-persistence is not taken into account, the effective standard deviation of the residuals may be seriously underestimated, resulting in a false claim for the statistical reality of the average variation.

The classical paper of *Bartels* [1935] comprises the most comprehensive work on the methods of analyses for testing the statistical significance of averaged variations, with original rigorous methods for dealing with data which contain

Paper number 2R0712.
0034-6853/82/002R-0712$05.00

nonrandom deviations from average. The importance of this seminal contribution to geophysics cannot be overestimated, in view of the fact that the standard textbooks on statistics deal only with random deviations. However, many geophysical time variations contain deviations which are neither random nor sequentially independent, because they can include trends caused by more than one phenomenon.

Since the methods developed in the present paper are extensions of the ideas first suggested by Bartels, we have taken the liberty of adopting the same title as that of his original paper (hereinafter designated paper 1) of which the present work is a natural sequel, thus emphasizing the importance of Bartels publication almost half a century earlier.

It is obvious that the Chree analysis result represents an averaged variation. Hence in this paper we shall first extend the established methods for the analysis of variance to provide procedures for testing the statistical reality of an observed average variation. These methods will further be applied to test the statistical significance of the average wave of a given frequency fitted to the average variation above. Finally, we will show that in the harmonic dial for the average vector (except for quite small samples), the method described here and one first developed by *Bartels* [1935] provide identical measures for the statistical significance of an average wave.

The *F* test, which is essential in the analysis of variance methods, takes proper account of sampling variations insofar as they affect the ratio of variances. The probabilities of obtaining an average as large as or larger than that actually obtained from the two alternative methods are compared. These probabilities may be crucially increased if the residuals are quasi-persistent rather than random and sequentially independent. In paper 1, Bartels showed how to derive an effective standard deviation (or variance) for quasi-persistent residuals that could be used to obtain a reliable measure of the statistical reality of the average. We will show how these considerations also apply to the *F* test for the analysis of variance. Procedures will also be described for determining the effective variance of residuals which are quasi-persistent.

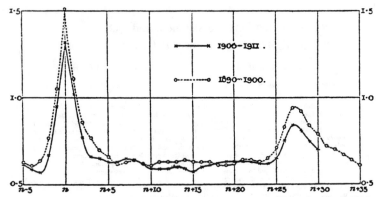

Fig. 1. Original analysis by *Chree* [1913], in which the epoch date *n* represents the 5 days each month on which the daily range of the horizontal magnetic force was greatest. For the period 1890–1900 the data are observations at Kew Observatory, whereas for 1906–1911 they are international character figures.

The measure of statistical significance from application of the analysis of variance to an observed average will be compared with the corresponding measure obtained from the application of the analysis of variance to the harmonic coefficients defining a wave fitted to the same observed average, in order to provide some basis for identifying the circumstances under which one or the other procedure may be preferable. Tests both for the homogeneity of variance that is required for the validity of the *F* tests and for the effect of quasi-persistence on these tests will also be performed.

ANALYSIS OF VARIANCE TO DETERMINE WHETHER AN AVERAGE VARIATION IS STATISTICALLY SIGNIFICANT

The method of Chree analysis procedure has become an important and widely used tool for evaluating the statistical relationship between events of two different types, for example, the occurrence of intense solar flares and variations in the intensity of cosmic rays. For studies of this nature the days characterized by intense solar flare events are defined as key days (day 0). The cosmic ray flux on days before $(-1, -2, \cdots)$ and after $(+1, +2, \cdots)$ each key day constitute an epoch. In many studies of solar-terrestrial relationships the duration of each epoch is set at 27 days, the nominal rotational period of the solar surface, although the choice could be arbitrary. The average (columns) of these epochs (rows) provides the basic result, or signal revealed by the method of superposed epochs. By this process of superposition the random noise is diminished, and hence the significance level of the signal is enhanced as the number of epochs increases.

The detailed statistical procedure based on the analysis of variance for testing the significance of the Chree analysis result has been described elsewhere [*Forbush et al.*, 1982]. To best illustrate the significance and application of this procedure, it was considered advantageous to use synthetic data. Here we will summarize the results of the analysis of variance and the determination of the contribution by non-random trends in the data in order to demonstrate the pitfalls of ignoring quasi-persistency. Thus the average 27-day varia-

tion in Figure 2 was obtained from Chree analysis of synthetic data tabulated in the manner shown schematically in Table 1 for 150 epochs (or rows) of 27 daily values each. Means $\bar{R}_1, \bar{R}_2, \cdots, \bar{R}_{150}$ for each 27-day epoch are indicated in the column headed M_R. Column means $\bar{C}_1, \bar{C}_2, \cdots, \bar{C}_{27}$ (Table 1) are the ordinates in Figure 2. In Table 1 a given entry for row R, column C may be represented by the sum of the mean for row R, plus the mean for column C, plus a residual Δ_{RC}.

These residuals may be entered in a table corresponding to Table 1. The variance of single residuals Δ_{RC}, designated S_Δ^2, may be obtained directly from the 150 × 27 residuals, in our example, with

$$S_{\Delta_1}^2 = \sum_1^{4050} (\Delta_{RC})^2/3874 = 0.593 \tag{1}$$

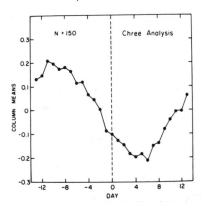

Fig. 2. Chree analysis of the data with 150 epochs (rows), each consisting of 27 days (columns). Contrary to the normal practice, no error bars are shown on the averaged variation (column means), since in the presence of quasi-persistency in the basic data, this can lead to erroneous conclusions.

458

TABLE 1. Matrix Used for Testing the Statistical Significance of Results Derived by Superposed Epoch Analysis

	Day					
Epoch	1	2	...	26	27	M_R
1						$\hat{R}_1$
2						$\hat{R}_2$
3						$\hat{R}_3$
.						
.						
.						
150						
M_C	$\bar{C}_1$	$\bar{C}_2$	...			M_T

in which 3874 is the number of degrees of freedom (i.e., 149 × 26).

From the 27 column means $\bar{C}_i$ we may obtain a second estimate $S_{\Delta_2}^2$ for the variance of single residuals with

$$S_{\Delta_2}^2 = [\Sigma(\bar{C}_i - M_T)^2/26] \times 150 = 2.94 \qquad (2)$$

in which 26 is the number of degrees of freedom for this estimate. Since there are 150 rows, (2) gives an estimated variance for single residuals. These estimates from (1) and (2) would differ only as a consequence of sampling variations if there were no persistent 27-day variation, and consequently, each column mean is an average of 150 random, independent residuals from a normal population with variance σ_p^2 for single values.

To determine whether to accept the hypothesis that the column means contain a real, persistent 27-day variation, the ratio F is computed with

$$F_{3874,26} = 2.94/0.593 = 4.96 \qquad (3)$$

in which 2.94 is the value of S_Δ^2 from (2) and 0.593 that from (1); 3874 and 26 are the degrees of freedom for the numerator and denominator, respectively.

From available F tables [Beyer, 1966], one can find the probability $(1 - P)$ that the observed F will be equaled or exceeded if the two variances used to obtain F are from samples from the same normal population. If this probability $(1 - P)$ is small enough, then we may conclude that the average variation is statistically significant. However, it must be emphasized that this conclusion is valid only provided the residuals Δ_{RC} are random, sequentially independent, and normally distributed and the variance of residuals is homogeneous among columns and rows. The large value of F in (3) is not covered by standard tables. Consequently, the probability $(1 - P)$ is calculated from [Beyer, 1966]

$$P = \int_0^F \frac{\Gamma[(m + n)/2]}{\Gamma(m/2)\,\Gamma(n/2)} m^{m/2} n^{n/2} \chi^{[(m/2)-1]} (n + m\chi)^{-(m+n)/2} d\chi \qquad (4)$$

where m and n represent degrees of freedom for the numerator and denominator, respectively.

For $m = 3874$ and $n = 26$ it is found that

$$(1 - P) \approx 3 \times 10^{-15}$$

Since this probability is infinitesimal, it would ordinarily be concluded that the 'signal' in Figure 2 is most definitely real.

Let us now examine the change in $(1 - P)$ which results from taking proper account of quasi-persistence in the residuals. The crucial test for randomness and sequential independence was first used in paper 1 to determine whether the vector deviations from the average vector in Bartels' harmonic dial were random and sequentially independent. To test this, he designated the standard deviation for single vector deviations as $M(1)$ and that for means of h successive nonoverlapping vector deviations as $M(h)$.

If these sample deviations are from a random normal population, then

$$M(h)h^{1/2}/M(1) = C(h) \approx 1 \qquad (5)$$

The approximation in (5) is in recognition of sampling variations in finite samples from the parent population.

It was shown in paper 1 that for nonrandom quasi-persistent deviations, $C(h)$ when plotted as a function of $h^{1/2}$ increases from unity to an asymptotic limit which Bartels designated $\sigma^{1/2}$. He further demonstrated that the effective standard deviation $M_e(n)$ for the mean of n observations is given by

$$M_e(n) = \sigma^{1/2}M(1)/n^{1/2} \qquad (6)$$

Thus, multiplying by $\sigma^{1/2}$, the observed standard deviation $M(1)$ obtained for single deviations gives the effective standard deviation $M_e(1)$ obtained as though the n quasi-persistent deviations were random and independent. Bartels designated σ as the 'effective length of sequences' to indicate that with quasi-persistence of the h vector residuals, the standard deviation behaved as if every successive σ individuals are the same and only h/σ of them are independent.

This method may be applied to determine whether the residuals Δ_{RC} used in (1) are quasi-persistent. These residuals may be tabulated in 27 columns for 150 rows in our example of synthetic data. Let $V(1)$, $V(2)$, $\cdots$, $V(h)$ indicate the pooled residual variance of the 27 column means for single rows, for means of two rows, for means of three rows, $\cdots$ and for means of h rows, respectively. If we let

$$\zeta(h) = [V(h)]^{1/2} \cdot (h)^{1/2} \qquad (7)$$

then in analogy with (5) the characteristic $C(h)$ is

$$C(h) = (h)/\zeta(1) \qquad (8)$$

It should be pointed out that if h is not an exact fraction of 150 (say $h = 7$), it is proper to obtain only 21 samples of $V(7)$ for calculating the pooled variance. Figure 3 shows that when plotted as a function of $h^{1/2}$, $C(h)$ approaches the asymptotic value 2.4, indicating quasi-persistence with an effective standard deviation of $2.4 \times \zeta(1)$. Thus the effective variance of residuals is $(2.4)^2 = 5.76$ times the value 0.593 obtained from (3) under the assumption that the residuals were random. Consequently, using the effective variance, $5.76 \times 0.593 = 3.92$, for the denominator in (3), F becomes 0.86, indicating that there is no reason for rejecting the hypothesis that the two estimates of variance in (3) pertain to the same population. Thus the average 27-day variation is not statistically significant.

The application to the simulated data of Bartlett's test [Hald, 1952; Dixon and Massey, 1957] for homogeneity of variances reveals no reason for rejecting the hypothesis that the variances within columns are homogeneous [Forbush et al., 1982]. It should be emphasized here that since Bartlett's test is based on the assumption that the basic data are random and sequentially independent, it is important that the

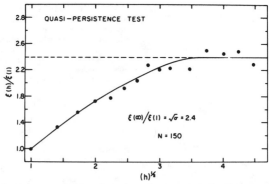

Fig. 3. Analysis of variance procedure for evaluating the quasi-persistence in the data. The asymptotic value $\zeta(\infty)/\zeta(1)$ ≈ 2.4 indicates that the equivalent length of sequences, σ, is 5.8 epochs.

column variances are separated in chronological order by at least $\sigma = 5.76$, or 6 units.

Strictly speaking, the analysis of variance is valid only if the data are normally distributed. Of course, it could be determined by application of the chi-square test whether the distribution of individuals departs significantly from normality.

However, according to *Dixon and Massey* [1957], the result of the analysis are changed very little by moderate violations of the assumptions of normal distribution and of equal variance. In addition, the central limit theorem insures that means from samples of N from arbitrary distributions are very closely approximated by a normal distribution even if N is quite small, say 10 or so. Since many causes combine to result in the observed residuals, the central limit theorem also indicates a near-normal distribution of single deviations.

ANALYSIS OF VARIANCE TO TEST THE SIGNIFICANCE OF AN AVERAGE SINUSOIDAL VARIATION OF A SINGLE FREQUENCY

Suppose we have a number of epochs or intervals with, for example, 12 observations in each epoch. Let the harmonic coefficients for the wave of frequency 1 in the interval 1 be a_1 and b_1. These coefficients may be listed in the format of Table 2.

The departures from means $(a_1 - \bar{a}_1)$ and $(b_1 - \bar{b}_1)$ may also be tabulated in a table analogous to Table 2. Analysis of variance may be used to determine the statistical reality of the amplitude of the average wave, i.e., $C = (\bar{a}_1{}^2 + \bar{b}_1{}^2)^{1/2}$. In this case,

$$F = \frac{n(\bar{a}_1)^2 + n(\bar{b}_1)^2}{S^2(a_1) + S^2(b_1)} = \frac{n(\bar{a}_1{}^2 + \bar{b}_1{}^2)}{2S^2(\bar{a}_1)} \quad (9)$$

where $S^2(a_1)$ is the variance of a single residual Δa_1. We assume that $S^2(a_1) = S^2(b_1)$, which can be tested; a significant difference between these variances may indicate that perhaps linear trends in the data have not been effectively eliminated; otherwise the variances may be homogenized. Then $2S^2(a_1) = S^2(C)$, where C is the amplitude of the wave. Referring to paper 1, we note that Bartels' parameter for

testing the statistical reality of the average wave is given by the ratio

$$\kappa = \frac{C}{M(n)} = \frac{h^{1/2}[(\bar{a}_1)^2 + (\bar{b}_1)^2]^{1/2}}{M(1)} \quad (10)$$

where C is the amplitude of the average wave and M is the expectancy obtained from observed values for means of n:

$$M(n) = \frac{[\Sigma(\Delta a_1)^2 + \Sigma(\Delta b_1)^2]^{1/2}}{n} \quad (11)$$

From (9),

$$F = nc^2/M^2(1) = \kappa^2 \quad (12)$$

so F from the analysis of variance is the same as Bartels' κ^2. In the analysis of variance above for vectors in the harmonic dial, the number of degrees of freedom for the numerator is 2, and n is that for the denominator. From F tables it can be determined (let $n \to \infty$ in the formula for F) that

$$F(2, \infty) = e^{-\kappa^2} \quad (13)$$

so that for large n the analysis of variance and Bartels' $e^{-\kappa^2}$ give the same probability P_c of obtaining by chance (from a population with no wave) an average wave with amplitude as large as or larger than that which was actually obtained.

For quite small values of n the values of F from harmonic coefficients result in correct values of P_c which are somewhat larger than those from e^{-F}. Even for values of n greater

TABLE 2. Harmonic Coefficients

Interval	a_1	b_1	c_1
1			
2			
3			
.			
.			
.			
n			
Mean	$\bar{a}_1$	$\bar{b}_1$	

than about 15 or 20 the difference between these two probabilities is not of practical importance.

If tests for randomness and sequential independence of deviations indicate quasi-persistence, then in (9) and (11) the effective expectancy and variance, respectively, must be used. These are obtained by multiplying the observed expectancy by the asymptotic value for $C(h) = C(h)h^{1/2}/C(1)$ as given in Figure 3.

Assuming that the quasi-persistence effects have been removed, let us examine the power of the two procedures described here when the average variation is fairly well fitted by a wave.

Let

σ_r^2 — variance of residuals;
S^2 — contribution of signal to variance of column means;
σ_r^2/N_R — variance of column means for N_R rows;
$(\sigma_r^2/N_R) + S^2$ — variance of single column mean from residual variance and signal.

$$F = \frac{((\sigma_r^2/N_R) + S^2)N_R}{\sigma_r^2} \quad (14)$$

$$F = 1 + \left(\frac{S^2}{\sigma_r^2}\right)N_R \quad (15)$$

For the harmonic analysis procedure, κ^2 (see (10)) can be written as

$$\kappa^2 = \frac{C_1^2}{M^2(N_R)} = \frac{C_1^2 r N_R}{4\sigma_r^2}$$

where

C_1 — amplitude of first (or other) harmonic;
$M(N_R)$ — expectancy for mean from N_R rows;
r — number of ordinates or columns.

For $\sigma_r^2 = 1$ and $r = 6$ columns,

$$\kappa^2 = \tfrac{3}{2}C_1^2 N_R$$

$$F(5, N_R) = 1 + S^2 N_R = 1 + (C_1^2/2)N_R$$

since for a single wave signal, $S^2 = C_1^2/2$.

In Table 3, starting from a given probability $e^{-\kappa^2}$ that the signal has appeared by chance, the corresponding significance level of the F distribution is listed. It can be noted that for the F test the probability $(1 - P)$ that the signal has appeared by chance is, in every case, larger than the corresponding probability $e^{-\kappa^2}$ in the harmonic analysis procedure. Conversely, for example, if the probability that the signal has appeared by chance is 0.001, we find from Table 3 that the amplitude of the signal, C_1, must be larger

(22–59%) for the F test to reveal the same level of significance as does the harmonic analysis procedure.

Another interesting way to compare these procedures is to assume that the original signal S consists of several harmonics, but only the first harmonic is used in the analysis (for the F test, of course, the total signal is effective). For simplicity, we consider only two harmonics; the amplitude of the second harmonic, C_2, has been calculated to make the probability $e^{-\kappa^2}$ that the signal has appeared by chance equal to the corresponding significance level $(1 - P)$ of the F test. For all three cases the amplitude C_2 is comparable to the amplitude of the first harmonic, C_1 (Table 3). Thus the two procedures give identical significance levels even when significant higher harmonics are present and neglected. In other words, the power of the harmonic analysis procedure is greater than that of the F test.

DISCUSSION

We have shown how the established procedures used in the analysis of variance may be extended to test the statistical reality of an observed average variation such as that obtained from Chree analysis. However, extreme caution is required when, as is usually the case, the residuals are neither random nor sequentially independent. This quasi-persistence may lead to a gross underestimate of the standard deviation of residuals. We have described an extension of the analysis of variance that can be utilized for evaluating the degree of quasi-persistence and for testing the statistical reality of an observed average variation.

With the help of an example (simulated data) it has been shown that by ignoring the quasi-persistent nature of residuals, the average variation can masquerade as a significant 'signal.'

The aforementioned procedure has also been applied to test the statistical significance of a wave of a given frequency fitted to the average variation. It is found that for vector averages in a harmonic dial from moderately sized samples this method and that developed by *Bartels* [1935] are not practically different. For very large samples the two results are identical.

Although the analysis of variance of an observed averaged variation is equivalent to the analysis of variance of all the significant waves that can be fitted to the same averaged variation, the latter procedure may require a complicated multidimensional analysis for the final test of the reality of the observed average variation. However, if the hypothesis to be tested is limited to only one frequency, for example, in case of a theoretical prediction of a single periodicity, Bartels' procedure is advantageous, since it is not affected by the noise from other frequencies.

Acknowledgment. This research is supported by the National Science Foundation's Division of Atmospheric Sciences Section under grant ATM-8005866.

REFERENCES

Bartels, J., Random fluctuations, persistence, and quasi-persistence in geophysical and cosmical periodicities, *J. Geophys. Res.*, 40, 1, 1935.
Beyer, W. H., *Handbook of Tables for Probability and Statistics*, The Chemical Rubber Co., Ohio, 1966.
Chapman, S., and J. Bartels, *Geomagnetism*, vol. I, Oxford University Press, New York, 1940.
Chree, C., Some phenomena of sunspots and of terrestrial magne-

TABLE 3. Significance Levels and Probabilities

$e^{-\kappa^2}$	r	N_R	C_1	F	$(1 - P)$	C_2
0.001	6	120	0.196	3.30	0.008	0.137
3×10^{-5}	6	120	0.239	4.42	0.001	
0.001	12	120	0.138	2.15	0.02	0.128
3×10^{-6}	12	120	0.188	3.13	0.001	
0.001	25	120	0.096	1.55	0.07	0.119
3×10^{-8}	25	120	0.153	2.40	0.001	

tism at Kew Observatory, *Philos. Trans. R. Soc. London, Ser. A*, *212*, 75, 1912.

Chree, C., Some phenomena of sunspots and of terrestrial magnetism, II, *Philos. Trans. R. Soc. London, Ser. A*, *213*, 245, 1913.

Dixon, W. J. and F. J. Massey, *Introduction to Statistical Analysis*, McGraw-Hill, New York, 1957.

Forbush, S. E., M. A. Pomerantz, S. P. Duggal, and C. H. Tsao,

Statistical considerations in the analysis of solar oscillation data by the superposed epoch method, *Sol. Phys.*, in press, 1982.

Hald, A., *Statistical Theory With Engineering Applications*, John Wiley, New York, 1952.

(Received January 26, 1982;
accepted April 28, 1982.)

STATISTICAL CONSIDERATIONS IN THE ANALYSIS OF SOLAR OSCILLATION DATA BY THE SUPERPOSED EPOCH METHOD*

S. E. FORBUSH, M. A. POMERANTZ, S. P. DUGGAL[†], and C. H. TSAO

Bartol Research Foundation of The Franklin Institute, University of Delaware, Newark, Delaware 19711, U.S.A.

Abstract. Although the method of superposed epochs (Chree analysis) has been utilized for seven decades, a procedure to determine the statistical significance of the results has not been available heretofore. Consequently, various subjective methods have been utilized in the interpretation of Chree analysis results in several fields. The major problem in the statistical treatment of Chree analysis results arises from the fact that in most studies of natural phenomena, data are neither random nor sequentially independent. In this paper, a statistical procedure which takes this factor into account is developed.

1. Introduction

For investigating the possible relationship between two sets of geophysical observations, Chree (1912, 1913) introduced a procedure for analyzing one set of measurements during epochs which were selected on the basis of a specific type of feature in the second set of measurements. The method of superposed epochs can also be used for investigating basic periodicities in time series of data (see e.g., Chapman and Bartels, 1940). This version of the superposed epoch technique is germane in the present context, since it has been utilized by several groups for investigating certain aspects of global solar oscillations (Severny *et al.*, 1976; Scherrer *et al.*, 1979; Grec *et al.*, 1980).

However, unfortunately, despite its long history, a proper statistical test for evaluating the significance level of the results obtained by superposed epoch analysis has not been available heretofore. This lack of a quantitative 'figure of merit' of the results of applications of the Chree procedure has led to controversial situations arising from different interpretations of the reality of an apparent signal. The fact that proper statistical methods have generally not been available for assessing Chree analysis results arises from a fundamental problem: Data representing natural phenomena are neither random nor sequentially independent. Consequently, the basic criterion for the application of standard statistical procedures is, in fact, violated.

The pitfalls of ignoring the non-randomness of data representing observations of natural phenomena were first demonstrated by Bartels (1935; see also Chapman and Bartels, 1940). He introduced the concept of quasi-persistency and developed a procedure for calculating the standard error by evaluating the extent of its effect. In this

---

* Proceedings of the 66th IAU Colloquium: *Problems in Solar and Stellar Oscillations*, held at the Crimean Astrophysical Observatory, U.S.S.R., 1–5 September, 1981.
† Shakti P. Duggal died, July 11, 1982.

Solar Physics **82** (1983) 113–122.   0038–0938/83/0821–0113$01.50.

paper we will describe a statistical procedure based on analysis of variance that takes into account the quasi-persistency and is suitable for testing the significance of Chree analysis results. (For a complete review, see Forbush *et al.*, 1982.)

It should be emphasized that the purpose here is not to discuss previous analyses of solar oscillation data by superposed epoch analysis. Rather, we wish to issue a caveat to the solar-physics community that conclusions based upon superposed epoch analysis must be viewed with extreme skepticism unless it is unambiguously demonstrated that both the nature of the statistical tests that are applied, and the assignment of error bars or other indices of the probable reality of a signal are strictly legal.

2. Chree Analysis

Let us assume that on the basis of some observational criterion (e.g., an unusually high value of a particular geomagnetic index) N key-days are associated with some variation in the data under investigation (e.g., the cosmic ray intensity). In the method of super-posed epoch analysis, each key-day is designated as the center of an epoch (day zero), the length of which is selected on the basis of a physically plausible period (e.g., the 27-day solar rotation period). We then list the data in the form of a matrix in which the rows r_j represent the epochs, and the columns c_i represent days before and after the individual key-days c_{13} as in Table I.

The column averages of this matrix, which will invariably show some variations, represent the Chree analysis result. We will refer to this result as the signal. The objective of the procedures described in this paper is to determine its significance level (i.e. the probability that it did not occur by chance).

Classical statistical procedures may be (and ordinarily are) followed for evaluating the apparent significance level of the variance attributable to the signal. However, as will become clear later, this is grossly erroneous because the data for any epoch are not sequentially independent. In general, there are real effects, in addition to the one under study, which can cause the measured phenomenon to vary over different time scales. This is the nub of the problem.

3. Statistical Test

A. ANALYSIS OF VARIANCE

Table I represents the data matrix in a typical Chree analysis. In this example, we assume that there are 150 epochs ($r = \sum r_j = 150$) each comprising 27 days ($c = \sum c_i = 27$). The statistical test of the resulting signal can be performed as follows:
   (1) Remove the linear trend, if any, from each row r_j.
   (2) Calculate the variance of the population S_c^2 from the column means $\bar{c}_i$:

$$S_c^2 = r \sum_{i=1}^{c} \frac{(\Delta \bar{c}_i)^2}{c-1} \,, \tag{1}$$

where

r = total number of rows,

c = total number of columns.

(3) Calculate the variance S_r^2 of the population from the row means $\bar{r}_j$:

$$S_r^2 = c \sum_{j=1}^{r} \frac{(\Delta \bar{r}_j)^2}{r-1} \,. \tag{2}$$

(4) Calculate the total variance S_T^2 from individual data points (x_{ij}):

$$S_T^2 = \sum_{i=1}^{c} \sum_{j=1}^{r} \frac{(\Delta x_{ij})^2}{rc-1} \,. \tag{3}$$

(5) Calculate the residual variance of the population S_R^2:

$$S_R^2 = \frac{[(rc-1)S_T^2] - [(c-1)S_c^2] - [(r-1)S_r^2]}{(c-1)(r-1)} \,. \tag{4}$$

(6) Test whether the variances of single rows and columns are homogeneous. One possible test is described in Appendix I.

(7) At this stage, let us first assume for simplicity that there is no quasi-persistency in the data. Under this assumption (which in most cases is not valid) the signal variance which is represented by S_c^2 can be compared with the residual variance S_R^2 of the data by using the F test with $(c-1)$, $(c-1)(r-1)$ degrees of freedom, df. If this test reveals

TABLE I

Data matrix that is generally used in Chree analysis. x_{ij} represents a single data point for column i and row j. $\bar{c}_i$ and $\bar{r}_j$ represent the averages for columns and rows respectively. Day 0 is termed key day.

CHREE MATRIX – $\alpha_{ij}(1)$

	DAY →				
EPOCH ↓	−13	−12	------------------------ 0 ------------------------	+ 12	+ 13
	1	2	---------------------- 13 ----------------------	26	27
1					$\bar{r}_1$
2					$\bar{r}_2$
. . . . .			x_{ij}		$\bar{r}_j$
150					
	$\bar{c}_1$	$\bar{c}_2$	$\bar{c}_i$	$\bar{c}_{26}$	$\bar{c}_{27}$

that the signal is not significant, there is no need to proceed further, since the determination of quasi-persistency leads only to an increase in the residual variance.

B. QUASI-PERSISTENCY

In order to provide a physical picture, let us assume that the signal in each row can be represented by a sine wave. In this case, following Bartels (1935), we define quasi-persistence as a periodicity which repeats for a certain number of epochs with approximately the same phase and amplitude, each such sequence ending more or less abruptly without any relation to other sequences. An example of quasi-persistent vectors derived from simulated data (see Appendix 2) is shown in Figure 1. To determine the

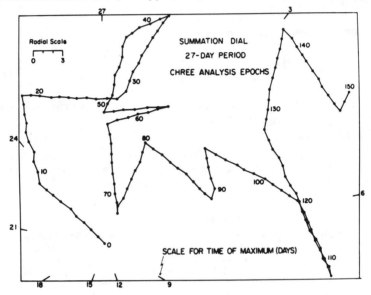

Fig. 1. Summation dial for the 27-day period. Vectors represent the 150 epochs of the simulated data. Every 10th epoch is marked at the end of the corresponding vector.

realistic residual variance, S_R^2 for quasi-persistent data, we transform the original Chree analysis matrix α_{ij} (1) into a new matrix in which the first row represents the mean of the first two rows of the original matrix, i.e.:

$$X_{i1}' \text{ (new matrix)} = \frac{(X_{i1} + X_{i2})}{2} ,$$

where the columns i extend from 1 to 27.

Similarly, the 2nd row of the new matrix α_{ij} (2), represents the average of the 3rd and 4th rows of α_{ij} (1). Following the same procedure, we construct h matrices:

α_{ij} (1) – Original data matrix with r rows.

α_{ij} (2) – Each row represents average of 2 consecutive rows of matrix α_{ij} (1). Total number of rows = $r/2$.

α_{ij} (3) – Each row represents average of 3 consecutive rows. Total number of rows = $r/3$.

α_{ij} (h) – Each row represents average of h rows. Total number of rows = r/h.

For each matrix, we repeat the first five steps described in the last section to obtain the residual variances $S_R^2(1), S_R^2(2), S_R^2(3) \ldots S_R^2(h)$ corresponding to the afore-mentioned h matrices. It is evident from Appendix III that if the data contain quasi-persistency, and if it is assumed that there is no persistent signal in the epochs, the ratio

$$\frac{S_r(h)h^{1/2}}{S_r(1)} = \frac{\zeta(h)}{\zeta(1)} \tag{5}$$

will exhibit some relationship with $h^{1/2}$. Note that in the analysis of variance, the signal is eliminated from the residual variance (4). However, this relationship will break down at some limiting value: $\zeta(h)/\zeta(1) = \zeta(\infty)/\zeta(1)$. In fact the quasi-persistency is negligible beyond $\sigma = [\zeta(\infty)/\zeta(1)]^2$ rows.

In the above discussion, it has been assumed that the chronological order of the epochs has been maintained in all the matrices $\alpha_{ij}(1), \alpha_{ij}(2) \ldots \alpha_{ij}(h)$.

C. Data Analysis

To clarify the procedure described thus far, and to demonstrate the pitfalls of using ordinary statistical tests for evaluating the significance level of Chree analysis results, let us consider the vectors shown in Figure 1. The data corresponding to these vectors (Appendix II) consist of a matrix with $c = 27$ and $r = 150$. The Chree analysis results derived from these data, i.e., the column means of this matrix, are plotted as a function of the day (column number c_j) in Figure 2. At first sight, Figure 2 reveals an impressive trend. In general, there appears to be a significant difference between the column means before and after the key-day (0-day). In fact, the plot in Figure 2 has the distinct appearance of a sine wave. Now let us examine, by using an ordinary statistical test, whether the variation evident in Figure 2 is actually significant. In other words, let us ignore the quasi-persistency in the data and perform the calculations enumerated in the seven tests outlined in Section A above. The final results are shown in Table II (random data). Application of the F-test reveals that the probability that the signal in Figure 2 has appeared by chance is very low (3 in 10^{15}). On the basis of this result, the 'signal' would be accepted as real.

An examination of Figure 1 suggests that this result cannot be valid because the vectors show abrupt change in direction after each sequence. Let us now apply the new statistical test described in the preceding section. To evaluate the quasi-persistency, we plot $\zeta(h)/\zeta(1)$ vs $h^{1/2}$ in Figure 3. It is clear from this figure that $\zeta(h)/\zeta(1)$ shows a definitive relationship with $h^{1/2}$ up to a point ($h \approx 4$) where the relationship breaks down. The break occurs at $\zeta(h)/\zeta(1) = \zeta(\infty)/\zeta(1) \approx 2.4 = \sqrt{\sigma}$. Thus the equivalent length of sequences is $[\zeta(\infty)/\zeta(1)]^2 = 5.76$. In other words, the quasi-persistency lasts for about six rows, which is consistent with the appearance of the summation dial in Figure 1. On the basis of this derived equivalent length of sequences, the results listed in Table II are

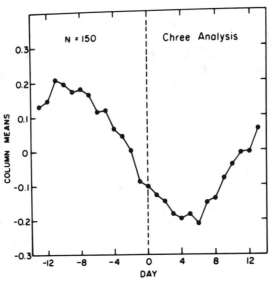

Fig. 2.   Column means ($\bar{c}_i$) representing Chree analysis results are plotted as a function of days (Table I). Total number of epochs for this analysis is 150.

TABLE II

Comparison of the signal variance in the Chree analysis result with the residual variance in order to evaluate the probability that the signal has appeared by chance.

Analysis of variance	Random data: quasi-persistency ignored	Non-random data: quasi-persistency included
Signal variance, df* = 26	2.94	2.94
Residual Variance, df = 3874	0.59	3.40 (0.59 × 5.76)
$F_{(26, 3874)}$	4.96	1.16**
Probability that the signal has appeared by chance $(1 - P)$	3×10^{-15}	3×10^{-1}

* df = degrees of freedom.
** Note that for this case, residual variance is larger than the signal variance, hence $F_{(3874, 26)}$ = (residual variance/signal variance) > 1, in accordance with standard practice (Hald, 1952).

obtained. The probability that the 'signal' has appeared by chance has increased to 0.33; hence, the apparent effect displayed in Figure 2 is *not* significant.

4. Discussion and Conclusion

For obtaining information concerning periodicities, or for understanding the relation-ship between two phenomena, superposed epoch analysis is unquestionably a useful procedure. However, this method of analysis yields meaningful results only if the

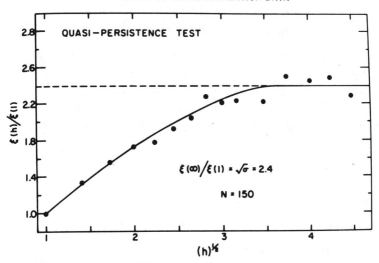

Fig. 3. Plot of the ratio $\zeta(h)/\zeta(1)$ vs $h^{1/2}$ derived from 150 epochs of Table I. The increase in the ratio for low values of $h^{1/2}$ indicates quasi-persistency in the data.

inherent quasi-persistency of natural phenomena is properly taken into account in the evaluation of the result. In fact, as was originally emphasized by Bartels (1935), the proper evaluation of quasi-persistency is of utmost importance in all types of analysis of problems in geophysics (and astrophysics). We have demonstrated here that the standard error can be grossly underestimated by ignoring the almost inevitable quasi-persistency in the data. In a statistical analysis of simulated data, we showed that ordinary (textbook) statistical tests led to the incorrect conclusion that the signal in Figure 2 is highly significant, whereas in reality it is not.

A new method, based on two-way classification analysis of variance, has been developed to determine the quasi-persistency (equivalent length of sequences) in the data. This analysis revealed that the effective (independent) number of epochs (rows) in the Chree matrix is much less than that determined under the invalid assumption that the data are strictly random. Thus, the standard error is modified, and the new effective standard error is found to be *larger* than the signal. It should be pointed out that an alternative method based on vectorial representation can be used to test the Chree analysis result. A study is in progress to determine the relative merits of the two procedures in various cases. It is hoped that application of the procedure developed here for including the effects of quasi-persistency in evaluating the statistical uncertainty of superposed epoch results will lead to more objective conclusions in future studies utilizing this powerful analytical tool.

Acknowledgements

This research is supported by the National Science Foundation's Division of Polar Programs under grants DPP–7923218–01 and DPP–7822467 and Atmospheric Research Section under grant ATM–8005866.

Appendix I: Test for Homogeneity

The hypothesis to be tested is that the variances of k normally distributed populations are equal. If there is no quasi-persistency in the data, Bartlett's test (see e.g., Hald, 1952; Dixon and Massey, 1957) can be utilized to test this hypothesis.

Let the variance of the ith sample of size n_i be given by S_i^2. Note that the sample size will take care of the fact that data in some rows or columns are missing.

Let

$$\eta = (N - k) \ln S_p^2 - \sum (n_i - 1) \ln S_i^2 \, ,$$

$$S_p^2 = \sum (n_i - 1) S_i^2 / (N - k) \, ,$$

$$A = \frac{1}{3(k - 1)} \left[\sum \frac{1}{n_i - 1} - \frac{1}{N - K} \right] \, ,$$

$$v_1 = k - 1 \, ,$$

$$v_2 = \frac{k + 1}{A^2} \, ,$$

$$b = \frac{v_2}{1 - A + (2/v_2)} \, ,$$

$$N = \sum n_i \, .$$

Then the sampling distribution of $F = v_2 \eta / v_1 (b - \eta)$ is approximately $F(v_1, v_2)$. It should be emphasized that this test is not valid for non-independent data. In case there is quasi-persistency, the equivalent length of sequences (σ) can be evaluated. Bartlett's test can then be applied to sets of k/σ independent samples.

Appendix II: Data Simulation

The simulated data $D(t)$ for each epoch, for the tests described in this paper, are generated from:

$$D(t) = R_q \sin [\omega t + \phi_q(t)] + \zeta(t) + \beta t \, ,$$

where R_q, $\phi_q(t)$ represent amplitude and phase of a quasi-persistent signal, $\omega = 2\pi/27$ and $\zeta(t)$ and βt represent random and linear effects in each epoch.

Harmonic analysis of these simulated data, after linear term corrections, yields the 27-day period vectors in the summation dial in Figure 1. Note that each vector represents a single epoch row in Table I.

Appendix III: Evaluation of Quasi-Persistency

Let

$M_i(h)$ = ith among N/h means of h consecutive means,

N = total number of $M_i(1)$,

$N(h)$ = total number of $M_i(h)$,

$r_i(1)$ = contribution of random effects to $M_i(1)$,

$r_i(h)$ = contribution of random effects to $M_i(h)$,

$q_i(1)$ = the quasi-persistent contribution to $M_i(1)$,

$q_i(h)$ = the quasi-persistent contribution to $M_i(h)$,

m = the contribution of the persistent wave to $M_i(1)$, constant for all i from 1 to N,

$c^2(1)$ = variance of $M_i(1)$,

$c^2(h)$ = variance of the means of h successive sequential means,

$$M_i(1) = [m + q_i(1) + r_i(1)],$$

$$c^2(1) = \frac{1}{N} \sum_1^N [m + q_i(1) + r_i(1)]^2$$

$$= \frac{1}{N} \sum \{[(m + q_1(1)]^2 + r_i^2(1)\}.$$

Since for large N, $\sum m r_i(1) = 0$ and $\sum r_i(1) q_i(1) = 0$,

$$c^2(1) = \frac{1}{N} \sum [m^2 + 2mq_i(1) + q_i^2(1) + r_i^2(1)]$$

(1A)

$$= m^2 + 2m\bar{q}(1) + S_q^2(1) + S_r^2(1).$$

Since

$$\frac{2m \sum q_i(1)}{N} = 2m\bar{q}(1),$$

$$\bar{q}(1) = \text{mean of all quasi-persistent steps}$$

and

$$\sum r_i^2(1)/N = S_r^2(1),$$

$$\sum q_i^2(1)/N = S_q^2(1).$$

Similarly

$$c^2(h) = \frac{1}{N/h} \sum_1^{N/h} [m^2 + 2mq_i(h) + q_i^2(h) + r_i^2(h)]$$

(2A)

$$= m^2 + 2m\bar{q} + S_q^2(h) + S_r^2(h).$$

471

Since

$$2m \sum_{1}^{N/h} q_i(h) = 2m\bar{q}.$$

Then

$$c^2(h)h = m^2h + 2mh\bar{q} + S_q^2(h)h + S_r^2(h)h , \qquad (3A)$$

$$c^2(h)h = h(m^2 + 2m\bar{q}) + S_q^2(h)h + S_r^2(1) .$$

As a special case, assume that the data contain no persistent wave, i.e. $m = 0$, then

$$c^2(h)h = hS_q^2(h) + S_r^2(1) . \qquad (4A)$$

For large values of h, the right-hand side becomes constant, i.e.

$$c^2(h)h = \text{const.} = c^2(1)\sigma , \qquad (5A)$$

where σ is defined as 'equivalent length of sequences' (Bartels, 1935).

Equation (5A) can be written as

$$c(h)h^{1/2}/c(1) = \zeta(h)/\zeta(1) \approx \zeta(\infty)/\zeta(1) = \sigma^{1/2} .$$

References

Bartels, J.: 1935, *Terrs. Magnetism Atmospheric Electricity* **40**, 1.
Chapman, S. and Bartels, J.: 1940, *Geomagnetism*, Vol. II, Oxford University Press.
Chree, C.: 1912, *Phil. Trans. London* **A212**, 75.
Chree, C.: 1913, *Phil. Trans. London* **A213**, 245.
Dixon, W. J. and Massey, F. J.: 1957, *Introduction to Statistical Analysis*, McGraw-Hill Book Co.
Forbush, S. E., Duggal, S. P., Pomerantz, M. A., and Tsao, C. H.: 1982, *Rev. Geophys. Space Phys.*, in press.
Grec, G., Fossat, E., and Pomerantz, M.: 1980, *Nature* **288**, 541.
Hald, A.: 1952, *Statistical Theory with Engineering Applications*, John Wiley and Sons.
Scherrer, P. M., Wilcox, J. J., Kotov, V. A., Severny, A. B., and Tsap, T. T.: 1979, *Nature* **277**, 635.
Severny, A. B., Kotov, V. A., and Tsap, T. T.: 1976, *Nature* **259**, 8.